Eureka!

An Illustrated History of Inventions from the Wheel to the Computer

Edited by EDWARD DE BONO

How and When the Greatest Inventions were made

Thames and Hudson

Acknowledgments are due to
Times Newspapers Ltd for the idea
on which this book is based. A
number of the articles appearing in
it were first published in the
Sunday Times Magazine.

First paperback edition 1979

Printed in Great Britain
by Severn Valley Press

Contents

Contents

Contents

List of Contributors

A.Bi.	Anthony Bird	J.C.	John Culshaw
A.Br.	Asa Briggs	J.G.	John Goulding
A.Bu.	Aubrey Burstall	J.I.	Janey Ironside
A.C.	Angela Croome	J.L.	James Laver
A.E.	Arthur Elton	J.M.P.	J. M. Pugh
A.K.	Alex Keller	J.N.	James Norbury
A.M.	Arthur Marwick	K.A.	Kenneth Allsop
B.D.	Bernard Dixon	K.G.	K. R. Gilbert
B.S.	Bryan Silcock	K.M.	Kenneth Macksey
C.G.-S.	Charles Gibbs-Smith	L.B.	Leon Bagrit
C.H.	Charles Hayward	L.D.	Len Deighton
C.L.	Christopher Lloyd	L.F.H.	L. F. Haber
C.M.B.G.	Michael Greenhalgh	L.M.	Leonard Maunder
C.StC.B.D.	C. StC. B. Davison	L.T.C.R.	L. T. C. Rolt
C.V.	Christian Verity	L.W.	Lynn White
D.D.	David Divine	M.C.	Margaret Costa
D.F.	Derek French	M.H.	Martin Hazell
Dd.G.	Donald Gould	M.McC.	Meriel McCooey
Dk.G.	Deryck Goodwin	N.G.	Norman Gibbs
D.K.	Donald King	P.D.	Percy Dunsheath
D.W.W.	D. W. Waters	P.J.	Paul Jervis
E.G.	Elizabeth Gundrey	P.Wh.	Paul Wheeler
E.H.	Edward Hyams	P.Wy.	Peter Wymer
E.J.	Elizabeth Jenkins	R.C.	Ritchie Calder
E.L.	Egon Larsen	R.J.	Richard Jones
E.S.T.	E. S. Turner	R.Wa.	Rex Wailes
E.W.	Edward Woods	R.Wi.	Raymond Williams
F.A.	Felix Aprahamian	S.F.	Sheena Ferguson
F.G.	Frank Greenaway	S.M.	Sylvia Miller
F.J.W.	F. J. Wilkinson	S.P.	Stuart Piggott
F.M.	Francis Maddison	T.C.	Trevor Crozier
G.B.L.W.	G. B. L. Wilson	T.J.	Timothy Johnson
G.E.E.	George Ewart Evans	T.L.	Tony Loftas
G.J.W.	G. J. Whitrow	T.M.	Tom Margerison
G.N.	Graham Norton	T.O.	Tony Osman
G.P.	George Perry	V.R.	Vasco Ronchi
G.R.T.	Gordon Rattray Taylor	W.B.G.	W. B. Gosney
G.T.	George Thomson	W.K.V.G.	W. K. V. Gale
H.C.	Hugh Casson	W.L.G.	W. L. Goodman
H.G.	Helmut Gernsheim	W.T.G.	W. T. Gunston

Preface

ıere is nothing more important than an idea in the mind of man. an's achievement is based on man's ideas. The fascinating thing out inventions is that often we can see how one idea in the mind one man at one moment in history changed the course of civiliza- n. There is no idea more personal than an invention which can traced back to a single man – as so many can. In the case of the vention, we can also see at once how the idea works, for very often product is still with us. This book is about all the most important ventions: how they came about, who made them, what effects they d. The personal element is most important. The book is not a story of technology, but a story of individual inventions and, ove all, of individual inventors.

What was the famous astronomer Edmund Halley doing sitting der 60 ft. of water for an hour and a half? He was trying out his n invention, the diving bell. With extraordinary bravery, the ventor of the first parachute jumped from a balloon 3,000 ft. up the air. He survived, but many other inventors concerned with ing killed themselves, like Otto Lilienthal, an early pioneer of ding.

It is not extraordinary to find an astronomer inventing a diving ll, for the history of inventions is full of examples of people who vented things outside their own fields. The first corn-reaper, for ample, was produced by an actor, who demonstrated it on a crop had 'planted' on stage. The first practical submarine was invented an Irish schoolmaster in New York, with the intention of sinking British Navy. The pneumatic tyre was invented by a Scots vet lled Dunlop to help his son ride his tricycle to school over rough bblestones. The hydrofoil was first thought of by a French priest, d the knitting frame by an English vicar who failed to convince ueen Elizabeth I of its potential. The principles of flying were orked out by a Yorkshire squire who sent his unwilling coachman oft in a glider, while the safety razor was invented by a 40-year-old lesman who was advised to invent something which would be ed once and then thrown away – as we would say today, something sposable.

The effect of an individual invention can be enormous. The vention of the simple cotton gin changed the whole economy of southern states of North America. Galileo's observation of the ndulum swing at once made accurate clocks possible; Bessemer's el process provided the world with its most important construc- on material; and the invention of gunpowder ended the feudal stem of the medieval world. The internal combustion engine was vented a little over 100 years ago, the motor-car more recently, d the aeroplane within the lifetime of many people still living, yet day they dominate the economy of the advanced countries. isha Otis's safety lift made skyscrapers possible, and Mario retic's fishing block caused a fivefold increase in the Peruvian h catch.

In spite of this, inventions are not always welcome. It seems believable that neither the Italian nor the British government owed any interest in Marconi's work on wireless, until one remembers that the British Admiralty had preferred to have men signalling to each other from hill-tops, rather than meddle with electric telegraphy. Percy Shaw's splendid catseyes were ignored until the wartime blackout forced their recognition. Only very rarely was an invention tested as thoroughly as when a paddle-driven sloop actually had a tug-of-war with a sloop propelled by the new-fangled screw. Establishments have always been conservative. Even the great Rutherford, father of nuclear physics, doubted whether the energy of the atom would ever have a practical use.

How do inventions come about? The stethoscope, which has, since its invention, been a basic tool of medicine, was first thought of when Laënnec noticed some children playing with a rod of wood. The gramophone was invented because Edison, playing about with Morse dots on a strip of waxed paper, noticed a hum when the paper passed rapidly beneath a needle. The important process of vulcanization, which turned rubber into the useful material it is today, was the result of some random experimentation by Charles Goodyear. Chance, observation and playing about have been the basis of many inventions. Radar was developed from the bizarre suggestion of a radio death-ray for shooting down planes. Instead of rejecting this idea, someone used it as a stepping stone to the concept of radar. Sometimes one thing follows on from another. C. T. R. Wilson started as a student of meteorology, and his interest in cloud formation led him to invent the cloud chamber, which became a basic tool of nuclear physics. In other cases, a new approach has been needed to a problem already under consideration. The Wright brothers succeeded in flying because they set out to build an inherently unstable aircraft, and so concentrated on the controls when everyone else was concentrating on stability. Inventions can, of course, result from individual problems: Mr Bissell was allergic to the straw scattered on the floor of the china shop where he worked, so he invented the carpet sweeper.

Some inventions have required technical or scientific knowledge, but others have been so simple that, with hindsight, they are only too obvious, as with the Chinese invention of the double canal lock. Three hundred years elapsed between the invention of lenses and the placing of one in front of another to make a telescope. Movable type was a simple idea, but a great improvement on the carving of a block for each page when printing a book.

There is nothing quite like the excitement of an idea. For instance, the whole field of photography, which in a way gave man an extra sense, was made possible in the instant in 1816 when Joseph Nicéphore Niépce substituted silver chloride paper for the ground glass in a camera obscura.

The extraordinarily successful xerographic copying process was invented by Chester F. Carlson in 1938 and the polaroid camera in 1947 by Edwin Land. Float glass was developed by Alastair Pilkington in 1959. The Wankel engine, a revolutionary change in engine design, was invented by Dr Fritz Wankel in the late 1950s and the hovercraft, an equally revolutionary change in transport methods, by Christopher Cockerell in 1955. The age of invention is not over.

EDWARD DE BONO
Cambridge, 1974

Acknowledgments

The Publisher's acknowledgments are due to the following for permission to reproduce illustrations:

Académie des Sciences, Paris; Allied Ironfounders Ltd; American Telephone and Telegraph Co.; Ampex Great Britain Ltd; Amplivox; Ashmolean Museum, Oxford; Associated Press; Australian Information Service; George F. Bass; Bavaria Verlag; Bayer AG; Bibliothèque de l'Assemblée Nationale, Paris; Bibliothèque Nationale, Paris; Bildarchiv Foto Marburg; Mme Blumer-Maillart; Bodleian Library, Oxford; Professor and Mrs Asa Briggs; British Aerosol Manufacturers' Association; British Broadcasting Corporation; British Esperanto Association; British Museum, London; British Railways Board; British Steel Corporation; Brooke Bond Oxo Ltd; Bryant and May Ltd; Central Electricity Generating Board; CERN, Geneva; Chicago Historical Society; Lady Cobbold; Cornell University, Ithaca, NY; Council of Forest Industries for British Columbia; the *Daily Telegraph*; Decca Record Co.; Deutsches Elektronen-Synchrotron, Hamburg; Deutsches Museum, Munich; the Duke of Devonshire; Charles Dollfuss Collection; Dowty Rotol Ltd.; Embassy of the Arab Republic of Egypt, London; *Farmer's Weekly*; F.M.C. Corporation, San Jose, California; Ford Motor Co.; Henry Ford Museum; Fototeca Unione; French Embassy, London; Professor O.R. Frisch, F.R.S.; Gabinetto Nazionale Fotografico, Rome; General Foods Ltd; German Embassy, London; Giraudon; Globe Photos Inc.; Guildhall Library and Art Gallery, London; André Held; Hermitage, Leningrad; Hughes Aircraft; Imperial War Museum, London; IBM; International General Electric Co. of New York; International Harvester Co.; Jagiellon Library; Kabul Museum; Kirkcaldy Museums and Art Gallery; Koninklijk Museum voor Schone Kunsten, Antwerp; Krauss-Maffei; Lawrence Radiation Laboratory, University of California, Berkeley; Leicester Museums and Art Gallery; Library of Congress, Washington DC; Linear Motors Ltd.; Tony Loftas; London Museum; Donald Longmore, F.R.C.S.; Louvre, Paris; Lund, Humphries; the Master and Fellows of Magdalene College, Cambridge; Department of Physics, Manchester University; Marconi Co.; Meteorological Office Library; MGM, Hollywood; Herbert Morris Ltd; Mullard Ltd; Musée des Antiquités, Saint-Germain-en-Laye; Musé Royaux des Beaux-Arts, Brussels; Museo Archeologico Naziona Cagliari; Museo Capodimonte, Naples; Museo Nazionale, Naple Museum of Fine Arts, Boston; Museum für Völkerkunde, Berli NASA; National Coal Board; National Gallery, London; Nation Gallery of Art, Washington DC.; National Maritime Museu Greenwich; National Motor Museum, Beaulieu, Hants; Nation museet, Copenhagen; National Museum of Rhodesia, Bulaway National Portrait Gallery, London; National Research Develo ment Council; Department of Psychology, University of Nottinghar Novosti Press Agency; Orbis Ltd., London; Ostia Museum; P American World Airways; C.A. Parsons & Co. Ltd; University Pennsylvania Museum, Philadelphia; Musée Pasteur, Paris; P kington Brothers Ltd; Plant Protection Ltd; Polaroid Corporatio Paul Popper Ltd; Popper-Handke Collection; Josephine Powe Press Photographic Agency; the editor, *Proceedings of the Ro Society*; Production Engineering Research Association of Gre Britain; Bernard Quaritch Ltd; Radio Times Hulton Pictu Library; Railway Gazette International; Rank Xerox Ltd; Re Automatic Co. Ltd; Rheinisches Landesmuseum, Trier; Rij museum Amsterdam; Ronan Picture Library; Royal Aeronauti Society; Royal National Institute for the Blind; John Rylan Library, Manchester; Scala; *Science Journal*; Science Museu London; Shell Chemicals U.K. Ltd; Short Brothers, Belfast; Edw Smith; Smith's Industries Ltd; Smithsonian Institution, Washin ton, DC; Staatliche Museen, Berlin; Stadtbibliothek Nürnbe *The Times*; Doreen Thomas; Tracked Hovercraft Ltd; the Mas and Fellows of Trinity College, Cambridge; Trinity House Libra London; Eileen Tweedy; Unilever Film Library; Union Paci Railroad; United States Department of the Interior; United Sta Information Services, London; Victoria and Albert Museu London; Walker Gallery, Liverpool; Waterman Pen Co. Lt Wellcome Institute of the History of Medicine, London; Woodlar Museum, Greenwich; Professor Yigael Yadin.

MAN MOVING

Transport

Communications

MAN MOVING

Transport

The wheel is so basic that we take it for granted. Yet in the Western Hemisphere the wheel was never invented. There were great and advanced civilizations in North, Central and South America – but no wheel. The various Red Indian civilizations, the Aztecs, Mayans, Toltecs and Incas, all did without the wheel.

In Mexico, wheels have been found on a child's toy from the pre-Columbian period but there is no evidence of their practical use. Does this mean that the invention of some ingenious designer remained confined to the world of toys and was never taken up in the adult world? Toy helicopters were in use in Europe in the 14th century but the adult helicopter had to wait many centuries for a suitable power unit: on the other hand, the wheel did not have to wait for anything.

The usual explanation is that the wheel was not developed because there were no beasts to pull carts until the Spaniards introduced horses at the time of their exploration and conquest of the New World. This may well be true. So may the explanation that much of the terrain is so rugged that transport by river and carried loads was more effective. But somehow these and other explanations are inadequate. I cannot help feeling that if the wheel had been developed its general use would have caught on very quickly. The real explanation, I think, is that no one needed it because they were quite happy with what they had: or perhaps no one thought of its usefulness.

Seven hundred years is a long time. Why did it take all that time from the development of the saddle to the invention of the stirrup in Western Europe? The stirrup is a very easy invention indeed, it requires no special material or advanced technology and it could have been tried out at once by anyone who happened to think of the idea. The advantage of the stirrup is that it gives the rider the same sort of stability he would have if both his feet were on the ground. This was a special advantage in warfare when the fighting man had to slash away with his sword or axe without falling off his horse. The Assyrian warriors seem to have had stirrups in 835 BC but it took fifteen hundred years for them to reappear in Western Europe. Why?

If someone had sat down deliberately to consider the problem of stability on a horse he would probably have come up with the idea of stirrups within a week. The likelihood is that no one actually noticed that there was a problem: indeed, riders prided themselves on the skill with which they controlled a horse and would have spurned any artificial aids beyond this skill. Besides, stirrups would have made the bad rider almost as effective as the good rider. So those who should have been most interested in the matter were probably the least interested. This sort of thing happens again and again in the history of invention. It was only after the stirrup had been developed and used that everyone noticed its immense advantage. Some historians have even gone so far as to suggest that the Asian penetration into Europe succeeded because the Asian horsemen used stirrups and the Europeans did not.

If people are happy with things as they are, they see no need for change and no advantage in it. The advantages can often be seen only after the change has come about. Neither the Italian nor the British government showed any interest in Marconi when he w developing radio communication – today, when we recognize t overwhelming advantages of radio communication, we find t incredible. Much earlier the British Admiralty had said they want nothing to do with the crazy idea of sending messages rapidly electric telegraph wire because they were quite happy with th existing system of posting men on adjacent hills and having th signal to each other by semaphore, so transmitting messages acr the country. No one now doubts the importance of aircraft in mod war and yet some years *after* the Wright brothers had flown a pla the US Congress actually passed a special Bill preventing the Ar from wasting its money experimenting with flying machines. N very long before, the US Patent Office was actually refusing pate for heavier-than-air flying machines on the ground that they w impossible. Space flight depends on rockets, as do the intercontinen missiles for delivering nuclear warheads: yet when Robert Godd was doing the pioneer work on rockets the US Government to no interest at all.

I am not suggesting that some inventions took a long time beca of official opposition or lack of support: this simply reflects the gene feeling of people inside or outside the government. Most peo are happy with things as they are. Even today, when we are us to change and invention, very few people actually set out to fi problems or produce ideas. In the past, when the world was sta for long periods, such people would have been even rarer. I beli that those inventions which did not depend on a scientific or techni advance elsewhere took a long time because no one could that there was a problem needing solution. In some cases inventi were actually made but never taken up because no one other th the inventor could see the problem. It is unfortunate that war h been one of the few situations that focus people's minds on proble and advantages.

Up to the development of the railway, animal transport and particular the horse was man's only means of movement on lan other than his own legs. For personal transport, this has been case right up to the development of the motor-car at the beginni of this century. So as far as transport on land is concerned, in ma areas the situation remained exactly the same for all generations humanity up to the oldest generation now living.

There were, of course, many improvements in saddlery, bridl stirrups and horse-breeding, but travel time by horseback ridi remained unchanged from the Roman era to the beginning of t century. While there was not much that could be done to a hor many improvements were made in the carts that horses could p Here war technology may have exerted its influence in the devel ment both of fighting chariots and also of support transport. Tw wheeled chariots replaced those with four wheels and eventually wheels were shifted further back so that part of the rider's weight w supported by the horse – thus giving him a smoother ride. Changes harness, yokes, wheels, suspensions continued to be made so that, the end, carriages were highly efficient and elegant. It is difficult see what else could have been done before the development of m rms of energy, but there are possibilities. For instance, the use of ind-power at sea might have suggested some form of land yacht. is also technically possible to design a horse-powered vehicle at actually moves faster than any horse.

If the great attention paid by the Romans to roads had continued, ese and other developments might have come about, since all of em would require wide roads with good surfaces. But when the ntralized authority that required good roads was gone, the need r maintaining them was lost. Small independent states had no terest in good roads and probably feared them, since they made tack easier. The lack of road technology may well have held back velopments in land transport.

Theoretically there has always been an alternative to the horse: an himself, at a faster speed and with a greater range. The bicycle, r example, uses no outside power or energy but by direct mechanical eans increases the speed and range of man's travel by a factor of out five. The bicycle was invented in 1790, but without pedals or ering; steering followed in 1817 and pedals in 1839. It may be gued that the bicycle had to wait for such developments as rather aborate metal work and a degree of precision engineering not quired in the building of carts. But metal-working among clock-akers, armourers and carriage-makers was certainly sufficiently vanced to have made the bicycle possible very much earlier than e 19th century.

At about the time the bicycle was being developed came an inven-n of an exactly opposite nature. Whereas the bicycle uses only man energy, the railway is a direct way of harnessing external rms of energy. The steam engine was an invention to make ailable the energy of coal. The first locomotive was a steam engine wheels. It is interesting to note that the first railways were veloped for use in mines. Steam engines themselves had originally en used in mines for pumping water out of the shafts; rails were ailable in mines for transporting material (in carts pulled by rses); and there was a need in mines for powerful cheap transport. is illustrates what very rarely happens in the history of invention: e combination of available technology and need.

In the case of railways we tend to look at the locomotives and get the rails, but it is doubtful if railways would ever have veloped if rail systems had not already been available. The mense advantage of railways would not have been recognized if am tractors had had to trundle over terrible roads. And yet rails emselves could have been developed long before. There is a ggestion that a primitive railway, consisting of parallel grooves which round balls of rock rolled, was in use in Stone Age Malta d from time to time thereafter the rail principle was used but never a large scale. It is interesting to speculate what new modes of nd transport might have emerged if rail systems had been in tensive use before the steam engine.

The advantage of steam-engined locomotives on rails was quickly ade obvious, for instance in the reduction of journey time between ndon and Edinburgh from 250 hours to 50. Today it is approaching e hours by railway. Once the railway had been invented, it went improving by a series of developmental steps. It is interesting to te how quickly it reached a pitch of efficiency and then stuck ere for many years.

The first locomotives were merely steam engines, complete with iler, on wheels. The first motor-car was also an engine put on eels in a light tricycle framework by Karl Benz. There the simi-ity ends. Once the principle of propulsion by an internal-com-stion engine had been shown, the path of development involved tting engines into carriages in place of the horse. The first motor-s really *were* horseless carriages.

Steam-engined automobiles had existed since 1770; it is strange that no attempt was made to refine and simplify them so that instead of being slow, lumbering tractors they approached the lightness of carriages. The general failure to simplify and lighten the steam engine is difficult to understand, except that everyone may have been so taken up with the immense power of the engine that they could only think in the direction of more power. There was probably no incentive to develop light steam-engined vehicles since the carriage was faster, lighter and could be owned by individuals (as contrasted with steam tractors). As soon, however, as the lightweight motor-car appeared – it did so at once and was not slowly developed – the possibility of horseless carriages became obvious. It was then that lightweight steam engines were actually developed and used in motor-cars.

There have been many key inventions in the development of the motor-car, but the surprising thing is that most of them occurred very early and that thereafter there was little fundamental change. In view of the vast size of the motor industry and its huge research budgets this seems surprising until we remember the basic principle that if something appears to be adequate there is little pressure to think of something better. The recent development of the Wankel rotary petrol engine could be called the only major change since the first motor-car. It is interesting to imagine what might have happened if the steam car or electric car – both of which were highly developed at one time – had been the winning choice. I think that by now we would almost certainly have non-polluting steam engines working off low-grade fuel, and also quite new concepts on electricity storage and generation.

The development of sea transport has been quite different from that of land transport. As we have seen, land transport (apart from the wheel and the horse) evolved little if at all until the invention of railways and then the motor-car. In contrast, sea transport developed steadily and surely. Boats became bigger and faster. They could carry heavier loads and travel longer distances. This type of develop-ment was made possible by several factors. The sea is a universal road, so no road-building was required. A source of energy was available in the wind. Because a boat is supported by the water, it was possible to build bigger and bigger boats without encountering the problems of support and friction that would have occurred in land transport. Finally, an excellent raw material, wood, was readily available and easily worked. From coracles made of a wooden framework covered with skins or tarred cloth to large galleons sailing the Atlantic, the process seems to have been smooth. Although sails were obviously better for long distances, they did not have the control and manoeuvrability of oars; so for naval warfare the development of oars continued alongside that of sail. Different types and rigs of sail were developed, some making it possible to sail into the wind or closer to the wind, others giving better directional control. Although these differences may appear to be small, in practice they would be very important.

Paddle wheels are a natural extension of human-powered oars or the water-wheel. As early as the Romans, at least the idea of a paddle boat powered by oxen seems to have occurred. When the steam engine was invented it appeared obvious that it should be put into a boat to drive paddle wheels. This was first done for boats that were to use canals or rivers: on such waterways the paddle boat would have an immediate advantage over the sail boat, which could not operate in a narrow waterway. It was only when the technology had been thoroughly worked out and when the con-denser made unnecessary a continuous supply of fresh water that the paddle boat moved out to sea. As so often, the next development was aided by war technology. Paddle boats are not suitable for warships because a single shot can wreck a paddle and render the boat useless. A screw propeller, being submerged and in any case very much smaller, does not have this disadvantage. Warships driven by screw propellers were first used in the American Civil War with some success. But the British Admiralty remained doubtful. So a classic tug of war was arranged between a paddle boat and a screw boat, and the screw boat pulled the paddle boat backwards through the water. Very rarely is there such an opportunity to demonstrate the advantages of a new invention.

Once the screw was established, ships continued to increase in size. Developments in engines and shipbuilding technology made this possible, and today tankers of one million tons are being built. The basic principles of propulsion remain the same, but there have been further innovations in sea transport. The first is the hydrofoil, which was invented by a priest. Two small wings are immersed in the water, and as they move along they lift the boat right up out of the water, so reducing drag and allowing very much faster speeds. In the case of the hovercraft the vessel floats on a cushion of air that is forced under the boat to lift it clear of the water. So far these developments are used only for relatively small boats but they do indicate a fundamental innovation in sea transport.

Inventions are difficult if people are satisfied with what they have, but not if there is a defined problem. For instance, the first chronometer was designed to win a prize offered for solution of the navigational problem. Unlike travellers on land, sailors could not tell where they were or where they were heading. A whole group of instruments had to be invented and perfected specifically to aid navigation: compass, sextant, chronometer. The compass was especially important when the sun or stars could not be seen. The north-pointing property of magnets appears to have been known for a long time in China and possibly by early Norse sailors. Until the invention of the modern gyroscope it is difficult to see what could have been substituted for the compass. Would history have been different if the magnetic properties of certain materials had not been available? The sextant solved the problem of accuracy and convenience, the chronometer the problem of an accurate timekeeper when the swing of a pendulum would be affected by the movement of the ship in rough weather.

Inland waterways presented no navigational difficulties and provided excellent roadways for bulk transport. It is true that the speed was slow but a fair-sized barge could be towed by a single horse. Where rivers were lacking, canals were built – often with much engineering skill. For a while it seemed that canals would rival railways for bulk transport and their usefulness may one day become apparent again. Inland waterways pose some problems of their own. The problem of silting up by mud and vegetation was solved by the invention of dredgers. Another much more difficult problem was that of going uphill and downhill by water: this seemed insuperable. The solution lay in the development of locks. The first type of single-gate lock was difficult to use but it was supplanted by the double-gate or pound lock which, like the bicycle, is an example of a superb idea which depended not on new technology or a new energy source but on pure creative thinking. It was designed in China in AD 938.

There have always been myths of men with wings and stories of men who made wings. If birds can fly, it is only natural to suppose that some gifted man could also fly. Unlike some other areas of transport, in which no one seemed inclined to attempt invention, people were always trying to fly. The feat itself was so remarkable that the thought of it inspired them.

The most obvious approach was to imitate birds and to use flapping wings. Many were the designs for flapping wings and many the broken limbs or deaths when the method failed. The idea of non-flapping wings probably derives as much from kites as from bir Kites are ancient, and the step from kite to glider is a big but not impossible one. The aerodynamics of gliders were worked out early as 1810 by Sir George Cayley. One hundred years late glider became a plane with the addition of a petrol engine and t Wright brothers' know-how. This is usually presented as an examp of a development that had to await technology – in the form of engine that had the right power/weight ratio. This is true but ignores two things. If energy had been devoted to developi lightweight and powerful steam engines for small land transp systems it is just possible that aeroplanes might have flown earli As it was, existing steam engines were too heavy to do more th lift a few eccentric flying machines off the ground for brief momer The more important point, which we must not forget, is that t Wright brothers did much more than simply put a petrol engi into a glider: their unique contribution was to consider the cont of flying machines. They realized that planes were unstable a would not fly themselves, so instead of trying to design a stable pla that would fly itself, they concentrated on finding ways by which t pilot could control the instability. It was their study of methods control, especially the idea of 'warping' the wings, that enabled th to become the first successful fliers.

Balloons, unlike aeroplanes, have no counterpart in natu In fact, the development of the balloon did not have to wait for a technological development beyond the availability of cloth a paper. Yet the first balloon was successfully launched many centur after these were available. When the Montgolfier brothers fl their hot-air balloon, they were only using the energy of fire, whi had been available for a long time. The use of hot air may seem ve obvious in hindsight but was not at the time.

When news of the balloon ascent reached Paris the physic J.A.C. Charles was commissioned by the Académie des Scien to make a balloon. He could think only of the gas hydrogen whi had recently been produced, and invented the hydrogen balloo thinking he was imitating the Montgolfiers. The story has t morals: the first is that using hot air was not at all obvious; t second is that when an inventor knows that something has be done he is much more likely to solve the problem – in the same o new way. The history of invention shows that on many occasic news of an invention sparked off another inventor. The hydrog balloon developed into the airship and the simplicity of the hot- balloon was forgotten, though today it is coming back into use f sporting purposes.

In the case of gas-filled balloons no energy is expended in keepi airborne, but rocket flight is dependent on the expenditure of ener at every moment. The rocket is an example of an invention that, principle, has been available for hundreds of years and yet practice had to await many developments before it could achieved. Crude rockets were used as display fireworks or war missi for many centuries but they could not be developed into spa vehicles until the discovery of suitable chemicals and cont methods, heat-resistant metals, gyroscope technology etc.

Yet another approach to flight is the helicopter principle. A helicopter was in existence in the fourteenth century and Leonar da Vinci's drawing of a helicopter design is well known. The idea a device which screwed itself up into the air is much more obvic than the idea of aerodynamic shaped wings to generate lift. Even ally the helicopter used the lift principle rather than the scr principle and it had to wait for the petrol-engine power unit. Ev so, helicopters are not the ideal method for slow vertical take- and landing and I suspect that a radical new principle of flying w soon be developed in this area.

Transport seems to be one of the most basic of human needs. an has always wanted to move around, and yet life has been ossible with very limited forms of transport. It might be said that an creates for himself the need to move. As a nomad, he may ant to move to new pastures. As an explorer, he may want to do e same out of curiosity, or population pressure behind him, greed for new trade or trade routes. As a warrior, he would want be mobile enough to defend himself or, more often, to attack hers. When man settled down and started growing crops, transport eds began to affect not only himself but his goods. Food grown in e place might have to be taken to another for consumption and orage. Food might also be transported in order to be exchanged r money or other goods in trading systems. Various substances ich as coal and iron ore) might have to be brought to one place so at out of their use together something new might emerge; then e finished product itself would have to be taken away and distri-ted. Today, a major use of travel is for pleasure. The travel stems were not developed because there was a strong need for em but because technology was available; the opportunity created e need, and people travelled simply because it was possible.

In some ways the sheer technical possibility of travel may be a advantage. It leads to congestion, pollution, badly distributed pulations and a great waste of natural resources both in land and energy. It is difficult to see how the process will be reversed. ice travel becomes a possbility it seems to become a necessity.

ommunication

ommunication is a use of transport. Until very recently the only ethod of communication was physical travel. If a person did not avel himself, he would send a message, carried by someone else. e excellent system of Roman roads was forged by the need to mmunicate over long distances (and to move troops). Today, en we can communicate by radio or telephone, communication d transport are no longer the same thing.

Modern communication is perhaps man's most astonishing hievement. It is amazing enough to know that man has gone to e moon, landed there, and come back; but what seems to me even ore incredible is that we can actually see the spacemen as they lk on the surface of the moon and hear what they say at the time ey are saying it, 250,000 miles away. Millions of people on earth ting in front of their TV sets can *all* watch and listen at the same ne. To a man born one hundred years ago this would have been mpletely beyond belief. Yet the moon is not the limit. The same ng could have been done at a distance a hundred times greater an that.

There are many different aspects of communication. First there the matter of distance. Two people can communicate with each er by speech over a short distance, but as the distance increases, eech is no longer sufficient – so the problem is one of overcoming tance or of making speech carry. Second, there is the multiplica-n aspect. For instance, by the use of radio or TV one man can mmunicate with millions of others at the same time. Third, there he recording aspect. Until the invention of wireless and telegraphy, long-distance communication involved recording the message d then physically carrying this record over the distance (by horse ship). Recording can also help multiplication, as when a news-per or book repeats a message and delivers it to thousands of ferent people. Recording has yet another advantage in that it n conquer time. The recorded message can be kept for an almost lefinite period (and with re-copying for an infinite period). Though the originator of the message may have been dead for hundreds of years his message can still be delivered. Consider a 6,000-year-old picture shown on TV today as part of an art series. An artist who lived 4,000 years before Christ is communicating simultaneously with millions of people today. Recording also offers a choice of access time: you can read a book or listen to a tape recording whenever you want to.

From a pointed stick used to scratch in the dust or on a clay tablet the transition to a pen is not a great one. It involves the invention of ink and paper. The next step was a pen which carried its own ink with it. This is like going back to the pointed stick, which needs no ink. A pencil can be regarded as a sort of solidified ink stick since the graphite in the pencil is related to the black colour in many inks. The first 'lead' pencils did indeed use lead but the switch to graphite came in 1795, and from that day to this pencils have remained exactly the same. From the pencil to the ball-point pen is an obvious step since the ball-point is just a pencil with semi-solid ink. Yet the ball-point was invented only in the 1930s. Why the delay? Probably because the fountain pen was so satisfactory that no one was interested in a new device. That may explain why the ball-point was first manufactured as a high-level 'writing-stick' for the air force. When ball-points were first marketed in the US the chief advantage publicized was that they could write under water; their cheapness was not a factor until *after* mass use made mass production possible.

Typewriters were first developed for the use of the blind. The earliest ones were slow and clumsy, and could not rival the efficient pen; only if you were blind and could not use a pen would there be any point in using a typewriter. The initial idea for the typewriter came from a machine that was used to number the pages of a book. The transfer from numbering to lettering was the sort of concept-transfer that can be found in many inventions: moving an idea from its original setting to a new one.

From a technical point of view, the step from personal writing to printing is not great: it is still a matter of applying ink to paper. But the effectiveness of the change is incalculable, for at once the multiplication factor is introduced, and there is no limit to the number of people who can receive the written message. As in many other cases, printing appears to have been invented first in China. In the West, its invention is usually credited to Gutenberg, though there were evidently others before him. At first the written page was regarded as a picture to be reproduced. A wooden plate was carved so that the letters stood out in relief and could be inked. Pressing the plate on to paper produced the printed page. This meant that a new carving had to be done for every page of a book. The amazingly simple step of using movable type that could be put together in different ways meant that the carving had to be done only at the beginning; thereafter it was a matter of assembly. The speed-up in the process was enormous. This has always been a classic example of a simple invention which anyone could have thought of – and yet profound in its effects. Ironically, with the advent of photo-litho printing we are going right back to treating each page as a picture on its own.

For three hundred years printing remained more or less the same until newspapers made necessary the introduction of mechanical high-speed presses. Even so, the basic process of making and assembling the type did not change until the recent introduction of the photographic methods mentioned above. It is interesting to speculate on what might have happened if inventors had been more active in those three hundred dormant years. For instance, the simple development of a cylinder rolled over many sheets of paper might have made a great difference, and no profound technology would have been required. But since knowledge was regarded as the prerogative of the élite there may not have seemed any need for developing mass printing techniques.

Long before written language, artists had drawn pictures on the walls of caves and elsewhere. Each picture in its place could be regarded as a message to the onlooker, just as a musical performance is a message to the listener. The purpose of the message may have been aesthetic, religious, magical or functional. The reproduction of pictures is not as easy as the reproduction of language since nothing like movable type is involved. The various methods developed (engraving, etching, litho, woodcut, silk-screen printing) all seek to prepare a surface so that it can be used to impart ink only to certain parts of the receiving surface. These processes tend to be slow, and until the coming of photography communication by picture could not rival communication by print.

The camera is a most remarkable invention, but a very simple one. The optical side of the camera had existed for a long time as the camera obscura, which projected on a screen a picture of the scene placed before it. All that was required was to capture that picture and make it permanent. Looking for a way to make this happen might have taken a long time, but a chance observation by Niepce in 1816 provided the method. He noticed that a piece of silver chloride paper on the laboratory bench retained the image of a spoon after the spoon had been removed. From that observation came the use of silver chloride paper to react to light and to give a permanent record and from then on it became merely a matter of refinement and improvement on the chemical and optical side. The remarkable thing about the camera is that with one single invention man became capable of recording scenes and pictures, and indeed any material that he could see (including print). It was as if man had been given an entirely new memory system. Some of the traditional functions of art at once became obsolete.

The step from the still camera to the movie camera was a much smaller one. The principle of moving pictures and various devices to activate them had been in existence long before the camera. All that was needed was to design shutters, sprockets and other mechanical devices to record and then project a series of still pictures in a succession rapid enough to give the illusion of movement.

Up to this point all the communicating devices mentioned have depended on recording. The essential process was to remove time. Once that had been achieved the message could be delivered at any pace. The coming of electronics made it possible to communicate over large distances without going through the recording stage. (Technically it might be said that the pattern of electric impulses in a telegraph wire or a modulated radio signal is a very brief form of recording.) This meant that it was possible to conquer distance directly without loss of time, so that a message could be spoken and received almost instantly even over very long distances (for instance, as we have seen, from the moon).

Before electronics the only means of transmitting messages rapidly over long distances was the visual relay system. An invasion at one point would be signalled by lighting a fire on a hill; this would be seen from a distant hill and another fire lit to pass the message on. This basic system was refined and developed by the French at the time of the Napoleonic wars with the use of semaphore arms which signalled messages by a change of position (rather like a railway signal). Systems of this sort were set up in France, England and the US. It is ironic that today, with microwave technology and lasers, we may be going back to this system of beaming messages directly from one hilltop station to another over vast distances – even using satellites in the process.

The development of electronic communication is a fascinating story of observation and invention, in which it was interested amateurs who made most of the advances. The usual dividing lir between scientist and technologist did not exist, for the man applyin the scientific principles directly to an invention was often the ma who had discovered the principle himself (e.g. Faraday, Marcon Edison). The telephone was invented by a teacher of deaf-mute The telegraph was developed by a retired Indian army man wh made anatomical models for a living. His name was Cooke and i partnership with Professor Wheatstone he developed a device i which a current transmitted over wires caused magnetic needles t point to different letters at the other end. This has remained th basic principle of the telegraph: a current altered in some way a the input end operates (via magnetism) some output device at th other end. The invention of the Morse code meant that the inpu and output devices only had to indicate an on/off signal (for instanc a buzzer or light would be on for a long period giving a 'dash' or short period giving a 'dot'). This excellent idea simplified telegraph – it was much easier than having magnetic needles pointing t letters. The Morse code is one of those elegant creative ideas whic arise directly from the mind without requiring any special technic knowledge.

A telephone is a very much refined telegraph which had to wa the development of microphones and loudspeakers. At the input en the microphone converts ordinary speech into alterations in a electric current and at the output end the loudspeaker changes th current back into the sound of speech. The development of the devices is a story of individual experimenters working with magnet coils and resistances, noticing results, testing them out, refining an developing ideas into a practical form. The microphone has gor through several stages of development but the loudspeaker is i principle exactly the same as when it was first invented.

Once the principle of transmitting electric current through wir was known, the development of the telegraph and telephone wa almost inevitable. Because we can think of wires as pipes or tube the business of transmitting a message by wire seems much le extraordinary than transmitting it without any wire at all. Th development of wireless or radio is a remarkable story becau only three people were involved. A much under-appreciate mathematician, James Clerk Maxwell, had worked out the behaviou of all electromagnetic waves from radio waves to X-rays, at a tin when neither of these was known. Much later Heinrich Her demonstrated radio waves in a primitive manner by showing how spark across a gap in a loop could excite another spark across a ga on the other side of the room. From this tiny observation Marco developed wireless to the point where he could transmit messag across the Atlantic and from ship to ship. To do this he had to mak a number of inventions along the way, though he benefited to son extent from the technology developed in the field of telegraphy.

What is extraordinary is that on one crucial occasion he succeede by being wrong. He was a good enough physicist to know that rad waves do not follow the curvature of the earth, so he should hav known that radio waves transmitted from Cornwall would trav out into space rather than reach America. Nevertheless he made h apparatus powerful enough to transmit radio waves to Americ and they were detected in Newfoundland. He had succeeded. Wh he did not know was that an ionized layer in the upper atmosphe had bounced his radio waves back to earth. The inventor sometim has to step outside the boundaries of the known and try things o instead of just theorizing.

The full development of electronic communication depended c two key amplifying devices which could magnify the tiny curren that were received: the triode valve and the transistor. Both can later and the latter is described under 'Key Devices' (p. 231).

Transmitting speech and sound were difficult enough, but easy hen compared with the complexity of transmitting a picture. The ectronic processes were much the same once the picture could be nverted into an electronic signal, but how was the conversion to ke place? There were two approaches to television. At first a echanical method seemed possible but eventually the electronic proach, using cathode ray tubes, turned out to be vastly superior. his is an example of how a false start may be sufficiently successful make the electronic process worth chasing.

Although pictures have always been more complex than speech or und, the direct recording of sound came much later than the direct cording of pictures in the camera. The gramophone is a mechanical cording device and does not depend on electronics, although these e now used to amplify and refine the sound. The gramophone was vented when Edison was recording Morse dots on waxed paper pe and noticed a hum when this ran past the needle too rapidly. om this he went on to experiment with waxed cylinders. The path development was fairly straight right down to the stereo records today. True electronic recording of sound involved the magnetic pe. Like the camera, this is basically a very simple invention, and ry effective. We think of it as recent but in fact it was invented in 00 by Vladimir Poulsen, who demonstrated the principle with a eel strip. Nothing significant was done about the invention because amophone recording seemed adequate. A big impetus was given to the emergence of magnetic tape recording by the computer, which relies heavily on this type of memory. Because the system is entirely electronic there is growing interest in tape recording and the basically mechanical gramophone system may eventually disappear.

Electronics have made a fantastic difference to communication, and yet the world might not have been very different without them. Apart from emergencies, rapid communication may be something of a luxury. Perhaps print technology would have been sufficient when coupled with an efficient transport system. But communication need not be only practical. Musical instruments can be regarded as an aesthetic communicating device for carrying a message from the composer, through the player, to the listener. Today the steps may involve written music, the player and instrument, the electronics of the recording studio, the disc and the gramophone to play it. The essential device is the musical instrument. The history of musical instruments is one of diverse developments and improvements. New families of instruments have arisen from time to time or been borrowed from other cultures but the basic instruments have remained the same for centuries. Perhaps this is not surprising because it is what the player does with the instrument that matters. Yet new technical developments in instruments – such as the electronic music synthesizer – have opened up whole new areas of musical interpretation and expression. More than anywhere else, musical instruments show technology as the servant of aesthetics.

Man Moving

Boat

Although the discovery that wood, or plaited reeds, will float on water and can be propelled is doubtless very old, few early boats have survived. At Star Carr, in Yorkshire, a wooden paddle has been uncovered from a peat bog which is dated *c.* 7500 BC: this must have been used to propel a dug-out canoe, of which we have an example found at Pesse, in the Netherlands (*c.* 6300 BC). Such dug-outs – hacked out with flint tools – could be large: that found at Brigg (Lincolnshire) is 16 m. long, and has a beam of 1½ m.

The early Egyptians and inhabitants of Mesopotamia would have built their boats of reeds, covered with pitch, for the navigation of canals or rivers. But it is not known how or when seagoing vessels were first built. Reed boats with sails were used by 3000 BC, and these, or wooden-hulled boats, perhaps with keels by 2000 BC, must have been essential to the trading nations of the eastern Mediterranean – for example the inhabitants of Crete.

Punting or paddling no doubt preceded the oar, but these methods of transport were unsatisfactory, as the legendary Babylonian hero Gilgamesh found out when he tried to cross the Ocean: at the waters of Death the ferryman told him to make an 18 ft. punt pole, but even 120 of these whipped together would not touch bottom, so Gilgamesh 'stripped himself, held up his arms for a mast and his covering for a sail'.

Two other types of boat which did not require the sophistication of a paddle-turned-fixed-rudder are very early, and survive today: the Irish 'curragh', or coracle – a frame of interwoven wicker covered with hide, then tarred – and the more simple inflated animal skin, which ancient reliefs show astute Assyrian soldiers using to carry themselves and their belongings across rivers. Animal skins still serve this purpose today in the Near East – as does the reed boat used on the Nile 5,000 years ago.

Two types of seagoing vessels emerge in Greek times: the large, solid and unwieldy merchantman and the much lighter ship of war which, from the 6th century BC, bore a large ram on its prow. Such vessels carried marines who, once two boats were interlocked, would fight a floating land-battle. Tactics in the Mediterranean changed little in 2,000 years; the battle of Lepanto in 1571 saw the forces of Spain, Venice and the Papal States combine to defeat the Ottoman Turks – the last contest of importance between oared boats (although Louis XIV still had war galleys 200 years later).

Long before Lepanto, the Vikings had fast war sailing-ships; but a newly developed form, the multi-masted, multi-decked and heavily armed galleon, was to aid the transference of power from the peoples of the Mediterranean to those of north-western Europe and Spain and Portugal.

C.M.B.G.

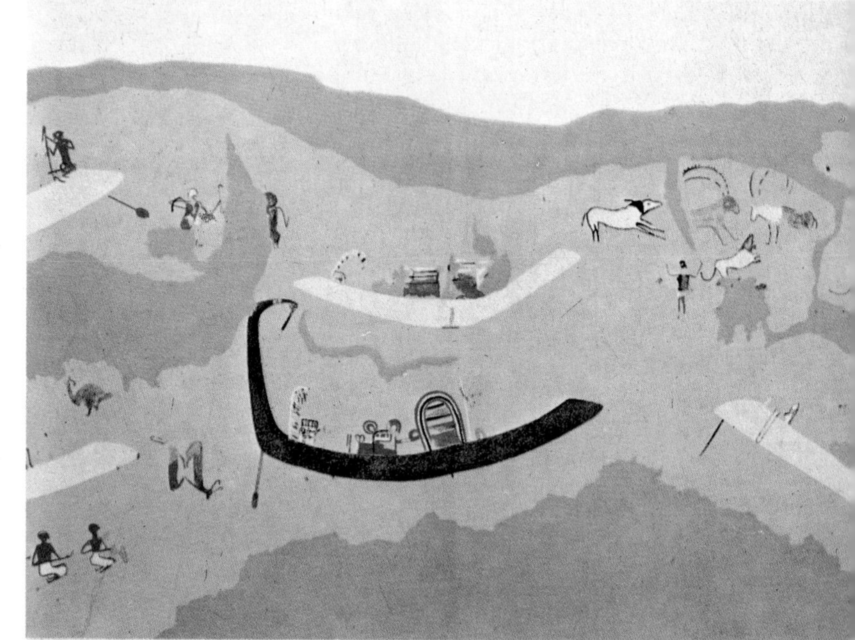

Kite

The kite is the earliest form of true aircraft in history, and was invented in China probably as early as 1000 BC. The earliest known kite in Europe dates from the 14th century. Kites in the Orient were used for religious purposes, for fishing, for military signalling and for man-carrying: in the latter, the man was spreadeagled on the upper surface of a plane-surface kite. In Europe, the kite first appeared as a military standard, and by the 16th century had become popular – in a form imported from the East – as a plaything for children.

The common kite played a vital role in the development of the aeroplane when in 1804 Sir George Cayley – the founder of the science of aerodynamics – made his first glider model of modern shape, using a kite as the main plane.

The most important modern development of the kite came in 1893 with the invention by Lawrence Hargrave of the stable box-kite, a type still flown today the world over. After some tentative man-carrying tests by Baden-Powell in 1894, the European man-lifting kite for naval and military observation was perfected in 1901–3 by S. F. Cody, the British Army adopting these kites in 1904. With this system a supporting train of box-kites was put up first, and the man-carrier then run up the line to whatever height was desired.

One of the most valuable uses to which kites were put in the last century and the early 1900s was to lift meteorological instruments to great heights: this too was generally carried out with a train of kites.

Glass vessel from Afghanistan, with a representation of the Pharos of Alexandria – most famous of the early lighthouses.

Opposite, above, *ancient Egyptian papyrus boat: fresco from the tomb of Neb-Aman. Thebes.* Opposite, below, *road-making in France. Painting by Joseph Vernet, 1774.*

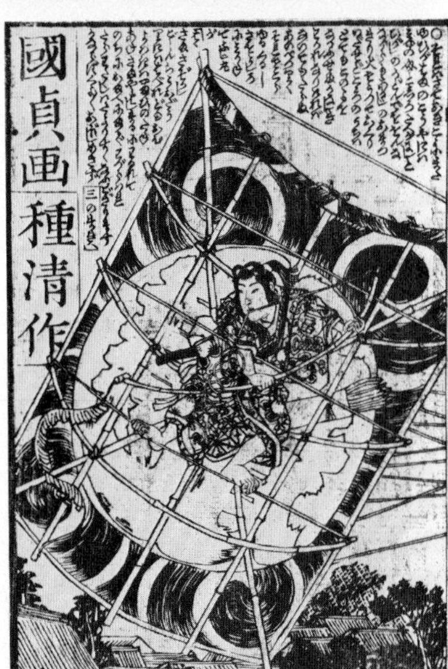

Man-carrying kite, from a 19th-century Japanese adventure book – the first known illustration to show how such a thing could be possible.

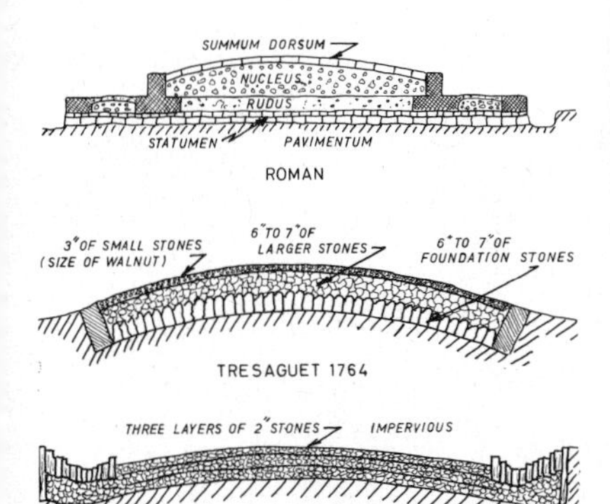

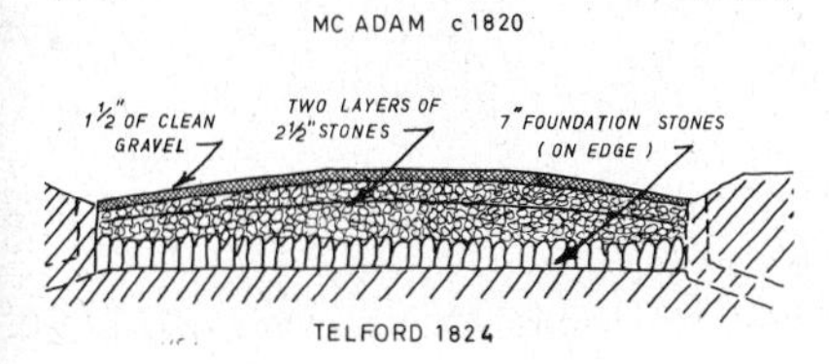

Roadmaking through the ages: four cross-sections.

In 1905, and later, the box-kite principle was applied to aeroplane wings, first on float-gliders, and then on powered machines, an array of large cells being used for the wings, and a smaller one for the tail-unit.

In modern times kites have been used not only as playthings but to hoist radio aerials from survival rafts and boats to save shipwrecked sailors and airmen brought down at sea.

C.G.-S.

Lighthouse

Beacons to indicate harbours at night were known to the Greeks of the fifth century BC and no doubt existed in the Mediterranean long before this, given that navigation, for any boat unable to hug the shore, was much easier by the stars than by daylight. But the first permanent constructions are Hellenistic and of these the most famous was the Pharos of Alexandria (built *c.* 300–280 BC): 'at the extremity of the isle is a rock, which is washed all around by the sea and has upon it a tower that is admirably constructed of white marble with many stories and bearing the name of the island' (Strabo). That other wonder of the ancient world, the Colossus of Rhodes, was also a lighthouse, popularly supposed in later times to have stood astride the harbour of Rhodes: for it was in the form of a statue of the sun-god, Apollo. It was certainly a colossal figure, built by Chares, a pupil of Lysippus, but it must have stood near the harbour mouth, not astride it. The statue was of bronze, and an earthquake destroyed it in 224 BC.

All early lighthouses – including the many built by the Romans, for example that at Dover built shortly after AD 43 to aid the invading forces – would have been towers of stone or rubble, with a wood or coal fire burning in a basket at the top.

It is not until the eighteenth century that improvements were made: the Eddystone lighthouse in England progressed from a timber structure (built 1698; lost in a storm in 1703), via a building of oak and iron (built 1708; destroyed by fire in 1755) to Smeaton's concrete tower of 1759. All three were lit by candles backed by simple metal reflectors.

In 1784, Aimé Argand invented an oil-burner suitable for lighthouses, and this was not surpassed until the end of the nineteenth century, which saw the introduction of acetylene gas and, before that (1862, on the Dungeness Light), electric carbon arc-lamps. Today, xenon high-pressure arc-lamps are used.

Argand's wick was used with a parabolic metal reflector, but in 1822 Augustin Fresnel improved lighting by producing a lens so ground that the beam of light was concentrated and projected through it. This compound lens is in use to this day.

C.M.B.G.

Roadmaking

Oddly enough, the wheeled vehicle for the transport of goods and passengers is thousands of years older than the road. The Greeks, who believed that woods and springs and other scenic features were inhabited by supernatural beings, were rather reluctant to interfere with nature, which is inevitable when a road is built. The few highways they constructed were rather avenues leading to places of worship than roads for transport; overland, carts had to carve their own wheel ruts on the country lanes.

The Persians, who had a formidable empire during the 500 years preceding the Christian era, were the first to build highways, which ran from Asia Minor to India, well dotted with inns and relay stations. These roads were not primarily meant to facilitate travelling and the inter-regional exchange of goods – they were military and administrative roads, a strategic necessity for moving troops and equipment, and for rapidly communicating orders and information between the seats of government and the outposts of the empire. The Chinese, too, built highways for the same purposes, but we know little about their roadmaking techniques.

We know more about the achievements of the Romans; their highway system was so excellent that many stretches of it are still in existence and have been archaeologically examined. They furnished their entire empire, from Hadrian's Wall to the Persian Gulf and from the Atlas Mountains to the Caucasus, with splendid roads, extending over some 50,000 miles at the height of their imperial power. Uninhibited by the Greeks' superstitions, the Romans cut their highways as straight as possible across the country, over steppes and mountains, draining marshes and building bridges by means of techniques which their engineers had developed. The Roman roads were built to last, to provide safe communication in all climates and seasons for marching legionaries and fast dispatch riders, for chariots and heavy covered wagons.

The materials they used were those they found locally. As a rule, the Roman engineers laid down a road bed of large stone blocks, and covered it with layers of small, broken stones and sand. The surface was made either of basalt blocks or, if none were handy, of gravel set in lime mortar. Where no suitable stone was available, particularly in marshy regions, they built wooden causeways on piles. Most roads had kerbstones, and the towns had raised footwalks on either side of the roadway; in regions where heavy rainfalls were likely, the roads were given a good camber to make the

Vision of St Eustace *by Pisanello. Not only the horse but also his harness was a status symbol in the late Middle Ages.*

Above, *the rigid padded collar which enabled the 10th-century European horse to haul heavy loads.*

water drain away. Labour was no problem – the work was done by prisoners of war or by forced labour collected from the surrounding country-side.

When the Roman Empire broke up, many of its roads began to decay through lack of use and maintenance. During the whole first half of the Middle Ages hardly a new highway was built on the European continent. England's roads were in a particularly bad state, and this was probably the reason why that country produced the first modern road-makers. The earliest of them was a blind man, John Metcalf (1717–1810). Despite his handicap, he built 180 miles of excellent highways, mainly in Lancashire and Cheshire, within 30 years. Like the Romans, he gave his roads a solid foundation of stone blocks, ramming in layers of chips and broken tiles on top of them; however, he did not regard an extra surface as necessary.

Next came Thomas Telford from Dumfriesshire (1757–1834). He used a double layer of small stones as a foundation, with even smaller ones filling the gaps; on top of this came a 7 in. layer of broken whinstones, and finally a surface of gravel. Telford's contemporary and compatriot, John Macadam (1756–1836), has bestowed his name to the surfacing of roads with granite chips or other durable stone, broken small enough to form a smooth surface; this became a particularly effective method when, 30 years after Macadam's death, the steam-driven road-roller came into use, invented by an ex-farmer from Kent, Thomas Aveling.

The coming of the motor-car early in our century made new road surfaces necessary; the most important of them were 'tar macadam' and asphalt (bitumen mixed with sand and pitch). The modern motorway, however, requires concrete foundations, usually with a reinforcing steel mesh embedded in them, interspersed with large sheets of waterproof paper to prevent 'water creep'. After the uppermost concrete has hardened, a surface of asphalt or tar macadam can be applied, but frequently this is done only after a few years' use and wear.

E. L.

Yoke

As late as *c.* 1420 BC in Egypt a T-bar was tied over the horns of beasts pulling ploughs, carts or chariots; the idea of fitting the yoke over their shoulders, making it comfortable by means of arched sections and pads, was a comparatively late development. But to keep it in place, the yoke also had to pass under the chest: this was a disadvantage, for the yoke then tended to ride up and choke the animal, and the more forcefully it pulled, the greater the discomfort. It has been surmised that the practice of using paired animals derives from ploughing – a practice which well antedates the invention of the wheel – although men and women probably pulled the first ploughs.

But the yoke is not merely a connecting-rod between puller and pulled; it can also enhance the suspension, the comfort and, therefore, the manoeuvrability of any two-wheeled vehicle, particularly the war chariot. Towards the end of the 2nd millennium BC in Mesopotamia, the axle on the two-wheeled chariot (itself a faster development of the two-axled vehicle) was moved towards the rear so that the warrior stood, not bearing directly on to the axle, but forward of it, with much of his weight taken by the yoke. And the increase in comfort for man and beast, and indeed efficiency, must have been great: the weight of the man kept the yoke from riding up and choking the beast, which could therefore pull more efficiently, and the man would not have every bone in his body shaken through standing on the axle. We can see the advance if we compare the four-wheeled chariots on the Standard of Ur (early 3rd millennium BC) with a one-axle chariot with yoke and harness shown on a stone relief of an Assyrian warrior in the Palace of Assurbanipal, Nineveh (7th century BC). The Greeks of the classical period used chariots, but only as 'taxis' to get them into battle; the Romans found the horse-cavalry more convenient. But the yoke remained, little changed, as an attachment for carts and ploughs.

C. M. B. G.

Saddlery

In Western civilization the horse early became a symbol of status, an essential part of an army as well as a beast of burden. It was originally a creature of the plain and it seems reasonable to assume that domestication was the work of nomads of the European Steppes, although it is not certain when this took place.

In Assyria the horse was certainly domesticated by 2000 BC, and Egypt first met it when the Hyksos swept down from the north to conquer the country in the seventeenth century BC.

Once the idea of domesticating the horse had arrived, it was necessary to find some means of guiding it, and very early Assyrian sculptures show a length of rope or leather with a loop round the horse's lower jaw. In a comparatively brief space of time, sculptures were depicting quite modern-looking bridles. The horse was controlled by a bit and the rider was able to vary the degree and direction of pressure by means of bars and straps. There seems to have been little further development until the Middle Ages, when bits became elaborate, with bosses and bars on the ends.

The whipple-tree, first depicted on the Bayeux Tapestry (11th century), enabled a pair of horses side by side to pull loads round corners without putting all the strain on one horse.

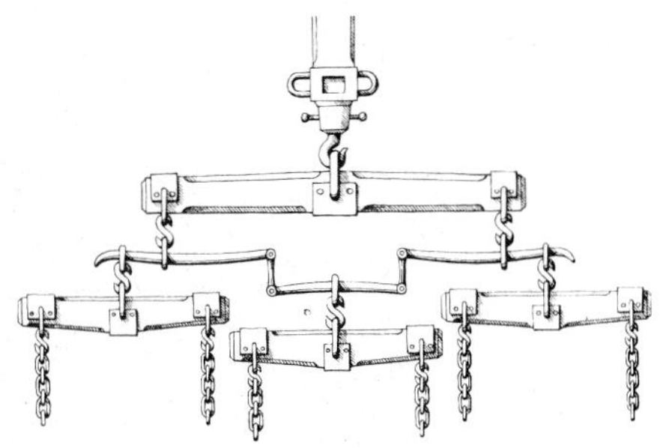

The bit, in a fairly basic form, goes back into pre-Christian times. Elaborations in the shape of bosses and bars can be seen in medieval paintings and sculptures.

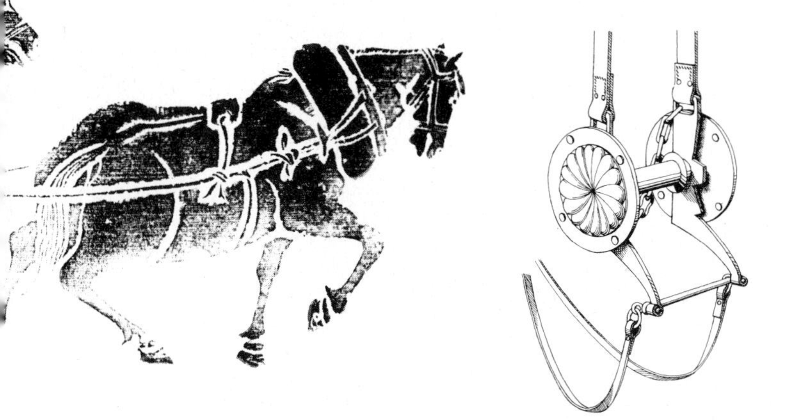

From the Standard of Ur, c. 2600–2500 BC, comes this picture of donkeys drawing a cart. The yokes would have pressed on their throats and reduced their hauling power.

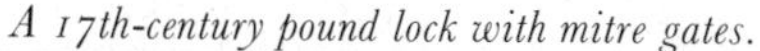

A 17th-century pound lock with mitre gates.

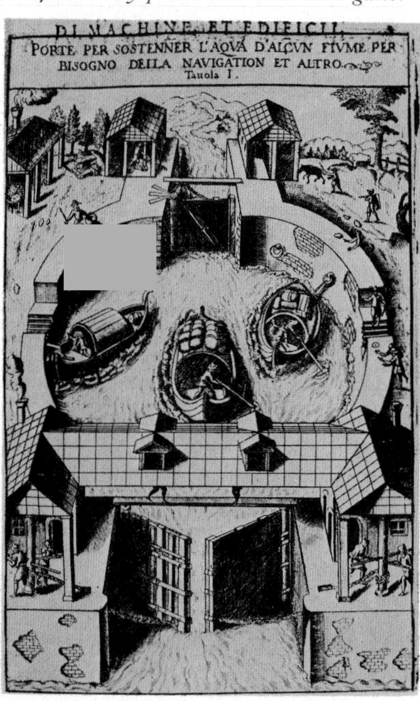

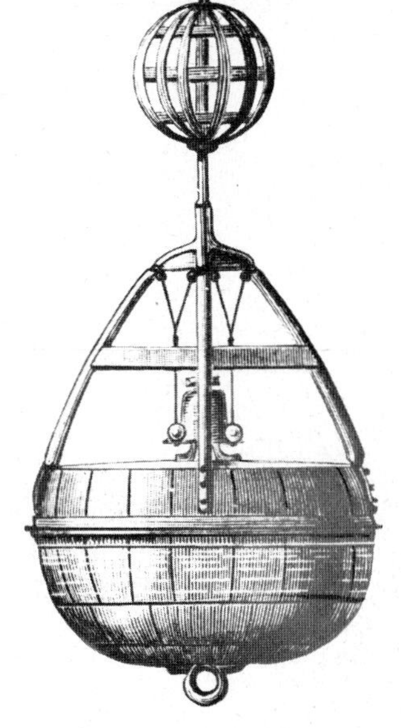

The breakthrough in the design of buoys did not come until the 1880s with the addition of warning lights and sound signals. This is an 1880 bell-buoy.

The second means of control was the spur, which was probably first used in prehistoric times. At first it was little more than a simple, sharpened attachment at the heel, but even so it was vastly more effective than mere pressure of the knees.

In western Europe, the earliest true form was the prick-spur – a U-shaped fitting round the ankle, from which extended a short shank with a spiked tip. Rowel-spurs, with a freely rotating spiked wheel, became common after 1325.

The earliest 'saddle' was simply a folded blanket or cloth. Presumably, some form of padding was the next step. It is not certain exactly when the wooden-framed saddle was developed, as it is difficult to decide whether saddles depicted in pictures are padded or framed, but it probably arrived in Europe during the first century AD. To ensure a good seat, the front (pommel) and rear (cantle) of the saddle were made higher than the seat. (For jousting, the cantle was normally very tall, with two curved arms which partly encircled the rider's waist, and the pommel had steel protective plates and projections to keep the rider securely in place.) Saddles were frequently decorated and fashioned from ivory or leather.

Despite the great value of the horse, its full potential was, until the Dark Ages, far from realization. A bareback rider was subject to many limitations since, in spite of his skill, his seat was not secure. He could not use a sword with maximum effort, he could not dodge easily and a heavy blow on his shield stood a good chance of dislodging him.

It is strange that it took nearly seven centuries from the introduction of the saddle for riders to devise an answer to the problem – the simple stirrup; even stranger because there is a form of stirrup depicted on Assyrian sculptures dated 853 BC.

Stirrups of modern form apparently originated in China and were in common use there by the fifth century AD. Knowledge spread slowly westwards, and it was not until the eighth century AD that the stirrup reached western Europe. With a stirrup secured to the saddle, the warrior now had a firm platform on which to deliver a hard, downward slash with a sword; he could block a thrust from a lance or dodge by moving himself in the saddle. As a result of this greater efficiency swords were made longer and armour became more important.

F.J.W.

Canal lock

Until the nineteenth century the canal lock played a more important part than the wheel in the development of civilization and commerce. For, unlike a wheeled vehicle, which requires some form of firm, man-made track, a boat needs only the natural element of water – provided it can be made to overcome differences of water-level.

The ancestor of the modern lock was a primitive and time-consuming device called a 'navigation weir' or 'flash-lock'. This consisted of a navigable gap in a masonry weir which could be closed at will by a single wooden gate with a sluice or similar movable device. Passage through such flash-locks was at once tediously slow and perilous, but nevertheless they had a very long life. They existed in China in 50 BC; and two were in use on the Warwickshire Avon in England until after the last war.

It was in AD 983 that a Chinese named Chiao Wei-Yo was struck by what now seems to us a blinding glimpse of the obvious. He had made two flash-locks only 250 feet apart on the West River section of the Grand Canal of China near Huai-yin. The short reach of river impounded between them acted as an equalizing chamber; delay and hazard were both eliminated; the pound-lock had been invented.

Knowledge of Chinese inventions rarely filtered through to the Western world. When, after an interval of four hundred years, the first pound-locks appeared in the West they owed their origin to a similar but quite unrelated brainwave. The earliest indisputable example was built near Bruges, Belgium, in 1396.

All the pound-locks had vertically rising or 'guillotine' gates. The swinging type of lock gate (or mitre gate) was invented by Leonardo da Vinci, when he was engineer to the Duke of Milan. Da Vinci built six locks with gates of this type in 1487.

L.T.C.R.

Buoy

The first recorded evidence of the use of buoys dates from about the 11th century, when European trade (as opposed to local trade) began to expand, mainly under the dominance of the Hanseatic League which took a keen and vigilant attitude in matters affecting the safety of ships. As early as 1066 buoys guided ships through the channels of the River Weser, and prior to the 17th century the most extensive use of seamarks was to be found along the German and Baltic coasts. More important, the German states were the first to evolve a method of directing ships through navigable channels by giving buoys specific colours and shapes for indicating port and starboard. The buoys used were either casks or staves of wood banded with hoops of iron and moored by chains to large stones.

When Henry VIII came to the English throne, he realized the importance of maritime trade to England and in 1514 – only five years after his accession – he formally established Trinity House,

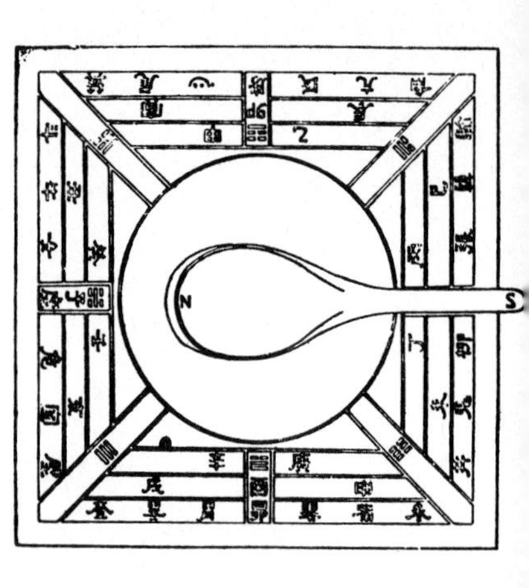

The directional properties of magnetite, discovered by the Chinese in the 1st century BC, gave early navigators that essential instrument, a compass. This Chinese model consists of a magnetite spoon resting on a polished copper plate.

which for over 450 years has provided essential aids to navigation for shipping. English buoys and beacons are shown as being in existence during the 16th and 17th centuries from books of sailing directions known as 'Rutters'.

In 1845 George William Lennox, whose firm supplied wooden buoys to the Trinity House Service, produced a design for the first iron buoy. Trinity House accepted the idea in principle and a prototype was built, 8 ft. high and barrel-shaped, of riveted iron construction. Today the trend is moving towards the extensive use of plastics, although in general buoys throughout the world are still made of either iron or mild steel.

Before the use of lights, sound signals were used on buoys. These usually took the form of a bell fixed to the top of the buoy, the sound being produced by clappers operated by the buoy's movement in the water. In the late 19th century an American, J. M. Courtenay of New York, developed an automatic whistle for use as a sound signal on buoys. The rise and fall of the waves compressed air in a tube fitted to the bottom of the buoy. The up-and-down motion of the buoy caused a piston to move in a cylinder, expelling air through a whistle attached to the top of the buoy and thus producing an audible sound.

G.T.

Navigation

Until the middle of the thirteenth century AD the overcast skies of autumn, winter and early spring had kept ships in harbour for half the year; without sight of the sun by day or stars by night, pilots quickly lost all sense of direction at sea. But in the last half of that century the trading season of Italian ships in the Mediterranean became progressively extended until, by about 1300, it lasted all the year round. And for the first time Mediterranean ships began trading regularly through the Straits of Gibraltar, past the Atlantic coasts of Spain, Portugal and France to the Low Countries and England.

A series of inventions had begun to transform the art of pilotage into the science of navigation. By about AD 1190 Italian pilots had begun to use an iron needle floated in a bowl of water and magnetized by a piece of lodestone or magnetite to check on their estimate of the direction of north when skies were overcast. By 1250 this had been developed into the mariner's or sea compass, a graduated circular card attached to the needle balanced on a pivot in a glazed box, which thus registered direction all round the horizon by day and, illuminated in a binnacle, by night.

Concurrently, the sand-glass had been developed to give the pilot the means to measure continuously equal intervals of time – usually half an hour or an hour – to enable him to estimate the speed of the ship in miles an hour and hence the linear distance sailed on the course steered over a given time. Methodically written sailing directions giving the magnetic compass direction and distance in miles between places and a geometrically constructed portulan chart, drawn to scale from compass directions and distances, enabled the pilot to select his course and to plot his progress.

The directional property of magnetite appears to have been identified first by Chinese necromancers about the first century BC; they used a spoon of magnetite in the shape of the Great Dipper to indicate the Pole on a polished copper celestial plate. By about AD 1090 Chinese pilots, when skies were overcast, were using a south-pointing needle floated in water. The origin of the European compass has not been established.

By 1450 ships had increased in size and better directional control had been achieved by the centre-line rudder and by hoisting sails on three masts. The aftermost lateen sail was used as a steering sail, while the main propulsion came from the traditional square sails of the north hoisted on the main- and foremasts.

Also by that year the Portuguese were pushing south down the Atlantic coast of Africa in search of Guinea's gold. They needed an aid to check their charted reckoning of course and distance sailed. Astronomers taught them to measure and record on a quadrant the altitude of the Pole Star at Lisbon and to convert the angular differences into miles sailed.

They thus learned to check their estimated position by scientific measurements independent of the effect of ocean currents and human errors.

D.W.W.

Dredger

'As the River Lez at Lattes (in Languedoc) is often choked with weeds and rubbish a dredging machine is constantly at work. It is a boat furnished with cables and iron rakes which, as they turn, pull all the plants out of the river and so maintain the depth and navigability of the river.' A Swiss student travelling in the south of France in 1595 gives this account of the earliest type of dredger. But even if this machine was new to him, dredgers pulling giant rakes had been used in the Low Countries for two centuries. The idea was to prevent plants retaining the silt, and to loosen it in the hope that the current would then carry it off and deposit it where it would not obstruct traffic. A better method was invented towards the end of the sixteenth century – or possibly adapted from a Chinese pump that dated from a least a thousand years earlier. The new dredger used an endless chain of paddles, like a waterwheel stretched round two axles. It was

Early 17th-century dredger in the Low Countries. Drawing by Roelant Savery.

Right, *the cross-staff, successor of the astrolabe and forerunner of the sextant.* Far right, *19th-century sextant.*

Halley's diving-bell with air barrel (left) and diver in 'cap of Maintenance'.

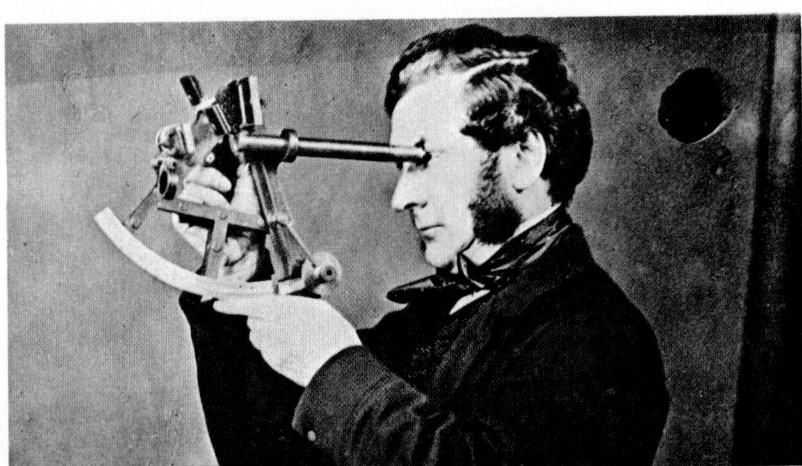

certainly already being used in Holland while the huge rake was still combing out the bed of the Lez, and was the ancestor of all modern dredgers. But the paddles still only disturbed the sediment, and left it to the stream to remove it. Where the current was not strong enough, some other method had to be tried. One way was to use buckets instead of paddles, carrying the silt up into the hold of the dredger. Another was to use scissoring grabs, similar to those now used in most earth-moving and excavating machinery; this method was first tried out in the lagoon of Venice.

A.K.

Diving bell

'Some good Effect has been of Diving; but if Mr Halley's should succeed, of which (were the Wars at an end and the Seas secure) he seems very sure . . . it would be very considerable.'

This is how John Houghton summed up the investment potential in the salvage activities of a company set up by Edmund Halley, a man better known to the world for his contributions to astronomy. The reason for the tip probably rested on the fact that Halley had devised a system which enabled men using a diving-bell to stay under water for hours on end.

Legend describes how in the fourth century BC Alexander the Great descended into the Bosporus in a large barrel to observe the life beneath the waves. No doubt the great general was only following a fashion already established in the warm waters of the Mediterranean. Aristotle, a more reliable chronicler than the scribes of legends, also describes a rudimentary form of diving-bell.

In Britain, where the climate – and the water – is not so kind, Sir Francis Bacon described the diving-bell in 1620 as a 'sort of metal barrel lowered into the water open end downwards'. And men using a diving-bell recovered guns from the *Florencia*, a sunken Armada galleon, in 1665.

Obviously Halley did not invent the diving-bell as such, but he did lay the foundations of the diving system which is used today. His system eventually included not only a means for renewing the air inside the bell, but also a way of providing air to men, who wore a 'cap of Maintenance', working in the water outside.

Halley's diving-bell, which was made of wood, had a capacity of sixty cubic feet. It was weighted with lead, boasted a plate-glass window, and had a bench and platform for its occupants. The whole thing was suspended from a spar attached to the mast of the parent vessel.

During their stay under water, the occupants would release the spent air from a tap at the top of the bell when new supplies came down from the surface. The fresh air was sent in barrels encased in lead. Each one had an open bung-hole at the base and a flexible leather hose at the top which was weighted so that the open end hung below the bottom of the barrel. Once the barrel had reached the diving-bell, the weighted hose was lifted inside the chamber. As soon as its open end was higher than the bottom of the barrel, water started to enter the bung-hole, pushing the air with some force into the diving-bell.

Like many other pioneers, Halley took part in his own experiments. He wrote: 'I myself have been One of Five who have been together at the Bottom, in nine or ten Fathoms Water, for above an Hour and a half at a time, without any sort of ill consequence.' Halley himself demonstrated the most advanced version of his system in the Thames in 1716.

Halley's company must have met with some success. In fact, earning from this source may have helped to repair the family fortunes: his father suffered serious losses from the Great Fire. In 1686, Halley had apparently been obliged to resign temporarily his Fellowship of the Royal Society in order to take a salaried position as its clerk. He gave up the position about ten years later. As early as 1691, however, he was being commanded by the Royal Society to tell of a successful salvage operation – in this case on 'a Guiney frigat'.

T.L.

Sextant

In order to find out how far north or south you are, it is necessary to have an instrument which can determine the latitude by measuring the angle between the horizon and sun at noon, or some fixed star. At first the mariner's astrolabe was used to 'shoot the sun', but this was too unwieldy for the heaving deck of a ship. It was replaced by the cross-staff, which remained in use for centuries until the introduction of the sextant.

The cross-staff was a piece of wood or bone on which was inscribed a scale along which a cross-piece could be moved to measure the altitude of the sun. A great improvement was made when an Elizabethan explorer and navigator, John Davis of Dartmouth, introduced the back-staff from the Arabs in the Indian Ocean because, by using an eye-slit in the horizon vane, the observer could turn his back on the sun and avoid being dazzled.

In 1731 John Hadley (1682–1744) invented a reflecting quadrant which superseded the cross-staff and was soon developed into the sextant, that is to say an arc measuring the sixth of a circle. Hadley was the friend of the better-known Halley, with whom he developed a reflecting telescope when he became a Fellow of the Royal Society. He then made an instrument for measuring angles at sea with which the observer looked at the horizon and at the sun at the same time by means of a mirror, thus bringing two objects together by reflection (an idea originally suggested by Isaac Newton). The angle between them could be measured on a graduated scale inscribed round the edge of the quadrant, on which there was a sliding vernier scale with minute subdivisions. When a small telescope or shade tube was fitted, observation became easier and sights could be taken independently of the roll of the ship.

The quadrant was tested on board a yacht by order of the Admiralty in 1732 and its readings were so accurate that it was officially adopted in the Navy. In 1757 Captain John Campbell extended the arc of this quadrant to measure angles up to 120 degrees instead of 90 degrees and the instrument became known as a sextant.

It consists of a triangular frame, one side of which is an arc on which there is a scale of degrees. An index pointer pivots across the frame and the arc, and a system of reflecting mirrors brings together the two objects whose angle is to be observed.

C.L.

Balloon

The balloon could have been invented, and men could have flown through the air, at any time back to five thousand years ago, when textile fabrics were closely enough woven, and could have been made reasonably non-porous by some primitive coating with gum or other substance. That the hot-air balloon had to wait until 1783 for its successful arrival on the world scene was a gratuitous accident.

The Montgolfier brothers, Joseph and Etienne, were paper-makers of Annonay, near Lyons; they had observed the light debris rising above the kitchen fire and experimented with paper bags, which floated up still more successfully. They made bigger and better bags float up until, on 5 June 1783, they gave a public demonstration in the square at Annonay with a model balloon of 110-foot circumference, made of cloth lined with paper and fastened together with buttons. They lit a fire of wool and straw beneath it and, before an astonished crowd, it ascended and flew away for about a mile and a half.

The news soon reached Paris, but without details of how the 'globe' had been raised. So the brilliant French physicist J. A. C. Charles was commissioned to build one. He could think of only one lifting agent, the inflammable gas isolated by Cavendish, which we now call hydrogen. So he built a smaller model, with fabric rubberized by the two brothers Robert, twelve feet in diameter; they filled it with hydrogen by pouring sulphuric acid over iron filings. This balloon was sent up from the Champs-de-Mars on 27 August 1783.

Meanwhile, the Montgolfiers had arrived in the capital, and after various mishaps sent up a huge, beautifully decorated hot-air balloon, carrying a cock, a sheep and a duck hanging in a basket; this took place at Versailles, in the presence of Louis XVI and Marie-Antoinette, on 19 September.

Then, in the grounds of the Château de la Muette, in the Bois de Boulogne, the world's first aerial voyage started on 21 November 1783 in a still larger Montgolfière, with two men standing in the gallery surrounding the neck; the pilot was the physician Pilâtre de Rozier, and his passenger the Marquis d'Arlandes. They travelled for some five miles over Paris and landed safely.

Charles, amazed by what he saw of the Montgolfier balloon, realized he had invented a new type of balloon altogether; and his man-carrying version was – except for a few refinements – essentially the modern gas balloon, complete with a valve in the crown. He and one of the brothers Robert then made the first voyage in a hydrogen balloon on 1 December and launched the modern history of ballooning. The flight started from the Tuileries Gardens, and ended at the village of Nesle, twenty-seven miles away.

Balloons have been in constant use ever since.

C.G.-S.

Parachute

The first human drop in a parachute from an aircraft took place on 22 October 1797, over what is today the Parc Monceau at Paris. The drop was made by Jacques Garnerin in a ribbed parachute folded like a modern sunshade, which was carried up beneath a hydrogen balloon and released at about three thousand feet. When he had attained the height he wished, Garnerin pulled a release-cord, which detached the parachute from the balloon and caused it to drop away, the canopy opening by the rush of wind into it from below. Owing to the lack of porosity, the Garnerin parachute oscillated alarmingly, making the occupant – in his little 'bucket' – very sick when he landed. Dollfus describes this event, with justice, as 'one of the great acts of heroism in human history'.

Although Garnerin's parachute derived directly from the parasol, the first known design for a parachute is that of Leonardo da Vinci, who sketched a tent-like device about 1485.

Parachuting in the nineteenth century became a standard variation of the aerial showman's act, when mere balloon ascents were apt to lose their excitement.

The first aerial voyage in a hot-air balloon invented and made by the Montgolfier brothers, 1783.

The Atlantic pioneers: the Mayflower *and the* Sirius, *which made the first steam crossing; the* Queen Elizabeth I *and an oil tanker are shown in silhouette for contrast.*

An ascent from the Tivoli Gardens, Paris, in 1817. The aeronaut was seated on a pet stag.

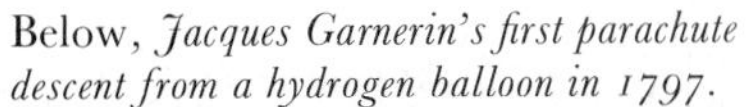

Below, *Jacques Garnerin's first parachute descent from a hydrogen balloon in 1797.*

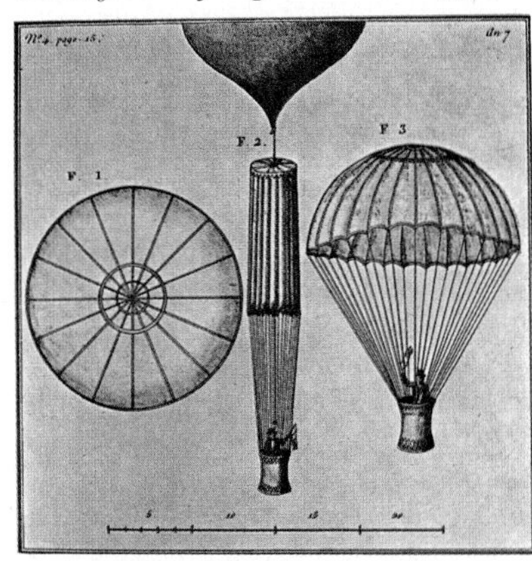

Patrick Miller's first steamboat, which made a speed of 5 knots on Dalswinton Loch, Dumfriesshire.

The aeronaut used a hot-air balloon, beneath which he hung on a trapeze; the parachute was lightly attached to the balloon, which was heated up over a trench. The ensemble would rise as high as the cooling air allowed, whereupon the jumper would simply let go the trapeze; this would pull the parachute away from the balloon envelope, and it would descend safely to the ground.

The first jump from an aeroplane, using a static line to pull out the canopy, was made in the USA on 1 March 1912, by Captain Albert Berry at St Louis.

The first free, rip-cord type of drop from an aeroplane was made in the autumn of 1912 in the USA, when F. R. Law dropped with a Stevens 'Life pack' parachute. It is one of the ironies of aviation history that when, in 1919, thousands of dollars were being spent on evolving the modern rip-cord parachute, no one thought of going back a few years and seeing what the tough old showmen had done, when the Stevens pack would have been discovered and improved upon. In the event it was not until 19 April 1919 that Leslie Irvin made the first descent (again in the USA) with his improved rip-cord type of parachute, the ancestor of most modern chutes.

C.G.-S.

Steamship

The search for a mechanized alternative to sail at sea began in Rome a millennium and a half ago. A low relief cut in the year AD 527 shows a Roman warship with six primitive paddle-wheels driven by oxen. More than a thousand years later Denis Papin, a Frenchman, developed a concept for a vessel driven by steam. A hundred years later another Frenchman, Perier, built a boat that moved experimentally on the Seine. Yet the germinal steamship was built by an Edinburgh banker, Patrick Miller, an eighteenth-century dilettante of science, on the Loch of Dalswinton in the deep valley north-west of Dumfries. Powered by an engine built by William Symington, it made a speed of five knots on its first passage; Robert Burns the poet was a member of its crew.

The succession is singularly clear. Miller, having fathered his steamboat, lost interest. Symington determined to perfect it. He found a patron in the Dundas family and designed a boat that he called the *Charlotte Dundas* to haul barges on the Forth and Clyde Canal. Too successful, the *Charlotte Dundas* threatened to bring down the banks of the canal, and was abandoned.

But Robert Fulton, an American jeweller, artist and engineer, came to Scotland to see it, was inspired, and went to Paris in 1803 to build a version of his own. Barely four years later he set up a steam packet line on the Hudson between New York and Albany.

It needed, however, the invention of a surface condenser in 1830 to solve the problem of freshwater for the boilers and take the steamship to sea as an economic proposition. The ship *Savannah*, using an auxiliary engine for part of the passage, crossed from New York to Ireland in 1818, the Dover-built *Calpe* crossed in 1827, and the *Royal William* six years later; but it was the tiny cross-Channel vessel *Sirius* which made the first passage under steam alone in 1838.

Even after the *Sirius*, two fresh advances in technology were essential before victory was complete – the iron hull and the screw-propeller. The *Aaron Manby*, backed by another Dundas and 'Mad Charlie' Napier, proved the feasibility of the first. The *Francis B. Ogden*, built by Ericsson on the Thames and rejected after trials by a purblind Admiralty, provided the second. In 1843 all three things were combined in Brunel's *Great Britain* – recently salvaged in the Falkland Islands to be brought back to her home port of Bristol.

What was the impact of the steamship on the sea world? The answer is complex. The short version is that it shrank it by more than a half, perhaps by as much as two-thirds; it eliminated dependence on the trade winds; it cut the seasons out of sailing. Only spherical trigonometry can compute its impact precisely but it can be illustrated by the practice of the East India Company, reckoned on a minimum of a year for a round voyage to Bombay; eighteen months was acceptable. A *Great Britain* at an average speed of 9·3 knots, with three coalings stops and an allowance for head winds, could have made three trips in eighteen months. From this kind of blunt fact developed the *Great Eastern*, the *Mauritania*, the *Queen Elizabeth* and the half-million-ton tanker.

D.D.

Railways/Railroads

The Age of Railways began with the opening of the Liverpool & Manchester Railway in 1830. This date marked the first wholly successful marriage of two key inventions – the steam locomotive and the road it ran upon – which enabled man to travel for the first time at a speed greater than the horse's. The railway immediately reduced the London–Edinburgh journey time from ten to twelve days to fifty hours.

The origin of the railway goes back to the primitive wooden tramways used in the mines of Transylvania at least as early as the sixteenth century. Similar wooden railways, which included primitive point work, were used in England to carry coal from local workings to loading depots on the River Tyne in the seventeenth century. As time wore on and traffic increased, these original wooden rails were progressively improved. The working life of their running surfaces was prolonged first by covering them with renewable strips of hardwood, then with metal ones and, finally, by making the rails themselves of cast iron. This last stage of evolution had been reached by 1804 when the first experiment of using a steam locomotive was made by Richard Trevithick on the Penydaren Tramway in South Wales.

Unfortunately, cast iron is a brittle metal and the weight of the first locomotives caused so many breakages that the owners of colliery tramways, who were then the only potential customers for the locomotive, rejected the 'iron horse' as uneconomic and went on using horses. Those who envisaged a national system of railways were ridiculed, and the only practical man with an unshakeable faith in the future of the steam locomotive on rails was George Stephenson, who had already built his *Blücher* at Killingworth Colliery, Northumberland.

There were two types of tramway at this time; the so-called 'plateway' with the flange on the rail and the 'edge railway' in which the guiding flange was on the wheel. Stephenson pinned his faith on the latter. Early in 1821, Stephenson was appointed Engineer of the Stockton & Darlington Railway. Here he proposed using cast-iron rails of his own invention, each three feet long. At the same time, however, John Birkinshaw, of the Bedlington Ironworks at Morpeth, had perfected a method of rolling wrought-iron rails in fifteen-foot lengths. Not only did this entail far fewer rail joints, but the new rails were not brittle and would not break under weight. Although it was not in his own interest to do so, Stephenson at once recognized their superiority to his own and had them laid on the Stockton & Darlington.

On that railway, early Stephenson locomotives shared the coal traffic with horses, and the only passenger service was worked by a horse-drawn coach. Locomotives were then only fit for very slow-speed haulage, largely because their primitive boilers could not produce enough steam for continuous high-speed working.

George Stephenson was determined that on his new Liverpool & Manchester line, which was then being built, steam locomotives would be used for all traffic. But the proprietors of that line thought otherwise. So unconvinced were they by the performance of the Stockton & Darlington locomotives that they proposed to haul their trains along with ropes from a series of fixed haulage engines at the lineside. Stephenson determined to give battle. 'Rely upon it,' wrote George's only son, Robert Stephenson, 'locomotives shall not be cowardly given up. I will fight for them until the last. They are worthy of a conflict.' He was as good as his word. At the Stephenson Works at Newcastle in 1828, Robert set out to design and build an improved locomotive that would convince the most sceptical that it was capable of far wider use.

The result was the *Rocket*, the first locomotive with a multi-tubular boiler of modern type. That steam locomotives were a practical proposition was shown once and for all in the famous trials at Rainhill in 1829, when the *Rocket*, drawing a load, averaged just under 14 m.p.h. for sixty miles and achieved a maximum speed of 29 m.p.h. Such a performance would have been impossible on the old cast-iron rails.

So, when it opened in 1830, the Liverpool & Manchester became the first railway in the world to rely exclusively on steam haulage. It was only thirty-one miles long, but from this small beginning railways quickly spread across the globe, reigning supreme in the transport world until the coming of the motor car.

L.T.C.R.

Railroad technology

Though interest in railways tends to centre on trains, these have increasingly come to be subordinate in terms of capital investment to the railway system along which they run. From the earliest days of rail the whole underlying reason for this mode of transport was the superior load-bearing qualities of metal wheels running on smooth metal track, and in particular the very much reduced rolling resistance (a result of the more intense wheel/surface loading) and thus lesser expenditure of energy in merely overcoming this resistance. Though monorail, air-cushion and magnetic-levitation 'railways' have all been developed, the classic steel duorail remains the dominant form.

By 1840 the form of rail had settled down into two main types, the bull-head supported in 'chairs' and the flat-bottom resting directly on the sleepers or ties used to link the two rails at the correct distance apart (gauge) and distribute the load into the ground. Bull-head rail is a rolled steel section with fattened top and bottom and a narrow centre. The rail is supported in chairs usually of cast iron, in which it is held by wooden wedges or, after 1940, steel clips of patented types. The chairs in turn are bolted through the depth of the sleeper. By 1960 this type of track was almost everywhere giving way to flat-bottom rail, which was used in crude form in early American railroads of the 1850s. The rail has a broad flat bottom which rests on the sleeper, or on a metal pad interposed, and originally was secured by a large spike with turned-over top driven vertically down by hammer blows. Today the sleeper is usually metal or reinforced concrete and the rail is secured by bolts, or by any of many patented fastenings which are usually of resilient steel, to force the rail under a heavy load against the sleeper while simultaneously preventing movement in the horizontal plane. Rails of up to 90 ft. lengths were joined by bolted fish-plates on either side of the narrow part at adjoining ends, but since 1950 continuous welded rail has been introduced in most countries to improve safety and reduce noise.

Today the dominant aspect of track design, and of rolling stock also, is to increase standardization and minimize need for maintenance. Most track today has sleepers embedded in ballast, which must be inspected, cleaned and periodically supplemented or renewed. Since 1947 this work has been increasingly taken over from expensive manual labour by specially designed machines, most of them produced to the designs of the Plasser company, which have mechanized most of the routine functions of track maintenance. Track construction is likewise a factory job, with complete lengths of new track prefabricated away from site and then swiftly set in place in a few night hours, with traffic starting next morning (though the need for reduced speeds during the following few weeks has yet to be fully overcome). Several rail administrations, especially the British, are seeking improved forms of track needing little or no maintenance over periods of many years (subject to occasional survey to check stability of the subsoil). One likely form uses prefabricated concrete sections laid in succession and fastened together; another has the rails laid on a continuous slip-form pavement of concrete specially designed to maintain stability at all temperatures with no cracking or deformation by rail expansion and contraction. Such track is likely to come into use by 1977.

The monorail exists in two chief forms, one laid close to the ground and straddled by the vehicles and the other laid high above the ground and with the vehicles suspended from it. Many monorail systems have been built since the 1860s, almost all of them over relatively short distances in urban environments where it is claimed such systems make fewer demands on available land. Monorail systems are in use with steel wheels running on a steel rail and with rubber-tyred wheels running on metal, concrete or other types of track.

Having established an alternative to the duorail, inventors have since 1958 devised further schemes. The TACV (tracked air-cushion vehicle) marries the hovercraft method of support with a smooth prepared track (exactly what the classic hovercraft does not need). Though absence of wheels means the TACV must expend energy merely in lifting itself into running position, the fact that it nowhere touches the track enables it to run faster, more smoothly and more silently than any previous railway vehicle, and with reduced incidence of defects and with little to wear out. The French Aérotrain was the first TACV to go into public use, various forms having been built since 1967 for

W. Hedley's Puffing Billy, *of 1813, ran on cast-iron rails.*

The machine that revolutionized transport: engineering drawing of a steam-engine, 1848.

Left, below, *American express locomotive, built in 1848 by Norris of Philadelphia.*

Above, *experimental tracked hovercraft. Propelled by a linear motor (q.v.), it has a design top speed of 300 m.p.h.*

Left, *experimental prototype of Britain's Advanced Passenger Train on its first main-line proving run.*

urban and inter-city use. These are supported on air pads and propelled by air propellers or linear motors (q.v.). The Aérotrain runs on track shaped like an inverted T, with the horizontal strips providing support and the vertical walls guidance, but a superior method (used by the British Hovertrain until its termination in February 1973) uses box-section track made of hollow concrete sections, bodily straddled by the 'train', which is guided by the portion overhanging the track on each side. TACV track can be made and prepared for use more cheaply than can steel duorail, but British experience shows that it is likely to be restricted to completely new undertakings. The same is likely to prove true of the two other alternatives to the duorail so far run in full-scale form. The Urba system, promoted originally by M. Barthalon in Lyons in 1965, uses suction suspension, in which a depression in a continuously evacuated chamber causes the suspended vehicle to lift slightly above its guideway. Two German systems, by MBB and Krauss-Maffei, were both demonstrated full scale in 1971 using magnetic levitation, in which attraction by unlike poles on the underside of a T-track is used to support the vehicle without physical contact. Both schemes use linear-motor propulsion. Extensive development is now in hand in Germany, the USA, France and Japan to transfer all this experience to a complete system capable of becoming the standard form of railway in the next century.

Motive power for traditional duorail systems is gradually becoming all-electric, the original experimental train by Siemens & Halske dating from 1879, though primitive 'electric locomotives' were run by Thomas Sturgeon in 1837, by Prof. Charles Page of Washington in 1839 and by Uriah Clark in 1840. Early electric railways operated under direct current at voltages seldom in excess of 660, usually supplied by means of a third rail between or alongside the running rails. By 1900 voltages were climbing to the 1,500 level, and supply by overhead cable suspended in the form of a catenary was becoming dominant. Today electric locomotives are running which can automatically adapt themselves in a fraction of a second to any of four kinds of current supply with voltages up to 25,000 (25 kV.), using mercury-arc rectifiers or thyristors and with precise and automatic power control.

Though likely to fade from the scene by the year 2000, diesel traction is today used more widely than any other railway motive power. The first IC-engined rail vehicle, a spark-ignition petrol-burner, ran in Württemberg in 1899; a diesel-electric passenger railcar ran in Sweden in 1913. Though little new technology was involved, the US railways dramatically changed their image in 1933–37 by switching from steam haulage to use of streamlined and brightly painted diesel-electric locomotives, invariably coupled in multiple under a single crew, for passenger haulage and later for freight. General Motors, the American Locomotive Co. and General Electric have since marketed a range of standard designs used today on more than half the world's rail systems. Use of small diesel engines mounted under the floor of a passenger coach led from 1920 onwards to the increasing use of multiple-unit (m.u.) trains in which groups of coaches furnish their own motive power and can be coupled into trains of any required length.

Various steam-turbine locomotives were built from 1925, and gas-turbine locomotives date from 1941. In 1965 gas-turbine m.u. trains were first used, and today gas turbines are used for some of the most powerful freight locomotives and the fastest non-electric passenger trains. A notable advantage of the free-turbine type of gas turbine is that it gives very high output torque with the drive turbine stationary (i.e. stalled), so there is no need for any form of gearbox or electric transmission.

The origin of the bogie is lost in antiquity, but a modern type of railway bogie was patented in 1812 by William Chapman, who saw the difficulty of negotiating bends with rigid-frame vehicles of a useful length. Since then the subject of rolling-stock suspension has governed both railway speed and riding comfort. The sheer mathematical difficulty of solving the equations of motion made the subject an empirical one – largely trial-and-error – until the British Rail Research Centre at Derby achieved complete solutions in 1966–69. The immediate result was that rigid-frame four-wheel freight vehicles, which previously had oscillated dangerously at speeds above 45 m.p.h., could safely be designed to run straight and true at 145 m.p.h. Bogie vehicles can run at speeds up to at least 250 m.p.h., the only limiting factors being the track and the fact that the power needed increases as the square of the speed. In the 19th century there were several half-baked proposals for passenger coaches suspended from above like a hammock to improve comfort. In 1964 British Rail, Canadian National and the French SNCF independently began designing high-speed trains with coaches tilting on bends to eliminate uncomfortable side-forces on the occupants. The British Rail APT (Advanced Passenger Train), first run in 1972, is unique in having coaches tilted under power by a system which senses lateral acceleration and smoothly imparts tilt to eliminate it. Like all high-speed trains the APT is constructed in ways which dramatically reduce weight, not only to reduce power required but also to reduce costly maintenance. Each bogie supports the ends of two adjacent coaches, a method proposed in 1818 and used a century later by Nigel Gresley for London suburban trains.

Man Moving

In early railways signalling was primitive, even in urban areas. The familiar semaphore arm was invented by Chappé in 1793 and adapted for railway use in 1841. In 1839 the first railway communications were laid, a telegraph being arranged to link all stations from Paddington, London, to West Drayton (and, later, all stations on the Great Western network). In 1844 W.F. Cooke of the Norfolk Railway introduced the concept of 'blocking' the track into sections separated by a stipulated time interval, and foolproof block-signalling followed in 1860. By this time points (switches) and signals were being mechanically connected, the first such interlocking being at Bricklayers' Arms, London, in 1854. In 1893 the Liverpool overhead railway introduced automatic signals set to danger by the passage of each train. Today rail systems are introducing computer-controlled operating and communications systems for safe running without the need for visual apprehension of signals by the driver. Such systems automatically allow for train stopping distance (varying with weight, adhesion, speed, gradient and other factors), sudden emergencies (such as line blockage) and all other eventualities. The objective for both inter-city and urban rail systems is total automation.

W.T.G.

Tramway/trolleybus/streetcar

After the coming of the steam railway, the urban tramway soon followed. New York was the first town to get it: in 1832 John Stephenson, a carriage-builder, laid tram rails right through Manhattan, whose streets were in an appalling condition at the time. Two horses were sufficient to pull a tramcar with 46 passengers, while the rail-less horse omnibuses, which were also introduced in this period, could carry only 26. Paris had its tramway by 1853 and London in 1861.

Early tram-rails protruded above the road surface and were a constant danger to pedestrians and vehicles, which often overturned; after some years the rails had to be removed, and the tramway got into its stride again only after the rails had been laid flush with the road surface. The first two lines of this kind were opened in London in 1870.

Most tramway lines were operated with horses, but there were a few towns with steam tramways and quite a number with cable tramways, among them San Francisco (which still has one), Birmingham, Edinburgh, and London (at Highgate Hill); the cable drums were driven by steam engines. The tramway had to wait for the development of efficient electric motors before it could dispense with horses. Werner von Siemens, the German industrialist and engineer, built the first electric tramway in a Berlin suburb in 1881. One of the rails carried the current, the other the return current, but this was found to be too dangerous for street traffic, and Siemens introduced the overhead contact wire. Surprisingly, it was the English coastal resort of Blackpool which first followed suit in 1884 with its sea-front electric tramway; London got one only in 1901.

Before the First World War there was hardly a large or medium-sized town in the world which had no electric tramways; many, particularly on the Continent, still have them although they are now rather inconvenient in crowded streets owing to their rigid routes and inability to give way. Paris, London and New York have done away with them.

There is, however, a half-way house between the tramway and the bus – the trolleybus. Developed around 1910 in the United States, this is a rail-less tramway or rather an electric bus, taking its current from overhead wires. It can move a few yards to right and left, which gives it a certain flexibility on the road, but it cannot operate where no overhead wires have been built. Another disadvantage is that trolleybuses, like tramcars, cannot overtake one another. For these reasons the heyday of the trolleybus, which began between the two world wars, did not last long. London, for instance, scrapped its trolleybus routes in the 1960s and replaced them by bus lines. Some towns and even capitals where roads are wide and traffic is not yet congested – such as Moscow – are still quite happy with their fast, economical trolleybuses. The first British city – and the last – to use trolleybuses was Bradford, Yorkshire, where they were introduced in 1911 and continued right up to 1972.

Screw propeller

Steam power was shown to be practical for navigation in the first decade of the nineteenth century. Experiments had been made in the last half of the eighteenth century, and three methods of propulsion by steam power were advocated and tested: paddles, jet propulsion and the screw.

In 1775 Benjamin Franklin saw that paddles would be inefficient, and suggested jet propulsion by a pump sucking in water at the bow of a vessel and forcing it out at the stern. In 1782 such a boat was tried on the Potomac, and as late as 1865 the Royal Navy built an armoured sloop, the *Waterwitch*, propelled in this manner.

In 1837 Sir Francis Pettit Smith (1808–74) experimented with his first screw-propelled steam launch; Captain John Ericsson, Swedish ex-army officer – later to become noted as an inventor and to build the famous screw-driven *Monitor* – was experimenting at about the same time. Pettit Smith's propeller was of wood; on the trials about half of it

First electric tram in Berlin, 1882.

Lyttleton's propeller, 1794.

Rennie's propeller, 1839.

Maudslay's variable propeller, 1853.

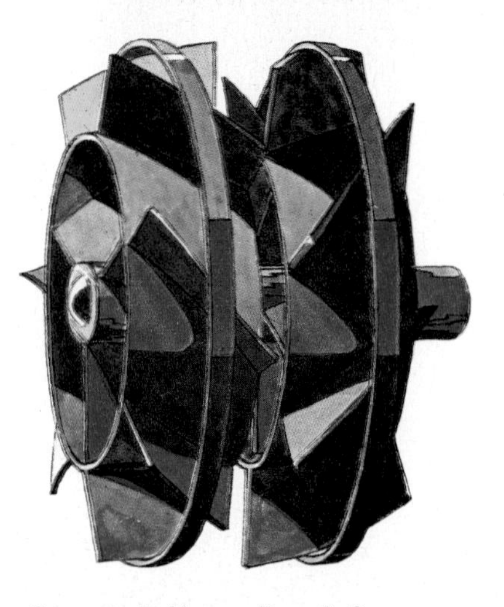

Ericsson's double propeller, 1836.

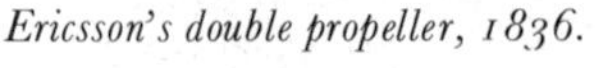

Francis Smith's propeller, 1836.

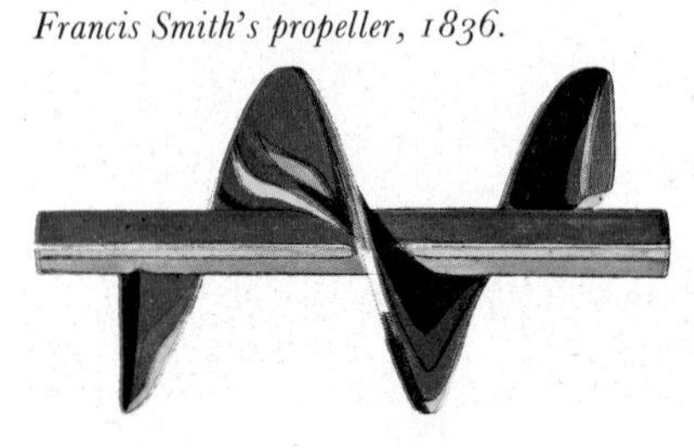

Top left, *early 19th-century célérifère.* Lower left, *1870s tricycle.* Right, top to bottom*: velocipede, forerunner of the penny farthing; early chain-driven 'safety' bicycle; 'reclining' bicycle.*

Right, *first closed diving-helmet.*

Below, *Assyrians swimming under water, breathing from inflated skins. Relief from Nimrud Palace.*

was broken off, and to everybody's astonishment this materially increased the speed.

Encouraged, he built the *Archimedes*, launched in 1838, at Millwall, a three-masted schooner fitted with an engine and screw-propeller. Brunel was at that time building the enormous iron ship the *Great Britain*. After testing the *Archimedes*, he scrapped his plans, already far advanced, for giant paddle-engines in the *Great Britain*, and substituted a huge propeller, with which she successfully crossed the Atlantic in 1845.

Brunel tried to persuade a reluctant Admiralty (which resisted most innovations at this time, fearing that a naval arms race would make Britain's wooden men of war obsolete) to equip a screw-propelled fleet. Paddle-powered battleships were out of the question – shot and shell would quickly smash paddle-wheels and put the ships out of action.

Smaller naval vessels, however, were being powered by steam. Paddle-tugs were used at the Crimea to tow battleships. In 1841, Brunel got the Admiralty to agree to experiments. An 888-ton sloop, the *Rattler*, then being built at Sheerness was adapted for propeller drive, and matched with a comparable paddle-sloop, the *Alecto*. Among other tests, there was a tug-of-war, which ended with the *Rattler* dragging the *Alecto* after her at 2·8 knots. This was conclusive.

Propellers needed a higher speed of rotation than the paddle-shaft; until engine speeds improved, geared drives using ropes, pitch-chains or tooth-wheels were used. The *Great Britain* used pitch-chains.

Although Brunel's next ship, the gigantic *Great Eastern*, used paddles as well as a propeller, by the end of the 1860s the propeller had triumphed.

G.N.

Diving apparatus

Aristotle tells of sponge divers and Livy of treasure divers; but even the best trained of them could not stay down for much more than two minutes. An artificial supply of air was, therefore, an obvious idea. The British Museum in London has two Assyrian bas-reliefs, dated about 900 BC, showing some swimmers under water with inflated goat-skins attached to their girdles, from which short tubes extend to the swimmers' mouths – probably a very early forerunner of our aqualung. The snorkel, too, is quite an old idea; nature itself invented it for some kinds of air-breathing insects which breed under water, drawing their air supply through the hollow stems of water-lilies; and Leonardo da Vinci made sketches of air-pipes leading from the surface to the divers' mouths. In German U-boats, where a large-scale model was used in the Second World War, it was jokingly called a *schnorkel*, because of the snoring noise it made.

Safe and efficient 'individual' diving began in 1837 with the invention of the closed diving-suit, supplied with air from a surface vessel, by Augustus Siebe, a mechanic from Saxony who had settled in England. It had a metal helmet which was screwed on to a breast-plate, a rubber collar, watertight cuffs, lead-soled boots and a lead breast-weight; air was pumped to the suit, which enclosed the entire body except hands and feet, through a hose from the surface, and the exhausted air escaped through a special valve in the helmet. The first diving apparatus with its own air supply was invented by a French naval officer and scientist, Auguste Denayrouze, in 1872; it combined a flexible, watertight suit and hood (which looked like a modern gas mask) with a compressed-air container, in the form of a small barrel which the diver carried on his back. But he had to be connected with a surface vessel by a lifeline for pulling him up.

E.L.

Bicycle

'Ah! How simple yet how sublime,' said a Turkish pasha on seeing Thomas Stevens dismount from his Ordinary, 'both the man and the machine!'

For about five thousand years man had used wheels singly, in pairs, in threes and in fours; it was not until 1690 that anyone joined two wheels in line and sat on the machine. A Frenchman, de Sivrac, sat astride what he called a *célérifère*, and trundled along, a leg on each side. His trundling was limited, because he had not invented a way of steering the machine. A steerable machine, the 'Draisine', was made by Karl von Drais in 1817.

The bicycle as a device that could be pedalled was invented by Kirkpatrick Macmillan in 1839 – the rear wheels were driven by cranks connected to pedals; at last, man could travel under his own power faster than he could walk. Yet not many people were attracted by the idea.

In 1861, a Monsieur Brunel, a French hat-maker, took his Draisine for repairs to Pierre Michaux, a coachmaker. His son Ernest, who presumably knew nothing of Macmillan's invention, suggested that the machine could be improved if the front wheel were fitted with cranks, like the handles of a grindstone, so that it could be pedalled around, and in this casual way he started the bicycle industry.

These 'velocipedes' moved along only one wheel-circumference per turn of the pedals, so gradually the front wheel became larger and larger until the penny farthing was invented. These machines, properly called a 'High Bicycle' or an 'Ordinary', were effective but unstable: any attempt at braking, particularly downhill, could send the rider over the handlebars, and a few tumbles were an accepted part of a day's run. People were undeterred, however. Sporting clubs were set up, and in 1884 the courageous Thomas Stevens rode, and walked, and at times carried, his Ordinary across the United States.

Attempts to make the front wheel smaller, and thus the bicycle itself safer, were made at intervals, but they were probably defeated by the engineering of the time – a reliable bicycle chain was a major invention. The chain-driven bicycle that actually ended the days of the Ordinary was the 'Rover Safety' of 1885 invented by John Kemp Starley. His innovations immediately caught on, but his frame design did not. Almost every imaginable way of joining the two wheels was tried in attempts to combine rigidity, steerability, mechanical efficiency and comfort.

The machines quickly became fairly reliable and effective, and the pneumatic tyre was reinvented by Dunlop in 1888. By the present time few radical innovations seem possible – a recent competition to find a better machine had no success.

T.O.

Underground railway/ subway

Perhaps the most obvious indication of man's biological affinity with the ape is the tremendous imitative ability of both primate species. Man in particular has repeatedly displayed such an intense propensity for imitation that he has very nearly duplicated, by one means or another the natural attributes of every other form of life which he has had sufficient opportunity to observe. The flying abilities of birds, the radar of bats, the engineering talents of spiders and numerous insects, the plant kingdom's life process of photosynthesis – all have been subject to the careful gaze of the human eye and the inquisitiveness of the human mind. So too, it seems, has the burrowing of the lowly mole, and in the hundreds of miles of underground railway track that wind their way deep below the streets and structures of the world's major cities, one sees the technique carried to a logical conclusion.

The first people to take this bold step within the history of transport were the English. In 1843 Charles Pearson proposed for the City of London the world's first underground rail system. It was, however, to be another ten years before Parliament authorized the construction of a three-and-three-quarter-mile stretch of railway between Farringdon Street and Bishop's Road, Paddington. In January 1863, using highly pollutant coke-burning steam locomotives, the Metropolitan District Railway opened for business; it carried 9,500,000 passengers in its first year of service. London's first electric underground line went into action in 1890, with a 'tube' ride to anywhere in the city costing twopence. Today, thanks in no small way to Charles Tyson Yerks, an American railway magnate of the turn of the century, London Transport operates eight primary underground lines, has a ninth under construction, and during every hour of peak service shuttles over a hundred thousand commuters within the city's belly. The average speed of the trains is only 20 m.p.h., but during rush hours this is more than double the pace of surface transport.

In 1896, the city of Budapest opened the first underground rail on the continent of Europe with a two-and-a-half-mile electric 'subway'. Four years later it was followed by the legendary Paris Metro, now the third largest system in the world and second only to the New York City subway complex in the number of passengers carried – more than 1,100,000,000 annually.

If the cleanest and most elegant underground in the world is the Moscow Metro, with spacious stations finished in polished marble and granite, decorated in mosaic and stained glass and embellished with statuary, then the most polite must be in Munich, the most disciplined in Tokyo, and the roughest in New York, where the trampling of old ladies during peak times at some of the city's 495 stations is not uncommon. All in all more than three dozen of the world's metropolises now utilize rapid underground transport, and, as populations continue to explode in number and surface traffic congestion increases accordingly, even more such systems are in the offing. The mole, if he could understand it all, would be amazed.

E.W.

Driving chains

In 1864 James Slater patented a driving chain which can be regarded as the first step in developing a precision chain strong enough to drive bicycles and other machines. His small factory in Salford, where he made textile-machine chains, was then acquired by a Swiss, Hans Renold, who patented the bush roller chain in 1880. The arrangement of bushes on this chain provided a much greater load-bearing surface than Slater's design.

The earliest known sketches of a driving chain are by Leonardo da Vinci, but it is not known whether the chains he drew were ever actually made. A chain-drive is also shown in a drawing of a water-raising machine of about 1588, from Ramelli's *Le diverse et artificiose machine*. This chain has square links which fit over projecting teeth on wooden wheels, and each square link is connected to the next one by three oval links.

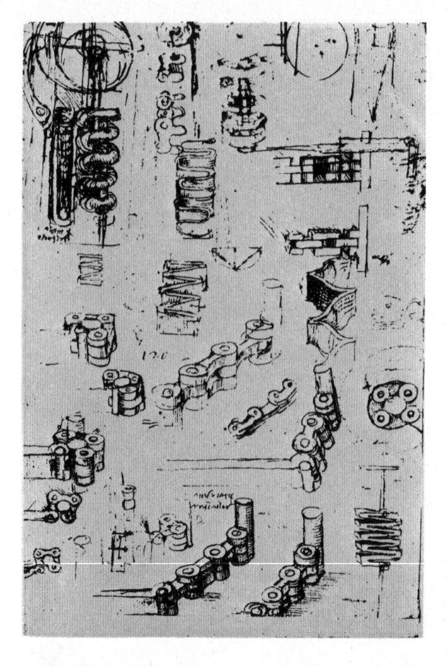

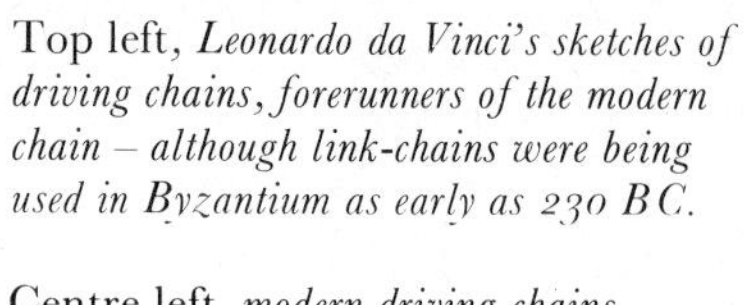

Top left, *Leonardo da Vinci's sketches of driving chains, forerunners of the modern chain – although link-chains were being used in Byzantium as early as 230 BC.*

Centre left, *modern driving chains.*

View of the Metropolitan Railway at Baker Street Station, London. Chromolithograph, 1868.

Below, *Irish schoolmaster John Holland in one of his early submarines – a one-man affair with paddles which the submariner operated with his feet.*

The scarcity and high cost of suitable metals, together with the absence of good tools to form and shape them, prevented a wider use of driving chains. It was not until well on into the nineteenth century, as a result of the Industrial Revolution, that they began to have a wider application.

In France, a 'Bicyclette' with the first chain-drive to the rear wheel was designed by André Guilmet and manufactured by Meyer et Cie in 1868. Although driving chains had been used for some time, mostly on textile machinery, the bicycle chain was still rather weak. Later another Frenchman, G. Juzan, developed his 'Bicyclette Moderne', in 1885. It had two wheels of equal size, with a chain-drive to the rear. Another advance was made by an Englishman, John Kemp Starley,, who made his 'Rover Safety' bicycle in 1885. This machine also had a driving chain to the rear wheel. Thus began the era of the modern bicycle, with Coventry as the centre of the manufacturing industry. And from the bicycle developed the application of the chain-drive principle to motor-cycles and cars. Nowadays precision chains are among the most reliable parts of industrial machinery.

C.StC.B.D.

Submarine

Scyllis of Scione and his daughter may or may not have destroyed the ships of Xerxes in the expedition against Greece, but submarine warfare and submarines have a long and respectable history. Alexander the Great with his glass barrel, Calluricus with his underwater gun, the messenger of Lucullus who slipped through the enemy's fleet in a goatskin may none of them have achieved what is claimed for them, but their stories bear testimony to two thousand years of passionate interest in the possibilities of undersea war.

Medieval enthusiasm for it dates at least from the Crusades, when the Christian encirclement of Ptolemais in 1150 was penetrated by a diver in a submerged apparatus; and forty years later there is a remarkable description of a leather underwater boat in the German poem *Salman and Morolf*. Modern interest begins, still with leather as the solution to its problems, with the '18th Devise' of the English mathematician William Bourne, who proposed in a book – *Inventions and Devises* – a wooden framework covered with impermeable leather.

He never built it, but Cornelius Van Drebbel did, and apparently rowed it under water from Westminster to Greenwich in 1615. Traditionally, King James I made a passage in it. The eighteenth century produced a crop of ingenious inventions which culminated in David Bushnell's one-man *Turtle*, and the splendid failure of its attacks on HMS *Eagle* in New York harbour in the War of American Independence.

Two things defeated the early inventors: the lack of an effective method of propulsion and of an effective weapon which their craft could use. Though the Confederate submersible sank the Federal corvette *Housatonic* off Charleston in the Civil War and proved the principle, the practice eluded the navies of the world until in 1875 the Irish schoolmaster John P. Holland began the series of vessels subsidized by the Fenian Society of New York to destroy the British Navy. Holland's success was irregular; one at least of his boats was stolen by rival Fenians, and it was not until he built the *Plunger* that he outstripped Garrett, the English clergyman, and Nordenfeldt, Zede, Simon Lake and the rest to secure the contract to build the US Navy's first submarine.

The Holland VIII had a gasoline engine, was propelled by electricity under water and was equipped with a Whitehead torpedo. It was the progenitor of modern submarines. Britain's first boats were Hollands, built by Vickers. The early German U-boats utilized the same basic principles. It made possible the submarine war that brought Britain to the edge of defeat in 1917 in the First World War, and failure to find an adequate answer to it brought her once more to the edge of defeat in the Second.

In January 1955 the frenetic technology of the post-war world produced the signal from USS *Nautilus*, 'Under way on nuclear power.' Five years later the first Polaris missile was fired from the USS *Washington* and a new era in warfare and a new dimension in human destructive capacity began.

D.D.

The submarine Resurgam, *built by an English clergyman, G. W. Garrett, in 1879.*

Motor-car

Nicholas-Joseph Cugnot's full-scale steam-truck moved under its own power in 1769–70, and many other automobiles were made during the next hundred years. They were mostly buses or tractors rather than cars, and were driven by steam, but from 1862 onwards attempts were made to harness the newfangled internal-combustion engine.

Karl Benz of Mannheim (1844–1929) just qualifies as inventor of the modern private motor-car by being the first man to sell light, self-propelled carriages, made to a set pattern, to the public. By 1880 he was running a small business making stationary gas engines, which worked on the pump-scavenged two-stroke principle to avoid infringement of the Otto and Langen four-stroke patent.

In 1884–85, Benz solemnized the improbable but harmonious marriage between a two-seat 'sociable' tricycle and a scaled-down gas engine adapted to burn liquid fuel by means of a surface carburettor. The Otto patent having been successfully challenged, the three-quarter-horsepower engine worked on the four-stroke cycle and had electric ignition.

Though it was frail and under-powered, the first car's performance encouraged Benz to make improved and more powerful three-wheelers, and in 1887 he actually sold one. The buyer, Emile Roger of Paris, was granted sole agency and assembly rights for France, and from 1888 onwards any member of the public so lost to all sense of the fitness of things as to want a horseless carriage could go to Paris or Mannheim and buy one for about £140. An 1888-type 'French' Benz survives in the London Science Museum, and in 1958 it was proved capable of covering the London to Brighton course at an average speed of $8\frac{1}{2}$ m.p.h.

After 1890, the unstable three-wheeler was redesigned as a four-wheeler, but thereafter the Benz design was altered little until 1902. The popular 'Ideal' model had an engine of 110 mm bore by 115 mm stroke, which developed $3\frac{1}{2}$ h.p. at 500 r.p.m. and gave the car a top speed of 14 m.p.h. Steep hills could defeat it, however, and after 1898 a 'crypto' emergency low gear was added to the two-speed belt and pulley transmission which enabled it to creep inexorably to the top of any gradient.

Benz has been called reactionary for clinging to his original design long after others had outstripped him, but his early cars were very simple and reliable. He was not interested in speed, and aimed at the pony and trap customers rather than the carriage gentry. Unlike the first Stanley steam-car the early Benz machines were not equipped with whip-sockets, but they were undeniably horseless carriages and they performed admirably.

A. Bi.

Automobile technology

Like the concept of the horseless carriage itself, the modern car is an assemblage of devices and techniques which can hardly ever be ascribed to one man or to one time. Many of them have their origins long before the age of the car: this is true of headlights, steering gear, warning horn, sprung suspensions and even winding windows.

Passenger coaches had sprung suspensions before 1580, though as late as 1805 Obadiah Elliot was able to obtain a patent for the elliptic and semi-elliptic spring assembled from superimposed laminations. This, plus the vertical coil, took care of most needs of motor-cars until Citroën introduced their self-levelling fluid system on the DS of 1956; yet this complicated concept was taken to a high standard of development half a century earlier by P. H. de St Senoch who devised a self-compensating four-wheel system in 1905 based on hydraulic fluid and an improved version in 1910 using compressed air. Again, the rubber suspension systems of Alex Moulton and Alec Issigonis rest on a background of at least 80 years of work in the same field.

Steering gears of primitive forms are at least 2,500 years old, but a major advance came in 1714 when Du Quet proposed a compensated gear that made allowance for the fact that the turn-radius of the inner and outer wheels is not the same (the inside wheel has to steer through a sharper angle). This gear was improved in 1818 and patented by Rudolph Ackermann, whose name is still used to describe the underlying principle of most car steering gears. Since then countless inventors have introduced improvements to give cheaper or more reliable construction, better dynamic qualities or reduced operating resistance; one example is the Burman gear introduced by J. G. Douglas in 1931. Even four-wheel steering, hardly ever used on modern cars, dates from not later than 1772. Various forms of power-augmented steering were used on both wheeled and track-laying vehicles from 1920 onwards, the tracked vehicle usually steering by stopping or reversing the inside track. The modern type of power steering, such as that devised by Ross, dates from as recently as 1940 and 1946. Its main advantage is to ease steering resistance at very low speeds, and facilitate effortless parking.

Brakes, as such, are centuries old, but in the 19th century various power-assisted brakes using steam, air under pressure, vacuum and hydraulic power were patented, and used on trains and other moving masses. Early cars typically used band or shoe brakes with a mechanical linkage to a handle or foot-pedal, and it was thought dangerous to fit them save on the rear wheels. In 1904, however, P. L. Renouf boldly patented a front-wheel brake; this was improved two years later by Allen and in 1909 fitted to an Argyll car (though for many years most manufacturers took care to ensure that the front brakes were almost ineffectual, to avoid the dreaded locking of the front wheels). Separate replaceable linings were proposed by Henry Ford in 1900 and soon produced in various forms matched to brake drums of steel or cast iron about $1\frac{1}{2}$ in. wide. In 1902 Louis Renault introduced the standard form of drum brake, with two hinged shoes forced apart by an interposed cam, and in 1917 R. Stevens patented a simple method of adjusting the gap to take up wear of the shoes. From 1919 onwards A. Dewandre and others took out a succession of patents on hydraulically assisted 'servo' brakes, and during the 1920s Bendix and others introduced duo-servo brakes for safer and more uniform action. Disc brakes,

Ford car factory at Dearborn, Michigan, in the 1920s.

Ignition: *dual system by battery and induction coil, as used on Lenoir gas engine, 1806: and H.T. magneto-generator devised by Simms and perfected by Bosch in 1898.*

Engine: *developed from the first practical car engine, built by Daimler in 1885.*

Cooling system: *water circulation system with air-cooled multi-tubed radiator, various forms of which were used from 1896.*

Steering: *Ackermann or radial-axle steering devised by George Lenkensperger, Munich, 1818: patented for him in England by Rudolph Ackermann, London print-seller.*

Lighting: *oil and acetylene gas lamps un c. 1910, when tungste filament enabled small electric bulbs to be powerful and robust.*

The $4\frac{1}{2}$-h.p. Panhard-Levassor – second in the 1896 Paris–Le Havre Race.

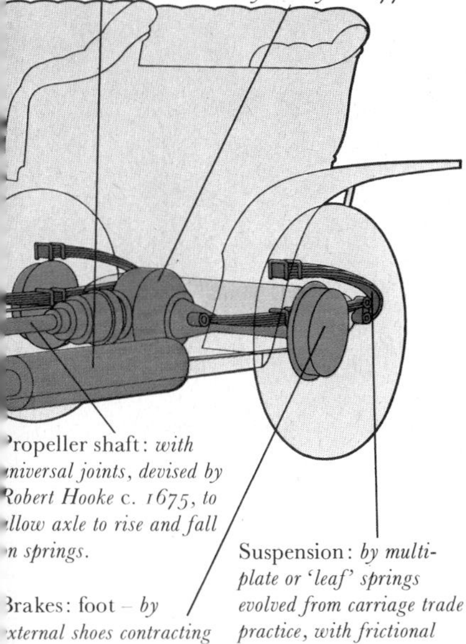

The motor car has taken many shapes in its relatively short life, and the constituent parts have been arranged in a wide variety of patterns. The drawing above shows some of these parts, as they appeared in the 1910 Rolls Royce 'Silver Ghost'.

identical in principle to those used today, were a feature of Darracq models of 1906 and were again used on AC cars of 1913–14.

Many early car manufacturers either believed, or affected to believe, that their customers would not demean themselves by touching their car except possibly to drive it; all servicing and cleaning was assumed to be the duty of the chauffeur. Henry Ford's mass production from 1908 made this belief demonstrably untenable, and among many other effects this highlighted the need for a self-starter. Over the next four years several forms were patented but the one that succeeded was that of A. H. Midgley of 1912, who bolted an electric motor to the crankcase, driving direct on a gear-ring surrounding the flywheel. V. Bendix simultaneously patented the familiar engaging mechanism comprising a gear pinion moving along spiral splines against a coil spring. Starting handles continued to be common until after 1945, and many modern owners consider their demise a retrograde step, on a cold morning with a flat battery!

No particular new invention was needed to create the first motor-car gearbox, because change-speed gears had been used in machine tools and many other power-driven devices. Some of the earliest cars used a slipping-belt drive, or belts linking pulleys of different ratios, but those using a chain or gear-train had to have change-speed gears in order to match engine speed to different speeds of the road wheels. The first car to have what could be called a gearbox – though the gears were completely unenclosed – was the 1890 Panhard, in which effecting a change was not merely difficult but potentially damaging to the gear-teeth. De Dion and Renault soon introduced improved schemes, and in 1902 the Premier car was marketed with gear-change motion effected electrically. But the giant advances all came in 1904, when Prentice and Shiels patented the synchromesh principle, De Dion invented the constant-mesh preselector and Wilson patented an epicyclic preselector. The preselectors relied upon gears that were never out of engagement but which, by arresting the motion of alternative parts of the system (such as, in the case of epicyclic gears, one or other of the peripheral annulus gears), could be made to transmit the drive with any of a series of different speed ratios. The synchromesh, on the other hand, did slide adjacent pinions into and out of engagement but overcame the dangerous fault of the early gearboxes in that gears about to be run together were first brought to matching speeds before engagement was made. Gears moved axially into engagement were also soon made with modified teeth with tapered ends, to facilitate smooth engagement.

From the earliest days of the automobile inventors sought other ways of linking the engine and road wheels. In 1896 Hall achieved the remarkable result of designing the first hydromechanical power transmission in history, with complete 'differential' or split-path qualities. It was a combination of a mechanical drive and a hydrostatic one, the latter being a closed hydraulic circuit comprising a pump and motor capable of transmitting the drive over an infinitely variable range of ratios between lower and upper limits (the lower limit usually being zero, the hydraulic fluid merely being pumped round and round with no output torque). From then on there were many 'automatic transmissions', some – such as Benson's of 1901 – merely being a fluid coupling in lieu of the friction clutch, and others – such as Coates's of 1924 – being complete automatic torque convertors fulfilling the whole function of clutch and gearbox. The fluid flywheel used on Daimler cars from 1930 was a 1926 design applied by Harold Sinclair. Numerous other inventors proposed different schemes, but the man regarded as father of the modern automatic transmission was Earl A. Thompson, who patented a semi-automatic drive in 1930 and finally realized the fully automatic 'Hydramatic' drive in 1939.

In the above the word 'differential' is the proper one to describe a split-path hydromechanical drive. There is, of course, a commoner meaning in which the word describes a gear system able to drive two shafts – typically the left and right half-shafts of a back-axle – at different speeds. The need for a differential gear was manifest as soon as power-driven vehicles began to steer round corners, when the outer wheel must rotate faster than the one on the inside of the curve. This kind of differential is usually ascribed to O. Pecquer, who patented the split-axle and bevel form in 1828. Many steam road vehicles used differentials, but the first car with a differential drive axle was as late as 1896, earlier models merely driving a live axle at one point and relying upon slight tyre slip to take up the different wheel peripheral speeds called for on bends.

Many advances made chiefly with automobile applications in mind were actually a surprisingly long time being introduced to production. Indeed, even today most cars sold in Britain retain a manually controlled gearbox, and there are several major advances devised to improve occupant safety in the event of a collision which are being introduced only with extreme reluctance, and often appear to need legislation to enforce their adoption. Perhaps the outstanding case of this kind is the collapsible steering column, to reduce the incidence of lethal chest injury to the driver in a frontal collision, which was developed more than 20 years ago, introduced on certain Chevrolet models in the mid-1960s and is still a rarity in Europe. Safety glass, invented primarily for use in vehicles in 1905, and commercially produced as 'Triplex' in 1910, was equally slow to catch on and as late as 1930 many makers were still using windows of celluloid which soon crazed and went increasingly yellow. Windscreen wipers, invented in 1920 (suction type) and 1922 (electric), soon became a standard fitment. On the other hand the many devices now available to reduce atmospheric pollution are ignored completely, except in states or countries that bother to draft legislation. 'Styling' – whatever that may mean – is considered far more important, both by the automobile industry and the public.

W. T. G.

Airship

As soon as the balloon was invented in 1783, its creators attempted to propel and steer it; and within three years a variety of hand-operated aerial oars, paddles, and winglets were tested – a small propeller was even applied by Blanchard to a spherical balloon in 1784. Elongated balloons were being built by then, but to no effect: the human animal was not nearly powerful enough to be a propelling engine. In 1785, the French Lieutenant J. B. M. Meusnier designed, but never built, an extraordinary and premonitory airship, with elongated envelope, interior ballonets to preserve the shape, and a long car suspended beneath with three air-screws to be worked by a team of men.

After many years of suggestions, and some good models, the great French engineer Henri Giffard succeeded in building and flying the world's first feasible – but only just – airship, propelled by the only existing prime mover of power and efficiency, the steam engine. This brave achievement was an elongated dirigible, with pointed ends, filled with hydrogen and with a 3-h.p. engine slung far beneath to avoid igniting the gas. On 24 September 1852 Giffard rose in his machine from the Paris Hippodrome before a huge crowd, and puffed away at some 6 m.p.h., finally landing triumphantly seventeen miles away. But Giffard's dirigible could not battle against anything more than a light breeze.

It was not until 1884 that the practical airship came nearer to reality with a large 'droop-snoot' ship, *La France*, designed and built by two Army engineer officers; Charles Renard and Arthur Krebs. She was 165 feet long, had a long car slung beneath, and one huge propeller at the front, driven by an 8½-h.p. electric motor. She took off from Chalais-Meudon on 9 August 1884, and returned safely to base after a flight of something under five miles at about 14½ m.p.h.

Man Moving

Petrol engines began to power experimental airships in the late 1880s; but the first practical dirigible had to wait until the Lebaudy brothers commissioned Henri Julliot to build them an airship. The Lebaudy was a semi-rigid ship – with a 40-h.p. Daimler petrol engine driving two airscrews, giving her speed of some 25 m.p.h. On 24 June 1903 the Lebaudy made her record flight of ninety-eight kilometres.

C.G.-S.

Motor-cycle

While Karl Benz, a railwayman's son in Mannheim, was developing the internal-combustion engine for propelling motor-cars, the middle-aged Gottlieb Daimler, a baker's son from Württemberg, was building the first motor-cycle. It was a neck-and-neck race between the two inventors: in the spring of 1885, Benz tried out his tricycle with a four-stroke petrol engine in the courtyard of his workshop; in the autumn of the same year, Daimler rode his motor-cycle for the first time in the backyard of his house in Cannstatt, near Stuttgart. Neither knew what the other was doing only 60 miles away, nor did they know of each other's existence, and they never met throughout their lives.

Daimler had been working as an engineer with August Nikolaus Otto in Cologne from 1872 while Otto was developing his stationary internal-combustion gas engine on the four-stroke principle. Daimler moved to his own workshop, with the idea of turning Otto's stationary engine into a movable one by using petrol vapour instead of coal gas, and by effecting its ignition not by a permanent flame as Otto had done, but by an electrical ignition system. His vehicle was a rough-and-ready structure, with wooden wheels and an exhaust placed immediately under the rider's seat; it was slow, for the engine made only 700 revolutions per minute. However, Daimler thought that his motor-cycle might be 'most useful for country postmen'.

After a slow start, the popularity of the motor-cycle grew so fast that for a time, just before and after the First World War, its lower price and running costs made it a serious rival to the motor-car; military police forces still use it extensively for its manoeuvrability. Between the wars the tendency was towards bigger machines, up to 1,000 c.c., often used with sidecars, and usually powered by four-stroke twin-cylinder engines. Around 1950, however, the small, light-weight, two-stroke motor-cycle pioneered by the Italians ('Vespa' etc.) began to gain popularity all over Europe; this type, which was developed during the Second World War for use by paratroops, is usually called a motor-scooter.

E.L.

Pneumatic tyre/tire

In 1845 a patent was issued to Robert Thomson, a civil engineer of Middlesex, for a tyre with a leather casing filled with air or horsehair. But John Dunlop, a Scottish veterinary surgeon living in Ireland, knew nothing of this when in 1888 his son asked him to think of some way to protect his tricycle from damage on the cobbled streets to school. Dunlop constructed a pneumatic tyre, a tube inflated through a valve and protected by a rubber-impregnated outer casing. This tyre was wrapped, mummy-like, on to the wheel of the bicycle. Punctures of the inner tube could only be repaired by soaking the casing apart with benzol and rebuilding afterwards.

The new tyres were ridiculed as 'mummy' or 'pudding' tyres, but racing cyclists found them faster as well as smoother and one, Harvey DuCros, set up a company to make them which became the Dunlop Rubber Company.

The tyre problem was not solved, however, merely by replacing solid rubber by an inflated tube. Charles K. Welch of Tottenham added to the tube the idea of wires embedded in the cover, and a dished wheel-rim to help keep the tyre in place. His patent was taken out in 1890, though there had been an American patent for locking solid tyres with the help of wires ten years before.

G.R.T.

Glider

The first glider constructed on modern aerodynamic principles was a model built and flown successfully in 1804 by Sir George Cayley, the 'father of aerial navigation' as he was rightly called. Of the various model and full-scale aircraft which Cayley built, the most famous was the triplane which – with its controls locked – carried his unwilling coachman across a Yorkshire dale in 1853.

Practical full-scale gliding was first achieved by the German Otto Lilienthal between 1891 and 1896 – when he was killed flying – in both monoplane and biplane hang-gliders, in which the only control exercised was the swinging of legs and torso to alter the position of the centre of gravity. Octave Chanute in the USA (1896) and Percy Pilcher in Scotland (1895–99) built improved, but still primitive, hang-gliders.

The first fully successful glider which was provided with proper three-axis flight-control – *i.e.* pilot-control in pitch, yaw and roll – was the Wright brothers' modified No. 3 glider, flown at the Kill Devil Hills, North Carolina, in 1902, and later. Many gliders were built thereafter in imitation of the Wright machines on both sides of the Atlantic.

Modern gliding, both for sport and for aerodynamic research, was the result of a remarkable forced growth which

La France, *designed and built by two French Army officers, was the first airship sufficiently fast to follow her pilot's directions – on a calm day. Her pioneer flight of some five miles took place in August 1884.*

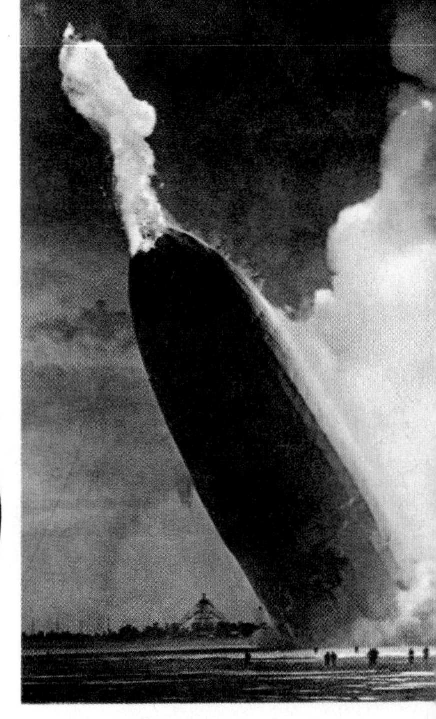

The Airship Age ended in 1937, when the Hindenburg *went down in flames in the USA.*

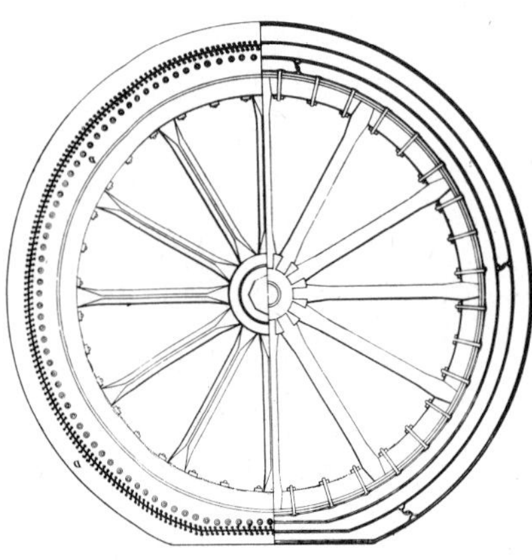

Pneumatic tyre: diagram from R. W. Thomson's Patent Office specification of 1845.

First motor-cycle built by Daimler, 1885.

One of the successful hang-gliders of Otto Lilienthal, photographed in 1896, the year he was killed.

The Wrights' No. 3 glider, with Wilbur piloting. Built in 1902, it was the first fully successful glider with three-axis flight-control.

appeared in Germany after World War I, and was created by the peace-treaty ban on her building powered aircraft. Since then, gliding and soaring have become world-wide in popularity, and glider construction has grown to be a large industrial enterprise.

The launching of gliders has always presented difficulties, and for many years gliders were launched by winches operating tow-ropes, the latter being cast off when sufficient height was attained. But today the towing-off by light powered aircraft has become universal, and has the great advantage that any height and any particular location can be selected before the glider pilot casts off the tow-rope and flies free.

C. G.-S.

The Butler petrol tricycle (1884), using a carburettor which anticipated the Maybach model by nine years, but was then abandoned.

Carburettor

Like many other things, the carburettor was 'invented' by several different people at different times. As early as 1830–35, an English industrial chemist named Donovan designed an evaporative device to enrich, or 'carburate', coal gas by guiding it over the surface of a volatile liquid hydrocarbon. In 1862 Étienne Lenoir used the same principle of surface evaporation to enable his patent gas engine to burn liquid fuel. Other pioneers, including Gottlieb Daimler, also used some form of surface carburettor.

The first cars sold as articles of commerce were those of Karl Benz, and he remained faithful to the surface carburettor until 1902. Between 1890 and 1900, some two thousand Benz cars were sold and the design was copied by other manufacturers. In essence the surface carburettor consists of a metal pot, holding anything from a quart to a gallon of spirit in its lower half, connected to the engine in such a way that air drawn in on the induction stroke passes over the surface of the fuel, 'licking up' vapour on the way; evaporation is assisted by an exhaust-heated by-pass.

Surface carburettors were simple, but they were very large and worked well only on very volatile petrol. Also, they evaporated the lighter elements of the fuel first and left behind a heavy residue. Lanchester's wick carburettor, used from 1895 to 1918, did not cause fractional distillation and gave excellent results.

In 1893, Wilhelm Maybach, Daimler's right-hand man, designed a float-feed, single-jet, spray carburettor which is generally regarded as the direct ancestor of the modern instrument. This device was immediately adopted by Panhard et Levassor, then building the world's most advanced motor cars.

In 1901, the Daimler Motoren Gesellschaft took action for infringement of patents against various manufacturers. On the evidence of Edward Butler's petrol tricycle of 1884, the English courts ruled that Maybach's carburettor was not the subject of a valid master patent. The Butler tricycle had a float-feed spray carburettor which anticipated the Maybach in principle, with the additional refinement of more automatic regulation of mixture strength to suit different speeds. Butler's carburettor included an air-cleaner, a refinement which did not come into general use for another fifty years.

However, Butler saw no commercial future for his 'petrocycle' and abandoned his experiments, leaving the field open to Maybach.

A. Fi.

Early attempt at an 'artificial road' – Boydell's steam-tractor of 1854, one of the forebears of the caterpillar track.

Caterpillar track

In 1770 Richard Edgeworth took out a British patent for 'a portable railway, or artificial road, to move along with any carriage to which it is applied'. He meant, in practice, several slats of wood connected to an endless chain and moving in regular succession so that one or more slats were always in contact with the ground. His aim was to spread the weight of the carriage over a greater surface than was possible with a narrow wheel, thereby enabling the carriage to cross rugged or soft terrain. But for most of the nineteenth century railways stole all the attention – and working capital – so neither Edgeworth's design nor the scores similar to it that were to follow got far from the drawing-board; yet each successive idea based on his original theme contributed to the caterpillar track.

It was from America, with its need to pioneer vast, roadless territories, that several promising patents for 'tracks' came, including F.W. Batter's of 1888 and Benjamin Holt's experiment of 1904. In 1906 a genuine caterpillar-tracked vehicle was built commercially and sold by the Holt Tractor Company to the Meadow Development Company for use in Louisiana – but even then Holts merely substituted tracks for the rear wheels of their existing steam-tractor.

Already, however, a far more sophisticated design, invented by David Roberts, had been tried out in Britain by the firm of Ruston Hornsby & Sons in 1904. Its claim to distinction lay in the use of separate links joined by lubricated pins and in 1907 the first petrol-driven caterpillar tractor was run on this track and demonstrated to the War Office. But military enthusiasm waned and in 1912 Hornsbys sold their patents to Holts. It took a major war, and the mud of Flanders, to bring caterpillar tractors to the forefront as weapon-carriers.

K.M.

Man Moving

Hydrofoil

Like the hovercraft, the hydrofoil is a 'skimmer', or boundary vehicle, moving in the region where the air meets the sea. While ordinary ships have to overcome water and air resistance – the 'drag' – the hydrofoil uses the air to lift itself as much as possible out of the water by means of wings or foils to reduce drag to a minimum; but it never leaves the water completely to fly like an aeroplane.

The first inventor to have worked on the idea was a French priest by the name of Ramus in the middle of the last century; he took his cue from the kite. Another Frenchman, the Russian-born Count de Lambert, understood that the petrol engine, then newly invented, would provide the necessary power for his 'hydroplane', as he called it. He tried his prototype out on the Seine in the 1890s, but it refused to lift its nose out of the water.

Now it was an Italian who took up the idea: Enrico Forlanini, an airship designer. He built a small hydrofoil boat in 1905, expounding the scientific and technical principles of the craft in his patent specification. Forlanini improved his subsequent models, demonstrating his last one in 1911 to a distinguished American visitor, Alexander Graham Bell, on Lake Maggiore. Bell proceeded to build a craft of his own design, based on Forlanini's patent, in Halifax, Nova Scotia; it established a water-speed record, at 71 m.p.h., in 1918. Further improvements were made by German inventors in the 1920s and 1930s, and later by a British organ-builder, Christopher Hook, during and after the Second World War. Commercial hydrofoil building, however, began in the 1950s in Italy with the 'Supramar' boats; they are also produced under licence in many other countries, while the USA and the Soviet Union have created their own designs for big military and passenger boats. Today, hydrofoil craft are maintaining regular passenger services all over the world, from the Thames to the Black Sea.

Nearly all types have a rather clumsy look – like giant insects with bow legs; their lack of elegance is due to the V-form arrangement of the foils, which means that the struts holding them stick out right and left from the hull. But the craft has proved most useful for fast commuter services, for fishery patrol, fire-fighting, harbour control, water-police purposes, and for air/sea rescue. Men working on oil-drilling platforms offshore can be taken to and from their artificial islands at top speed. The hydrofoil is faster than anything afloat, with the exception of speedboats; these, by the way, can be converted into hydrofoils at low cost, by fitting hydrofoil fins to the bows and the outboard engines.

E. L.

Aeroplane/Airplane

The first practical aeroplane in history was the Wright brothers' biplane, known as the *Flyer III*, of 1905. This remarkable machine, the outcome of three seasons of gliding experiments (1900–02), and of two previous powered aircraft (1903–04), was a biplane with a span of 40½ feet and a 16-h.p. engine driving two pusher-propellers. It was catapult-launched along a wooden monorail, and landed on skids, a technique which allowed the aircraft to take off easily, and to land on any ground, however rough. From late June to mid-October 1905, the *Flyer III* proved that it was robust in structure and under complete control: it could bank, turn, make figures of eight, circle and fly easily for half an hour at about 35 m.p.h.; it made more than forty flights at the world's first aerodrome, the Huffman Prairie, near Dayton, Ohio, where Wilbur and Orville Wright had set up their shed, and where they brought the powered aeroplane to maturity. Later, the *Flyer III* was modified to carry two men successfully. As a result of these achievements, the Wrights flew in public for the first time, in two new planes, in 1908, both in Europe and America, and revolutionized the then primitive work on flying that was going on on the Continent.

The essence of the Wrights' success lay in the full controllability of their machines in pitch, roll and yaw; the Europeans – until they saw Wilbur perform in 1908 – treated the aeroplane as a winged automobile, whose primitive control surfaces were used more for corrective purposes than to initiate control in the air. Most important was control in roll, in order to bank and turn, used in combination with the rudder. The Wrights also made a great stride forward by gearing down their propellers, and gaining maximum thrust with minimum engine power. The Europeans were at first bent on inherent stability, and their controllability was minimal: they adopted various devices, including the dihedral angle (i.e. the wings rising from the roots) which tended to bring the machine back to the horizontal when gusted out of it. The Wrights at first deliberately built their machines inherently *unstable*, as they thought this was the only way to make them fully controllable. But, after seeing the Wright control systems and how they worked, the Europeans progressed to the necessary compromise between inherent stability and full controllability.

Before the triumph of the Wrights' *Flyer III*, there had been many attempts to fly, from the brave men who threw themselves off towers and cliffs – often at the expense of life or limb – to the first time that a powered machine left the ground, in 1874, when the French inventor Félix Du Temple made a full-size steam-driven monoplane which ran down a ramp and was airborne for a

First amphibious hydrofoil vehicle, built for the US Marine Corps.

Cockpit of the Vickers Vimy (below) *and, for contrast, the instrument console of a Boeing 747.*

Orville (extreme left) and Wilbur Wright, and the 1903 plane in which they first attempted powered flight.

The X-15 aircraft mated to a B-52 'mother ship'. Of all manned aircraft, this has the highest performance to date – 4,530 m.p.h. (Mach 6·72).

second or two as a result of this accelerated take-off. Then a monoplane built by the Russian Alexander Mozhaiski did the same thing in 1884, with a steam engine made in Britain. In 1890 came the first aeroplane to take off under its own power, the steam-powered *Eole*, with its inventor Clément Ader on board; it rose and careered forward for about fifty metres; but this ingenious machine could not sustain itself, or be properly controlled. In 1894, the expatriate American Sir Hiram Maxim succeeded in getting his huge biplane test-rig – also steam-driven – off its rails for a few feet; but this, too, was only a creditable uncontrolled hop.

Inspired by the work of the great German gliding pioneer, Otto Lilienthal, the Wrights entered aviation in 1899 with a kite whose wings could be 'warped' (twisted) for control in roll. They then made ever-improving gliding flights, and in their third glider (1902) they learnt proper three-axis control in the air. It was not until they had mastered this flight-control in gliders that they built their first powered *Flyer*, and made four short hop-flights on 17 December 1903, the last of which covered 852 feet; this amounted to half a mile air distance, as they were flying into a stiff breeze. These four tests were the first in which a powered and piloted aeroplane had lifted itself off the ground under its own power; had flown under control; had sustained itself in the air; and had landed on ground as high as that from which it had taken off.

European flying was reborn in 1904 with Esnault-Pelterie, as a result of the Wrights' success with gliders which were copied in France; but progress was very slow, and the many pioneers lacked tenacity; and it was as late as 1906 when Alberto Santos-Dumont (from Brazil) just succeeded in making some short hop-flights.

In fact it was not until 1908–10 that, inspired by the Wrights' mastery of flight-control, the Europeans caught up with their masters, and the already well-known names of Henry Farman, Blériot, Levavasseur, Curtiss, Breguet and many others came fully into their own.

C.G.-S.

Aviation technology

As in most broad areas of technology, aircraft engineering makes progress as a result of countless daily advances made within huge organizations, the individual 'inventor' playing a minor role. Even such clear-cut advances as the jet engine, the 'swing wing' and carbon-fibre structure all stemmed from work of many people in many places.

In 1910, aeroplanes (or airplanes, whichever is preferred) were constructed of a wire-braced lattice of high-quality hardwood (especially spruce) with steel joints and a covering of taut doped fabric. By 1920 there were alternatives. Junkers used the new Duralumin, an alloy mainly of aluminium and copper, to build all-metal aircraft with a self-stabilized skin of corrugated sheet. Dornier and Short attempted to use a load-bearing yet smooth skin to reduce both aerodynamic drag and the weight of the underlying structure. Fokker's chief designer, Reinhold Platz, made efficient monoplane wings entirely of glued and pinned wood with a sufficient depth to give good lift at low speeds and need no exterior bracing. Numerous companies made the fuselage and tail from welded steel tube.

This diversity continued until after 1930. Then it became clear, notably in the United States, that the Dornier/Short concept, taken much further in 1925 by Rohrbach with the title 'stressed skin', was superior for all except light aircraft. In the Second World War almost all aircraft were made from Duralumin sheet and strip, formed very accurately and assembled in precise tooling (jigs) to give standard and interchangeable parts with smooth surface and no exterior bracing. An exception was de Havilland's Mosquito (chief designer, R. E. Bishop) which was made of plywood laminations and ply/balsa sandwich. Parts of the Mosquito were bonded by the adhesive Redux, which by 1945 was also being used to bond metal aircraft. By 1950 material thickness had become so great that a revolution in manufacture took place. Heavy slabs were machined to form light 'integrally stiffened' panels; chemical baths were used to etch slabs to predetermined patterns, with perfectly smooth surfaces free from tiny marks that could start a fatigue crack; giant presses 90 ft. high were squeezing large forgings into complex shapes, and various combinations of metals and non-metals were being used to make light yet rigid 'sandwich' and 'honeycomb' panels.

Since 1966 the discovery of the outstanding physical properties of stiff and strong fibres of carbon (graphite) and boron has led to a range of fibre-reinforced composite structures. Previously many parts of aircraft were made of glass-reinforced plastics; the new fibres differed in that their rigidity allowed lightweight composite to replace light alloy and even titanium or steel in the most highly loaded portions of the structure. By 1980 carbon and boron composite will be used for a major portion of the structure of new advanced aircraft.

Engineering developments have been innumerable. Pressurization of inhabited spaces, to improve human comfort at great heights, was experimented with from 1932, in parallel with the use of partly pressurized pilot suits. In 1938 Boeing flew the Stratoliner, a four-engined airliner pressurized to $2\frac{1}{2}$ lb./sq. in., giving an 8,000 ft. interior altitude at a true altitude of 14,000 ft. In 1949 the Comet flew at a pressure differential

Man Moving

The Short 'Silver Streak' – one of the first aircraft with 'stressed skin' construction.

of 8¼ lb./sq. in., giving an 8,000 ft. interior at a true altitude of 40,000 ft. Accessory power has been multiplied thousands of times. The biggest aircraft of 1920 had a lead/acid battery which was charged by a wind-driven generator. Today the horsepower of electric and hydraulic devices in large aircraft is measured in thousands, and more than 5,000 separate devices may be installed. Further power is applied in the form of compressed air, low-pressure airflows, heat (for example, for protection against icing), liquid-oxygen conversion (for breathing) and countless other services.

Hydraulic power is invariably used to move the flight controls in high-speed aircraft swiftly and without flutter (self-excited vibration), and to operate the landing gear, slats (or other high-lift device) and flaps. Retractable landing gear appeared on racer aircraft of 1918–20, slats date from Handley Page's work in 1919, and flaps of many types appeared in 1923–25 after having worked well on the S.E.4 designed at Farnborough in 1913–14. Piston-engine superchargers, to maintain power in the thinner air at great heights, were used experimentally in 1915; exhaust-driven turbo-superchargers were devised by Sanford Moss at US General Electric and in France by Rateau, both in 1918. Variable-pitch propellers, to match engine speed and power to pilot demand despite great variation in aircraft speed, were tested at Farnborough in 1916 and patented by Hele-Shaw and Beacham in 1924 (though commercial success waited for Hamilton in the USA in 1932). Low-drag air-cooled engine cowlings were devised by H. C. H. Townend in 1927 and at the US National Advisory Committee for Aeronautics in 1928.

High-subsonic aircraft with swept wings were designed by nine German companies in 1944–45, while Alex Lippisch devised the delta (triangular) shape and Messerschmitt the variable-sweep 'swing wing'. The first successful supersonic aircraft was the Bell XS-1 of 1946, designed by a team led by Roy Sandstrom. Mach 2 was reached by the Skyrocket, designed by Ed Heinemann, in November 1953. The highest-performance manned aircraft to date is the North American (now Rockwell) X-15 series, first flown in 1959, which has exceeded 4,530 m.p.h. and reached Mach 6·72. Ejection seats for emergency escape at high speeds were used in German fighters in 1945, but the most widely used types stemmed from work by Capt. (later Sir) James Martin in 1947.

The first successful marine aircraft was a glider, towed by a speedboat along the Seine in 1905, built and flown by the brothers Voisin. In 1911 Glenn Curtiss flew his Hydro-Aeroplane (though a year earlier Henri Fabre had successfully flown from water) and went on to build a flying boat from which sprang many fine marine aircraft. J. W. Dunne flew a tailless glider in 1907 and a powered machine in 1908, establishing a configuration for many later 'all-wing' machines. In 1907 Paul Cornu flew a helicopter, leading to jet-driven rotors by Papin and Rouilly in 1915 and the fully successful tip-drive Doblhoff of 1943. In 1955 Bell and McDonnell flew two types of 'convertiplane', one having tilting rotors used as propellers in forward flight and the other having separate rotors and propellers. Numerous other VTOL (vertical take-off and landing) aircraft of this period included the Hiller Flying Platform, in which the pilot stood on a small fan-lifted platform and leaned in the direction he wished to go, and the Bell Rocket Belt which is simply strapped on to the wearer and flies him on the thrust of small rocket nozzles or, in later (post-1963) versions, a small jet engine. Ordinary turbojets were first used to provide direct VTOL lift in the Rolls-Royce 'Bedstead' of 1953; the Flying Atar, by the French SNECMA company, was an even starker solution with a seat above an up-ended engine. The Bell VTO of 1954 had swivelling engines for lift or propulsion, the Short SC.1 of 1957 had a battery of specially designed lift jets used only in the vertical mode, and the Hawker P.1127 of 1960 had a single engine with four swivelling nozzles to thrust upwards or horizontally. There are many other ways in which engine power has been either used to give lift or else integrated with lift from the wing.

Elmer Sperry developed a primitive autopilot in 1911 (it was flown in a Curtiss seaplane in 1912) and the Sperry company brought out much-refined commercial models from 1929, when other types were also coming into use. US air-mail routes used bonfire beacons in 1927, with radio beacons introduced in 1928 and a complete system of 'airways' introduced in 1929 with radio beams heard aloft as a series of Morse A's on one side, as N's on the other and as a continuous note exactly in the centre of the beam. A somewhat similar scheme was used to give steering guidance in the ILS (instrument landing system) of 1948, with a separate 'glide path' beam giving vertical guidance and point beacons giving distance information. Numerous forms of radio navigational device (Oboe, Consol, Gee, Decca, Loran, VOR, DME, Tacan, Dectra and others) have come into use since 1940, supplemented by Doppler navigators, radar methods and self-contained inertial systems based upon gyros and accelerometers more precisely made than any previous fabrications in bulk metal.

W. T. G.

The Bell Aerosystems rocket belt, powered with hydrogen peroxide, has hand throttles for control of thrust and for steering.

Helicopter

The first power-operated aircraft in the Western world was a toy helicopter; the earliest known illustration can be dated

Paul Cornu's twin-rotor helicopter, the first machine to make a free vertical flight, 1907.

The first traffic signal, installed outside the British Houses of Parliament in 1868 by J.P. Knight, a railway signalling engineer. It had red and green gas-lamps for use at night. Unfortunately it blew up and killed a policeman.

American three-colour lights. This was the first set to incorporate the red, green and amber lights in a single unit.

to the early fourteenth century, so the toy was in use long before that. Directly derived from the windmill, which is a passive airscrew, this toy reversed its role and made the windmill's sails into an active – propulsive – airscrew. Thus it also embodied the first example of the aircraft propeller. The toy consisted of four rotor blades mounted on a spindle dropped into a holder; a long piece of string was wound round the spindle, then taken out through a hole in the holder; a strong pull caused the spindle to rotate rapidly, and sent it soaring high into the air. This toy has an unbroken history to this day.

Leonardo da Vinci was the first to make a helicopter model with its power-source on board (a watch or clock spring); the lifting surface was in the form of a helical screw.

A short time after the invention of the balloon in 1783, the helicopter screw was made to act horizontally; it thus became an airscrew in the modern sense. This was first (ineffectively) applied by Blanchard to his balloon in 1784, and was rotated by hand. From then on, the airscrew was seen in countless designs and models for aeroplanes. The first time an airscrew actually propelled a full-scale man-carrying aircraft was on Giffard's airship of 1852.

Helicopter toys became increasingly popular in the seventeenth and eighteenth centuries until, in 1796, the French savants Launoy and Bienvenu devised a twin-rotor model operated by a bow-drill mechanism. This was publicized by Sir George Cayley in 1809, and was reproduced so often thereafter that it became the main source of helicopter inspiration. Many helicopter models were successfully flown throughout the nineteenth century. Then in 1907, after the petrol motor had been perfected, two full-size machines succeeded in rising from the ground vertically with a man on board: first off the ground was that of the Breguet brothers, in September at Douai, but this fails to qualify as it was steadied from the ground; when Paul Cornu's twin-rotor machine took off near Lisieux in November, however, it was unaided.

Before the helicopter became a practical reality, there came the intermediate step of the Cierva Autogiro, which first flew in 1923. Cierva saw that a practical helicopter would present formidable problems in the way of power-units and rotor-blade control. In the helicopter proper, the machine is both lifted and propelled by its rotor blades; and it can not only land and take off vertically, but hover. So Cierva devised a normal airframe, with a tractor airscrew and engine in the nose; then, instead of wings, he mounted freely rotating rotor blades above the fuselage, which, when the machine was pulled forward by its airscrew, rotated and provided the lift. But, by a preliminary acceleration of the rotors, which were clutched into the engine and then released, the Autogiro could make a bound into the air after running for only a few feet. It could also land very steeply, but not vertically; it could not hover.

The first fully practical helicopter arrived in the shape of the Focke-Achgelis Fa 61 in Germany. Finally, with the Sikorsky VS-300, there appeared the first successful single-rotor helicopter; this heralded a long succession of machines which firmly established the helicopter as a new and essential aircraft.

C.G.-S.

Traffic lights

A traffic signal, invented by J. P. Knight, a railway signalling engineer, was installed outside the Houses of Parliament at Westminster as long ago as 1868. It looked like a railway signal of the time, with semaphore arms and red and green gas-lamps for night use. After a short period of operation it blew up, killing a policeman. The accident discouraged further experiment until the age of the internal-combustion engine.

The modern traffic light is an American invention. The first red–green lights were set up in Cleveland in 1914. Three-colour signals were installed in New York in 1918, worked manually from a 'crow's-nest' tower look-out in the middle of the street. In 1920 similar lights appeared on traffic towers in Detroit. Traffic lights came to London in 1925. The first were put up at the junction of St James's Street and Piccadilly. A policeman controlled them from a hut in the middle of the road, decked out in the style of a railway signal-box, with a switchboard display in front of him.

In the following year automatic signals, working on a time interval, were installed at a junction in Wolverhampton, where they remained until 1968. The first vehicle-actuated signals in Britain were put up at the junction of Gracechurch Street and Cornhill in the City in 1932. By some strange quirk of history a seepage of gas in the controller cabinet caused an explosion when they were switched on.

Today traffic lights throughout the world work on a three-phase red–amber–green pattern, although there are subtle variations, such as in Boston, where green and amber appear together. Boring standardization has replaced such eccentric specimens as the elegant gilded columns of Fifth Avenue, each surmounted by a statuette, and the traffic lights of Los Angeles which, not content with changing mutely, would ring bells and wave semaphore arms to awake the slumbering motorists of the 1930s. But the strangest traffic lights of all are surely the set in Venice which control a junction of two canals.

G.P.

Man Moving

Rocket

On 16 March 1926, an obscure American scientist called Robert Goddard posed in a snow-covered field at Auburn, Massachusetts, beside a contraption looking like a child's climbing-frame, for a picture taken by his wife. He then got on with his work: an experiment that marked the beginning of modern rocketry. Rockets had been in use for hundreds of years, as fireworks and in war, but they had all been powered by solid, gunpowder-like fuels. Goddard's was liquid-fuelled, the first of the breed that led to the Saturn V moon rocket, though on that cold March day it reached a height of only forty-one feet.

Goddard's pioneering work continued almost up to his death in 1945. He was the first to use a gyroscope coupled to movable vanes in the exhaust to steer a rocket, the first to use liquefied gases as propellants, the first to send scientific instruments aloft on a rocket. But he was a secretive man, partly no doubt because he tended to get called 'the crazy moon professor'. The US Government took no interest in his work, and his enormous contribution had little effect on the mainstream of rocket development, which took place in Germany.

There a group of amateur enthusiasts, inspired by dreams of inter-planetary travel, were conducting their own, occasionally fatal, experiments on liquid-fuelled rockets. In 1930 they were joined by an eighteen-year-old student called Wernher von Braun. Soon afterwards the Army started to support their work and in 1936, after Hitler had come to power, they were transferred to a new research station at Peenemünde on the Baltic coast, where a team led by von Braun successfully launched in 1942 the first V-2, the rocket that was later used to bombard London and Antwerp. ('The V-2 was a fine rocket. The only thing wrong with it was that it landed on the wrong planet,' said von Braun later.) The V-2 had an importance extending far beyond its technology. It brought the possibilities of long-range rockets dramatically to the attention of military and political leaders for the first time. The intercontinental ballistic missile was an inevitable development anyway, but without the V-2 it would probably not have been regarded as a serious possibility until many years later.

B.S.

Parking meter

By the late 1920s the car dominated the American way of life and parking had become a problem. Most big towns had parking ordinances, but enforcing them was another matter. In December 1932 Carl C. Magee of Oklahoma City applied to the U.S. Patent Office with a claim for the invention of a parking meter. The first meter looked like a loaf of bread on a short pole. A small glass side window showed a 'vacant' or otherwise sign, and the principle of a coin-operated meter selling a space for a specified period of time was established. Patent rights were granted in 1936, but in 1935 Magee had filed another application, refining his original device. This second prototype resembles in most particulars the cobra-shaped meters to be found on the streets of most of the world's cities today.

G.N.

Catseye

Before the British inventor Percy Shaw designed and made the first catseye in 1934, tramlines were the only means by which the motorist could find his way along roads in thick fog. The authorities doubted the eye's usefulness, and it took the war, with the slogan 'fifty thousand catseyes help to fight the blackout' for it to become established as a commercial proposition. It had the wartime advantage of reflecting light back the way it came; unlike other markers it would not scatter light upwards for aircraft to spot.

The reflecting road-stud is a carefully engineered prism and reflecting system, with a sealed, vacuum-deposited layer of aluminium forming the mirror, all mounted in a rubber pad which protects both the eyes and the vehicle tyre passing over them. The most ingenious part of the eye is the self-wiper which protects the stud. The inside wall of the rubber in which the glass eyes are embedded is cut so that when the eyes are depressed they are automatically cleaned; dirt is removed and rain washed away. As they are deeply embedded in their rubber case, the reflectors never come into contact with tyres. The device clips into an iron base with ramps to protect the pad, and is sunk into the road.

Percy Shaw, the catseye's inventor, owned a flourishing business repairing roads which took him all over the Yorkshire countryside. 'At night I had to drive home down a dangerous road from Queensbury to Halifax. One night it was so dark I couldn't even see the tramlines. It was then that I saw some reflectors on a poster by the road and I thought it would be a good idea to bring them down to road level. It was also the eyes of a cat sitting on a fence which helped the idea to develop.'

The rubber disc of the catseye has not been seriously superseded – the British Ministry of Transport still regard it as being the best stud of its kind.

C.V.

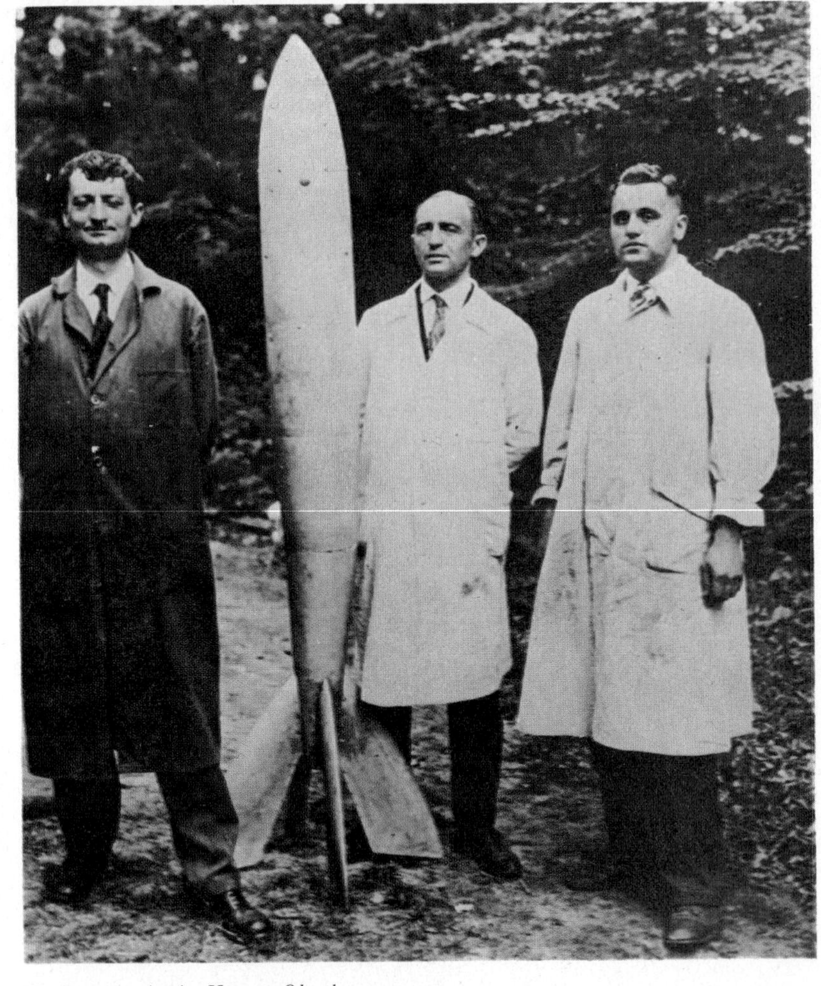

The German scientist Herman Oberth (left) poses with his 'postal rocket', forerunner of the V-2. It was said to be capable of carrying mail from Berlin to New York in 24 minutes.

Early American parking meter.

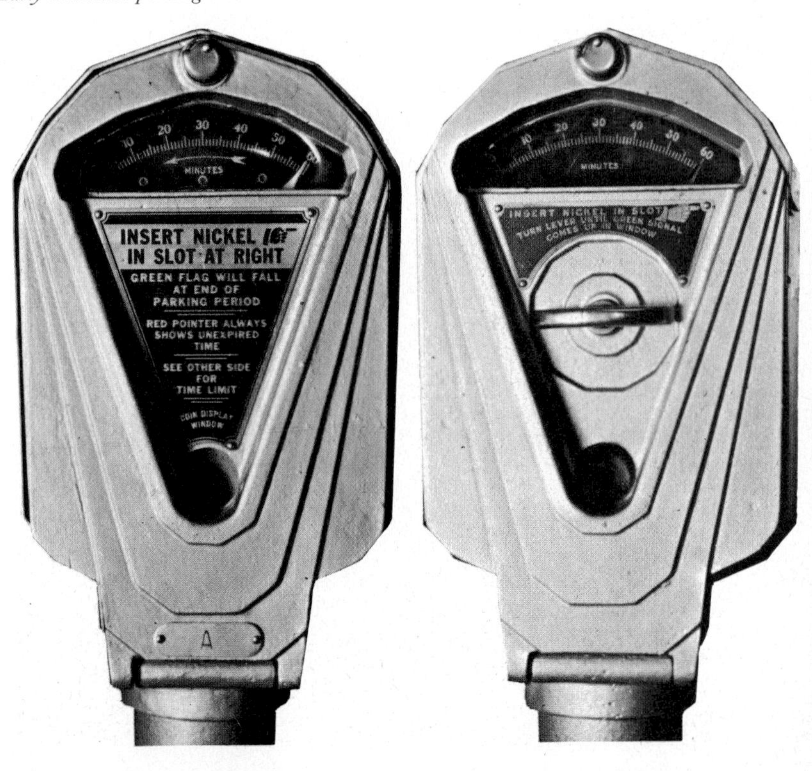

Germany established her lead in jet engines with the turbojet designed by von Ohain (above), and maintained it with the Messerschmitt 262 (top).

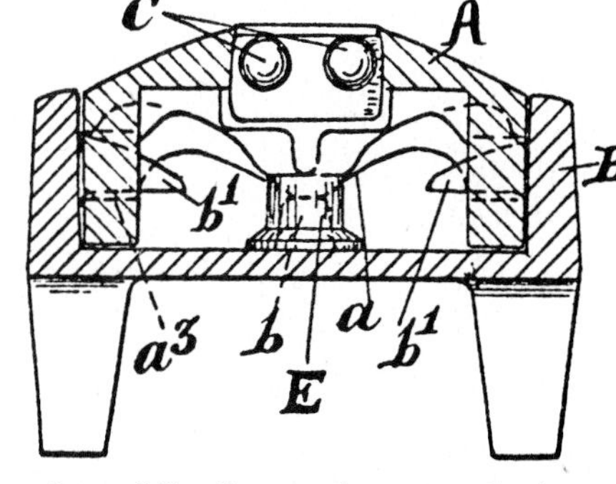

Patent Office diagram of a catseye, showing the reflectors embedded in the rubber disc.

Aqualung in use by an underwater archaeologist.

Jet engine

The first successful turbojet engine, and the first successful flight by a turbojet aeroplane, were achieved in Germany when the Heinkel He178 flew in 1939, powered by a centrifugal-flow jet engine designed by Dr Hans von Ohain. It used to be thought that von Ohain had seen and profited by patents taken out by Britain's Frank Whittle; but it is now known that this was not so. Whittle, working under great difficulties, saw the first machine powered by one of his turbojet engines take off and fly in 1941. But even then progress in England was very slow; and it was again Germany which, by the end of the war, had the only successful jets in action, led by the remarkable Messerschmitt Me262 fighter.

But British Whittle engines were to achieve a far greater historical fame than their German rivals; for Whittle engines directly gave rise to the whole of the vast production of jet engines in the United States and Russia, after examples had been delivered to them during the war, and became the true ancestors of all jet engines in those countries.

The first suggestion for reaction propulsion for full-size aircraft dates back to the first year of flying in balloons, and was made by Joseph Montgolfier at Lyons in 1783; his idea was that the hot air which supported his balloons could also be expelled rearwards from the envelope, and the aircraft thus made to travel in the opposite direction.

The first mature idea for a jet aeroplane came from a remarkable Frenchman, Charles de Louvrié, who designed his jet-propelled monoplane in 1865: it was not, of course, built, but was a fine and prophetic achievement. This was followed, the next year, by suggestions for delta-wing jet aeroplanes designed by the Englishmen Butler and Edwards. There were numerous other designs along the same lines throughout the last century, but no attempts to build such a machine.

Rocket propulsion for aircraft – which is also reaction propulsion – dates back to about 1420, when Joanes Fontana designed a model bird propelled by a rocket hidden in its body, an idea which was quite likely tried out in practice. The idea was again revived as a proposed method of propelling balloons in the 1780s, and continued to be a popular pipe-dream during the nineteenth century. But it was a rocket-propelled aeroplane piloted by F. Stamer which first took off and made a short flight in Germany in 1928, resulting from the combined efforts of Fritz von Opel, F.W. Sander and Alexander Lippisch. In 1944, the first successful rocket aeroplane went into action when the German Me163 Komet fighter flew against the Allies.

C.G.-S.

Aqualung

It was the perfection of a valve which made the Aqualung possible and put man back into the sea with all the freedom of a fish. For the first time it made breathing underwater fully automatic and routine. The swimmer was free equally of umbilical connections to the surface and of awkward hand-regulated pumps and devices that could prove very chancy in a crisis. The breakthrough was made in an obscure corner of occupied France in the middle of the Second World War.

Commandant Jacques-Yves Cousteau of the French Navy was part of the Mediterranean free-diving tradition – part sport, part science – that had flourished for some decades around Marseilles and elsewhere. The divers had tried using portable tanks of compressed air, but until the summer of 1943 it was not possible for them to regulate the flow of breathing gas as their needs varied at different depths. The problem was how to guarantee, simply and economically, a supply of air at the same pressure as the surrounding sea and so avoid lung collapse at depth. (Pressure increases by one atmosphere every thirty-three feet.)

Cousteau had tried out all the available equipment, and had had some close escapes. Through experience he isolated the central problem. Then Émile Gagnan, a specialist control-valve engineer, translated the diver's specifications into working hardware. His device supplied compressed air to the lungs, on demand and at the correct pressure to match the local depth. He achieved this by linking the diver's outward and inward 'breaths' by a diaphragm valve adapted from the regulating valve of a gas stove. The diver's 'exhaust' breath pressure opened an outlet valve which simultaneously cut off the input air supply. On breathing in, the outside pressure of the water closed the outlet valve and pushed in the diaphragm, switching the diver's supply back on. The air flowed until the pressures became equal on either side of the diaphragm, which then returned to 'neutral' and shut off the input.

The Aqualung, as the breathing-set – mask, valve unit and gas-tank – soon became known, has opened the undersea world to millions and has made Cousteau a household name.

For many years, Aqualung dives were limited to three hundred feet because of the effects of the nitrogen fraction in air under several atmospheres of pressure. But free dives on a mixture of oxygen and helium, using the basic Cousteau valve system, are now taking place to four hundred feet in the search for oil in the North Sea.

A.C.

Man Moving

Hovercraft

Two tins, a vacuum cleaner and an English engineer of rare detachment and persistence were the keys to the elements of the hovercraft.

It had been known for years that a special kind of 'lift' helps to support aircraft flying very close to the ground. A helicopter just clear of the ground may need no more than a quarter the power it needs to hover at height. Lindbergh made use of this 'ground effect' to save fuel on his famous transatlantic flight in the 1920s. Various people in several countries tinkered with the 'ground effect'. The Americans spent $3,000,000 in a fruitless attempt to apply the principle to naval craft.

In 1950, after a tough war in the demanding field of radio engineering, Christopher Cockerell turned boatbuilder on the Norfolk Broads and found that two factors greatly reduced his boats' performance – skin resistance and wave resistance. 'If I could make the skin of my craft a skin of air, i.e. introduce a film of air between hull and water, skin friction would be negligible.' He found that by blowing air through open-ended tins of different sizes downwards on to his kitchen scales, he achieved vastly different readings. The same mass of air through the narrow tin exerted up to three times as much thrust – which was contrary to the classical laws of physics. He was on to something.

A working model of Cockerell's hovercraft flew over the most illustrious carpets in Whitehall in 1955 – and was taken out of the inventor's hands and put on the Secret List where no one wanted it. It was finally prised loose through the enterprise of a far-sighted civil servant, R.A. Shaw, who authorized a small feasibility contract. The report was favourable and was the turning-point in obtaining the money necessary to transform a bright idea into a viable production item.

On 11 June 1959, a full-scale overwater craft, *SRN-1*, was shown off to the Press – and received rave notices across the world. On the fiftieth anniversary of Blériot's first Channel crossing in 1909, *SRN-1* repeated the performance in the opposite direction.

The key to practicality proved to be the manipulation of the air-jets beneath the craft and the refinement of the British-patented 'skirt' which keeps effective contact with the surface, while avoiding actual hard contact.

Cockerell's concept has produced not just a flying-boat, but a family of new transporters. Besides a revolution in short sea and inter-island fast ferry services and patrol craft, the hovercraft principle promises to make 300-m.p.h. trains a reality in a few years. The first eighty-passenger prototype has been tested on an elevated monorail outside Orléans and the more sophisticated British version is also in preparation. The British tracked hovercraft design is in view for the busiest travel route in North America, the Boston–Washington 'corridor'.

A.C.

Spacecraft

Jules Verne wrote his famous space-travel novel about the first trip to the moon in 1865; just over a hundred years later it became a reality. However, scientists did not take the possibility of space flight seriously until the 1920s, although a Russian scientist and schoolmaster, K.E. Tsiolkovsky, had already done some essential physical and mathematical studies of the problem as early as 1903. It was he who pointed out that only rocket propulsion would be suitable for vehicles leaving the earth's atmosphere, for two reasons: first, the rocket does not need air for propulsion, as the airscrew engine does, but acts like the recoil of a gun according to Newton's third law of motion, that to every action there is an equal and opposite reaction; second, the rocket carries within itself all the chemicals needed for combustion and for the generation of the gases which drive it through airless space: it needs no oxygen from the air. (The jet engine, invented in the 1930s, also works as a reaction engine, but it needs large quantities of air for combustion and is therefore unsuitable for spacecraft.)

The most important theoretical and practical work on rocket propulsion was done in Germany, where the physicist Hermann Oberth published his influential book, *Roads to Space Travel*, in 1923 while he was still a student. A few years later the motor-car industrialist Fritz von Opel tried out a rocket-propelled car near Berlin. Max Valier, another rocket pioneer, built a car which achieved a speed of 235 m.p.h. on a frozen Bavarian lake in 1929, using ethyl alcohol and liquid oxygen as a fuel; he was killed in a later experiment, which ended with an explosion.

Meanwhile, an American professor of physics, Robert H. Goddard, was doing much systematic rocket research. His first laboratory experiments resulted in a slim booklet, *A Method of Reaching Extreme Altitudes*, published in 1919; a few years later, he successfully tried out liquid propellants – petrol and liquid oxygen – in a series of test launchings, achieving heights of 7,500 ft. and speeds of over 700 m.p.h. with his rockets. He was also the first to build gyroscope-controlled rockets. His work was backed by the Smithsonian Institution and the Guggenheim Foundation. He died, 63 years old, in 1945, a decade too early to see the beginnings of manned space flight.

During the last year of the Second World War, the Germans bombarded southern England with 'V2' rocket bombs, largely designed by Wernher von Braun; their engines, fuelled with

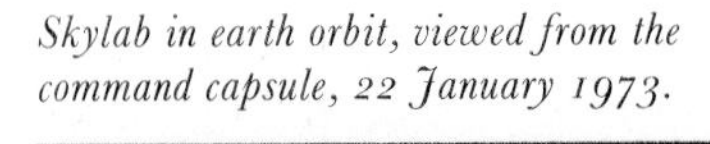

Reconstruction of Cockerell's original hovercraft experiment, with kitchen scales, the works from a vacuum cleaner, and some tins. His intention was to find a way of reducing friction round the hulls of conventional craft.

Skylab in earth orbit, viewed from the command capsule, 22 January 1973.

Top, *the Krauss-Maffei magnetic levitation test vehicle. The linear motor reaction plate can be seen, and the traction current conductor rails at the side of the track support.*

Above, *a small linear motor used in the fabrication of aluminium sheet. Working against the transporting rollers, it keeps the sheet in tension and thus prevents warping and wrinkling.*

alcohol and liquid oxygen and later with nitric acid and hydrazine, developed a thrust of 55,000 lb., carrying the vehicle up to an altitude of 60 miles and over distances of about 650 miles. The top speed during descent was 3,700 m.p.h. Most of the postwar development of space flight can be traced back to the V2.

The Russians achieved two 'firsts' in space exploration. In October 1957, a Soviet rocket carried a small *sputnik* (satellite) 560 miles up into space, where it began to circle the earth at a speed of 17,000 m.p.h., which means enough centrifugal force to counteract the gravity pull of the earth. Some additional velocity is needed to let a space vehicle escape from the earth's pull altogether and shoot out into outer space, which subsequent Russian and American unmanned spacecraft have done many times, reaching the moon and planets in our solar system.

The other Russian 'first' was the manned spaceflight of a 4½-ton rocket-launched craft carrying a man, Yuri Gagarin, into orbit and around the earth in 89 minutes at a speed of 18,000 m.p.h. in April 1961. That first manned spaceflight was followed by a second one by Titov, who completed 17 orbits, and a third by the American Glenn.

The greatest space spectacular was achieved by America with the first landing on the moon on 20 July 1969. The three-stage, 44-ton Apollo 11 rocket – liquid-fuelled, gyroscope-steered, computer-controlled, with 56 independently working systems – carried the three astronauts Neil Armstrong, Edwin Aldrin and Michael Collins after a three-day journey into an orbit around the moon. While Collins continued to circle it, the other two men descended to the surface of the moon on board the 'lunar module' or space ferry, which had been detached from the main craft; the ferry idea had been worked out by John Houbolt, a senior technician at the National Space Agency (NASA). There were some anxious moments before the ferry lifted off the moon again and as it 'docked', or linked up, with the command ship; but all went well, right to the splashdown in the Pacific 195 hours after the start. Technically, the successful moon trip was man's most brilliant achievement, although it did not reveal very much the scientists had not yet known about the moon.

Subsequent moon flights, until the end of America's lunar programme in 1973, were bound to come as a series of anticlimaxes; but many of the techniques and experiences of these flights have proved to be of great importance in the whole field of space exploration, particularly the docking process, which makes it possible to assemble permanent manned space stations orbiting the earth just outside its atmosphere. These stations, put up and worked jointly by the USA and the Soviet Union, will enable the scientists to probe deeper into the still unsolved mysteries of the universe than from the earth's surface.

E. L.

Linear induction motor

It so happens that, at present, the term 'electric motor' instantly conjures up a picture of a machine which turns a rotating shaft. In fact there is no reason why the output should not travel in a straight line, and the traditional induction motor – today responsible for at least 95 per cent of the total horsepower of the world's electric motors – obtains its rotation because its magnetic field is forced to travel in a circular path. The linear electric motor, often called a linear induction motor or LIM, is one of those inventions which were known more than a century ago, virtually ignored and then suddenly blossomed forth in the modern world in a score of variations for a wide range of purposes.

Around 1830, when electrical machinery was non-existent, there was no reason to suppose that rotary machines would be any more important than linear ones. William Sturgeon, one of the first engineers to construct an electromagnet (in 1825), built several 'magnetic engines' from 1832 onwards of which several gave a linear output motion. Other electrical experimenters, such as Froment in 1860, used solenoid-driven oscillating rods and then added pivoted beams (just as in James Watt's early steam engines) to convert what was essentially a to-fro motion into a rotary one. Today's LIM is not a solenoid device. Instead it consists of two elements, usually called a 'stator' and a 'rotor' despite the obvious misnomer, free to move past each other with a constant small air-gap between them. Alternating current is supplied to one of the elements and passed through one or more rigid coils to produce a magnetic field travelling in a straight line. The moving field cuts the secondary member, which can consist of further coils, rows of bar conductors or, in the most common form, simply a strip or tube of aluminium. The reaction between the moving field and the secondary member tends to cause linear relative motion.

What can cause confusion is that usually the secondary reacting member, or rotor, is actually the part that stays still, while the current is supplied to coils in the stator which is the moving component. This terminological nonsense is a heritage from the time when LIMs were explained by starting with a rotary induction motor and then hypothetically unrolling it. If this is done, obviously the two parts will thrust against each other for only a few inches before shooting apart. This is all right if the LIM is wanted merely to serve as a short-stroke actuator, a duty that cannot always be done by hydraulic or pneumatic rams. Where the LIM has no rival is in its ability to give a smooth, silent push for as long as either member is constructed, which may be 1,000 miles or more. The ideal application is propulsion for high-speed guided land transport, such as trains running on wheels or air cushions, or magnetically levitated. In view of their speed, such trains are at present thought unlikely to comprise more than one vehicle.

The LIM appears to be nearing the point where it will take over from other propulsion methods in an increasing part of the world's guided land transport. When the first edition of this book was written the favoured scheme was to use double-sided stator coils operating on each side of a vertical aluminium strip. Since 1969 the use of single-sided coils running over a flat track of composite aluminium/steel has become important, because this can be made to give lift as well as thrust. Since 1973 Professor Eric Laithwaite and his team at London's Imperial College have refined this system still further to give thrust, lift and guidance. The vehicle collects a.c. or d.c. from pick-up shoes running on supply rails. The current in the coils creates a linear travelling field which constantly reacts with the aluminium secondary track, causing relative motion. The only limits on speed are set by wind resistance and other economic factors; there is no physical contact, noise or wear. To brake the vehicle, the track may be made to generate current in the coils, as in conventional regenerative methods, the kinetic energy of the vehicle appearing as heat in the affected length of track.

Even electrical engineers often do not appreciate the wealth of varieties of electric motor possible. There are far more than 200 structurally different types of linear motor alone, divided into families with short primary or short secondary windings, or arranged in flat or tubular forms, or single or double sided, or having the magnetic flux transverse or longitudinal. One family comprises 'synchronous' machines in which the moving material is not the conductor coils but the magnetic iron. The only thing that can be predicted with assurance in this field is that engineers have as yet only scratched the surface, and the term 'electric motor' will in future be as unacceptably vague as the word 'engine' is today.

W.T.G.

Wankel engine

The design of the engines that drive our motor-cars has hardly changed since the days of Benz and Daimler in the 1880s. The conventional internal-combustion motor still works with pistons which are forced up and down in cylinders by gas explosions, and their reciprocal movement is then transformed into a rotary one which turns the wheels.

It was only in the late 1950s that Fritz Wankel, a German engineer from Lindau on Lake Constance, found a solution to the problem of how to cut out the need for transforming one kind of movement into another – by replacing the up-and-down movement in the cylinder by a rotary one. His 'cylinder' is not oblong but round, and his 'piston' is a triangular rotating disc whose sides are slightly curved so that throughout its rotation there is always some space for the moving and expanding gases on at least two of its three sides. The rotating piston-disc also opens and closes the inlet and exhaust valves automatically.

The result of his revolutionary design is that the engine is much smaller, works more smoothly, and has fewer moving parts; its disadvantage is that it consumes slightly more fuel than a conventional engine. The four strokes of the latter – suction, compression, ignition and exhaust – are completed during the rotation of the piston-disc; but there are *three* power strokes during each rotation. In fact, a one-rotor Wankel engine does the work of a conventional three-cylinder motor, and a two-rotor Wankel that of a six-cylinder one, developing more than 100 b.h.p. gross. The average Wankel engine speed is 5,500 rotations per minute, with a maximum of 7,000 r.p.m.

The rotary-piston engine has only two moving parts: the rotor and the output shaft, leading to the gearbox. It needs a carburettor and spark plugs but no piston rods, crankshaft or complicated valve control; this makes the Wankel engine about a quarter less heavy than a conventional engine, and it is cheaper to build in mass production.

The first Wankel car was produced by NSU in Germany, followed by Japan's Mazda in 1968. General Motors in the USA marketed their small Chevrolet Vega Wankel car in 1974. Major companies which have acquired Wankel licences are Mercedes-Benz, Volkswagen, Toyota, Rolls-Royce, Alfa Romeo and Citroën, and two dozen others in many countries.

General Motors estimate that by 1980 eight out of every ten cars produced by them will be powered by Wankel engines. There are also efforts to design aircraft engines (with propellers, of course) and stationary electric generators based on the rotary-piston principle.

E.L.

Disc drive for an electric crane. Powered by a linear motor, it ensures jerk-free movement of the loaded crane.

Diagram of the Wankel engine, showing two triangular rotors.

Musicians at a Tartar feast, from a 14th-century Genoese manuscript. The drummer is playing a pair of kettledrums.

Woman playing a lyre: Pompeian wall painting.

Drum

The first drum of all was doubtless the human body, slapped by the hands in rhythm; sticks beaten on the ground, or clappers, must have followed, preceding the hollow vessels of wood with a vibrating membrane of uncured or cured skin which we know as drums. Even today, in primitive tribes such as the Bantus or the Australian Aborigines, no such vessel is used.

The earliest drums found are of pottery (perhaps a little later than 4000 BC), and come from sites including the Upper Nile and Bohemia: these are of hour-glass shape, with lugs on to which the skins – at either end – would be tied and tensioned. These are perhaps the ancestor of the side-drum or bass-drum; other classes are the open-ended type (the tambourine) and the kettle-drum (a cup-shaped drum of wood, metal or pottery).

Everyone seems to have known and used the drum; pre-Columbian America had drums of clay, or tree-trunk, with anything from llama-skin to human skin for covering; they were played with the hands, or with one stick. Large drums were used in temple ceremonies in Mesopotamia, where they were known from at least 3000 BC; a Sumerian temple drum of great size is shown on a relief from Ur of *c.* 2500 BC. The instrument was also used by the Greeks and the Romans, especially for the cults of Dionysos and Cybele; the Romans used it to march to, and it was to their steady beat that the famous 18th-century general, the Maréchal de Saxe, attributed their military success. The tambourine, a wood or metal ring with one skin, and hung with bells or discs, is also of great antiquity, and had a religious function before it was adopted for military use.

Of all the drums mentioned, only the kettle-drum has a definite musical pitch: the rest can only be used for rhythm. It probably originated in the East, though at what period is uncertain; the Greeks and Romans knew of it as an instrument beaten by barbarians, but seem to have had no close familiarity with it, and its introduction into Europe may date only from the 14th century – through the Moors, to judge by their name of 'nacairs' or 'nakers'. Kettle-drums were early used in churches, and saw regular military use by the time of the Restoration, being carted into battle on a chariot. Mounted in pairs on a horse, they are still used by, for example, the British Life Guards.

Side-drums – of which Egyptian examples have survived – were used by the English in battle as early as the battle of Halidon Hill (1333), an account of which notes that 'Englische mynstrelles beaten their tabers and blewen their trompes and pipers pipened loude and made a great schowte upon the Skottes.' It seems possible, however, that the snares, which give the side-drum its characteristic rattle, were not introduced until the 16th century.

C. M. B. G.

Stringed instruments

The harp is probably the oldest stringed instrument. It appears as a type of lyre in the cemetery at Ur, which is about 4,500 years old. Homer mentions lyres frequently, and numerous inscriptions and works of art from the classical world testify to its continuing popularity. From such lyres the harp developed, leading in turn, via the psaltery and the dulcimer (a kind of harp in which the strings were struck by small hammers), to the harpsichord and piano.

Early harps, however, were of limited use in ensemble music since their strings were tuned to a fixed scale. It was not until 1720 that a Bavarian musician from Donauwörth, Simon Hochbrucker, produced a harp with seven pedals attached to various strings which, when depressed, shortened them and so raised their pitch a semitone. This basic principle was refined by Sébastien Erard, a Parisian, in 1792, and it is Erard who is credited with fixing the form of the modern harp.

The lute seems to have originated in Persia, and by the time of the Renaissance had spawned a large family, one member of which, the cittern, was to become the 'gittern' and eventually the guitar, of which two examples made by Stradivarius in the early 18th century survive today. The guitar continued to assume a variety of shapes and sizes, however, and varied numbers of strings were tried: the 'lyre guitar', 'harp guitar' and 'harp lute guitar' could all be found in Napoleonic France and Regency England. In 1783 Christian Krauss, a German living in London, devised a 'keyed English guitar', complete with piano keys, but it failed to

The making of ink in China, from carbon and glue.

Opposite, *St Mark dips in his pen in the ink to write the 'Harley Golden Gospels'. 9th-century manuscript.*

attract attention. It was Antonio Torres who, in 1854, began to make guitars in what is today accepted as the standard pattern.

By contrast with these early stringed instruments, bowed ones are a relatively modern conception. Possible forerunners of the bow include the 'java', a long wooden plectrum used in India for plucking the strings of a lute-like instrument in one sweep, but the modern type of bow, intended mainly for playing on a single string and with adjustable tension and concave shape, dates only from about 1780 and is mainly the result of the work of the master craftsman François Tourte, a Parisian watchmaker. (John Dodd, however, is thought to have arrived at a design similar to Tourte's at almost the same time.) Bowed guitars – viols – were played between the legs, originally with frets on the fingerboard and with six strings. For the immediate ancestors of the violin, therefore, we should look to the rebec and the Renaissance 'fiedel'. The violin's basic shape is attributed to Gasparo da Salo (1540–1609); during the 16th and 17th centuries the violin was perfected by Andrea Amati, his sons Antonio and Girolamo, Girolamo's son Nicola and Nicola's pupil Antonio Stradivarius, who died in 1737.

P. WH.

Wind instruments

There are three groups of wind instruments – flutes or flue pipes; trumpets or lip-reed pipes; and reed pipes. Primitive flutes were always associated with magic, and even in the 4th century, the Greek *aulos* was said to be able to cure sciatica if played over the affected part of the body. The earliest flutes, in which the performer had to blow vertically against the sharp edge of a pipe, had no finger-holes, but were made in various sizes and bound together – the pan pipe. 'Fipple' flutes, ancestors of the recorder, had a plug (a 'fipple') slightly smaller than the aperture, thus creating a wind-channel which produced the sound.

The transverse flute, held across the player's face, was well known in the East long before Christ, but was not popular in the West until very much later. In the 17th century, Jean Hotteterre and his colleagues are credited with designing the recorder, making it in three joints, with a conical foot, as is familiar today. In 1677 Lully introduced an improved transverse flute but it was not until after the middle of the 18th century that the transverse flute was considered superior to the recorder. It was Theobald Boehm (1794–1881) who devised, in 1847, the reformed flute that we know today, and whose key-system still bears his name.

The first lip-reed instruments, played by passing air through pursed lips into the instrument, were the conch and the hollowed horns of various animals. The trumpet, as with all primitive instruments, was blown for ritual purposes, basically in war and the hunt, or to scare away evil spirits. The 'Oliphant', an often elaborate ivory horn, was a badge of knighthood (as mentioned in the *Chanson de Roland*), and came from Byzantium in the 10th century. In the late Middle Ages, the trumpet was still associated with regal fanfares; the slide trumpet, and the horn, whose unwieldy length was curved back upon itself for convenience, were developed. In the Renaissance, the trombone, also called 'sackbut', appeared, and has not substantially changed since. A bass cornet emerged at this time, which was fashioned in the shape of a double 'S' and appropriately called the 'serpent'. The keyed trumpet is ascribed to the Viennese Anton Weidinger (1801); the most revolutionary refinement of it was the 'rotary valve', attributed to Joseph Riedl of Vienna in 1832, which made the length of the tube, and hence the range of notes playable, more variable and controllable than ever before.

The Greek *aulos* seems to have been both an early flute and also an early reed instrument which had two pipes played simultaneously. Two reeds were bound together in such a way that when they were placed in the mouth they vibrated together and the air then passed down their stems into the pipe. It is not known whether these double pipes were played in unison or whether one acted as a drone. The Greek players had a technique of blowing out and breathing in at the same time. This put a great strain on the cheek muscles, which is why players of the period are depicted with straps around their cheeks. This technique led to the development of the bagpipes, where the bag takes the place of the player's cheeks.

The earliest form of oboe was called a shawm, the earliest record of which is in 12th-century Saracen Sicily. The bassoon, known as the dulzian or kortholt, appeared in the 16th century, as did many other instruments, including the rackett, a cylindrically bored tube curled round many times into a small cylinder, with finger holes bored through both the outer and inner casing. It was at this period that the crumhorn, whose double reeds were embedded in a wooden chamber into which the player blew, also appeared.

The invention of the clarinet, with a single reed fully controllable by the lips, is attributed to J.C. Denner (1655–1707) in Nuremberg around 1700. In 1839, Auguste Buffet's clarinet won a medal in the Paris Exhibition, and at the same Exhibition, Henri Brod showed the first cor anglais. Adolphe Sax patented his saxhorn in 1845, and his saxophone in 1846.

P. WH.

The Australian aborigines' 'didgeridoo', seen (left) *at a corroboree at Alice Springs and* (above) *playing for dancers in the Northern Territory, is one of the most primitive of wind instruments.*

Right, *15th-century organ, still more akin to the Pan-pipes than to the sophisticated instrument played by Bach.*

Ink

The use of ink seems first to have appeared in Egypt, where from *c.* 3200 BC black inks were prepared from lamp-black suspended in an aqueous solution of vegetable gum, and red inks (from much later on, in the Hellenistic period) from red oxide. The Egyptians also knew how to make inks with a base of iron oxide from *c.* 2800 BC, but these were apparently only used for marking linen, and not for writing. The Chinese, too, had red ink, probably much earlier than the Egyptians, for they were using sulphate of mercury to produce it during the 2nd millennium BC. Black ink was prepared with sulphur of iron, mixed with a varnish obtained from the sap of the sumach tree; made up in block or stick form, and used with a moistened brush, this was similar to modern Indian ink.

The Romans used a purplish ink called *encaustum* (something which 'burnt in'), from which our word 'ink' derives. They also made frequent use of a dark brown liquid, sepia, obtained from the dried and powdered ink-sacs of cuttle-fish. Throughout history many coloured juices and extracts of plants and animals have been used as inks.

Iron-gall inks – made from galls and a salt of iron – are first described by Bishop Isidore of Seville in the 7th century AD, but they did not come into general use until the Renaissance; they are, unfortunately, prone to serious fading. In the 17th century a new writing ink was introduced into Europe. It was coloured by a mixture of tannic acid, obtained from the bark of trees, and iron salt. Modern inks basically use the same ingredients, but a blue dye is added to heighten the colour. (Ballpoint pens, however, use ink with an oil base so that it transfers easily from the reservoir to the writing tip.)

Other writing inks include coloured inks, made by dissolving dyes in water – the first British patent for making them was issued in 1772 – indelible inks, and the various 'invisible' inks. During the Indian Mutiny (1857–58), for example, the British used a solution of milk and lemon juice to write secret messages with; the writing would become visible when exposed to heat. Other invisible inks, such as rice water, were developed, which became visible when iodine was applied.

Printing inks are of a totally different kind. While inks suspended in a solution of water were quite adequate for the hand impression of wooden blocks onto paper, metal types did not hold them so successfully. The answer was found in a base of linseed oil, boiled until it resembled glue and added to finely ground lampblack.

This method has changed little, except that today a drying agent is also added to the solution – previously the dampened paper had to be taken off the press and hung up on a line to dry. For very high-speed presses heat-set inks are used, which dry quickly when exposed to heat; others can be dried by steam. The composition of coloured inks has, however, changed considerably, for synthetic compounds have been substituted for the original vegetable and mineral pigments.

Today ink is mass-produced by machinery, and a wide variety of special inks is made. Reading, writing and printing have so expanded in this century that the average amount of ink required to print all the newspapers in the United States is now about 700,000 lb. per week.

M.H., C.M.B.G.

Organ

The early ancestor of the organ was the *syrinx* or Pan-pipes, which figured in ancient mythology. Here human breath supplied the wind, wind-chest and control. Then an Alexandrian engineer, Ctesibius (*c.* 265 BC), invented or improved the *hydraulus*, a hydraulic organ in which water was used for stabilizing the wind pressure. The following millennium saw the evolution of keyboards to control the valves admitting wind to the pipes and 'stop' mechanisms which controlled (or 'stopped') the various ranks of pipes at will. St Dunstan (925–88) is said to have directed the building of an organ at Glastonbury.

The next 500 years saw the instrument develop into a more recognizable precursor of the classical organ. The 16th to 18th century organ-builders who contributed to the kind of instrument for which Bach wrote were the Niehoffs (Netherlands), the Antegnatis (Italy), the Cliquots (France), Schnitger and the Silbermanns (Germany), Dallam, 'Father' Smith and Renatus Harris (England). In 1712, another English builder, Abraham Jordan, provided one division of the organ with a sliding pedal-controlled shutter, and so is credited with the invention of the swell box, though this was being developed in Spain at about the same time. During the 19th century, the organ developed in size, and direct mechanical control of the valves admitting wind to the pipes was superseded by pneumatic and, later, electrical action.

F.A.

Book

Egyptian papyrus being quite pliable and written on one side only, the sheets of it could be glued together end to end to form long rolls. Since it was hardly practical to write all the way along, the scrolls were divided into columns; the longest that has survived is 65 ft. in length and has 116 columns, but rolls for normal use were not more than a third of this size. They were rolled round a stick, or if long, two sticks, and in order to be read had to be gradually unwound. This made it very hard to use one for reference, because to find a particular passage near the end of the roll the whole thing had to be unwound, and once you lost your place it was no easy matter to find it again.

When parchment superseded papyrus, the roll format was taken over. The Scroll of the Law (i.e. the Pentateuch), as appointed to be read in synagogues, to this day preserves the two-stick roll format. Anyone who has remarked how large and unwieldy this is, although the letters are not outsize, will appreciate that rolls had normally to be kept as small as possible. That is why many works of ancient authors which now fit comfortably into a single volume are formally divided into 10, 20 or more 'books'.

For taking notes, school exercises, tradesmen's accounts and the like, the Greeks had used little rectangular wooden or wax tablets. Two or more of these could be tied together to form 'notebooks'. As it had begun as a more economical (though less flexible) alternative to papyrus, parchment eventually took over the wax-tablet format. Each sheet was folded over to form a 'signature' and these were then tied together in a protective cover to make a *codex*. This practice seems to have begun in the first century AD, when the poet Martial acclaimed the new method, so handy and compact especially for the more voluminous authors, who would hardly fit into his library as scrolls, but could now be compressed into 'fist-books', in 'many-folded skin'. But it took roughly the whole duration of Imperial Rome before the codex ousted the roll completely. Because of its seniority, no doubt, the roll retained the dignity proper to royal proclamations, municipal charters, honorary degrees and diplomas (all conveniently short).

Papyrus books had had no cover but their cases, and codices were probably at first covered only in blank sheets. The oldest leather bindings are Coptic Egyptian of the 6th century; they are already decorated with the tooled patterns which later became so popular. The codex provides a compact shape, good for storage, joining so many more sheets together that books could be much longer. But their prime advantage is ease of reference, especially once they began to be paginated, so that it is no trouble to find a particular place. Those who imagine that microfilm will now supersede the codex ought to remember that it is read like a scroll, unwound from a spool mounted on one spindle on to a second. And it is just as hard to find your place.

A.K.

Woodcut

The idea of producing a block of wood or metal bearing a design in either relief or intaglio and capable of being impressed into a receptive material such as clay is very old. The Egyptians used relief stamps to make impressions on their pottery, as did the inhabitants of Mesopotamia before 3000 BC. Yet neither civilization took the discovery further. This is understandable for Mesopotamia, where records were kept on clay tablets; but in Egypt, where papyrus was used, it is odd that attempts were not made to impress a pattern on to a piece of papyrus.

The use of wood-blocks to produce repeatable patterns (in which the parts required to print white were cut away, leaving the design in relief, to be inked and applied), is found in Egypt in the 6th century AD for the printing of textiles, but the first appearance of printing on paper was in China at about the same time. Impressions showing the Buddha are to be found in rolls, and some actual stamps, datable before AD 800, have been found in eastern Turkestan.

In Europe, textiles were printed from wood blocks from the 6th century AD onwards, but again no one seems to have thought of other applications until the early 15th century – coinciding with the beginnings of large-scale paper production in Europe. There is a record of payment in the accounts of the Charterhouse of Dijon, dated 1393, for 'John Baudet, carpenter, for having cut blocks and tablets for the chapel of my lord at Champmol called the chapel of Angles', but this is quite possibly for textile hangings there.

Early woodcuts on paper were for two main purposes: playing-cards (which were being produced from blocks at Bologna as early as 1395) and, more importantly, devotional pictures. One of the earliest dated prints is the *St Christopher* of 1423, but undated examples exist which undoubtedly date from several decades earlier. Probably from *c.* 1400, individual sheets were sewn together to make books mostly of a devotional character; these are known as 'block-books', because not only the pictures but the text as well – usually very limited – was cut from blocks. Famous types of 15th-century date (but based on much earlier medieval manuscript traditions) include the *Ars Moriendi* – which gives every Christian essential information on how to die properly – and the *Biblia Pauperum* – a picture book of well-known Biblical scenes. Block-books were, in effect, the first printed books, for they appeared before the invention of movable type.

Unlike engraving and, particularly, etching, the making of woodcuts tended to be an art executed at second hand: an artist made designs which were then handed over to a trained craftsman for transfer to the block. The quality of the

St Jerome distributing the Vulgate, recognizably bound as a book. From a 10th-century manuscript.

The Buxheim St Christopher. South German woodcut of 1423.

Above, *11th-century Buddhist text made up of wooden strips carved with Chinese characters: first step to movable type.*

Below, *detail from an engraving by Jean Duvet, 16th century.*

Above, *engraving,* c. *1600, of a printshop, showing the setting of type by hand.*

Below, Landscape, *etching by Hercules Seghers,* c. *1630.*

woodcuts which bear the monogram of Albrecht Dürer is probably solely due to the strength of his original designs, for there is no evidence that he had a hand in the cutting of his blocks.

C. M. B. G.

Movable type

New worlds are not made by technical devices, but by what we use them for, and by all the other activities in which we are also engaged. The invention of printing is a case in point. We usually say that in about 1440, in Strasbourg or Mainz, Johann Gänsefleisch (or, using his mother's family name, Gutenberg) invented printing. In the last 20 years an attempt has been made to date the modern world from this invention: to see what McLuhan called 'the Gutenberg galaxy'. But printing was invented several times, and with several different effects. In each case, what happened to it was profoundly affected by a complex of other social factors.

The earliest printed book that has survived came from China in 868. Printing from wooden blocks on which text and illustrations had been engraved was then widely practised and developing fast in China. A different method, printing from movable type, also originated in China in the 1040s, and was widely developed in 13th and 14th-century Korea. Since the languages normally printed had an ideographic rather than a phonetic or alphabetic script, movable type had fewer advantages over block printing than it was to have in Europe. Chinese movable type for an alphabetic script was made in words rather than letters, and a Korean movable type tried to combine phonetic symbols and ideographs.

The earliest printing in the West was also from blocks. But in the course of the 15th century a movable letter type was invented and improved, and its application to alphabetic scripts quickly transformed the reproduction of books.

Through a series of inventions and developments and their subsequent expansion the importance of printing became central in human history, making possible the accurate and rapid multiple transmission of writing. But this importance can only be gauged in its social context: what kinds of writing, and to what kinds of readers, the print carried.

The invention of movable letter printing (supported by improvements in presses and in papermaking) came in Europe when the demand for books was from a growing community of scholars. But the period was also, and decisively, one of expanding knowledge and trade, and of the emphasis on new literatures written in the vernacular rather than in the classical languages. Within that complex, printing was a major agency in the spread of open, accessible and native experience and knowledge. This is also why it was repeatedly opposed and controlled by various forms of licensing and censorship. But still its humanism and openness on the one hand, and its commercial viability on the other, brought it through as a means by which men could extend and alter their world.

R. WI.

Engraving and etching

An engraving is made by taking a thin sheet of soft metal, usually copper, and gouging lines in it with a bodkin-like tool (the 'graver') to produce a design in reverse. The plate is then wiped with an inking pad, which forces the ink into the gouged lines; the surface of the plate is wiped clear of ink, and the plate is then printed under great pressure on dampened paper. A copper plate, depending on how it is handled in printing, will give from 1,000 to 2,000 good impressions before it begins to wear.

Etching is a more effortless and versatile technique, but the process is similar: a plate is covered with wax, and the design is drawn through the wax by the artist. The plate is then immersed in acid, which attacks the plate only where there is no resistance from the wax – that is, the lines drawn by the artist. By trial and error the artist can achieve effects much more subtle than those possible with engraving, and for this reason etching has been preferred to engraving for producing what are in effect multiples from the artist's own hand; engraving, on the other hand, had its place as a reproductive technique – serving as photographic illustrations do nowadays. Both processes played an important role in communicating ideas as well as style to men in other cities, often in other countries. They have been to the artist what the book is to the writer.

The engraving of designs into metals (although not for reproduction onto some other material) is ancient, and one of the first techniques learned by metalsmiths for decorating their wares. But it is only in the 15th century AD, in Europe, that we find engraving for transfer onto paper or vellum; the earliest dated engraving known is of *Christ Crowned with Thorns*, of 1446. And it is also in Germany, slightly earlier, that we find the Master of the Playing-cards using the new technique to produce playing-cards which, we may imagine, were distributed all over Europe through the big trade fairs. The process might also have developed separately in Italy.

As for etching, this was the field of the armourer before it was used by some ingenious person for producing multiples on paper. The first dated etching is Urs Graf's *Girl Bathing her Feet*, of 1513; Dürer's almost complete avoidance of etching (perhaps because Germans used iron, not copper, and this gave a clumsy

Waiting for The Times *by B.R. Haydon.*

effect) left the field clear for Parmigianino in Italy (1520 onwards) to demonstrate the potentialities of this swift and expressive technique.

C. M. B. G.

Pencil

The earliest description of a pencil in the sense in which we understand the word today occurs in a treatise on fossils which Konrad von Gesner, 'the German Pliny', published in 1565. Gesner, who was in fact a German Swiss and who wrote important works in fields as diverse as botany, linguistics, zoology and theology, described a writing instrument which consisted of a piece of lead held in a wooden casing. There are other early references to the existence of pencils but it was not until the start of the 19th century that their use could be said to be anything like universal, and they never threatened to supplant the quill pen as the pre-eminent writing instrument.

Among the reasons for this must certainly have been the fact that, at least until the 1760s, when the famous Faber family set up their large factory in Nuremberg, pencils were only sporadically effective. The breakthrough came in 1795, when N.J. Conte first produced pencils made of graphite which had been ground, formed into sticks and baked in a kiln – the process fundamental to pencil manufacture to this day.

Coloured crayons were an obvious development and became widespread after the introduction of aniline dyes in the 1850s had made its contribution to their manufacture. Today several hundred types of pencil are available, their uses ranging from the marking of metal ingots to the delineation of female eyebrows.

J. G.

Newspaper

Wherever people gather there is a market for gossip, and the first published news bulletins designed to meet this demand was probably the *Acta Diurna* posted up daily in Rome's public places when Julius Caesar was consul. In China a similar series of handwritten reports was made in the 7th century, and by the Middle Ages most societies had evolved systems where news was dispensed by town-criers and wandering minstrels (though this was not a new idea – the Homeric epics were doubtless originally designed to serve as news bulletins, written in verse for greater memorability), while governments relied on posted proclamations to inform the people of decrees or special events when they occurred.

Gutenberg's inventions in the field of printing of course gave a new stimulus to the dissemination of news and information, but the preparation of a news-sheet was still a time-consuming and laborious job when in 1529 the Viennese authorities published the earliest newspaper-style broadsheets appealing to 'all Christendom' for help against the besieging Turks outside their city gates.

The first true newspaper in our sense of the term was probably the *Mercurius Gallobelgicus* published by Micael of Isselt in Cologne in the 1580s, which provided the inspiration for the first regular English newspapers, the *English Mercurie* (1588) and the *Mercurius Britannicus* (1632). The latter organ was the first newspaper to fall foul of authority, for the British Star Chamber edict of 1632 banned 'the printing of news from foreign parts' and a struggle ensued which ended with the victory of the Press – and the abolition of the edict – in 1641. The freedom of the Press has been an issue to arouse strong passions for longer than one might think.

During the short period of Press censorship in England a journalist named Benjamin Harris escaped to Boston in America. Subsequently to become known as the 'father of American newspapers', Harris produced in 1690 the first issue of his *Publick Occurrences Both Foreign and Domestick*, which was to appear monthly 'or, if any Glut of Occurrences Happen, oftener'. But the journal was suppressed by the authorities four days later. Boston, however, was nevertheless to be the scene of the publication of the first continuously running American newspaper, John Campbell's *Boston Newsletter*, of which the first issue appeared in 1704. Boston also spawned the first news agency in the world. In 1811 the new owner of Gilberto's Coffee House and Marine Dairy, one Samuel Topliff, observed the avidity of his customers' appetite for news and decided to cater for it directly. He changed the establishment's name to the 'Merchant's Reading Room' and began to collect news for his patrons from correspondents in foreign countries. Soon local newspapers were publishing stories from 'Mr Topliff's correspondents'.

S. M.

Piano

The mists of antiquity shrouding earlier keyboard instruments lift with the invention of the piano. The credit for this belongs to a Florentine harpsichord maker, Bartolommeo Cristofori (1665–1731), who in 1709 published the diagram and description of the earliest piano. Cristofori's action even had an escapement mechanism which, after slinging the hammer up to the piano string and causing it to vibrate, immediately freed it, so that it returned to its rest position and was ready to be acti-

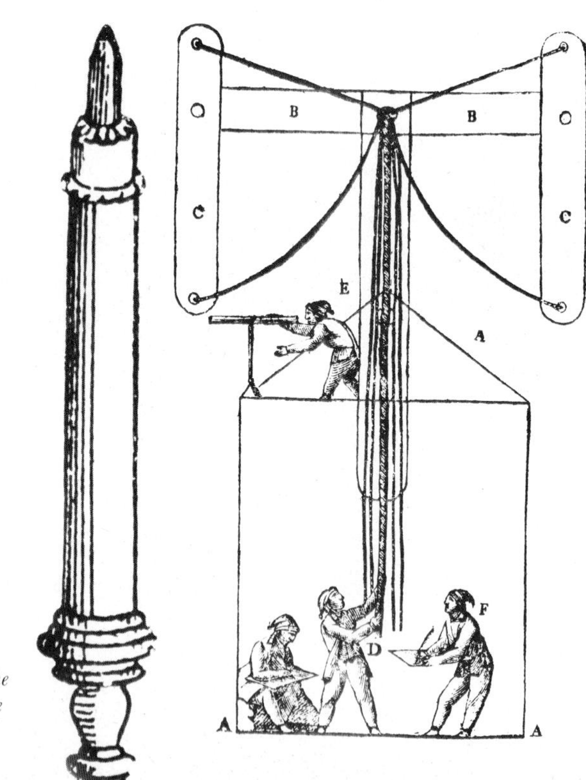

Right, *Konrad Gesner's first pencil (1565).* Far right, *Claude Chappe's 'télégraphe'.*

Below, *piano made by the German Johannes Zumpe in 1767. Bach played a piano of this type when he was in London.*

Aloys Senefelder (above) *and two of his presses. Both pictures are lithographs.*

Below, *lithographic poster:* Divan Japonais *by Toulouse-Lautrec.*

vated again, even though the player's finger still rested on the key. The piano-player had direct control over the volume of tone he produced, for this now depended on the force with which the key was struck. (A harpsichordist could not vary the volume of tone, for the strings of his instrument were plucked.) Piano-making was developed in Germany by the organ-builder Gottfried Silbermann (1683–1753), who knew Bach. It was Silbermann's pianos that the elderly composer tried when he visited Frederick the Great at Potsdam. For the next few decades, pianos were made only in Germany. Johann Stein (1728–92), an excellent musician as well as a fine piano and organ-builder, made instruments that were played and praised by Mozart and Beethoven, and further improved the piano's mechanism. In 1783, John Broadwood (1732–1812) took out the first patent for a piano pedal. This enabled the player's feet to operate the mechanism, formerly operated by the hands or knees, which lifted the dampers, leaving the vibration of the piano strings unchecked (*forte*), or additionally damped the strings (*piano*). In 1821 Sébastien Erard (1752–1831) of Paris, who had built the first French piano in 1777, created the definitive form of piano-action with a double escapement. The final step in the piano's evolution was the over-strung scale, a new arrangement of the strings. In 1855, Steinway gave this its present form.

F.A.

Semaphore

The idea of relaying visual signals over large tracts of land is extremely ancient. Most peoples seem to have employed it at some time or other, and military commanders from Agamemnon to William the Conqueror used beacons of smoke or flame to signal the news of victory. In 1588 England was warned of the approach of the Armada by similar means and a little under a century later no less a personage than King James II devised an early set of naval signals for ship-to-ship communication when he was Lord High Admiral of the British fleet.

Telegraphy in the modern sense is the product of the Napoleonic Wars, during which the French were fighting on several fronts at once and speedy communication between the different armies became of vital importance. In 1792 the French Revolutionary Council approved an idea for an optical telegraph designed to solve the problem. This was the brainchild of an engineer named Claude Chappe, an ardent supporter of the Revolution who had, with his brother Ignace, invented a system of communicating messages via an upright post with a transverse bar fastened at the top, from each end of which hung a smaller arm mounted on a pivot. Chappe had devised a code whereby the position of these arms could be read as different words or letters and he proposed that a series of them be set up at regular intervals of about 10 miles, each signalling machine to be manned by an official with a telescope who would receive messages from his neighbour and pass them on down the line. In 1793 Chappe was made *Ingénieur Télégraphe* and instructed to set up a line of his semaphore stations between Paris and Lille. The first message – announcing the capture of Le Quesnoy – was sent in August 1794.

By the end of the century links between Paris, Brest and Strasbourg had been set up – ultimately the Chappe network in France covered over 3,000 miles. The British Admiralty copied the system and telegraph routes ran from London to Deal and Portsmouth (whence the ubiquitous 'Telegraph Hills' of the Ordinance Survey maps, then first being prepared). A similar system was also put into use in the Boston area of the United States.

In suitable conditions a short message could be flashed from London to Deal in a single minute. But Chappe's system was extremely extravagant in manpower as well as being vulnerable to climatic conditions, and even before the arrival of the electric telegraph made it obsolete it was beginning to fall into disuse. His signalling code is preserved, however, by many of the world's navies and by the Boy Scouts, using flags as the medium of communication, while the signalling system used with railways descends directly from his idea. Chappe himself suffered so greatly from aspersions cast on his originality in making the invention that he is said to have committed suicide in a fit of depression in 1805.

J.G.

Lithography

The history of picture reproduction is older than that of printing; the invention of the woodcut preceded that of movable type. Etching and engraving came next, but the modern age of picture-printing began with the discovery of lithography at the end of the 18th century. It was by chance that 25-year-old Aloys Senefelder, the son of a strolling actor from Prague, discovered what he called 'a new way of printing'. His family had come down in the world; they moved to Munich, and Aloys's mother had to take in washing. One day, she asked Aloys to write an urgent laundry list, and the young man, who had been experimenting with a slab of polished limestone and all kinds of chemical ink, looked in vain for pencil and paper. So he wrote the list on the slab with a grease ink he had made of wax, soap, dripping and lampblack. Then it occurred to him that he might try to etch the stone with *aqua fortis*, nitric acid; this would eat some of the surface of the stone away, leaving the grease letters in high relief.

This was indeed what happened. The letters stood out slightly, and after applying a solution of gum arabic to the stone he inked it with printer's ink – which only the letters accepted, but not the rest of the stone. The laundry list came out perfectly sharp on the paper he pressed on the slab – in mirror writing, of course.

All this happened in 1798, and Senefelder's invention was very soon taken up by the printers. The first to do so was a publisher of Mozart's works, and the musical world benefited immediately from the new process. The Bavarian Elector granted him a patent in 1799, and a year later he obtained an English one.

E. L.

Printing technology

For three and a half centuries after the invention of printing with movable type, the technology of this first and most important form of mass communication remained largely unchanged. But with the development of the Press in the 18th century the process of composing type by hand and making prints by hand-operated presses became inadequate to satisfy the demand for speedy information. Obviously, the new prime mover, the steam engine, had to be harnessed for newspaper production.

A German printer, Friedrich König, and his chief mechanic, Friedrich Bauer, came to London in the early 1800s and built, with the aid of an enterprising financial backer, the first steam-operated mechanical printing-press. In 1812, John Walter II, son of the founder of *The Times*, saw the prototype and recognized it as the first major advance in the art of printing since Gutenberg and Caxton. Walter ordered two double machines, and on 29 November 1814 *The Times* was printed on them for the first time.

The König-Bauer machine was based on a simple idea. Until then, every sheet to be printed had to be positioned by hand on top of the 'forme', which had been inked by hand-operated rollers; the press bar was pressed down by hand or foot. König mechanized these various movements by making the 'forme' pass under an inking cylinder; only the feeding of the sheets was done by hand – the machine put them automatically on top of the forme, and another cylinder pressed them against the type which moved underneath. The printed sheets were delivered into the printer's hands while the forme moved back to receive another coat of ink. Up to 1,200 sheets an hour could be printed in this way, as compared with 300 sheets printed by hand.

For 50 years, this machine printed the major newspapers in Europe and America; but then it was superseded by the rotary press. William Bullock, an American, built the first rotary press in 1863, and it is still the standard machine in newspaper production. It holds giant rolls of newsprint, which are fed into the machine in a continuous process. While König's machine worked with a 'flat bed', Bullock introduced the cylindrical forme; now everything was on rotating cylinders – paper, ink, type – thus speeding up the process enormously.

Type-setting, however, was still being done laboriously by hand; the time was ripe for mechanization. In 1876 Ottmar Mergenthaler, a young German mechanic who had emigrated to Baltimore, was asked to develop a type-composing machine with a keyboard like that of the typewriter. Ten years' intensive work was necessary to solve the innumerable problems involved. In 1886, the *New York Tribune* was the first newspaper to order thirty of Mergenthaler's 'Linotypes'.

This machine, still in use in the world's newspaper and printing shops, produces lines of metal called 'slugs', each made to the width of the paper's columns. When the operator presses a key on the keyboard, a matrix is released from a magazine – a small rod with the die of a character on its vertical edge. This matrix falls into a line-composing box, where the spaces between the words are automatically adjusted by little wedges. After a line of matrices has been set it is carried off to be cast in metal, while the matrices return to the magazine, each character in its compartment, ready to be used again. The cast lines are assembled in the forme, hand-set headings and pictures are added, and the stereotype plate can be cast from it.

For other type-setting jobs, especially books, the 'Monotype' is more practical; this was invented in the 1880s by an American, Tolbert Lanston. It has two separate parts: a keyboard machine which perforates paper rolls in certain patterns, each representing a character, and a 'caster' in which the perforations release matrices from which the characters are cast in single metal types (not in complete lines). The types are assembled automatically to form lines, which are then made up into pages.

Recent developments in printing technology aim at speeding up the setting process. Very fast line-casting machines work from punched tape like the Monotype; computers do the job of equalizing the lines and hyphenating words that have to be separated at the end of the lines. An entirely new concept, however, is photo-composition, which works with several keyboard units on which the copy is typed and turned into perforations on paper tape. The tapes are sent into a photo unit, where the perforation patterns are changed into type on film or paper, which is then checked and corrected. Another unit, the 'composer', makes up the page on film or paper, and the page goes to the platemaker.

Caxton's printing-press – a Victorian representation.

An embossed picture with Braille title, early 20th century.

The Camera Club Outing – late 19th-century American devotees.

Five-needle telegraph patented by Cooke and Wheatstone in 1837.

Original Daguerreotype whole-plate camera.

Negative of a latticed window taken in 1835 by Fox Talbot.

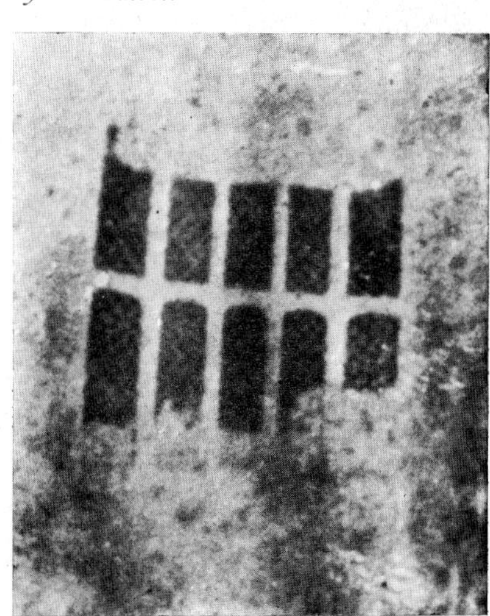

Type-setting at a distance is no longer a problem. The same issue of a newspaper can be produced in two or more towns simultaneously. The perforation patterns of the copy are transmitted by cable or radio, and reproduced by a 'reperforator' unit, to be fed into the line-casting machines. Alternatively, the 'Whole-Page Facsimile Transmitter', a British innovation of the 1960s, serves provincial publishing centres from a central office. It works with a television-type camera which scans the proof copy of a whole page within $12\frac{1}{2}$ minutes and transmits it to a receiving unit, where it is reproduced as a stereotype plate.

E. L.

Camera

Unlike many other inventions the photographic camera was merely an adaptation. It developed from the camera obscura, known to artists and scientists since the 17th century as a portable drawing instrument, complete with lens, variable stops to sharpen the image and a ground glass on which it formed. It would have been a simple matter to convert the model illustrated by Johann Zahn in 1685 to photographic purposes had the chemical side of photography been as advanced as its optics.

The first person to make the simple adaptation – by substituting silver chloride paper for the ground glass – was the French amateur scientist Joseph Nicéphore Niépce in the spring of 1816. He even succeeded in fixing the pictures partially, but he was unable to print positive impressions from his negatives. Niépce's first successful and only surviving camera picture, which he submitted to the Royal Society in 1827, had been taken the previous year using a pewter plate, made light-sensitive by asphalt. It measures 6×8 in. and all his cameras were correspondingly large. One consists of two wooden boxes, the rear part with the ground glass sliding within the front part containing the lens panel, and measures 30 cm. square when closed. In another model of the same size, the rear and front parts have been connected by a locally made accordion-like square bellows. Still another is fitted with a variable metal leaf iris diaphragm to sharpen the image. The bellows and iris diaphragm were novel features introduced by Niépce.

L. J. M. Daguerre employed for his early photographic experiments, from 1830 on, similar camera obscuras made by Charles Chevalier and fitted with an achromatic meniscus lens. Those made for the daguerreotype process, after its publication in 1839, were produced by his brother-in-law Alphonse Giroux bearing his guarantee and Daguerre's signature. Like Niépce's cameras they consisted of two wooden boxes and were of similar size to produce wholeplate pictures. The lens cap acted as shutter. Until the introduction of the gelatine dry plate in 1880, exposure times were far too long to require a shutter.

W. H. Fox Talbot also arrived at photography via the camera obscura, though he used for his experimental work in 1834–35 locally made 6 cm. square wooden boxes which his wife called 'mousetraps'. With these he produced miniature negatives.

Though smaller and much larger cameras were constructed during the following decades, camera design remained essentially the same as that used by the pioneers of photography.

H. G.

Braille

The first book to be printed in raised letters was produced in France in 1784 by Valentine Haüy (1745–1822). His biographer says that he got the idea from seeing a blind Viennese pianist distinguish the keys of her piano by touch. According to another and more likely story, his first pupil, François Lesueur, showed how he could 'read' several letters on a card which had been heavily indented by the printer. Haüy then seized a pen and wrote several letters pressing heavily, which Lesueur was able to read. After some experiments with different sizes of letters and type-faces, he fixed on a design and soon showed that blind children could be taught to read with their fingers. A new organization, the Royal Institute for young Blind Persons, was created, led by Haüy.

Louis Braille (1809–52), inspired by a system devised by a Capt. Ch. Barbier, invented his sytem of embossed dots in 1829. It was adopted in France as the standard system in 1854, though the École Braille was not founded until 1883. In 1932, a form known as Standard English Braille was agreed on for worldwide use.

G. R. T.

Electric telegraph

'Telegraphs of any kind are wholly unnecessary and no other than the one in use will be adopted.' So said the British Admiralty, on being offered the electric telegraph by Francis Ronalds in 1823.

The 'one in use' was a kind of semaphore, installed in 1796 in imitation of a French version. The idea of signalling by torches (or arms) placed in different positions is very old; it is said that the fall of Troy was communicated to Argos by this means.

The first attempt to make an electric telegraph dates from before the discovery of electromagnetism. As early as 1808, after the discovery of the electric battery, Sömmering, the German anatomist, had suggested putting 36 electrodes in acidified water, each marked with a letter or number, and each attached to wires which could be connected to a battery at the sending station. Gas would then evolve at the electrode which corresponded to the letter being sent. The system was of course hopelessly slow and complicated. It was an Englishman, Francis Ronalds, who reduced the number of wires to two and set up a complete system in the grounds of his house at Hammersmith, using eight miles of wire.

But the first telegraph based on current electricity was essentially the work of W.F. Cooke and Professor Charles Wheatstone, who worked on the idea from 1837; it was the needs of the expanding railway system which provided the social motive.

Cooke, a mechanically minded man who had retired from the Indian Army because of ill health, was supporting himself by making anatomical models. In 1836 he happened to see some telegraphic experiments which Prof. Munke of Heidelberg was making with a needle galvanometer, and was seized with the idea of making a commercial success of this idea. He therefore devised an improved version, consulting first Michael Faraday and then Professor Wheatstone, at King's College, who had been conducting scientific experiments with telegraphic apparatus. Cooke and Wheatstone decided to go into partnership.

The telegraph which they patented in 1837 used five magnetic needles arranged to point at different letters, and a telegraph of this type was installed on the Great Western Railway in 1838, between London and West Drayton. It involved five lines of wire. By 1838 they had reduced the number of needles to two by introducing the idea of a code.

On New Year's Day, 1845, the telegraph was used to catch a murderer who had taken a train from Slough to London. The police followed him from Paddington to a lodging house in Cannon Street, where they arrested him. He was subsequently tried, convicted and executed, and this success drew general attention to the telegraph.

Cooke and Wheatstone were soon on bad terms, with Cooke protesting at the other's oft-repeated claim to be the sole inventor of the electric telegraph. Charge and counter-charge were published. Nevertheless, by 1852, 4,000 miles had been installed.

An American, Samuel Morse, is often credited with the invention of the telegraph in 1832. Though his code was universally welcomed and though he contributed to the introduction of the telegraph in America from 1843 onward, he can hardly be considered the inventor. His first private demonstration was in 1837 and he did not install a commercial telegraph until 1844.

He did, however, introduce relays – devices to reproduce or amplify the signal *en route* – which increased the range of the telegraph. Further, his code was an early example of what is now known as 'communication theory'. The most frequently used letters are the most quickly sent.

G.R.T.

Typewriter

On 7 January 1714, Queen Anne granted a patent to an engineer named Henry Mill, who, it was stated: 'hath by his humble peticion, represented unto us, that he has, by his great study, paines, and expense, lately invented and brought to perfection an artificial machine or method for the impressing or transcribing of letters singly or progressively one after another, as in writing, whereby all writings whatsoever may be engrossed on paper or parchment so neat and exact as not to be distinguished from print.'

No drawing or model of Mr Mill's contraption survives. There are typewriter *aficionados* who consider that it may have been some form of stencil, yet Mill is generally accepted as the father of the typewriter – at least, of the idea which was eventually developed into the typewriter as we know it. It did not exactly soar off the ground: the 18th century was not fidgeting for the arrival of a typewriter; secretaries, like Napoleon's Meneval and Bourrienne, could take down the great man's dictation, at normal talking speed, for hours without pause, and legibly and accurately. During the next 100 years, innumerable essays were made with the notion of a mechanical recorder approximately in focus – mostly with the blind in mind.

The first American patent for what was recognizably of the typewriter genus was granted in 1829 to William Austin Burt, of Detroit, for 'Burt's Family Letter Press'. Four years later Xavier Projean of Marseilles produced his *Machine Criptographique*. He claimed proudly that it would go 'almost as fast as one could write with an ordinary pen'. Credit for the invention of the prototype of the modern typewriter – for getting action out of a mechanism with an inked ribbon, which was capable of being manufactured and marketed – is thus corporate, but the main impetus came from Christopher Latham Sholes.

In the Kleinsteuber machine-shop in Milwaukee, Sholes and Carlos Glidden were developing a machine for numbering book pages consecutively. Glidden asked one of those blindingly simple questions which are the basis of most inventions: 'Why cannot the paging machine be made to write letters and words, and not figures only?' And Sholes, with the aid of Glidden and a printer named Samuel Soule, proceeded to supply the answer with a wooden mock-up. Although it had no shift-key and printed only in capitals, this was the best yet of its kind, a fact that was recognized by two businessmen, James Densmore and George Washington Newton Yost. They bought Sholes's patents for $12,000 and talked the Remington Fire Arms Company, then diversifying from its Civil War weaponry into sewing machines, into trying the new line. The manufacturing contract was signed in March 1873.

The business community did not rush to seize the stylish efficiency improver. Indeed, the Sholes typewriter's first serious appearance in public was at the 1876 Centennial Exposition, and it drew small interest. It was pushed aside by another knick-knack making its début at the Exposition, the Graham Bell telephone. Remington promoted the typewriter by lending its machines to several hundred firms, however, and after that competition livened up. There have been more than 300 different makes of typewriter: up-strokes and front-strokes, clock models with type-wheels and those of the type-bar principle. Today's low-slung electrified elegance is a far step from the stately baroque of the originals, yet what remains almost wholly unvaried is the Sholes keyboard.

The apparently mad scattering of the alphabet was so arranged to prevent jamming – he disposed the letters commonly occurring together as far apart as possible. Sholes – a man of 'modest and retiring disposition' – like most inventors faded from sight once his brainchild had gone forth and multiplied. But in one of his last letters he reflected: 'Whatever I may have felt, in the early days, of the value of the typewriter, it is obviously a blessing to mankind, and especially to womankind. I am glad I have had something to do with it. I builded wiser than I knew, and the world has the benefit of it.' He was certainly right about the typewriter's special meaning for women. It can be persuasively argued that the typewriter's most significant social effect was its provision of a new role for the new working woman, and that by providing employment for literally millions of women in the past 80 years it has been a powerful instrument in their emancipation – though the individual shorthand-typist might argue that she has not been liberated but enslaved.

K.A.

Telephone

Though Alexander Graham Bell was the first to demonstrate, in 1876, that people could talk to each other over a distance by electricity, rudimentary transmission of sounds in this way had been demonstrated in 1863 by Philipp Reis in Germany.

By the mid-19th century the electric telegraph, on which messages could be sent by various codes, was operating successfully – even across the Channel – but it was not until Bell's demonstration in 1876 that people could talk to each other over a distance.

The first receiver used by Bell in 1876, and (above, left) *his first transmitter.*

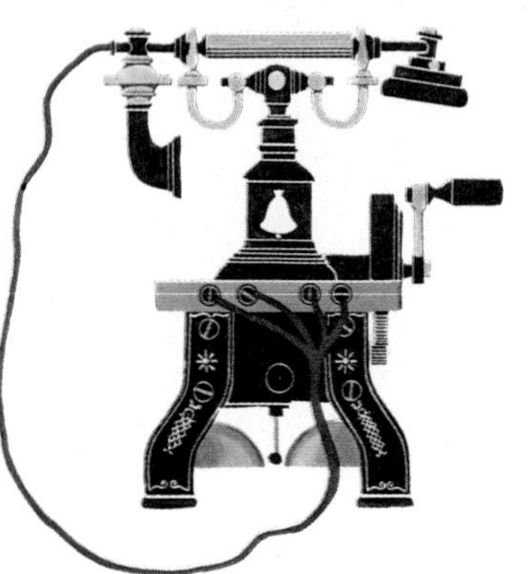

The Ericsson 'skeleton' magneto phone – the first one-piece instrument.

Below, *Burt's 'Typographer' of 1829.*

Sholes using one of his own typewriters.

Developed from his early phonographs (right), Edison's 'Ediphone' of 1911 (above) achieved rapid success as a dictating machine.

The earliest records were on wax cylinders: discs came later.

Early Swedish half-tone block: an immediate improvement on the older non-mechanical methods.

Bell's telephone consisted of a metal diaphragm suspended near coils wound on magnetizable iron cores. When the sound waves struck the transmitter diaphragm, its movements caused changes in the magnetic field through the coil and thus produced varying electric currents corresponding to the sound. These currents passed through the receiver coil which changed the magnetic field in sympathy and caused the receiver diaphragm to vibrate similary, re-creating the original sound.

Telephone earpieces are similar in principle today, but the mouthpieces use more sensitive microphones. A great contribution of Edison was the carbon microphone. This is basically a 'button' of packed carbon granules, the electrical resistance of which varies when the pressure on them is varied by a diaphragm actuated by sound waves.

Bell was born in Edinburgh in 1847. After graduating as a doctor and spending two years teaching deaf children with his father, he became Professor of Vocal Physiology at Boston University. Besides teaching the deaf and dumb he was absorbed in the scientific study of sound, and also in the possibilities of transmitting it by means of electricity or light.

While he and his assistant were transmitting musical notes produced by vibrating metal reeds, his keen, trained ear detected that the harmonics of the notes were also being transmitted. He realized that the harmonics in speech could also be sent over the circuit. After some adjustments he summoned his assistant, some rooms away, with the message, 'Mr Watson, come here – I want you', which were the first words transmitted by telephone. This was in March 1876.

Bell's telephone was a great attraction at the Centennial Exposition held in Philadelphia that year to celebrate the 100th anniversary of the Declaration of Independence. The visiting Brazilian Emperor Pedro II was apparently so surprised that he dropped it, crying out: 'It talks.' The vast telephone industry began, and in less than 25 years one in every 50 people in America had a telephone. Bell was rich and famous at 30.

P. WY.

Gramophone/ Phonograph

When Thomas Edison set out to record messages in Morse on a waxed paper tape so that they could be sent out later at a high speed, he found, not surprisingly, that a spring resting on the embossed paper gave off a musical note when the paper was pulled rapidly past it. He was able to invent the gramophone because, in his mind, he connected this sound with that produced by the telephone which had just been invented. He argued that a mouthpiece diaphragm connected to a stylus would cut a pattern that a second stylus attached to an earphone diaphragm could decode, and he made a device based on this principle. In July 1877 he shouted 'Halloo' into it and convinced himself that a very low-fidelity 'Halloo' was shouted back.

It took him a few months to decide that, apart from telephony, reproducing the human voice was worth while, but by 6 December he had designed a 'phonograph' which his mechanic Kruesi built. The message was cut into a tinfoil cylinder by a stylus that was moved along it by a screw. This stylus was then removed, and a second one, attached to a hearing tube, was placed on the 'record' so that the message could be reproduced. 'Mary had a little lamb' was the entire contents of the first ever gramophone record made in 1877. From then on the problem was to increase the power of reproduction and its accuracy, driving the machine and amplifying electronically the vibrations of the stylus.

T. O.

Half-tone block

An American, Frederick E. Ives, invented the photomechanical half-tone process of printing in 1880. The first half-tone block ever printed was in the New York *Daily Graphic*; the new reproductions were a tremendous improvement on the previous line woodcuts; the speed and quality of their production made an immediate impact on the newspapers of the day. The process is named after the finished photo-engraved block which reproduces not only blacks and whites, but also the intermediate shades or half-tones.

The fundamental principle is the photographing of the original through a glass screen engraved with opaque black crossed lines. The tones in the original are translated through the negative to a print on the surface of a metal sheet. After etching, the surface of the block is made up of varying sizes of points and hollows. It blends into solid colour in the darkest part of the picture.

Silk-screen printing

Although the silk-screen method of printing, as we know it today, is a 20th-century development, its ancestral line reaches back to antiquity. In both China and Japan, straight stencil printing was in common use by AD 500, while 15th-century Europe saw the consistent use of stencils for colouring playing-cards and religious subjects which had first been printed from wooden blocks. Two hundred years later, the French were

Man Moving

employing the same method for producing elegant wallpapers, and in the United States stencilling was extensively used after 1800 for decorating walls, floors, furniture, clocks and trays. During the same period, stencil printing became such a fad among the well-to-do that it was considered a 'genteel accomplishment' for young ladies to paint neat patterns on velvet by this means.

In 1907, Samuel Simon of Manchester received a patent for an early modern silk-screen process, although there is considerable uncertainty as to who first developed this method and where. One story dates the invention to the 1880s in Germany, after which it was taken to Japan and from there to the United States. In any event, by the late 1920s silk-screening had come into limited use for commercial printing and is today the most widely used method where limited runs are involved.

E.W.

Silk-screen printing: spreading the colour with the 'squeegee'.

Stereophony

On 30 August 1881 Clement Ader was granted a patent in Germany for 'improvements of Telephone Equipments for Theatres'. Two groups of microphones were placed either side of the stage at the theatre and these were relayed directly to two telephone receivers which subscribers held to their ears. This invention was first displayed at the Paris Exposition that year, where it 'broadcast' presentations from the stage of the Paris Opéra. It was a great success, and may be considered as the début of stereophonic sound. An inventor called Ohnesorge used a similar device to Ader's in the Crown Prince of Prussia's palace at about the same time.

The most striking feature of stereophonic sound is that one can locate the source of a sound much more easily than when one hears a monophonic, or single-source sound. This phenomenon is similar to the way in which one can judge distances better visually with two eyes than with one. This aspect of stereophony was exploited during the First World War by 'binaural receiving trumpets' which were used to locate enemy planes. Two large horns, similar to the ones used on early phonographs, were connected by rubber tubes from their thin ends to the ears of the operator, whose sense of aural direction was thus greatly improved.

The idea of stereophony being linked with telephone systems of communication continued during much of its early development, and the Bell Telephone Laboratories, under the direction of Harvey Fletcher and others (and with the advice of Leopold Stokowski) were the driving force behind the development of stereophony in the early 1930s.

In the winter of 1932, 'Oscar' was the main instigator of research at the Bell Laboratories. Oscar was a tailor's dummy who had two microphones built into his ears, thus reproducing the conditions of human hearing as closely as possible. On 27 April 1933, a concert performance in Philadelphia was stereophonically transmitted by telephone line to Washington DC; this was the most widely publicized of Bell's experiments.

As early as 1925, the WPAJ radio station of New Haven, Connecticut, gave 'binaural' broadcasts by broadcasting the same programme on two different wavelengths, picked up individually by two receivers – one for each of the listeners' ears. The first stereophonic disc was patented by A.D. Blumlein of Electric and Musical Industries in 1930. The first stereophonic sound film track accompanied Walt Disney's *Fantasia* in 1941: this project once again involved Leopold Stokowski.

P.WH.

Reception through two earpieces of an opera performance, by Ader's stereophonic system.

Fountain pen

The first British patents for reservoir pens were taken out in 1809, but in these early models the ink did not flow freely. The writer started the flow of ink by pressing a plunger, which he had to press again at intervals to keep it going.

It was not until 1884 that a satisfactory capillary feed was devised, by an American insurance salesman, Lewis Edson Waterman. The end of his pen could be unscrewed, and ink was squirted in with an eye-dropper. The first self-filling pens came early in the 20th century, and used a piston to draw up the ink. When the rubber sac was introduced, a coin had to be pushed into a slot to squeeze the sac and suck up the ink. The idea of a lever followed in 1908.

The Schnorkel pen, with a tube which extends into the ink, dates from 1952, and the capillary pen from 1956.

G.R.T.

Cover of Waterman's magazine, Pen Prophet.

Cinematograph

Many men contributed to the discovery of how moving pictures could be made. Roget's Theory of the Persistence of Vision in 1824 codified a phenomenon observed for several centuries – and is the scientific basis of the moving picture. For the rest of the 19th century men tried to make it work, initially with simple toys such as the Phenakistiscope of Plateau, Horner's Zoëtrope and Beale's Choreutoscope. The invention of photography was the spur; and in the 1850s parlour audiences were entertained by magic lantern shows of photographed tableaux, arranged in narrative stories.

In England J.A.R. Rudge had by 1875 invented a magic lantern projector that could show seven phases of action in rapid time. In California Eadweard

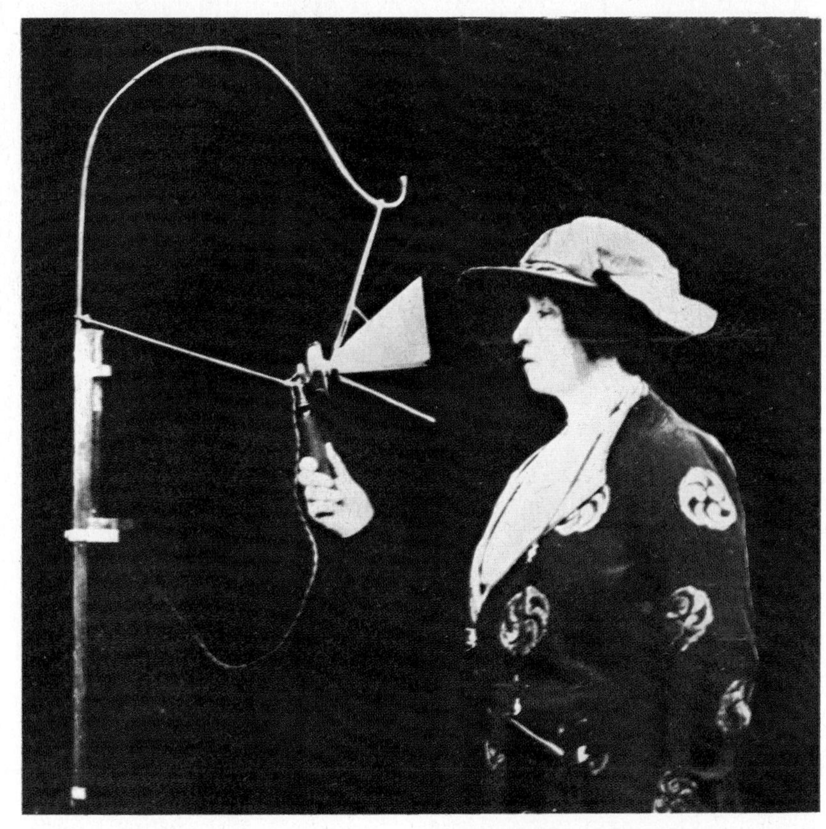

Dame Nellie Melba giving a broadcast recital on 15 June 1922.

A hand-cranked Pathé movie camera of c. 1920.

SURPRISE FOR CRIPPEN.

"NO SUSPICION."

CAPTAIN'S MESSAGE TO "DAILY MAIL."

INSPECTOR DEW'S ARRIVAL.

ARREST EXPECTED TO-MORROW.

Crippen and Miss Le Neve, on board the Montrose, are now in the Gulf of St. Lawrence, and due at Father Point, near Rimouski, 170 miles from Quebec, at 10 a.m. (3 p.m. Greenwich time) to-morrow.

Inspector Dew, in the Laurentic, arrived at Father Point at 4 p.m. (9 p.m. Greenwich) yesterday, and awaits the fugitives.

CAPTAIN'S TELEGRAM.

LIFE OF THE COUPLE IN THE MONTROSE.

By MARCONI WIRELESS.

Radio traps a murderer: Daily Mail, *30 July 1910.*

Muybridge rigged up a row of cameras in a famous demonstration to prove that a horse's feet left the ground in a gallop (and although Muybridge never tried it, the sequence of pictures that were produced could have been animated). Then a Frenchman, Jules Etienne Marey, used a single dry plate on which to shoot a dozen consecutive pictures and, later, paper film to shoot 120 with exposures as short as a thousandth of a second.

His work was observed by Thomas Alva Edison from his Menlo Park laboratory. Seeking a way of adding pictures to his recent invention, the phonograph he assigned an assistant, William K.L. Dickson, to work with celluloid film supplied by George Eastman. Dickson built a peepshow apparatus called the Kinetoscope which used a continuous loop of film. It was ready by 1891.

But it was still not the cinema as we know it. Edison did not believe in the future of the projected image; film viewing was a solitary pleasure as far as he was concerned. Edison's legacy to film that persists to this day is the 35 mm. gauge and the positioning of the four sprocket holes that advance each frame of film.

The pioneers had to find a way of achieving intermittent movement. It was no use the film moving continuously – it had to be stopped and started at least a dozen times a second, each frame stopping at the precise moment when the shutter was open. There were many ingenious methods, most of them unsatisfactory. Several inventors worked with the Maltese Cross movement, which is still used in modern projectors. In England Friese-Greene, Donisthorpe, Acres and the mysterious Frenchman Le Prince strove to make projected images work, and at the same time Armat, Jenkins and Latham in America, and Max and Emil Sklandanowsky in Germany, were also conducting experiments.

But the credit for the very first reasonable demonstration of the projected image must go to the French brothers, Louis and Auguste Lumière. Their apparatus substituted a claw movement for the Maltese Cross and the same box not only took the pictures but later projected them. On 28 December 1895, in the Salon Indien of the Grand Café on the Boulevard des Capucines, Paris, an incredulous crowd watched moving film of a train entering a station, a rowing boat leaving a harbour and workers coming out of the Lumière factory at Lyons. On that night the cinema was born.

G.P.

Radio

Half-British, half-Italian, Guglielmo Marconi knew by 1895 that he had a method of transmitting messages without the need for wires. The possibilities were endless. He was just 21.

Since the age of 16 he had conducted scientific experiments, indulged by his mother, Annie Jameson, who had, in spite of his Roman Catholic baptism, raised him as an Anglican. His Italian father was far less encouraging. The Marconis lived in a country house, the Villa Grifone, outside Bologna, and Marconi set up a wireless laboratory in the two top-storey rooms, which were previously used for storing trays of silkworms.

He began his radio experiments in 1894, and brought a fresh and ingenious mind to a field where most of the theoretical research had already been done. Marconi was able to convert this work into a practical and useful device.

James Clerk Maxwell had said in the 1860s that it should be possible to generate electromagnetic waves that would travel at the speed of light. Twenty years later, Heinrich Hertz managed to demonstrate this 'radiation' (whence the word 'radio'). He used a high-voltage alternating current to generate sparks between two metal balls, and found that the sparks produced an electromagnetic radiation that could be detected by a loop with a small gap. A spark would jump this gap whenever the loop received radiation from the 'transmitter'.

The electric telegraph was by now well established, and there were many attempts to use 'Hertzian waves' to carry messages in a similar way. They were helped by E. Branly's invention of the 'coherer', a glass tube filled with iron filings and connected to a battery and some device for detecting a current. Normally, none could flow through the iron filings, but they became conductive if they were struck by Hertzian waves. This gave a way of amplifying their effect and Branly succeeded in detecting these waves at 150 yds.

Marconi first transmitted his messages across the silkworm rooms, and then, when he discovered that the distance could be increased if he used an aerial and an earth on both transmitter and receiver, he increased the distance.

Marconi's brother carried the receiver further and further away from the house. Reception was signalled by a white handkerchief. But the most dramatic and proving moment came in September 1895. Marconi attempted to transmit beyond the line of sight with a hill between transmitter and receiver. The shot from his brother's hunting rifle which echoed up the valley was the sound of success.

The Italian government refused to help develop the invention, and Marconi turned to the British authorities and his family connections in London. With its maritime might, Britain ought to have been interested. The Post Office was helpful, and Marconi applied in June 1896 for the world's first radio patent. The Government, however, did not follow up the Post Office initiative, and in

Man Moving

July 1897 Marconi and his British relatives formed the Wireless Telegraph and Signal Co., which by 1900 became the Marconi Wireless Telegraph Co. It concentrated on installing radio in ships – the first was in the German liner *Kaiser Wilhelm der Grosse*. Its use was dramatically demonstrated off New York in a collision in fog between the *Republic* and the *Florida*, when 1,650 passengers were taken off the two liners by the *Baltic*, summoned by the *Republic*'s Marconi operator. On 12 December 1901 the three dots of the letter S were sent from Marconi's station at Poldhu in Cornwall and picked up by him in Newfoundland on an aerial supported by a kite. He had spanned the Atlantic.

G. N.

Teleprinter

Frederick George Creed (1871–1957), the accredited pioneer of the teleprinter, was born near Canso in Nova Scotia. Canso was distinguished only by the fact that the transatlantic cable lines terminated there, and Creed became a telegraph operator as a young man. In the machines which Creed operated in his youth, the operator punched Morse signals on to a paper tape by means of three plungers, one for dot, one for dash and one for space, which were struck by hammers to perforate the tape. This seemed a laborious process to Creed, and he set his mind on simplifying it.

It was not until Creed had travelled to Glasgow in 1897 and was working for the *Glasgow Herald* that he developed his first teleprinter in a shed rented for 5*s*., basing his experiments on a Barlock typewriter bought in Sauchiehall Street for 15*s*. This typewriter was an obsessive mascot of Creed's which he guarded all his life.

Creed's first instrument was designed to cut out the process of the operator encoding letters into Morse Code and then punching this code on to a tape, by making the machine encode the letter automatically when the operator depressed a typewriter key. Twelve copies of this instrument were bought by the British Post Office in 1902, but for the next 20 years or so he had difficulty in persuading people to accept his invention, perhaps mainly because although it was much quicker than existing machines, its introduction made so many trained operators of the older machines redundant.

In the early 1920s, a machine was produced in America by Morkrum Company, who had been developing their 'teletype' machine independently since Charles L. Krumm produced a prototype in 1907. Siemens-Halske of Germany also developed a machine at this time. These both employed a new system of coding based on permutations of a five-unit binary code, itself developed by Jean Maurice Emil Baudot and Donald Murray, a New Zealand farmer. Five keys were aligned vertically across a tape, all or any of which could be made to mark the tape. If, for example, the first, fourth and fifth keys marked the tape, this constituted the code for the letter 'B'. When that particular line of marks passed the head of the transmitting machine, an appropriate electrical pulse was generated to the receiver which then decoded the letter and printed it. This code is still in use today.

The *Daily Mail* adopted Creed's system in 1912 to publish simultaneous regional editions of the paper.

Creed was a large man with strongly puritanical religious convictions: employees had to sign a pledge of teetotalism before working for him, and on 8 March 1930 he resigned as chairman of his own company at the age of 59 because his employees insisted on playing on the sports field on Sundays. After his resignation, Creed continued with his inventions: his most notable was probably his floating airport or 'Seadrome'; less successful was his irremovable hair dye, only used once, on his own beard, which was consequently dyed in an indelible rainbow pattern.

P. WH.

Operator receiving a message on a Morse printing telegraph, 1887.

Magnetic tape-recording

Valdemar Poulsen invented the magnetic tape-recorder in 1899. In 1900 his 'Telegraphone' was successfully demonstrated at the Paris Exhibition. This device recorded messages on a moving magnetizable steel tape. It was the forerunner of today's familiar tape-recorder, but had to wait for electronic amplifiers to make it practicable.

Poulsen conceived the Telegraphone originally as a message-taking device for use with telephones. As well as using steel tape, he also had the idea of using wire and a strip of flexible material covered with a magnetizable powder, thus anticipating the modern recording tape, which is a coating of a brownish iron oxide (which can be magnetized) on a flexible plastic base.

Today the tape-recorder is almost as familiar as the gramophone, and it has revolutionized the recording of sound in films, broadcasting and the gramophone industry. TV programmes are now often recorded and replayed by this means. Musicians – and others – in recording studios need no longer suffer the tension of having to make a four- or five-minute 'take' on wax which one mistake would spoil. Today, individual notes can be corrected and spliced into the tape.

P. WY.

The printing telegraph, a 1907 version of the teleprinter.

This sturdy tape-recorder of 1934 held two miles of tape which could be run off in 35 minutes.

Triode

There are two parallel developments in the history of modern communications.

Fotos Type R triode. Without this modulating device, sound transmission would have been impossible, either by wire or by radio.

Below, *the Monopol Disc musical box, 1900, forerunner of the juke box.*

The electric telegraph required only a simple circuit over which signals – such as the Morse dots and dashes – could be sent by merely switching the current on and off; but in order to transmit natural sounds like speech and music by telephone, much more subtle and sensitive instruments had to be developed. Similarly, wireless telegraphy operated by means of the on/off signalling system, but to make the electromagnetic waves transmit natural sounds it was necessary to develop new equipment. 'Wireless telephony' – or radio, as it was called later – had to wait for that achievement.

Three men worked on it independently from each other: an Englishman, an Austrian and an American. Professor (later Sir) Ambrose Fleming (1849–1945), a Lancastrian who had helped Marconi with his early experiments, discovered in 1904 that a vacuum tube with two electrodes, one heated and one cold, had the effect of a 'detector' of alternating wireless waves arriving 'by air'; it allowed them to pass in one direction only, with the flow of electrons that were emitted from the heated cathode, towards the anode, or 'plate', maintained at a positive potential. Fleming called his tube 'thermionic valve' (from the Greek *thermos*, warm).

In Vienna, Robert von Lieben (1878–1914) was working on the problem of amplifying telephone signals. He realized the great importance of Fleming's valve for sound transmission by wire as well as by wireless, and improved it decisively in 1906 by placing a third electrode, a perforated 'grid', between cathode and anode; it was fed with the sound impulses coming from the telephone transmitter, and it had the job of modulating, like an invisible brake, the stream of electrons flowing between cathode and anode, controlling it very finely. In this way the arriving weak impulses could be amplified as much as required, with complete accuracy. The new valve with its three electrodes was called 'triode'.

Lieben died too young to take part in the development of radio, but the American physicist Lee de Forest (1873–1961) perfected his own modification of the Fleming valve, calling it 'audion'. He arrived at the same solution as Lieben, with a control grid as the third electrode. After a few years of research he and his team discovered that the triode could also be used to generate high-frequency, continuous-wave oscillations for a radio transmitter. The latter sends out a continuous 'carrier wave', with the impulses coming from the microphone superimposed on it by means of the grid. In the receiver, the carrier wave is 'filtered off', the modulated waves are amplified by another set of thermionic valves, and translated back into audible sound in the headphone or loudspeaker.

The triode reigned supreme in radio transmission and reception until about 1950, when receivers were gradually equipped with that new electronic miracle – the transistor.

E. L.

Juke box

In intention if not in type, the juke box is the child of a whole series of automatic musical instruments, starting with the musical box (popular and fairly cheap from the 18th century), and continuing through the more complicated automata of the later 19th century.

These instruments reached the peak of popularity in the first decades of this century, and the main centres of production were Germany, Austria and America. The customers of the Aeolian Company of New York could, from 1913, buy a Duo-Art Pianola Piano; this would play both automatically and as an ordinary piano, and one with Steinway action in a Louis XVI case cost $2,950. But they were good; Paderewski wrote to the makers, 'I congratulate you again. . . . I shall be glad to have my playing reproduced with such fidelity.'

There were a host of similar contraptions, some of which played violins or whole orchestras (for example the Cremona Theatre Orchestra, which could manage violin, piccolo, flute, bass, sax, 'cello, and xylophone). The company which made the Cremona, the Marquette Piano Co. of Chicago, was manufacturing coin-in-the-slot pianos by 1905, for amusement arcades rather than for home use.

The juke box differs from all previous record-playing devices in that it can play a selection from anything up to 300 discs. It works as follows: after passing the coin-tester, the coin releases the selector mechanism; depression of the required keys causes a micro-switch to energize a magnet (of which there is one for every key combination) which actuates a contact tongue on the pre-selector unit. This 'counts' down the pile of records until it comes to the one selected, the unwanted ones being made to fall down the stacker column.

Juke boxes were first produced in America in the early 1930s, but they did not become popular in Britain until after the Second World War. A report in *The Times* (4 January 1958), entitled 'Juke Box Boom', outlined the high profits (up to £12 per week from a £300 investment) to be made by speculators, and estimated that there were already 3,000 in Britain and that the total was increasing at the rate of 260 a month.

C. M. B. G.

Television

Television is one of those things like the motor-car which had a multiplicity of inventors ranging from the university egghead to the entrepreneurial eccentric to the big corporation team. And, like the motor-car, it had its false trail, its Neanderthal or steam-car form, which lives in the memory through the legend of John Logie Baird and the flickering shadows of his strange mechanical television sets.

In fact, the history of television goes back to 1862 and the Abbé Caselli, an Italian-born priest who discovered fame in France through inventing a method of sending pictures, or to be more precise, shadows, over a telegraph line. His experiments were supported by Napoleon III, and he set up several commercial stations in France for sending handwritten messages and drawings over the telegraph lines. Alas, he was not commercially successful. Other messages on the line interfered with his pictures, which were often liberally sprinkled with dots and dashes.

This was, however, the beginning from which the Baird approach to television derived. By 1881 an Englishman, Shelford Bidwell, was demonstrating his electric distant vision apparatus, which analysed the picture to be sent with a selenium cell mounted in a box, moved up and down by a cam.

In 1908 Bidwell wrote to the scientific magazine *Nature* on the problems of electric vision and suggested that the way to get a good picture was to connect 90,000 photoelectric cells at the transmitter, each through a separate wire to a corresponding point in the receiver. This letter interested the Scottish-born electrical engineer Alan Archibald Campbell-Swinton, who started to think of ways of carrying all the information on a single wire.

Campbell-Swinton had already had a successful career. He was one of the first to experiment with X-rays and used them to take photographs. He had launched Marconi on his career with a letter of introduction to Sir William Preece of the Post Office. He knew the work of Braun in Germany on the use of the newly discovered cathode ray as an oscilloscope to make visible the wave form of alternating currents.

He saw that an electronic switch was needed to turn on, in turn, Bidwell's 90,000 photoelectric cells, and that Braun's cathode ray oscilloscope could be used to do just this. Further, the same device could be used as a receiver, creating a patch of light corresponding to the amount of light falling on each photocell. In 1908 he described the idea in a letter to *Nature*. By 1911 he had patented the basis of the television system which eventually came into service from Alexandra Palace in 1936.

It was an amazing piece of scientific clairvoyance, comparable perhaps to Charles Babbage's anticipation of the principle of the computer. It included the most significant principle on which all modern television cameras work, and which was not fully appreciated until many years later. Campbell-Swinton's camera stored the amount of light falling on each photocell element during the whole period, so that when that cell was switched on by the cathode ray beam, the message sent to the receiver was the sum of all the light that had fallen on the cell since it was last switched on.

'It is an idea only,' wrote Campbell-Swinton modestly, 'and the apparatus has never been constructed. Furthermore, I do not for a moment suppose that it could be got to work without a great deal of experiment and probably much modification.' He was right. Success could only come after the invention of the electronic amplifier, the development of much-improved photocells and cathode ray tubes. Campbell-Swinton died in 1930, only a few years before the final vindication of his distant electronic vision.

In fact, he had not been the first to think of the cathode ray oscilloscope as a possible picture receiver. At the Technological Institute in Petrograd, Professor Boris Rosing built himself his own electric vision apparatus in 1907, using a mechanical transmitter similar to one developed some years earlier in Germany, and a cathode ray oscilloscope as receiver. It did not work very well, although with a little imagination it was possible to see a pattern on the screen of his picture tube. Rosing had no amplifiers, and the signal from his photocell was just not strong enough to produce more than a faint image.

But the experiment fascinated one of Rosing's students, a young physicist from Mourom, Vladimir Zworykin. Zworykin went to Paris to study under the famous physicist Paul Langevin; he spent the First World War in the Russian Army Signal Corps, and in 1917 he joined the Russian Wireless Telegraph and Telephone Company. By then, quite independently of Campbell-Swinton, he had come to the same conclusions about the best way of achieving distant electric vision. Rosing was right about the receiver, but wrong about the transmitter.

In 1919 Zworykin left Russia for the United States, and took a job with Westinghouse, who were no more receptive to his ideas than the Russian Wireless Telegraph Company had been. He left for another company in Kansas, working on the television idea in his spare time. In 1923 he patented his camera tube using the storage principle. 'I had considerable difficulty getting patent protection', wrote Zworykin, some years later.

But the patents served him well. By now all the big electrical manufacturers were becoming interested in the idea of television, and demonstrations of the mechanical system with its spinning discs and whirling mirrors were being given all over the world. Sam Kintner, head of the Westinghouse Laboratories, was keen to get into the business. Zworykin was persuaded back to Westinghouse, giving them an exclusive option to purchase his television patents.

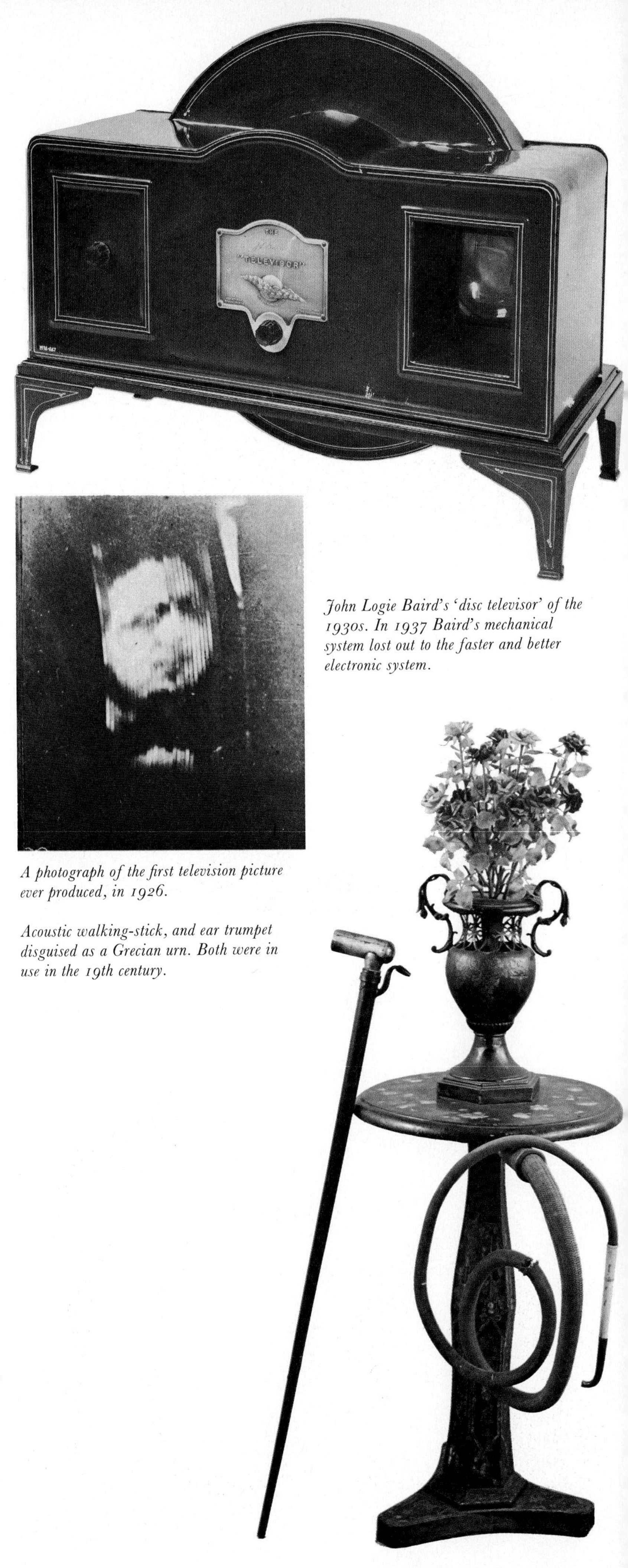

John Logie Baird's 'disc televisor' of the 1930s. In 1937 Baird's mechanical system lost out to the faster and better electronic system.

A photograph of the first television picture ever produced, in 1926.

Acoustic walking-stick, and ear trumpet disguised as a Grecian urn. Both were in use in the 19th century.

Chester F. Carlson with his prototype photo-copying machine.

'Acoustic throne' made for King John VI of Portugal in 1819.

Below, *19th-century ear trumpets.*

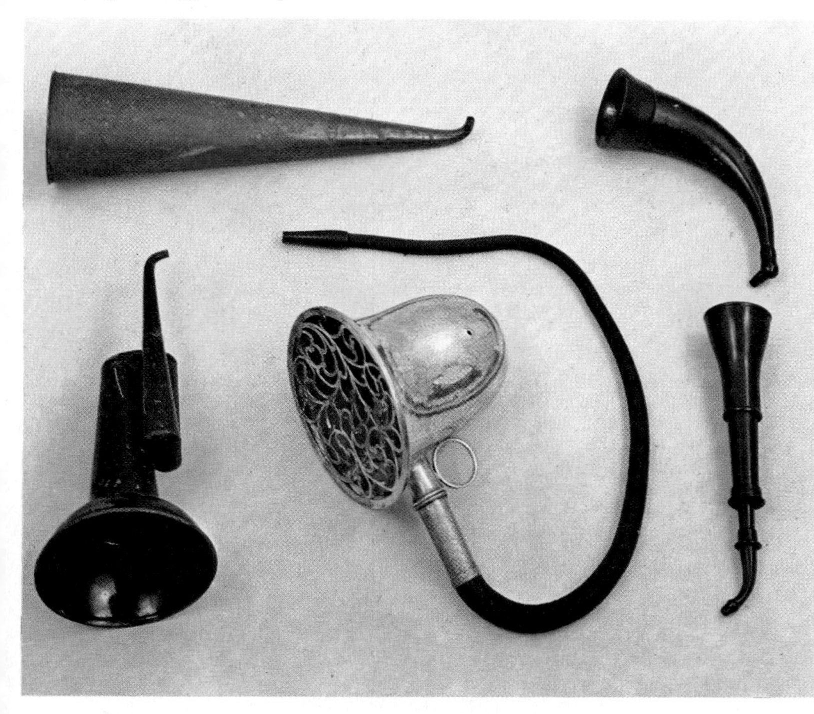

But Zworykin soon found that things were not so easy. He was asked to demonstrate his invention and did so in 1924. But the pictures were little more than hazy shadows: not much better than Rosing's and certainly no match for the mechanical system which was developing fast.

So Zworykin went back to the laboratory and worked on new and more sensitive photocells, and applied them not only to his new camera tube, but also to sound movies. At last, in 1928, the new camera was ready. Zworykin called it the 'iconoscope'. David Sarnoff, vice-president of RCA, recognized its importance and at once started work on strengthening the patent situation.

Between 1930 and 1939, RCA spent more than $2 million on patents and legal battles and a further $7 million on television development. By that time television, almost as we know it today, had established itself both in Britain (through the brilliant work of Isaac Schoenberg and his team at EMI) and the United States.

T. M.

Hearing aid

The biggest hearing aid ever built was surely the 'acoustic throne' supplied to King John VI of Portugal by the London firm of Rein in 1819. Kneeling courtiers spoke into lions' mouths in the hollow arms and resonators fed the sound through tubes to the royal ears. Now there are hearing aids weighing one-fifth of an ounce.

Originally, man cupped his hand to his ear to increase the sound-catching area, as when a dog pricks its ears. Animal horns may have been the first ear trumpets. Later, chair-to-chair speaking tubes canalized the sound and prevented its dissipation. The Victorians produced mock urns and vases, with catchment areas to carry conversation by tube to deaf guests. But the deaf man wanted a wearable aid. So inventors gave him an acoustic top-hat, an acoustic walking-stick and an under-the-beard receptor, all with tubes to the ears; they gave his wife a feathery acoustic bonnet, receptors disguised as tiaras, fans and reticules. For the not so shy there were big chased silver resonator trumpets. Another idea was a curved 'fan' for holding in the teeth, the sound gathered by the fan travelling by bone conduction to the ear.

The scientific advance came when Alexander Graham Bell, striving to invent a hearing aid, came up with the telephone. Its principle was then adapted to the original quest; the carbon microphone turned sound into voltage which was amplified and turned back into sound. But batteries were cumbersome. In 1923 the Marconi Company launched the valve-operated Otophone; the equipment was packed in a case weighing 16 lb. In the 1930s miniaturized valves made possible electronic hearing aids the size of box cameras, weighing 4 lb. Wearable? Nearly.

The claim to have made the first really wearable electronic aid, weighing 2$\frac{1}{2}$ lb., is made by A. Edwin Stevens, who produced it in 1935, the year he founded Amplivox. By the 1950s the arrival of the transistor led to undreamed-of miniaturization, with micro-circuits the size of a pin-head, and the latest aids are all but invisible. Yet a visible aid has one advantage: it encourages people to speak louder and more clearly.

E. S. T.

Xerography

In the xerographic process, an electrostatic charge is induced on an insulating photoconductive surface in darkness. This 'xerographic' (from *xeros*, dry, and *graphein*, to write) surface is then exposed to light reflected from an original (as for a photographic plate). The resultant charge pattern is dusted with a coloured powder suitably charged to be attracted to the pattern but rejected by the background. The powder pattern (the image of the original) is transferred to ordinary paper with the further application of an electrostatic charge, and 'fixed' on the paper by heat or chemically. Finally, the xerographic surface is cleaned and ready for re-use. In an alternative, and simpler, system used today, copy paper can be used which itself forms the photoconductive surface and consequently becomes the final copy.

Before the invention of xerography, copy methods had mostly relied on photographic processes or systems involving chemical changes in specially prepared papers by the action of light or heat. These methods were, however, subject to a number of limitations, including the use of wet chemicals, restrictions on types of originals that could be used, and the need for relatively skilled operators. It was these limitations and the ever-growing requirement for a simple and versatile copying method that led Chester F. Carlson, a physicist working for a patent firm in New York, to conclude that the real answer was the combination of electrostatics and photoconductivity. He filed a provisional patent application in this sense in 1937.

Working in a small back room in Astoria, New York, he found his experiments fraught with difficulties. Apart from acute shortage of money, there was considerable danger of fire, and from the fumes of the materials (sulphur and anthracene) with which he was trying to make a photoconductive plate. Not the least of his problems were the complaints of the other inhabitants of his apartment block.

He persevered, however, and produced the historic first xerographic copy, '10–22–38 Astoria', on 22 October 1938. His first patent was filed on 4 April 1939, and he followed this with a patent for an automatic copying machine on 16 November 1940.

Carlson tried to get industry interested in the process, but to no avail. However, in 1947 a small family firm, the Haloid Company (later to become Xerox Corporation) of Rochester, New York, acquired the rights of the Carlson patents and gave the first public demonstration of xerography at the annual meeting of the Optical Society of America in Detroit on 22 October 1948 – 10 years to the day since the first 'Xerox copy'.

Ballpoint pen

A printer's proof-reader, fed up with blotting the galleys and incessantly refilling his fountain pen from a bottle of ink, was driven to invent the ballpoint. He was a Hungarian named Biro. Actually there were two Biros. The proof-reader – he was also a journalist, hypnotist, painter and sculptor – was Ladislao and his chemist brother was Georg. They fiddled with preliminary experiments in Budapest in the late 1930s, emigrated to the Argentine in 1943 and there found a backer, a British financier named Henry Martin. With Frederick Miles, Martin set up a factory in England to manufacture non-leaking high-altitude writing sticks for the RAF. They took over Swan, became a million-pound mass-production industry, and were in turn taken over by the French company of Bic, who developed an ever cheaper throw-away ballpoint.

All this was neck and neck with what was going on in America. During the Second World War, Milton Reynolds of Chicago was on a sales trip to Buenos Aires. He was intrigued by Biro's nib-less pen, brought some specimens back to America, and found there were enough legal fissures to develop his own variation without infringing the Biro patents. This he did by converting the liquid-supply system to gravity feed, whereupon the American government bought his foolproof marker by the hundred thousand to issue as combat kit. Reynolds was able to invade the field because some digging revealed that (as is repetitively the case) the idea was not new. In fact another American, John J. Loud, had invented and patented a ball-point pen in 1888. It was a crude device, designed for marking warehouse bales, and did not oust the steel nib, which had then been scratching away for 80 years. In any case, Loud's new tool was overshadowed by the then much more practical fountain pen, also a child of the 1880s. Four years after Loud's leather and fabric brander came a similar try-out by an inventor named Evans, a pen which replaced the ball with a tiny wheel, lubricated by revolving against an ink pad within the housing. Both were fuelled with the standard writing ink. The Biro brothers thought of the vein-like tube inside the barrel which moistened the ball by capillary attraction. But what made the ballpoint the 'mod. con.' it now is was the production of a viscous fluid similar to that used by printers. It was mixed in his Californian kitchen by Franz Seech, a chemist from Austria. The instant-dry dye forms a skin on the surface when exposed to air.

The post-war civilian market was attacked in New York with the slogan that it was the only pen which could write under water. Demonstrators sat in the windows of Gimbels scribbling in tanks of water. Five thousand customers who seemingly had always longed to write under water swarmed into the store and paid £3 each for the novelty item. Subsequently the ballpoint thrust, in a flying wedge, between the vanishing pencil and the élitist fountain pen.

K. A.

Polaroid camera

It was an American, Edwin Land, who recognized that many camera-users would like to be able to produce 'instant photographs' and who, in 1947, invented the Polaroid camera, which can produce positive prints without the negative having to be taken out of the camera. The negative is developed by chemicals which are released immediately the picture is taken, and which then carry unused silver salts to a second piece of paper to give a positive print within 10 seconds of the picture being taken. Within the last few years a similar colour system has been developed. Polaroid photography is now widely used by the police for identity pictures. The one disadvantage of the system is that the negative can be used only once and only one print of each picture can be produced unless a special film is used.

Long-playing records

The long-playing record was announced by Columbia Records (CBS) at a press conference in New York in June 1948. Several earlier attempts to increase the playing time of records – notably by Neophone (England) in 1904, and by RCA (USA) in 1931 – had found only a limited commercial application. The CBS research team, under Dr Peter Goldmark, began work in 1944 and came up with a record made of unbreakable plastic instead of shellac, revolving at 33⅓ r.p.m. instead of 78 r.p.m. This micro-groove record accommodated about 250 grooves per inch instead of about 80 on the old kind of disc, and

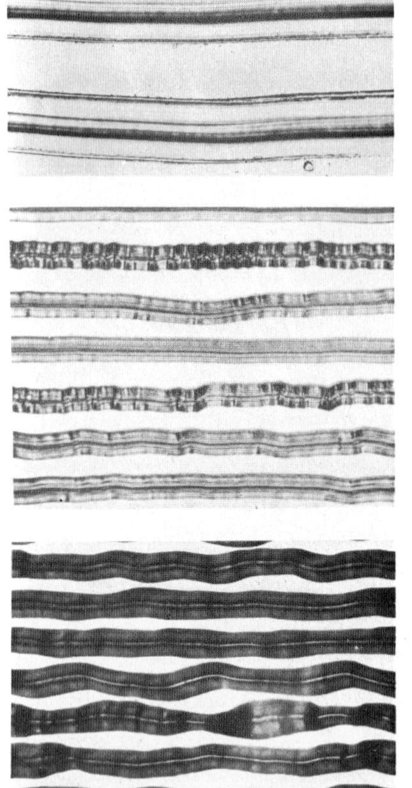

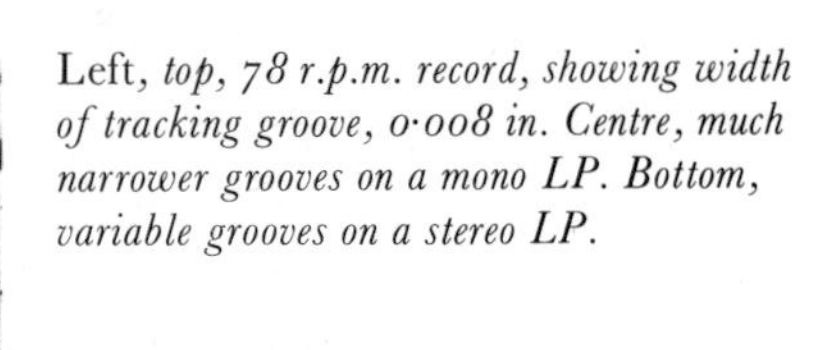

Left, *top, 78 r.p.m. record, showing width of tracking groove, 0·008 in. Centre, much narrower grooves on a mono LP. Bottom, variable grooves on a stereo LP.*

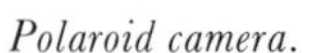

Polaroid camera.

How the ballpoint pen works.

Early experiments in the analysis and synthesis of sounds were made in the 1850s on this apparatus by the German physicist Hermann von Helmholtz.

The first videotape recorder, demonstrated by Poniatoff in 1956.

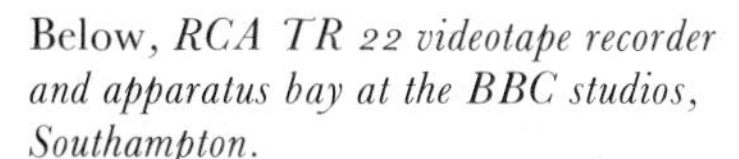

Below, *RCA TR 22 videotape recorder and apparatus bay at the BBC studios, Southampton.*

each side played for 20 minutes instead of five. The principal snag in 1948 was that new equipment was needed to play the new records.

Then began the 'battle of the speeds'. CBS offered to share their invention with their biggest competitors, RCA, and with all other interested parties. But RCA replied with the announcement of a 7 in. record, to be played at a speed of 45 r.p.m. Its playing time was about the same as the standard 78 r.p.m. disc, but a 'quick-change' apparatus was said to reduce the interruption between sides to seconds.

Meanwhile, British Decca (under their US trade name of London Records) began exporting their own long-playing records to America in August 1949. In June 1950, Decca launched the LP in the United Kingdom.

By 1953 the 78 r.p.m. format was in its death spasms. Over the next five years the recorded repertoire expanded beyond all predictions.

J.C.

Music synthesizer

On 5 August 1953 the RCA Synthesizer at Columbia University in America completed tapes of Chopin's Polonaise in A flat (opus 53) and other selected pieces. These were interspersed with tapes of famous musicians playing the same pieces and played to various people who were asked if they could distinguish the difference between the synthesized versions and the human performances. Interpreting the results by standard statistical methods it was concluded with 70 per cent certainty that only one in four people can tell which is which.

A synthesizer is an instrument which generates sound electronically and can modify it almost indefinitely. Once any sound has been analysed it can be reconstructed on the synthesizer. The machine contains a 'white noise' generator (white noise is a combination of all frequencies at once, in the same way that white light is a combination of all colours), and two or three oscillators as sound sources. The wave forms and frequencies of the oscillators can be controlled and fed through a number of filters, reverberation devices which modify the sound; the 'envelope' shape (the arc from growth to deterioration, or 'attack' to 'decay') of a sound can also be controlled. Many synthesizers also incorporate a device called a 'ring modulator' which produces a rather unpleasant discordant sound and is probably included because it is economical to produce.

An electronics professor, Robert Moog (pronounced to rhyme with 'vogue'), has, for many people, given his name to the synthesizer. Moog was the first person to make a model accessible to the general market, but Don Buchla and Al Perlemon also produced early synthesizers with this in view. Opinion as to the merits of their various versions is still divided, and there is, for example, disagreement as to who first invented the voltage-controlled oscillator, an acknowledged milestone in the development of the synthesizer.

The exact application and scope – and indeed the definition – of the music synthesizer has been under discussion ever since RCA first announced its version in the 1950s. At that time it was generally viewed as a development designed to replace conventional musical instruments, and this concept was reinforced when Walter Carlos produced *Switched on Bach* and *The Well-Tempered Synthesizer*, reconstructions of J. S. Bach's works on the synthesizer – a laborious process in that each note had to be recorded separately and spliced together. Current attitudes, however, seem to have progressed through imitation ('I've tried all the currently available synthesizers and none of them matches a decent Steinway or Baldwin when it comes to subtlety', Carlos now says) and even through concentration on sounds which can be obtained only with a synthesizer (and there undoubtedly are such sounds) to regarding the machine as nothing more than its name implies: 'in the same way that a recording studio should be seen as an instrument, because it transmutes sound,' says Eno, who played synthesizer with the pop group 'Roxy Music', 'a synthesizer should be seen as a studio, and used in co-operation with other instruments.'

P.WH.

Videotape recording

John Logie Baird made the first video recordings in 1928. They were made on the 78 r.p.m. discs used on contemporary radio-gramophones and fed through his Baird 'Televisor'.

Accelerating television technology made these recordings obsolete very quickly. The requirement for a tape-recorder able to record and reproduce both picture and sound efficiently, together with the growth of a standardized television industry after the Second World War, meant that a lot of money was spent on research. In 1956 Alexander M. Poniatoff demonstrated a machine at the National Association of Broadcasters Convention in Chicago, and, using it, the first coast-to-coast network television broadcast pre-recorded was 'Doug Edwards and the News' on 30 November of that year.

Poniatoff, a Russian-born engineer, had founded a small company producing electrical equipment for airborne radar systems in 1944. The company's name was composed of his initials, AMP, and EX for 'excellence': Ampex. Two years later, a large American equipment manufacturer, RCA, announced their video recorder; it was fully compatible with the Ampex machine, and thus the track patterns of these two machines became the standard of the broadcasting industry for tape-recording.

The development and application of videotape recording has accelerated ever since, and it may still be too early to define its significance.

Two early problems were that a large bulk of tape was needed for a short recording, and that editing was very difficult to synchronize and consequently clumsy. These were greatly helped during the 1960s by the 'helical scan' system (whereby instead of the magnetic tape having signals longitudinally along it and passing over a head which scans the information, the information is on the tape in vertical lines and passes right round the head helically), and by the development of the 'CMX' computerized editing system, which makes editing easy and precise.

In 1967, Ampex developed a specialized recording system that uses a large metal disc as a recording medium rather than a reel of tape. Although the disc can only record 30 seconds of programming at a time, which it does continuously, it can locate and prepare to replay a desired action in less than four seconds. This is probably the best-known use of video recording: the 'instant replay'.

Less visible to the viewer, but highly important to the broadcaster, are new cartridge-loading videotape recorders that permit the automated airing of short programme segments such as news coverage and commercials. Also new are high-speed contact duplication systems that for the first time make videotape copies available rapidly and economically. These developments, too, come from Ampex.

In 1966, Ampex combined computer technology with videotape recording techniques to produce a new system capable of filing and retrieving documents. The Videofile system permits compact filing of up to 200,000 documents on a single reel of tape. Videotape machines are also used in education, enabling a student to rerun a demonstration, for instance, in his own time; home-to-store video communication to order a pound of cheese has also been envisaged. But perhaps the real videotape machine has not yet been invented, and Poniatoff will be dismissed with Baird in the videobooks of the future.

P. WH.

Communications satellites

Easily the most practically valuable of all the space-travel efforts made since 1957, when the first Russian *sputnik* was launched, have been the communications satellites. The idea of using some kind of celestial telephone exchange for relaying wireless signals goes back to 1946; American scientists then suggested the moon for that role. Ten years later, a leading US electronics corporation undertook to study the possibilities of transoceanic communications by means of man-made satellites; and only another seven years after that, the first one was launched.

It was called *Telstar* and went into orbit from Cape Canaveral in July 1962, circling the earth at an altitude of 500 to 3,000 miles at a speed of 11,000 to 18,000 m.p.h. – that is, it completed each trip round the planet in 160 minutes, receiving and retransmitting telephone calls and television signals from America to Europe and vice versa. The signals coming from ground stations were amplified and sent back to earth by means of storage batteries and solar cells, which provided the necessary power.

Telstar and several other satellites positioned in orbits around the northern and southern hemisphere were of the 'rising and setting' type, comparable to the moon's orbit; they were in range of their transmitting and receiving stations on earth for only a few hours per day. The first 'second-generation' satellite, *Early Bird*, which was launched in 1965, inaugurated the age of the 'synchronous' type: apparently stationary above a certain point in mid-Atlantic, halfway between the transmitting and receiving stations, *Early Bird* in fact keeps pace, at an altitude of 22,300 miles, with the rotation of the earth, relaying signals between stations on the earth for 24 hours a day. These high-frequency signals travel, like light, in a straight line, with only slight deviations which can be corrected by computers controlling the directions in which the aerials face.

Today, these *Intelsat* communications satellites – to give them their technical name – carry half the world's telephone calls and many of its television programmes from continent to continent. There are now about half a dozen of them in continual operation all over the globe, and every year one or more is added. The busiest earth station is the British Post Office installation at Goonhilly in Cornwall; it carries about 500 telephone calls at a time across the Atlantic and at the same time 200 to the Middle and Far East, but the latest *Intelsat IV* type can carry 9,000 telephone circuits or 12 colour TV channels, apart from Telex circuits and data transmissions – the fastest-growing telecommunications service of all. The costs are shared by 83 member countries of the International Telecommunications Satellite Consortium.

E. L.

The best-known use of videotape recording – the 'instant replay'.

MAN TAKING

Energy

Materials

MAN TAKING

Energy

When we think about energy we usually have in mind its basic sources: muscles, water, wind, wood, coal, oil and the nuclei of atoms. Yet the history of energy really consists of the ways man has been able to exploit it – the mechanical devices and physical processes he has invented to convert raw energy into power which he can use directly.

Coal itself may be a powerful source of energy but the heat of burning coal has to work through steam and an engine in order to produce useful power.

We can start at the wrong end of history and look at a very modern form of energy use: the nuclear reactor for producing electricity. The energy used is the most basic form possible – the splitting of atoms so that mass is converted directly into energy. This process produces heat – always the most basic of man's methods for obtaining energy, and the form which he has learned how to handle best. In the nuclear reactor the heat is used to produce steam. (This may seem surprising since steam has been such an old-fashioned way of getting power.) The steam is used to drive a turbine. (Here again, the turbine is really an advanced version of the simple waterwheel which man has been using for centuries.) Finally the turbine drives an electric generator, itself a mechanical device which uses movement in a magnetic field to produce electricity. So in the nuclear reactor we have atomic energy, heat, steam, the mechanical device of the turbine, the mechanical device of the generator, and electricity. The electricity is carried over wires and then used as light or heat; or it may again be put through the mechanical device of an electric motor to produce power.

The mechanical devices are simply converting devices, but without them there would be no practical energy. At first man's energy was limited to his own muscles, and of course this remains the only source of energy at the disposal of animals today. Some time later man learned to use the energy of animals. To do this he had to invent a number of mechanical devices to turn animal (for instance, oxen) energy into something useful. The first energy devices may well have been used to raise water from rivers for irrigation. There are many of these still used throughout the world – powered by animals or by man himself. It is often suggested that the reason civilization seemed to start in those areas where irrigation was a problem was that this problem had to be overcome.

This required, on the one hand, mechanical devices and ways of using energy and, on the other, organization, centralized bureaucracy and an integrated society. Irrigation and crops also meant that a people was fixed in one area and therefore had more chance of erecting permanent buildings. It is also possible that because water was scarce there was more incentive to make war and conquer other people's lands – and war has often been an impetus to invention.

The article on the waterwheel notes that many of the basic mechanical devices were started in connection with raising water: piston, gear, screw, cylinder, rotor, endless chain. The step from a wheel that raises water to one that is turned *by* the water is a very small one: anyone might have observed that when a stream was running fast it turned the water-raising wheel. Although the step is a tiny one, to make it was of immense importance. At once a pow ful new source of energy becomes possible. Instead of energy bei expended beside the stream, energy is now obtained *from* the strea Furthermore, unlike man or animals, the new source of energy nev gets tired, never needs feeding and can run itself. So a simp mechanical device made of a readily available material (woo makes possible a new source of energy. It is interesting that at fi the waterwheel uses energy to produce an effect (raising water) b in reverse it gives out energy – like the electric motor, which us energy to produce an effect but in reverse can actually produce ener as electricity.

A windmill is a waterwheel in the air. Wind has the advanta that it is more readily available than a stream and many mo people can use it at the same time, but the disadvantage that it less constant and can come from different directions. It is surprisi that the windmill took so long to develop: fabric was available, was woodworking technology. Perhaps it required some genius discover how to make the sails so that they revolved in the win This is rather more difficult than a waterwheel, only part of whi is immersed in the water. Windmills seem to have originated Persia in the 7th century. These were horizontal, rotating paral to the ground, which meant that they would work no matter t direction of the wind. Four hundred years later the windmill appea in Europe as the vertical windmill we know today. This is mo effective since the whole, instead of only half, of the wheel is used capture the energy of the wind. But it does have to be turned in the wind each time the wind changes direction. At first this was do by hand but, later, automatic devices worked by the wind itself we used – a very early instance of feedback.

The windmill principle turns up again in the water turbine, t steam turbine and the gas turbine. The general rotary principle of course, also used in the electric motor and recently, though in very different way, in the Wankel rotary petrol engine. But for while man's use of energy took a very different path: the path pistons that went up and down inside cylinders.

The first steam engine was not a steam engine at all but a vacuu engine. It was invented by Thomas Newcomen in 1712 to pum water from Cornish tin mines, but it needed so much coal that it w only possible to use it in the coal mines themselves. (It is interesti to note that the impetus for both steam engines and railways car from the mines. In a sense, the mines were the descendants of ma first irrigation problems: a lot of energy was needed to raise hea and bulky material and to carry it away.) In this first steam engine piston was pulled up a cylinder. The space behind it was filled wi steam as it rose. The steam itself did nothing to raise the piston b simply acted to fill the space; it was very low-pressure steam. A j of cold water then cooled the steam, which at once condense This created a partial vacuum and so the air pressure above t piston forced the piston down. This energy was transmitted to t pump, but some of it was stored in a weight or flywheel to pull t piston up again. Later on James Watt and others improved the engi and used steam to push the cylinder down, thus using the stea ectly as an energy source. At first this was not possible because re were no boilers which could produce high-pressure steam, so e one invention had to wait for another. This may have been the e with the next development in the use of steam power – the steam bine, which came about two hundred years later.

The idea of using steam to rotate a wheel had been suggested and nonstrated on many occasions in history, but only as a toy, since y little power was produced. It is suggested that the full development of the steam turbine by Parsons had to await special materials l bearings that would allow the high speeds that were necessary produce power. Parsons's own contribution in devising a method extracting the full energy from the steam by making it pass from fan to the next must not be forgotten. In fact his steam turbine n example of an invention which was brilliantly successful right m the beginning. Moreover, Parsons envisaged his turbines ng used to drive electric generators, and today most electric erating plants (even the nuclear ones, as we have seen) do in t use steam turbines.

Steam engines had to have boilers, and fire to heat them, and t meant that they tended to be bulky. The petrol engine or internal nbustion engine started as a gas engine. If gas could be made to plode inside a cylinder then it should act just like steam in raising piston, and there would be no need for boiler or fire. The invention of the internal combustion engine took place over quite a short e (1860–1900) and involved many different individuals working their own. These people were often self-taught mechanics working their own backyards or small workshops; hearing about each er's successes, they were spurred on to be the first to produce a werful, reliable engine.

The principle of self-combustion under high pressure was gested by de Rochas in a pamphlet (he never made it) and then gotten until rediscovered by Rudolf Diesel who gave his name to diesel engine. The four-stroke engine, also suggested by de chas, was eventually re-invented by Otto. Later on, the carettor was invented, neglected and then re-invented. The basic tures of the petrol and diesel engines had been worked out by beginning of the 20th century. There have been many improvements but no radical changes since (until recently). This is exactly allel to what happened with the steam engine.

t was inevitable that the gas turbine should be developed to do the internal combustion engine what the steam turbine had done the steam engine. This windmill principle works well enough in gas turbine, and many are in use for various purposes. Often it seemed that the gas turbine might take over in the motor-car, t it has never happened. There are many advantages but some advantages, such as a higher fuel consumption; but the real reason robably that the petrol engine is adequate and the cost of change ot worth while. The gas turbine has achieved its triumph in the , for a jet engine is really a marriage between a gas turbine and a ket. The advantage here was obvious though it was based not on power of the engine but on the way that power could be used to pel an aircraft.

The Wankel rotary petrol engine was not invented until the 1950s, t now seems likely to be used in a large way for motor-cars. This curious engine because it is really neither a piston engine nor a ndmill' type. It is somewhere in between – a piston that rotates. is notion was probably difficult to develop because people's nds would tend to be pulled along the separate tracks of either nventional piston engines or turbines.

It is not difficult to see the path of development from waterwheels windmills, turbines, pumps, steam engines, internal combustion gines. The basic principles remain the same. Petrol or gas is just a substitute for steam, and a turbine is a refined windmill. But in the case of electrical energy, the development from two completely chance observations was direct and very swift.

It seems a huge jump from a pair of dissected frogs' legs to the most powerful and convenient source of energy but the connection is straightforward and the steps amazingly few. Luigi Galvani noticed that the legs twitched when he happened to touch them with two pieces of metal, and thought the electricity came from the frog. But his friend Volta was more interested in the two pieces of different metal, and from this interest arose the first battery or voltaic cell. This invention had a dual importance. First of all, it started the line of development of the chemical generation of electricity, which has led to batteries and the fuel cell – a line which is likely to develop much further in the future. But, perhaps more important, the voltaic cell made possible a practical source of electricity. Once this source was available, people could experiment with electricity and see what happened.

Not long afterwards, the Danish professor Oersted observed that a magnetic needle on a pivot was deflected when a wire carrying a current was put near it. Directly from this very simple observation Michael Faraday developed electro-magnetism, which showed how mechanical movement could link up with electricity. So Faraday made the first electric motor. Run in reverse, an electric motor can be a generator which provides electricity. Though the basic principles have remained the same, there have been many further developments in the field of electricity, perhaps the most significant being the linear induction motor developed for surface transport by Eric Laithwaite. Here the electric motor has no rotating parts, since the moving body is pulled along by a magnetic field always just in front of it (in the same way as the rotor of an electric motor is pulled round).

Electricity is a remarkably convenient source of energy because it can be carried long distances by nothing more substantial than a wire. It is possible to build electric motors so tiny that they fit inside toy cars, and so large that they drive the biggest locomotives. Electric motors are powerful and clean. They involve no large boiler and no heat, and they can be started and stopped instantly. As a source of energy, electricity has the further advantage that it can be stored in accumulators. This is a unique advantage because storing energy allows it to be used exactly when wanted.

The discovery of nuclear energy is a story of a sequence of scientific discoveries interspersed with theoretical jumps. There was one key discovery which turned atomic science from a description of the basis of matter to an actual producer of energy, at first in the atom bomb and later in nuclear power stations. This was the discovery that nuclear fission in an isotope of uranium could produce both energy and a chain reaction producing further energy. Intensive work is now going on to try to develop even more powerful methods which do not rely on rare uranium. Instead of splitting the atom, the new methods seek to fuse hydrogen atoms into helium atoms. This is the reaction produced in the sun; if man succeeds in making it work he will have reached the ultimate source of energy and tamed it for his own uses.

Materials

Man takes many materials from the world around him for his own use. There is no attempt to cover all these materials here; those included are the ones which have been made available only by man's inventiveness. Wood is a widely used material but man did not have to do much to it except devise tools for working it, and these are described elsewhere. Gold is an important metal but it is not essentially changed by man's treatment of it.

Man is often described as an animal that can use tools. Tools are an extension and an amplification of his own power, with which he can get the soil to work hard for him in agriculture and produce the food he needs – instead of having to move around to find it as animals have to do. With tools he can kill animals that would otherwise have eaten him, and he can also kill animals for food. With tools man can kill animals at a distance or animals that would otherwise have escaped from him, with tools he can fight those other people who are using tools to fight him. Tools enable man to make bigger tools to make still bigger tools. A supersonic airliner is the result of ideas, raw material and tools. The raw material is turned into the shape and function of the idea by means of the tools used.

But before man can use tools he has to make them, and for this he needs suitable material. The type of material available determines the type of tools he can make: for instance, you can make things of steel which could never be made of copper or wood. That is why archaeologists and anthropologists call the early ages of man by the basic materials available in each age: Stone Age, Bronze Age, Iron Age. It is perhaps sad that in many cases man has devoted his most ingenious toolmaking to designing devices to kill other men. At least the need to defend and the apparent need to attack have provided one of the major pressures for the development of new materials. You will survive if your plough is good enough for you but not quite as advanced as your neighbour's, but you would not survive very long with Stone Age weapons against Bronze Age ones. Fortunately the material that would make good weapons was exactly the material that would make good tools in general.

The search was for a material that was hard, sharp (to take an edge) and durable, but at the same time could be easily worked. Copper was easily worked but not hard enough and no good at taking an edge. The discovery that adding a little tin to copper produced bronze led to the great civilizations of the Bronze Age. The step to bronze was a significant one because bronze was much harder and could also be cast much more easily. It was later found that bronze could be hammered cold, and this gave an even harder edge.

Two other important alloys are based on copper: brass and pewter. Composed of copper and zinc, brass was more difficult to produce than bronze, and so it was expensive and used mainly for luxury items until fairly recently. The great advantage of pewter, which contains copper, antimony and tin, is that it can be worked cold by hammering, much as can silver. For this reason it was very suitable for domestic vessels. It could also be worked by individual craftsmen on their own without the need for large foundries.

The next major tool material after bronze was iron. Considering how iron has come to dominate man's materials – though now slightly threatened by aluminium – both for tools and also for construction (ships, skeletons of buildings, cars, etc.), it gave man a tremendous amount of trouble. Iron was never easy to obtain. There was plenty of it in ores, but transforming it into a usable material with the right properties has made great demands on man's inventiveness. At one time iron was so rare that it was as precious as gold.

First of all, iron requires a high furnace temperature, so it had to await furnace technology and especially the invention of the bellows. Removal of the impurities from the ore was a tedious process in which the 'bloom' was periodically taken out of the furnace and hammered until the impurities were reduced. This was done again and again until in the end the iron was more or less pure. The whole process required a great deal of effort, skill and luck. Luck seems to have been involved also in the partial development of steel. So natural iron ores have the right carbon content to produce ste and it seems that some people learned how to produce small amou of steel for special weapons. But the process was never relia enough to establish steel as a material. Processes like quenchi tempering and hammering were also probably developed in t weapon-maker's workshop.

The next big step in iron-working was the development of t blast furnace and the use of coke instead of charcoal as fuel. T blast furnace provided a much quicker way of separating out t impurities, and produced cast iron which was suitable for many u other than those which required a high tensile strength. It is doubt if railways could have developed if the rails had continued to made of cast iron for they were always fracturing, and only wroug iron solved the problem.

Wrought iron was made in rather a haphazard way, by putti oxides into the furnace along with the iron and allowing the oxyg to burn off the excess carbon that made iron brittle. Then Bessem came along with one of those truly important technological develo ments that suddenly change the whole picture. His process involv blowing cold air through the molten iron and removing carbon th way. The beauty of the process was that it was simple, reliable a easily controlled. In fact, instead of producing wrought iron actually produced mild steel. The next step was obvious. If y removed all the carbon from the iron you could put back exac as much as you wanted, to form different types of steel. The Bessem process was the crucial invention in the use of iron. Before it sk was more important than technology, but after it iron could produced in large quantities, in a reliable and predictable mann and with greatly improved properties.

The metal that is beginning to rival iron in some areas is al minium. Compared to the antiquity of bronze, brass, silver, gol iron and copper, aluminium is a late starter since it appeared or at the beginning of the 19th century. Even so, the first producti method was so costly that, like iron in its early days, aluminiu was regarded as a semi-precious metal. It was only in 1886 tha new economical way of extracting aluminium was discover separately by two individuals in the US and in France. This d covery was similar to that of Bessemer, for at once it changed t whole availability of the metal.

Metals are used not only for tools but also in structures. Ships a bridges can be made from iron, aircraft are made from aluminiu Alongside man's search for tool material has been his developme of structure materials. By far the oldest of these is pottery. Ma other materials were undoubtedly used in antiquity (wood, bo skins, gourds), but pottery is the only one likely to survive centuries until discovered by archaeologists. The basic ingredie of pottery, fire and clay, were readily available to early man. fire built on clay soil would harden the patch underneath; a lump clay tossed into a fire would be found hard in the embers. There a many ways in which pottery could first have been discovered, b its history is one of development rather than invention. A continuo series of chemical and technical improvements were made over t centuries. New kilns, new methods of firing, different clays a different mixes, new methods of glazing, new types of decorati were continually being developed or discovered. Some of the i provements were functional: for instance, ways would be develop for making vessels watertight. But once the functional proble had been solved, all further improvements were aesthetic – efforts make the vessel more beautiful or more interesting. This is an i portant point because most of the inventions in this book have functional purpose: to solve man's problems or achieve his purpo is interesting to observe that aesthetic exploration or aesthetic ssures were just as effective in stimulating innovation as the ving of practical problems.

The development of glass is similar in many of these respects to t of pottery. Glass does have a unique function when used as a ndow or a lens, but its use for drinking vessels and ornaments was entially aesthetic – perhaps even more so than pottery, for it is gile and cannot be used for cooking. Glass is such an old material t it is difficult to realize that a fundamental advance in glass-king occurred very recently. This was the method of making float ss by floating liquid glass on a bed of molten tin. In this way ets of glass with no blemishes could be made. Before this innova-n, such sheets had to be ground and polished.

Leather is a good example of how the development of a particular ocess can completely change the nature of a material. Animal ns and hides were probably used from the beginning of man's tory, but the development of the process of tanning turned skins o leather. Leather held a unique position: it was stronger than ric or even rope and yet much more flexible than metal. In its ction leather was the forerunner of all modern plastics: it is pervious to water, it has some rigidity but also flexibility, it is ong and durable, it is easy to cut and shape and easy to join to er pieces. The actual process of tanning has changed extra-linarily little over the centuries. Why should it change if it is equate?

n the case of rubber, too, a change in processing made all the ference to the properties of a material. Rubber was known in -European civilizations many centuries before it found its way Europe, because in certain countries it occurs naturally in trees. t natural rubber is not a satisfactory substance, particularly cause it changes its nature with changes in temperature. The vital ovation was the discovery of the process of vulcanization by arles Goodyear, who deliberately set out to make a more usable m of rubber and succeeded by accident when he overheated a xture of rubber, sulphur and white lead. We cannot know whether had specific reasons for putting these substances together in the t place or was simply acting at random. Vulcanized rubber was vastly superior to natural rubber that it was as if a new substance l been invented, with the elastic properties of rubber but much der and much more durable.

Rubber might have been used for a great many more purposes if stics had not been invented. Synthetic plastics are a vast area in mselves. They are materials which man has made entirely for nself and by his own inventiveness. Where other materials were cerned it was a matter of treatment, of extraction, of mixing, in the case of plastics the very materials themselves were created man. For instance, many plastics are made originally from some-ng as insubstantial as a gas. The properties of plastics are also arkable: nylon is so strong that it can rival metal in many gineering applications, yet at the same time it can be used as a ead and made into clothes. Polythene film and cellophane are a t of flexible glass and acrylic plastics are in fact much more nsparent than glass. Plastics can be brightly coloured without ing to be painted. Not only are the properties of plastics remark-y versatile but the methods of manufacture (extrusion, spinning, ction moulding, vacuum forming, etc.) are themselves so versatile t their cost is very low. Compare the easy moulding in a few seconds of a plastic bucket with the hand fabrication of a metal one. Imagine photography with only glass plates and no cine film. The history of the invention of plastics is one of chemical discoveries by accident or by design. It is a story of sudden inventions or gradual development, a story of individuals and teams.

When looking at the major materials used by man, it is easy to overlook materials that played a small but very important part in his technological and, indeed, his general history. Pitch is such a material. It can be obtained from vegetable sources as wood tar or as a mineral from surface deposits. It is not of much use by itself, but it has proved immensely useful in waterproofing. Wooden boats could conceivably be made with such precision that no caulking at all is required, but this becomes more difficult the bigger the boat, since the water pressure increases. Pitch applied directly or as tarred rope made it possible to waterproof such vessels. It is difficult to see what other material would have done as well. Even today this use of pitch is virtually unaltered. It is only a slight exaggeration to suggest that had there been no pitch there would have been no boats.

The impression may have been given that man would use a good material as soon as it became available to him. This was not always the case. Today concrete is used in vast quantities in skyscrapers, bridges and buildings of every sort. Concrete was known to the Romans and used by them, but then the knowledge was lost or at least dormant for many centuries. It was not widely used as a building material since bricks and stone seemed adequate and aesthetically more satisfying. For certain special purposes, such as the building, in the 1880s, of the underwater base of the Eddystone lighthouse, concrete was excellent, but not many lighthouses were being built. As in the case of steel, aluminium and rubber, it was a crucial invention that changed the nature of concrete. This was the idea of reinforcing the concrete with iron bars – an invention first made by an Englishman in 1854 and then by a Frenchman in 1892. Concrete could now be used for large spans, as well as for pillar-type support – it had acquired a tensile strength which bricks and stone could never have had. The later development of pre-stressed concrete improved this quality even further. Nevertheless, what really made concrete the favoured building material it is today was not these technical improvements but a combination of high labour costs, the style and size of modern buildings, pre-fabrication and speed of building, and steel skeleton support.

The history of the development of man's materials is full of events that took place so long ago that we cannot be sure exactly what happened. It is possible that discoveries were made independently in different places at different times. A process may have changed by slow, almost imperceptible steps, or there may have been sudden jumps (as was true of iron, aluminium or rubber). The jumps may have been entirely accidental. For instance, the value of quenching may have been discovered when a piece of red-hot iron was accidentally dropped into a bucket of water. Or it may have been a matter of trial and error, as with a workman trying out all sorts of different materials during his tanning attempts. So long as something is happening, the reasons behind what is happening are not so important. What is important is that an effect be noticed and then followed up. The process of observation, repetition, search and development can only get started once something has happened. It is the initial happening that is so difficult to generate. That is why invention is often the outcome of chance and observation.

Waterwheel

All our basic mechanisms, gear, screw, the piston and cylinder pair, began life as devices used in raising water; so did the endless chain and above all the first rotor, the waterwheel. The crucial events in the story took place during the period of Greek rule in the Middle East, even though the Greeks are always said never to have bothered their heads with technology since they had plenty of slaves to do their work for them. Ancient civilizations certainly knew of the pulley, by means of which a man can exert the whole weight of his body downwards to hoist a pail of water up a well shaft. But it was almost certainly left to a Hellenistic artisan to use a wheel as a perpetual water-raising lever, by fastening his pails or pots around the circumference of a large wheel. Oxen cannot be trained to haul up a bucket. They can, however, be made to walk in a circle round a vertical axis. But the wheel of pots has to rotate about a horizontal one. The answer was the oldest form of gear: a face wheel or crown wheel, just an ordinary wheel with pegs at right angles to the rim set in the shaft of the wheel of pots.

Within a few generations the pots lashed to the wheel's rim became boxes shaped to its circumference, raising much more water for the same diameter. Then somebody noticed that if the waterwheel was used to raise water from a fast-flowing stream, the current itself would turn it. If projecting boards were set around the rim to catch the current, the river's power could turn the wheel quite fast. But the wheel did not have to raise water. The whole transmission could be reversed. Let the water turn the wheel which rotated the shaft: that way any rotating instrument could be water-powered. An obvious candidate was the millstone that ground flour. By the 1st century BC, watermills were widely known, even if not widely used.

There was a simpler version, a horizontal waterwheel. This is a much smaller affair, and less powerful, but it also needs less timber, and less water to drive it. In the 1st century BC a Hellenistic poet composed an ode in its honour: 'You girls who worked so hard in grinding corn – you can go back to bed and leave the birds to their dawn chorus . . . now the nymphs of the stream leap down on the wheel and turn the axle, and with it the millstone.'

All the basic components of water-powered industry were in existence by the time Rome took over what was left of the Hellenistic world. But medieval Latin Christendom was the first civilization to make real use of machinery. By the 11th century AD, watermills were to be found in nearly every parish. Waterwheels were beginning to perform other functions – driving grindstones and turning rolling millstones, or edge-runners, which were used as olive-presses, or for mashing apples for cider, perhaps crushing linseed or rape for oil, woad for dye, and later pulverizing charcoal for gunpowder.

A. K.

Windmill

The earliest mention of a windmill is in AD 644, when a Persian 'builder of windmills', Abu Lu'lu'a, who had assassinated the Caliph Umar ibn al-Khattab, was captured. A little more than 200 years later there are specific references to the remarkable windmills of Seistan (on the borders of Iran and Afghanistan), which derive from the horizontal waterwheel of Asia Minor first known in the 1st century BC. The sails, mounted on a vertical 'windshaft', turn in a horizontal plane.

Similarly, the vertical windmill of the West derives from the vertical watermill described by the Roman Vitruvius between 22 and 11 BC. This type of windmill, known as the 'post mill', had appeared in France by 1180 and in England by 1191. As the sails had to be faced square into the wind at all times, the wooden mill body containing the stones and gearing was mounted on a suitably supported upright post on which it could be turned into the wind by a long lever or 'tailpole' at the back. The earliest illustration of one of these mills is in the so-called Windmill Psalter, probably made in Canterbury in about 1270.

Not long after, a new variety of this Western mill was evolved which was to be found in France by 1300. The 'tower mill' consists of a fixed tower containing the stones and gearing, while only the roof or 'cap' which carries the sails is turned into the wind. Sometimes this was effected by a tailpole, at others by a lever inside the cap. There are far fewer medieval illustrations of this type; but

By the 15th century, waterwheels were in use for grinding and polishing. Detail from Virgin and Child with Saints *by Hans Memling.*

An illuminated initial from a 15th-century French Book of Hours, showing a windmill of an early type.

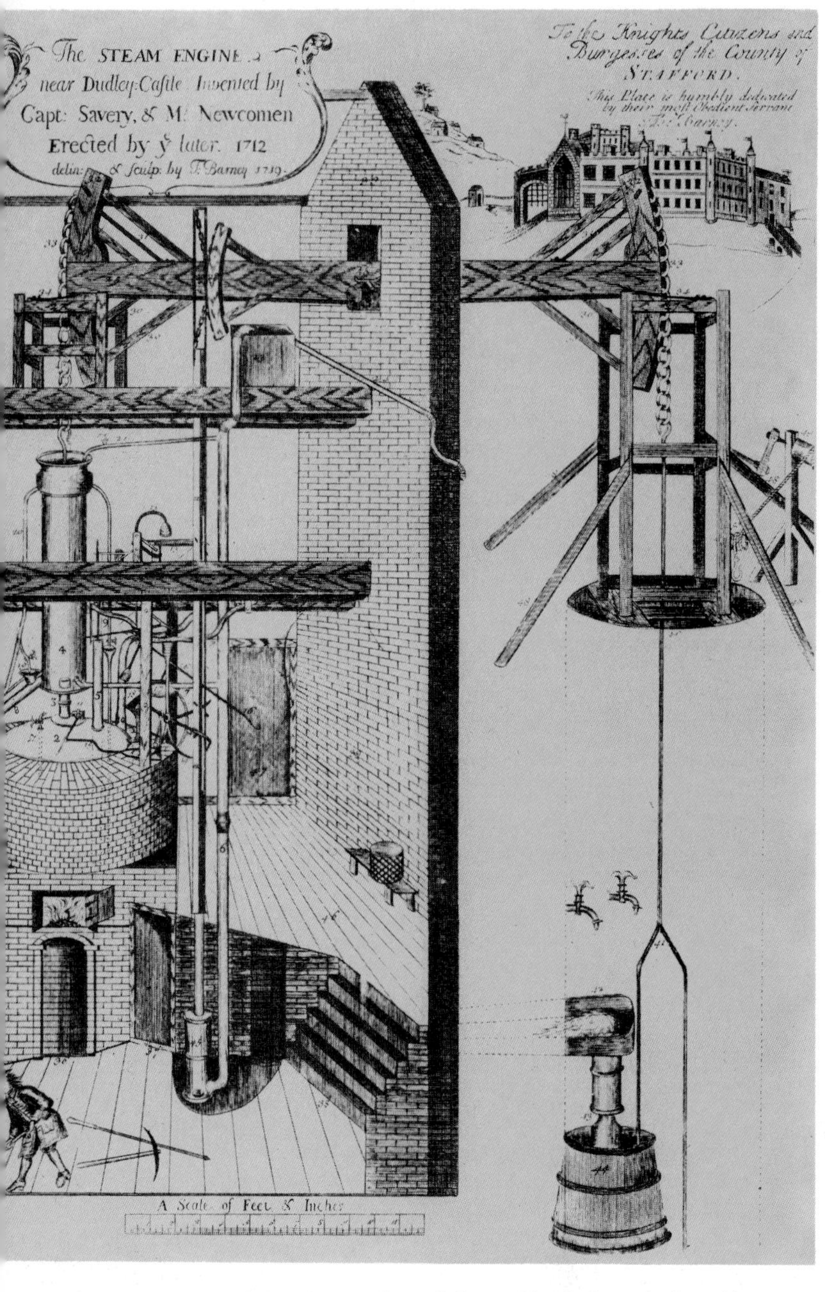

Newcomen's steam engine, which he erected at Coneygre Colliery, Staffordshire, in 1712.

Below, *pithead of a coal mine with steam winding gear,* c. *1820.*

one is shown in a stained glass window dating from *c.* 1470 in Stoke-by-Clare Church, Suffolk.

The windmill was a blessing to all European communities which had no stream that could drive a watermill the year round; it also made it possible to dispense with hand querns and cattle mills, especially if there was a neighbouring watermill. But the cost was high and two competing mills could seldom be worked economically.

R. WA.

Steam engine

The world's first practical working steam engine was erected by its inventor, Thomas Newcomen, at Tipton, Staffordshire, in 1712. It pumped water from the workings at the Earl of Dudley's Coneygre Colliery for more than 30 years.

Thomas Newcomen was a Dartmouth ironmonger, familiar with the tin mines of Devon and Cornwall and with their besetting problem – how to remove the water from their ever-deepening levels. Newcomen determined to find a solution, but it took him 10 years of experiment before he arrived at a satisfactory answer. And even then his engine was unacceptable to the Cornish miners because it had a voracious appetite for coal, which they had to import by sea at great cost. It was the mine-owners of the Midlands who, faced with similar water problems, became the first users of Newcomen's steam engine.

The vertical cylinder of the engine was open at the top and the piston-rod was linked by chains to one end of a pivoted wooden beam. The pump-rod was connected by chains to the other end of the same beam. As the pump-rod descended, its weight drew the piston up the cylinder. At the same time a valve opened to fill the cylinder, below the piston, with steam from the boiler. When the piston reached the top, the steam supply was cut off and a second valve opened to admit a jet of cold water. Condensing the steam created a vacuum in the cylinder and caused the piston to be driven down again by the atmospheric pressure on its upper surface. This power stroke raised the pump-rods and so lifted water out of the mine.

Strictly speaking, Newcomen's invention was an 'atmospheric engine', using steam merely as a convenient way of forming a vacuum. Nevertheless, he was the first man to harness power successfully by means of a piston in a cylinder, and his engine of 1712 was the undoubted parent of all subsequent machines of this type. And it played a vital part in Britain's Industrial Revolution by enabling more coal to be mined from deeper levels.

James Watt did not 'invent' the steam engine as is commonly supposed. What he achieved was a great improvement in efficiency by condensing the steam in a separate closed vessel instead of the cylinder itself. He also closed the top of the cylinder and used low-pressure steam instead of cold air to drive the piston down. This improved engine of Watt's still depended on the creation of a vacuum and used steam at extremely low pressure. It could not be otherwise when boilers were little better than brewers' coppers, unable to withstand high pressures.

The Cornishman Richard Trevithick was the first to harness 'strong steam', as it was then called. Yet the Cornish beam engine, which was the outcome of this third stage of development, still closely resembled Newcomen's engine.

L. T. C. R.

Battery

In 1801 Alessandro Volta was summoned from Italy by Napoleon to give a demonstration of his invention of a year or two earlier, now the talk of scientific Europe: the electric battery. Napoleon's summons showed that he had an instinct for the scientific as well as the military breakthrough, for the invention of the battery, the first continuous source of electric current, was the take-off point for electrical research.

Volta's discovery followed from a chance observation by a fellow countryman, Luigi Galvani, a Bologna anatomist, that dissected frogs' legs twitched when touched with certain metals. After a series of experiments Galvani concluded, wrongly, that there was a source of electricity in the muscle. Volta came to the right conclusion: electricity was produced when two different metals touched the leg simultaneously, and it came from the metals, not the frog.

By trial and error Volta found copper and zinc to be a good combination, and his experiments led eventually to the famous Voltaic pile, a stack of discs, alternately copper and zinc, interleaved with cardboard or cloth moistened with brine. It was an easily reproducible source of continuous current.

Voltaic piles were soon in use all over Europe, but it was almost 20 years before the discovery that an electric current flowing in a wire would deflect a compass needle demonstrated the fundamental relationship between electricity and magnetism and raised the possibility of producing continuous mechanical motion from electricity.

B. S.

Electric motor

On 22 September 1791 Michael Faraday, the son of a blacksmith, was born in rooms over a mews coach-house near

Man Taking

Manchester Square in the West End of London. Through his personal character and insistence he secured, at the age of 21 a junior appointment in the laboratories of the Royal Institution in nearby Albemarle Street.

One discovery that interested him was that of the Danish Prof. H.C. Oersted who noticed by accident that a pivoted magnetic needle was deflected when a wire carrying a current was brought near to it. Faraday gave much thought to this effect and became convinced that there was a clear connection between electricity and magnetism. After many experiments, he produced his 'electrical rotations' which we see today as the earliest electric motor.

In the most important of his various arrangements he suspended a 6 in. length of copper wire from a hook with its lower end dipping into a bowl of mercury in the centre of which was a bar magnet fixed vertically. When he passed a continuous current from his battery through the hook down through the wire to the mercury, the wire began to rotate and continued to do so as long as the current flowed. The year was 1821 and the electric motor had been invented.

P.D.

Matches

Attempts to produce a match began in 1680, soon after the 'discovery' of phosphorus by Robert Boyle, but the match as we know it did not appear until 1834, when 'Congreves' were invented, and named after the rocket expert. Boyle's assistant, Godfrey Haukewitz, got very near the answer, using bits of wood dipped in sulphur which, with the aid of phosphorus and friction, produced a light. But this device was smelly, dangerous (phosphorus being highly poisonous), inconvenient and expensive.

At the beginning of the 19th century, the normal method of striking a light was to use a flint and steel to ignite tinder. This was then used to light a wick. The idea of substituting small pieces of wood (known as splints) dipped in sulphur emerged about 1800, and it was not long before the sulphur was being mixed with chlorate of potash and sugar to improve inflammability.

The idea was first used by one Captain Manby to fire rockets of life-saving apparatus, and was converted to a domestic form about 1830 by a chemist named Jones, whose shop was in the Strand, in London. His 'Promethean matches' were made of a roll of paper, which had some of the above mixture in one end with a small hermetically sealed tube containing a minute amount of strong sulphuric acid. By crushing the end of the tube with a small pair of pliers, sold for the purpose, the sulphuric acid was made to come into contact with the mixture and ignite.

In 1826 the first friction matches appeared, invented by John Walker of Stockton-on-Tees; he never patented the invention, however, although he was urged to do so by Faraday. They were known as 'lucifers' and were coated with sulphide of antimony and chloride of potash, made into a paste with gum and water. They were lit by being drawn between the two surfaces of a folded piece of sandpaper. The final step was to substitute phosphorus for the antimony, thus creating the 'Congreve'. These, at a shilling a box, were relatively cheap, and rapidly became popular in several countries, the price falling to a halfpenny a box.

'Safety matches' based on the less dangerous amorphous phosphorus were introduced in Sweden in 1852. Eventually, after 1900, the use of yellow phosphorus was banned in a number of countries.

Apart from the 'mainstream' development of the match there was also the 'light-syringe', a squirt which ignited tinder with the heat developed in compressing air; Volta's 'instantaneous light', which used electricity and hydrogen; and Debereiner's hydrogen lamp, of 1823.

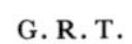

G.R.T.

Dynamo

It was 10 years after his invention of the electric motor in 1821 that Faraday devised the dynamo. Instead of producing mechanical motion from electromagnetism, he generated an electric current through the application of mechanical power. During the 10-year period he had made many attempts to produce a current with magnets and lengths of conducting wire, but, as his diary records, his time was much occupied by his chemical experiments. It was only at intervals that he was able to concentrate on electromagnetism, and his first experiments failed because he did not appreciate the difference between starting with a mechanical motion and starting with an electric current. He came very close to success in 1824 and 1825, but it was not until 17 October 1831 that, using a very simple piece of equipment, he succeeded in generating the current mechanically. The first electromagnetic current generator consisted only of a cylindrical coil (solenoid) and a bar magnet which could be slipped into the coil by hand. Faraday's original bar and coil are today exhibited among other treasured relics at the Royal Institution, London.

Faraday had at first thought that there would be a current while the magnet lay still within the coil, but he soon discovered that while his galvanometer registered a current when the magnet was being inserted there was no current at all when it was stationary. Further, when he withdrew the magnet,

John Walker (1781–1859), the first man to put a 'friction light' (above) *on the market.*

Tiger, Pearl, Runaway and Elephant – four of the many brands shown in Bryant & May's diary and almanac of 1874.

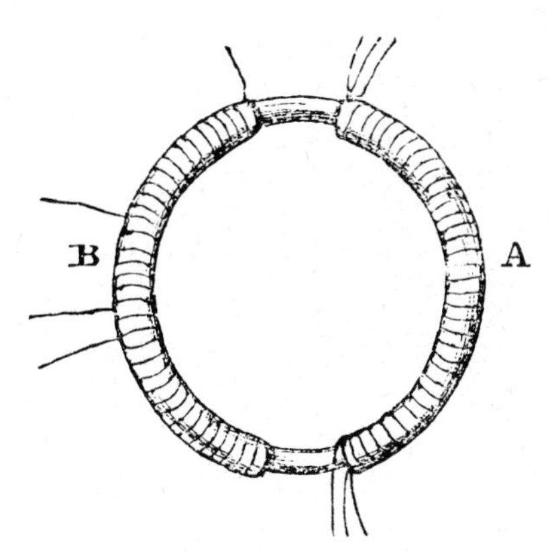

Faraday's first transformer – two coils of wire, A and B, wound round an iron ring.

Michael Faraday, the greatest name in electricity. He invented the electric motor in 1821, the dynamo and transformer in 1831.

Horizontally operated water turbine designed by James Thomson in 1889.

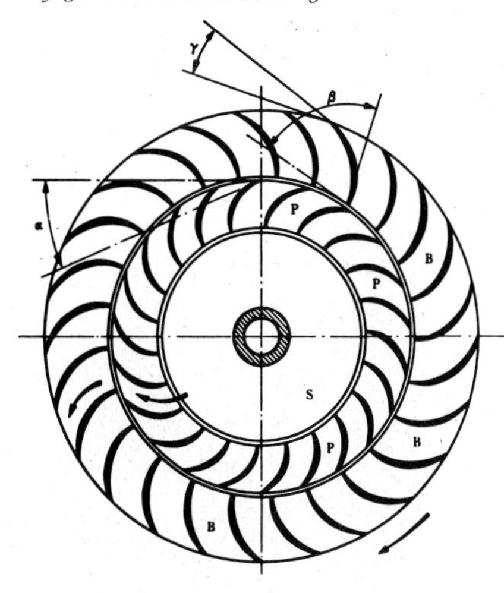

the current started again in the opposite direction.

Faraday had thus produced an alternating current, but it was his ambition to generate a continuous current similar to that produced by a battery. He quickly drew a further important conclusion from his results: if relative movement of a magnet and a conductor produced a current, then a current should also be produced by moving a conducting plate in a magnetic field. This led him to construct what we now see was the first continuous current machine, the dynamo. He set up a copper disc on a spindle, with a handle by which it could be rotated between the poles of a powerful magnet. The result was exactly as he had predicted: two wires taken to a galvanometer – one rubbing on the spindle and one on the rim of the disc – produced a steady deflection, showing a continuous, direct current.

The generation of eddy currents in the copper disc made this first dynamo very inefficient. But a year later Hypolite Pixii of Paris fixed a pair of wire bobbins from a wooden gantry and rotated a horseshoe magnet below them. This invention started a long train of further ideas using coils in various forms which have led up to the present highly efficient dynamo.

P. D.

Transformer

Electricity is a marvellously flexible form of energy. It can be generated wherever it is convenient, and transmitted to the exact point of use – a motor on a work-bench, a cooking-stove or a radio loudspeaker. It is very rarely convenient to transmit it at the voltage at which it emerges from the generator, or to use it at the transmitted voltage, and the usual device for altering the voltage is a transformer.

It was invented by Michael Faraday in 1831, his most productive year, although he did not see it as a way of altering voltages. He found that if the current in a wire wrapped around an iron rod were interrupted, it would generate a current in a second wire wrapped around the rod. To him the excitement of the discovery lay in the fact that electrical energy was transferred between two circuits.

The transformer is particularly valuable for power transmission, but needs an alternating current – one that regularly reverses its direction. Somewhat surprisingly, the use of alternating current was at first opposed as dangerous by, for example, Edison who pointed out, accurately but irrelevantly, that it was used in the electric chair.

T. O.

Water turbine

Benoît Fourneyron took his interest in water turbines from Claude Burdin, his professor at the School of Mines at St Etienne. He started work on it in 1823 and in 1824 Burdin suggested the word 'turbine' from the Latin *turbo*, whirlpool. Fourneyron got it to work in 1827. Such a machine was badly needed, for the steam engine was satisfactory only where coal was cheap, and the waterwheel was cumbersome, inefficient, and not sufficiently powerful for the growing needs of industry. A prize of 6,000 francs was offered by the Society for the Encouragement of Industry in France to the first builder of a successful water turbine. There were four competitors, and in 1833 the prize was awarded to Fourneyron, who had entered a 50 h.p. machine.

Fourneyron's machine was an outward-flow reaction turbine. The rotor had 30 curved vanes to which water was admitted from the inside, and revolved at the then incredible speed of 2,300 revolutions per minute. Following its success, more than a hundred machines of different sizes were built, and one of the largest (800 h.p.) was installed in the Paris Waterworks at Pont Neuf in 1855.

Fourneyron showed his brilliance at a very early age. His college entrance examination papers were so remarkable that he was awarded a special dispensation to enter at the age of 15. Seven years later his first industrial task – the erection of a rolling-mill – was so successful that his salary was immediately doubled. He was not only a professional engineer and designer, but also a metallurgist and a mechanic clever enough to make a success of Burdin's original idea. He found out how to support and lubricate the bearings of the annular rotor, spinning at high speed while surrounded by the stationary blades in the casing. He was secretive about his designs, especially about the curvature of the blades.

Fourneyron's turbines were very efficient. One made in 1837 had an efficiency of more than 80 per cent, nearly four times that of the undershot waterwheel.

A. BU.

Hydro-electricity

If we disregard the beaver, which has been practising the art of building dams for millions of years, the history of hydro-engineering may be said to date from the Third or Fourth Egyptian Dynasty, some 5,000 years ago. That, at least, is the age of the oldest dam we know of, the Sadd el-Kafara or 'Dam of the Heathens', discovered by the German explorer Georg Schweinfurth in 1885. It was quite a formidable structure, nearly 350 feet long, straight across the Garawi river valley, and it created an artificial lake for the irrigation of a large agricultural area, as well as a constant freshwater supply for its people. All over the Middle East, in China and India dams were being built for these purposes in later centuries, though the idea of power from water did not occur to those ancient civilizations which always had plenty of energy at their disposal – that of the slaves.

It was only in Roman times that the waterwheel was developed as a prime mover. Roman engineers began to build watermills; large wheels with many small paddles or shovel-shaped blades fixed to their outer rims were being turned by the force of a flowing river. Later, waterfalls were also harnessed, and occasionally dams were built to channel the power of the falling water so as to increase its impact on the mill wheels. Some waterwheels were fitted with buckets instead of paddles or blades, in order to raise water from the river for irrigation. Around AD 200 the Romans built what may be called the earliest power station: a combination of 16 waterwheels near Arles in southern France. The wheels turned 32 millstones, which produced 30 tons of flour a day.

The Moslems, too, were ingenious technicians. Throughout their Empire, which at its height reached from India to the Pyrenees, they made great use of the waterwheel while Central Europe and England forgot all about that prime mover when the Romans had gone. It returned gradually only after the 9th century and was eventually used universally, providing power for flour-mills, saw-mills, fulling-mills, sledge-hammers, foundries, and pumps in the mines. The Domesday Book (1086) lists 5,624 water-mills in England, many of them operating in connection with small dams; the monasteries often provided the incentive and the know-how for the construction work.

The Industrial Revolution did not begin with the steam-engine: it began with the water-mill, and as late as in the 1830s about one-third of England's cotton-mills were still working with water power. Of course, that power – like its aerial counterpart, the power of the wind – was tied to the spot where it was available; there was no way of 'sending' energy where it was needed, except the latent energy of coal after the perfection of the steam-engine. The great breakthrough in energy utilization came only towards the end of the last century with the development of the electric generator, the 'dynamo'; now electricity could be produced at one point – for instance where water power was available – and distributed to many others for lighting, heating, various industrial purposes and, most important, for driving the electric motors in factories and trains.

Basically, water power for generating electric current is 'free' – the rain and the melting snow cost nothing. But that is not really true. In fact, a hydro-electric power station needs more expensive equipment than a coal- or oil-fired one (though it is still cheaper than a nuclear power station). Even the water is not just a gift of Nature. As a rule, dams have to be built and reservoirs constructed to make the best use of the available water power, and to ensure that the supply of water is constant and does not dry up. Waterfalls cannot be utilized in their natural state; they, too, have to be harnessed and conducted into the power station.

The water turbine (q.v.) took decades to develop. Back in 1827, long before the generation of electricity from mechanical power was even thought of, a young Frenchman, Benoît Fourneyron, won the first prize in a competition for the most effective design of a water turbine; his solution was to use two wheels, one inside the other, the inner one being the power wheel. The turbines which are still in use today were all designed in the last two decades of the 19th century. There are three types: the *Pelton* wheel, or 'impulse turbine', in which two high-pressure water jets are directed against the rim of a wheel fitted with large, cup-shaped paddles – the most efficient type where the head of water is at least 500 feet above the turbine, i.e. where waterfalls or high-level storage lakes are available (France, Switzerland, Norway, California) so that high water pressure is created by the altitude; the *Francis* 'reaction turbine' for falls of water at medium pressure in which the water is directed, by means of an outside casing with guide vanes, simultaneously against all 16 blades of the turbine; and the *Kaplan* turbine, which works best with low-pressure water. It looks like a ship's propeller and has four very large blades, whose pitch can be altered during the actual running so as to get the best results from varying water pressures. Thus there is a turbine type available for every kind of water supply.

Two of the world's first hydro-electric power stations, both begun in 1891, were as different as can be. One was situated under a small waterfall in the river Neckar and equipped with a 16,000-volt turbo-generator (designed by C. E. L. Brown and built by Oerlikon) to supply Frankfurt, 100 miles away, with electricity; the other one at Niagara Falls, was put into operation a few years later with an output of 5,000 h.p. (today, over 8 million h.p.). It was among the first stations to use alternating current, although the sceptics predicted that this would never work; the power was generated at high voltage, and transmitted by overhead cables to the surrounding towns, especially Buffalo, where it was 'stepped down' to lower voltages for industrial use and domestic appliances.

Direct current (DC) travels only in one direction through the wires, from the generator to the user and back again. Consequently, it must be produced at the same voltage at which it is used. But in transmitting electric current over long distances while keeping the loss of energy to a minimum, voltage must be as high as possible. Therefore another form of current has to be used, alternating current (AC), in conjunction with transformers which step the voltage up or down without affecting its frequency, the number of vibrations per second.

Alternating current, after reaching a maximum in one direction, decreases quickly, then reverses and reaches a maximum in the other direction; by means of secondary coils it can be stepped up or down. Thus, a turbo-generator may produce current at, say, 11,000 volts, which the step-up transformer increases to 22,000 volts. This current is sent by overhead cables to users in a wide radius, and sub-stations at the population centres step the voltage down to 110 or 250 volts so that the current can be used safely in the homes and factories.

Nikola Tesla, born in 1856 as the son of a clergyman in Croatia (now part of Yugoslavia), was an undergraduate at the University of Graz in Austria when he saw the first Gramme generator. His brilliant technical intuition told him that this type of dynamo and the direct current it produced would never fulfil the great variety of tasks which he saw lying ahead in electrical engineering. Most electrical engineers of the time thought that high-tension AC was too complicated and potentially dangerous, but Tesla was convinced that it was practicable. In the early 1880s Tesla emigrated to America – arriving, so the story goes, with no more than a few coins in his pocket – went straight to Edison, and was taken on as his assistant. A few years later he was able to take out his first patents for high-frequency AC generators and transformers.

Tesla's ideas were the basis of the Niagara Falls hydro-electric station; they had been accepted as sound by George Westinghouse who designed it – the former railway engineer who had invented the compressed-air brake as a young man and who had devoted much of his research work to the problems of electric-power distribution from 1884.

It was not all glory and success in the early days of the hydro-electric power stations. In 1893, for instance, the Colorado river was dammed near Austin, Texas; the dam was 65 ft high and nearly 1,300 ft long, built of masonry with a concrete core. The station had an initial output of 15,000 h.p., three times as high as the Niagara station. In the early spring of 1900, heavy rain made the Colorado rise rapidly, and soon a flood of water 11 ft high was rushing over the dam. The whole central section gave way and crashed downstream. The accident created a sensation all over America,

Turbine in the first Niagara power station. A contemporary engraving gives a vivid impression of the force involved.

Above, *the hydro-electric power station at Maentwrog, North Wales. Overhead cables link it in a symbiotic relationship with the nearby nuclear power station at Trawsfynydd* (below).

The Aswan Dam, Egypt, under construction.

apparently confirming the doubts which many people had about electrical power stations in general.

As early as 1902, the first dam was built across the Nile in order to wrest large areas from the desert and turn them into fertile land by irrigation. The point chosen was near the town of Aswan in upper Egypt. After the second World War, a new Aswan dam scheme with an electric power station generating 10,000 million kW hours a year was begun, but in the summer of 1956 the USA and Britain, which had promised to back the scheme financially, both withdrew from it, and Egypt reacted by nationalizing the Suez Canal. The consequence was the Suez crisis, Anglo-French intervention, and hostilities between Egypt and Israel. The Soviet Union stepped in and undertook to finance the major part of the Aswan scheme.

Construction of the new Aswan High Dam began in 1960, and by 1964 the diversion of the Nile was completed. 'Lake Nasser', an enormous reservoir, began to take shape; here, the problem was that of rescuing the unique ancient temples and monuments of Abu Simbel. This was done with a great deal of international help, but further south, in the Sudan, thousands of people had to be resettled because their livelihood was threatened by the engineers' interference with the Nile.

Another gigantic scheme, the hydro-electric station 'Dnieproges' on the Dnieper in the Ukraine, then the largest in Europe, was completed in 1932. The dam was destroyed during the second World War in the face of the German advance, but completely restored by 1950, with a power capacity of over 650,000 kW. Smaller hydro-electric power stations have since then been added to the Ukrainian network.

Water power derived from the rise and fall of the tides is an old technical idea. Many experimental schemes have been tried out in various countries, but only one seems to have been worth the effort – that on the Rance estuary, based on designs by Georges Claude. Situated between St Malo and Dinard in Brittany, it was completed, and ceremonially opened by President de Gaulle, in 1966. This tidal generating scheme has a capacity of 85,000 h.p.; the turbines work on the rising as well as on the falling tides. During the daily periods when the demand for current is low, the generators work as motors pumping water into the tidal dam, where it is stored until the demand rises. Rance is a combination of dam, tunnel, lock and a system of turbines coupled to AC generators; the barrage is 700 yards long. The problem of corrosion has been tackled by using stainless steel and aluminium bronze for the turbines. The difference between low and high water levels can be up to 44 feet; the maximum flow of water is half a million cubic feet per second. Still, the expensive and sophisticated installation of Rance supplies only enough electricity for the needs of an industrial town of 250,000 inhabitants – half a million megawatts per year.

One of the essential features of Rance, the pumping back of water during hours of slack demand, has also been incorporated in a nuclear power station in Britain, Trawsfynydd in North Wales. It has been built on the shore of an artificial lake, created in the 1920s as a storage reservoir for the Maentwrog hydro-electric power station, with which the nuclear power station at Trawsfynydd is linked by overhead cable. What happens is that during off-peak periods the nuclear station uses its power to pump some of the water back into the reservoir, thus providing additional power for Maentwrog. In a country with limited hydro-electric potential such as Britain, this symbiosis of two different kinds of power station is a sound technical concept.

E. L.

Paraffin/Kerosene

The credit for the discovery of this particular hydrocarbon cannot be given to one man alone, for several scientists were working on the problem at about the same time: Carl Wilhelm Fuchs had identified solid hydrocarbons in oil from the Tegernsee (Germany) in 1809, and the German chemist Eduard Buchner had separated them by 1819, in a high state of purity. In 1830, Auguste Laurent obtained the substance by distillation of bituminous schist, but it was the German scientist Carl Reichenbach, in an article published in the same year, who coined a name for a product with such 'few and weak affinities' (*parum affinis*), paraffin; he himself had obtained it from wood-tar. Dr Christison, of Edinburgh, about the same date, suggested it be called *petrolin*, but this received little support.

Until the late 1830s, no one seems to have thought of a use for paraffin, but in 1839 Ure wrote: 'Paraffine is a . . . solid bicarburet of hydrogen; it has not hitherto been applied to any use, but it would form admirable candles.' This was not, of course, a practicable procedure until the manufacture of paraffin could be placed on a commercial basis; this was achieved by Dr James Young's patent of 1850, 'for obtaining paraffine oil, or an oil containing paraffine, and paraffine from bituminous coals'. The process entailed distillation: broken into small pieces, placed in retorts, and heated, some coals would give as much as 120 gallons of crude oil per ton, to be separated into naphtha, domestic oils, lubricating oils, heavy fuel oil and paraffin wax. Oil-shales, first worked by Robert Bell in 1859 at Broxburn in Scotland, yielded much less. Both methods became unprofitable when the

Man Taking

Lenoir's coal-gas-fired car of 1860 – the first internal combustion engine that worked.

American crude petroleum industry developed in the later 19th century.

From about 1854, paraffin wax was indeed used as a constituent in candles, supplementing spermaceti (obtained from the sperm whale, and in use since the middle of the previous century). In 1861, the *Annual Register* noted that 'There has lately been introduced, for the purposes of light, an oil called "paraffin"'; by 1874, the use of paraffin lamps was widespread, as a cheaper and safer substitute for candles.

C. M. B. G.

Internal combustion engine

In 1860 Etienne Lenoir, a self-taught mechanic, revealed to the world the first internal combustion engine that worked. For several decades, engineers had realized both the low efficiency of the steam engine and the desirability of a motor which would burn its fuel inside the cylinder instead of using it to produce steam as an intermediary.

Inventors were tantalized with the prospect of an engine that would do away with wasteful furnace, boiler, condenser and all the valves, pipes and heat loss they involved. Many were designed but Lenoir's was the first to pass the experimental stage. It worked with illuminating gas – derived from coal – which was mixed with air and drawn into a cylinder by withdrawal of a piston: at midpoint the mixture was ignited by an electric spark, so that only the second half of each stroke was powered.

However, Lenoir's engine was double-acting, so fuel entered either side of the piston in turn. In short, the design kept closely to the traditions of the steam engine. But if this new motor worked, it did so very inefficiently, consumed great quantities of its expensive fuel, and, at about 100 r.p.m., it could only deliver something over 1 h.p. It was also very rough-running, producing violent shocks at each explosion even though Lenoir tried to tame it by the use of springs and other devices to absorb the shocks.

In 1862 a railway engineer named Alphonse Beau de Rochas published a pamphlet about improvements in locomotive design, in which he suggested compounding steam engines with gas engines. The gas in the engine should, he said, spontaneously ignite under high compression, which was to be done by making it work in four stages: '1, intake during one whole stroke of the piston. 2, compression during the following stroke. 3, firing at the dead point and expansion during the third stroke. 4, expulsion of the burnt gases from the cylinder at the fourth and last return.' Thus he had grasped the principle of the four-stroke engine. Yet he himself never tried to construct an engine on these lines, nor to publicize it.

Quite independently of Beau de Rochas, Lenoir's invention was taken up by Nikolaus August Otto, a German travelling salesman even more self-taught as a technician than Lenoir himself. As Otto saw it, the problem was to control the richness of the gas-air mixture so that the engine would run smoothly and efficiently. By the mid-1860s, he had evolved an engine in which the expulsion drove the piston up in a vertical cylinder, where the contraction of the spent fuel as it cooled produced a vacuum into which atmospheric pressure and gravity forced the piston back – thus reverting to the technique of the low-pressure engines of the 18th century. The result was still noisy and wasteful, but a marked improvement on Lenoir.

One day in 1875 he was watching smoke emerge from a chimney and his imagination was caught by the way that it came out thick and then was gradually dissipated in the air. In the same way he supposed it would be possible to have a rich mixture of fuel at the point of ignition, which would yet be cushioned from the piston by a much thinner layer pervading the inert air next to it. This notion of stratification was almost certainly a red herring. But to produce it, Otto reinvented the four-stroke cycle, just as de Rochas had envisaged it (but henceforward called the 'Otto cycle'), and embodied it in his 'Silent Otto' of 1876, which was an immense success, giving 3 h.p. at 180 r.p.m. – though it still kept the dangerous flame ignition (where Lenoir had used a spark) and illuminating gas (although the Austrian inventor Siegfried Marcus in 1867 had invented a carburettor to convert liquid petroleum into a flammable gas).

A. K.

Dynamite

Dynamite was no chance discovery but the result of a logical search by an able Swedish technologist. Alfred Nobel (1833–96) was the son of Immanuel Nobel, the inventor of plywood. Immanuel was particularly interested in the use of gunpowder, then the only practicable explosive known, and he passed on this interest to his son.

One very powerful explosive, nitroglycerine, had been discovered by the Italian organic chemist Ascanio Sobrero in 1846, but it was so sensitive and difficult to control that it was useless. In 1865, Alfred Nobel made his most important invention, a detonating cap incorporating mercury fulminate with which nitroglycerine could be detonated positively and at will. This detonating cap is the basis of all subsequent explosives technology, but although it made possible the use of nitroglycerine on a large scale, it did nothing to make this devilish oil any easier to transport or handle.

Parsons's Turbinia*, built in 1884, steaming at an incredible $34\frac{1}{2}$ knots.*

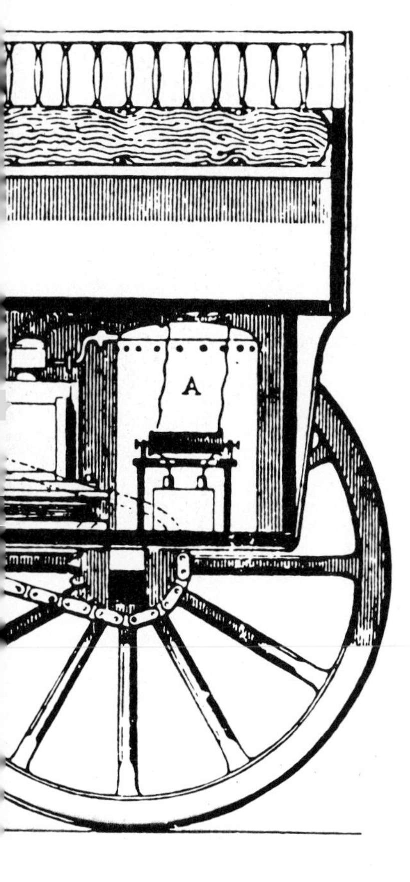

Alfred Nobel of Sweden: peace prizes from the man who invented dynamite.

Below, *an 1897 advertisement for Rudolf Diesel's engine.*

Nobel went into the explosives business in a big way, but it was a precarious way of life, partly because of the murderous risks run by all his customers and partly because of the opposition of well-entrenched gunpowder firms like Du Pont.

All the same, Nobel saw that his future lay with nitroglycerine and sought ways of taming it. An obvious solution was to turn it into a manageable solid by soaking it up in something porous. He tried all manner of porous, non-explosive inert substances: paper, wood waste, brick dust, dry clay, until (in 1866–67) he hit on an unusual though by no means rare mineral, kieselguhr. The dough-like explosive, although 25 per cent less powerful than pure nitroglycerine, was still far more powerful than gunpowder, and was easily manageable.

The story got about that Nobel had hit upon his invention when some nitroglycerine accidentally leaked into kieselguhr used in packing some containers. This story irritated him, and he very firmly stated that his discovery was no accident, but the result of a systematic search based on an appreciation of the technical problems.

Nobel knew his own merits, and recognized those of others. He gave secure employment to Sobrero, who had discovered the essential ingredient in the new explosive. Nobel realized that although this ingredient was well known, his preparation was the starting-point of a new epoch and therefore deserved a new name: he christened it 'dynamite'.

Nobel later invented gelignite (a mixture of nitroglycerine and guncotton or nitrocellulose) which was more powerful and easier to use even than dynamite. He made a great fortune and bequeathed it to a foundation for awarding prizes for contributions to science, literature and the promotion of peace. The committee has never lacked for candidates for the first two. It would have saddened Nobel to see how often they have sought in vain to award the third.

F. G.

Steam turbine

The steam turbine is an example of a very ancient idea which was made practicable only by an advance in technology – in this case the widespread introduction of steel and its alloys in the second half of the 19th century.

To convert the linear motion of a piston in a cylinder into rotary motion by means of a connecting-rod and crankshaft seemed a roundabout way of doing things. It would be much simpler, surely, to direct a jet of steam on to the vanes of a wheel, like water against a waterwheel. This, the principle of the impulse turbine, was suggested as long ago as 1629 by an Italian, Giovanni Branca. The principle of the reaction turbine, that of turning a wheel by jets of steam emerging from its rim like sparks from a Catherine wheel, is even older. It was suggested by Hero of Alexandria in the 2nd century AD. Yet ever since the beginning of the Industrial Revolution many inventors, including such men as Watt and Trevithick, had experimented – unsuccessfully – with the 'steam wheels' or 'whirling engines'.

For there was a big snag to this deceptively simple idea. Before it could absorb even a useful fraction of the energy of the steam, the wheel would have to revolve with unheard-of velocity. If it failed to do so, the device was woefully inefficient compared with a piston engine. It was the Hon. Charles Parsons, sixth and youngest son of the Earl of Rosse, who solved this knotty problem, early in his engineering career at Newcastle. He did so by placing a series of vaned wheels on a single shaft and making the steam pass from one to another, the wheels increasing in diameter as the pressure of the steam dropped. In this way, each wheel absorbed part of the steam's energy. Even so, the speed was high. Parsons's first turbine, patented in 1884, revolved at 18,000 r.p.m. At no earlier date could technology have coped with the demands made by such an invention.

The turbine was subsequently applied to marine propulsion and the Cunarders *Carmania, Lusitania* and *Mauritania* were the first large liners to be propelled in this way. But Parsons's original idea was to use his invention for generating electricity and he also patented a design for a dynamo capable of being driven at high speed.

Subsequently, rival turbines appeared on the Continent and in America, but Parsons was undoubtedly the pioneer. With the possible exception of Newcomen's steam engine, few inventions have emerged so fully fledged from the brain of one man and few have had social consequences more far-reaching. Nowadays we refer disparagingly to 'the steam age' as if it were a thing of the past, ignoring the fact that an overwhelming proportion of the electricity we use is still generated by steam turbines. These machines are the true descendants of the engine Parsons fathered 86 years ago.

L. T. C. R.

Diesel engine

'Find out more about this!' wrote the 20-year-old student Rudolf Diesel in his notebook during one of his professor's lectures at the Munich Technical College in 1878. The professor was talking about the poor thermal efficiency of the steam engine, which can turn only 6–12 per cent of the latent fuel heat into power; he had also explained the theory of the French physicist Nicolas Carnot, the 'father' of thermodynamics, who had stated around 1830 that he could visualize an ideal heat engine which would convert nearly all the latent energy into power. Only an *internal* combustion engine – where the fuel is burnt inside the cylinder – would be able to approach that ideal system, said the professor. 'That idea kept following me,' recalled Diesel. 'I used every moment I could spare to enlarge my knowledge of thermodynamics.'

A Cologne engineer, Nikolaus Otto, had already invented an efficient internal combustion engine in 1872, but it ran on town gas and was therefore stationary. However, it was from Otto's engine with its four 'strokes' that two other German engineers, Gottlieb Daimler at Cannstatt and Karl Benz at Mannheim, developed independently of each other the petrol engine for the automobile and the motor-cycle, which need an ignition system to make the fuel burn inside the cylinder to move the piston.

Diesel worked on somewhat different lines. He aimed at keeping the temperature and pressure in the cylinder fairly constant during combustion so that much more of the heat thus created would turn into power. It took him 14 years before he could write a slim booklet on his engine – which existed only on paper – and take out a patent. Big German engineering firms, including Krupps, enabled him to build his first model in 1893. In the diesel engine, which does not need refined petrol but works with cheaper heavy oil, the heat of the piston-compressed air is increased to such a degree that it ignites the fuel without a special ignition system; the liquid fuel enters the cylinder gradually, so that temperature and pressure are maintained in it throughout the 'power stroke' of the piston. Thus the diesel engine converts 35 per cent of the latent fuel energy into power, compared with 28 per cent in the most efficient petrol engine. Its disadvantages are that it is heavier and noisier and that heavy-oil exhaust gases are a great nuisance.

Rudolf Diesel lived to see no more than the beginning of the enormous success of his engine, which today powers lorries, buses, taxis, small ships and power stations, and – mainly in the form of the diesel-electric system (in which the engine produces current for electric motors) – also railway locomotives. Diesel disappeared without trace from a Channel steamer in 1913, driven to suicide by his desperate financial situation.

E. L.

Electronics

During the last decades of the 19th century, when electricity began to play an important part in modern civilization, scientists everywhere tried to answer the question of what it actually was, how it travelled, from where it derived its power. One of the essential instruments used for research in this field was the Crookes tube, so called after its inventor, the London physicist and chemist Sir William Crookes (1832–1919), but now better known as the cathode-ray tube. This was an oblong glass vessel from which the air had been pumped out, with two metal plates called electrodes in it, one at each end. When these were connected to a battery there was an invisible discharge of electricity flowing from the negative electrode, the cathode, to the positive one, the anode. All that one could see was the glow of the heated cathode while the discharge took place.

It was with the Crookes tube that Röntgen discovered X-rays (q.v.) in his laboratory at Würzburg, Bavaria, in 1895. These X-rays, as we now know, are in fact electromagnetic waves of a much shorter wavelength than visible light, produced when the cathode rays strike a material object – in this case a metal shield in the tube. But what those cathode rays actually were remained an unsolved riddle, though only for a year and a half.

At the Cavendish Laboratory in Cambridge, Professor Joseph John Thomson (1856–1940) and his young New Zealand-born assistant Ernest Rutherford (1871–1937) repeated Röntgen's experiments, mainly in order to establish the nature of the cathode rays. They used various types of cathode-ray tubes, the largest being a Braun tube. (Professor Ferdinand Braun, an Austrian physicist, had invented a tube in which the inside of the end wall was coated with a substance which fluoresced under the impact of the rays – the grandfather of our television tube.) In one of the most important experiments in the whole history of physics, Thomson, having outlined the path of the rays more clearly by interposing a barrier with a small hole in it between the terminal and the fluorescent end of the tube, showed that a magnet applied at right angles to the beam deflected the beam as a whole. Thomson came to the conclusion, which he reported at a meeting of the Cavendish Society, that 'cathode rays are particles of negative electricity'. And he named them 'electrons'.

That was in the summer of 1897, and from that discovery sprang the development not only of electronics but also of atomic research. The former found its most important implement in the triode valve (q.v.), invented in the first decade of the 20th century; the latter came of age one day in 1911 when Rutherford, then working at Manchester, and his Danish assistant Niels Bohr presented their picture of the atom, with its mass concentrated in the nucleus, which has a positive electric charge, surrounded – like a miniature sun with its planets – by electrons revolving in various orbits round it. Thus the positive and negative

William Crookes demonstrates the deflection of cathode rays by a magnet.

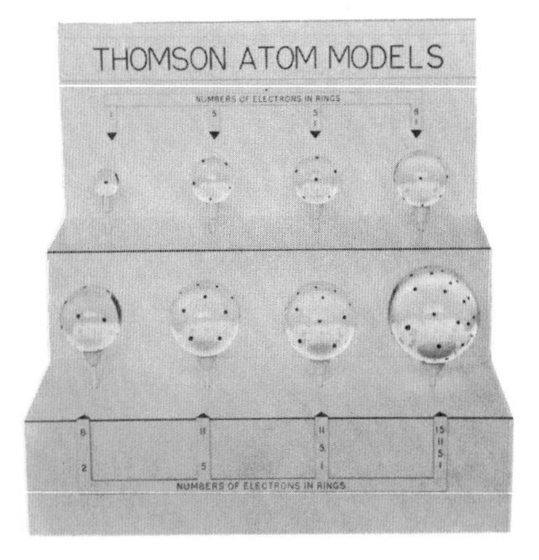

An early visualization (by J. J. Thomson) of the structure of the atom, with the electrons embedded in it like currants in a bun.

Three common encapsulations of integrated circuits (shown much enlarged). The top device is shown before encapsulation.

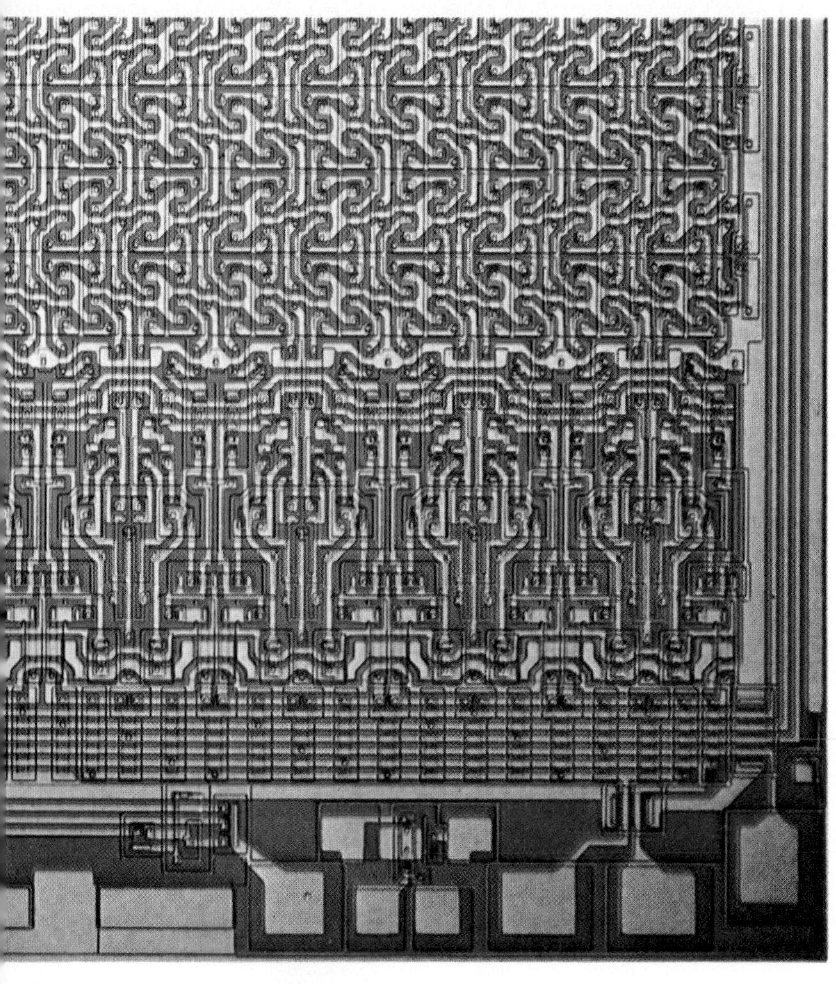

Part (about a quarter) of a memory circuit containing 1,244 transistors and about 1,200 other devices. The whole circuit occupies about the same area as the word 'the' in this caption.

charges cancel each other out in the atom, which is electrically neutral. 'A beautiful model of the electronic structure of the atom', a fellow scientist called the Rutherford-Bohr concept.

Just as electrical engineering had made great strides even before the nature of electricity was explained, so electronics had begun to develop before the term was coined. In 1873 the first head of the Cavendish Laboratory, the Scotsman James Clerk Maxwell (1831–79), presented his theory that light waves were in fact waves of electric and magnetic forces. Eight years after his death the German physicist Heinrich Hertz (1857–94) demonstrated the existence of electromagnetic waves, by transmitting them across his laboratory without any wire connection, and radio (q.v.) was born, with Guglielmo Marconi as the midwife.

The miracle of wireless telegraphy had hardly become reality when the general public and the technicians began to ask when the wireless waves would be able to carry not only Morse signals but also voices and music. But this was rather a complicated matter as it required the transmission and reception of finely modulated electrical impulses. These impulses, coming from the microphone, had to be superimposed on a powerful electromagnetic 'carrier wave', and then changed back into sound in the receiver. The triode valve made this possible, but the triode itself was based on an earlier invention, Ambrose Fleming's 'thermionic' valve. From the thermionic valve and the triode began the stupendous development – still continuing – of electronic communication, details of which can be found on other pages, in articles on radio, television and radar.

In 1924, the French physicist Louis de Broglie put forward the paradoxical notion that the electron was both a particle and a wave. In its wave manifestation it has a wavelength much shorter than that of visible light, and so an electron beam can be used to resolve much finer detail than can be revealed by visible light. This is the basis of the electron microscope (q.v.), which has made a whole new world of almost infinitely small things visible, showing detail as small as 5–10 atomic diameters.

The technical trend in electronics is towards smaller and smaller equipment. Even transistors (q.v.) are not yet tiny enough for the technical designers; their watchword is 'miniaturization'. There are three ways of reducing the size of components: the 'solid circuit', the 'micromodule' and the 'microcircuit' techniques. The solid circuit incorporates the semiconductor crystals of the transistor, the resistors and the capacitors, all in one solid piece or 'chip', ready for insertion in, say, a computer. Both the micromodule and the microcircuit use very thin film or glass wafers on which the components are mechanically bonded, complete with all interconnections. The 'printed circuit' overcomes the problem of assembling minute parts, of wiring and soldering, and speeds up production: the entire circuit is first outlined in acid-resistant ink on a copper film, and then the copper parts which are not protected by the ink are washed away in an acid solution, leaving the circuit standing. There are no soldered joints to become loose, no wires to break. No wonder that these little technical miracles have prompted the development of electronic 'bugs' – miniature transmitters and receivers for spies of all kinds, industrial, political and military.

E. L.

The world's first gas-turbine car, made by the Rover company. Registration number JET 1.

Gas turbine

If a basic turbine could be worked by steam, as Hero of Alexandria had suggested in the 2nd century BC and Charles Parsons had demonstrated in 1897, why not then by any other force producing greater power? It was a logical question, and by the 1930s scientists, mainly in Switzerland, had begun to consider the gas turbine as an economic proposition.

Basically a gas turbine is a kind of bladed wheel, the blades of which are forced to move by a powerful jet of gas directed on and between them. The blades are fixed to a shaft and thus turn the shaft round with them; the power is taken off the turbine shaft, which belts or gear-wheels can connect to any type of machinery. Air for the gas turbine is drawn in by a compressor and joined in the combustion chamber by injected oil fuel, which is then ignited. The fuel takes fire, expands, and blasts hot gases down into the turbine, thus driving the shaft round and producing power.

At first the advantages of the gas turbine, namely the reasonable initial cost, small size and weight, relatively simple build, and quick starting, were overshadowed by the great disadvantage that special metals were required to withstand the very high temperatures involved, and by the problem of heavy fuel consumption. Apart from Holzwarth's practical gas engine of 1908 it was not until the Swiss firm of Brown-Boveri began their work on gas turbines that gas could be looked on as a possible source of power for propulsion. In the mid-1930s, in the oilfields of the United States, some 6,000 kW. gas-turbine generating units were being used, and a few years later gas turbines were increasingly in use to provide power for electricity generating at peak periods. However, at first this form of power was not too efficient and it was used sparingly.

After the Second World War gas turbines increased in size and became more efficient. In Britain the first stand-by gas-turbine electricity generator was in operation in Manchester in 1952. In

Russia experiments to cut the cost, and waste, of fuel consumed have been successful. A 20,000 kW. turbine has used cheap piped underground gas, and in 1962 the excess heat generated by a gas-turbine unit was first used to heat water. Today probably the largest single gas turbine in the world is a 50,000 kW. unit in Russia.

This form of power is also used in transport. In 1941 the Swiss Federal Railways produced the first gas-turbine locomotive, giving up to 2,200 h.p. The turbine was found to be more efficient than steam power. The first gas-turbine car was made by the British Rover Company in 1950, and later this type of car performed well in motor racing. In 1963 a gas-turbine car, entered in the gruelling Le Mans 24-Hour Race, covered 2,553 miles at an average speed of 109·7 m.p.h., with only nine pit stops. Successfully powered gas-turbine fire-engines, and also light portable turbine-driven fire-pumps, have been used for some years now in the United States. At sea a number of navies have favoured gas-turbine engines, at least in conjunction with other forms of power, and in aviation, too, gas turbines have formed part of the machinery of jet-propulsion engines, as perfected by Britain's Sir Frank Whittle.

Although in transport the gas-turbine engine would seem to offer reliability, lightness, long life and low maintenance, high fuel costs have prohibited its widespread use on railways and motorways. Gas-turbine engines are, however, increasingly being used in industry to power machinery.

M. H.

Stirling engine

In the early years of the 19th century the muscular toil of men was at last being eased by the versatile power of the steam engine. Unfortunately the technology of steam engines was still highly imperfect, with the result that efficiencies were very poor, skilled engineers were constantly having to effect adjustments and repairs, and – by far the worst – boilers frequently exploded, causing great destruction and loss of life. Such tragedies were particularly common at sea, where machinery weight was cut to a minimum and where boilers were often subjected to stresses in storms that their designers had not allowed for. A Scottish engineer and inventor, Robert Stirling (1790–1878), spent part of his teenage years trying to devise a better kind of engine that did not need a boiler.

Before long his thoughts were centred on the use of air as the working fluid, instead of steam. The Frenchman Sadi Carnot had published his idealized thermodynamic cycle, but had no idea how such an engine could be constructed. Stirling experimented with hardware, while simultaneously poring far into the night over paper calculations. Eventually, in 1817, he filed a patent for the engine which bears his name. He was never able to build a really good working example, and the idea was undoubtedly ahead of its time.

The basis of the Stirling cycle is that a volume of gas is continuously oscillated in a volume enclosed by a cylinder (plus a heat source and a 'regenerator') and reciprocating pistons. On each cycle the gas is heated at constant volume, expanded isothermally (constant temperature) to drive the output piston, cooled at the now larger volume and then compressed isothermally (at a lower constant temperature). It was fundamental to his engine that the heat was applied from outside. One of the reasons why he could not build a useful engine himself was that merely trying to heat and cool the outside of the cylinder was doomed to failure. The strong metal wall could not be heated and cooled fast enough to drive an engine at a useful speed.

By the mid-19th century Lenoir and others had built gas engines where the heat was generated by the working fluid itself inside the cylinder. By 1900 Otto and Diesel had made engines burning liquid fuels, and these are used by the million in the modern world. But in 1938 the Dutch firm of Philips NV, seeking an engine to drive small electric generators in remote locations, gradually came to the conclusion that the Stirling was potentially of great importance, and had been unjustly neglected. Today many firms are running engines of the Stirling type, often under Philips's licence, most of them having a clever rhombic linkage to provide an output drive from two oscillating pistons which move independently along the same cylinder. The gas, which may be air, helium or hydrogen, gives up 99 per cent of its heat on each cycle to a porous metal regenerator (heat store) and takes it back on its return flow. The source of heat can be anything hot enough to be useful – it could be a garden bonfire – and one unusual feature of the engine is that portable models can run on high-octane petrol, crude oil from the well or even salad oil!

As the fuel burns continuously and externally it is possible to minimize production of unwanted pollution in a way that is impossible with engines of the internal-combustion type. The Stirling can also be made extremely quiet and smooth-running. Already engines suitable for a wide range of ships, road vehicles and other applications have been demonstrated to have power, economy, reliability and life at least as good as those of alternative engines. There seems little doubt that this 160-year-old principle is on the verge of wide use, hastened by the increasing concern for what Otto and diesel engines are doing to the environment.

W. T. G.

A Stirling engine built according to Stirling's original specifications. Modern versions may be the answer to the pollution problem.

The core of the Herald research reactor at Aldermaston. The eerie blue glow in the shielding water is given off by escaping fast electrons.

Nuclear reactor

Coming from the man who first split the atom, Rutherford's famous doubts as to whether the energy in the atomic nucleus would ever be put to practical use are often regarded as a typical example of a scientist's short-sightedness about the implications of his apparently harmless experiments. In fact it was a perfectly reasonable thing to say until, well after his death in 1937, Otto Hahn discovered nuclear fission in uranium.

In his experiments Rutherford had always had to use more energy to split an atomic nucleus than he ever got out of it. The process discovered by Hahn was quite different. He found that the nucleus of certain uranium atoms could split into two approximately equal halves, releasing energy and several particles called neutrons which could split further uranium atoms in their turn.

Hahn published his observations in January 1939. The implications were worked out by his former colleague Lise Meitner and her nephew Otto Frisch. The possibilities were obvious and, particularly after the outbreak of war, ominous too. In both Britain and the United States refugee scientists were active in drawing the governments' attention to the dangers of an atomic bomb in Hitler's hands. Research began in both countries.

An important first step was to demonstrate that a chain reaction could indeed take place in uranium. This was made the responsibility of an American team led by the refugee Italian scientist Enrico Fermi. At 3.45 p.m. on 2 December 1942, in the squash court of the University of Chicago, they were successful. While two assistants stood by with buckets of a chemical solution which could damp down the chain reaction, the cadmium control rods were slowly withdrawn from the pile of graphite blocks and uranium fuel. The Geiger counters began to click as the radiation built up, and a cryptic telegram went out to those in the know: 'The Italian navigator has entered the new world.'

Fermi's 'pile' was the first nuclear reactor, the prototype for the nuclear furnaces in the atomic power stations of today, though it was almost 14 years before the first practical atomic power station started up at Calder Hall, Cumberland, in 1956. It was 125 years since Faraday had invented the dynamo. Now this device could be driven by the most efficient source of energy yet discovered – nuclear power.

B.S.

Electrolysis in reverse: a battery of fuel cells. More research is needed to achieve economy, small size and useful power.

Fuel cell

Sometimes a new technical idea is 'a solution without a problem' – but invariably the problem turns up not very long after the solution has been found. A case in point is the fuel cell.

There are three basic ways of producing or storing electricity: by generator, by battery, or by accumulator; the fuel cell is a fourth. Its history goes back to Sir Humphry Davy who, early in the last century, experimented with electrolysis, suggesting that when an electric current is sent through water it splits the water up into its constituents, oxygen and hydrogen. Then, in 1842, another British scientist, Sir William Grove, succeeded in reversing this process – that is, producing electric current from the interaction of the two gases, hydrogen and oxygen. However, the amount of current produced in this way was negligible, and the idea of 'electrolysis in reverse' lay dormant for 90 years.

But in 1932, a young English chemist at Cambridge, Francis T. Bacon, had another look at it; and 27 years later he was able to demonstrate what he called the 'fuel cell'. In fact it consisted of a whole battery of cells, each with two electrodes, porous flat plates made from nickel powder; they were suspended in a 40 per cent solution of potassium hydroxide, and fed with hydrogen and oxygen at a pressure of several hundred pounds per square inch and a temperature of a few hundred degrees centigrade. The result was a 5 kW. current of 24 V., sufficient to operate a circular saw or a welding apparatus. That result was encouraging, and Bacon hoped that the efficiency of the fuel cell could eventually be raised to 80 per cent, i.e. that 1 lb. of gas would produce over 1 kWh. of current.

Large corporations with their vast research capacity took up the idea in the USA; General Electric succeeded in streamlining the arrangement for use in vehicles, and an experimental tractor, powered by 1,008 fuel cells and an electric motor, was tried out in the field. It used propane and oxygen, and 15 kW. of electricity was generated – sufficient to pull a plough. The Chrysler Corporation built a fuel-cell car with four electric motors, one attached to each wheel so that no gearbox, transmission, differential, drive shaft, or rear axle were needed. The Russians, too, are working on similar lines, for now the problem to which the fuel cell is the solution has become evident – the demand for a silent car with no exhaust fumes.

The first practical application of the fuel cell, however, was in an American space satellite, where it generated current for the radio transmitters. But a great deal of research will still have to be done before an economical electric car or lorry, powered by the fuel cell, can be developed, or even small power plants serving factories or local areas. In the distant future, we may have fuel-cell trains or ships.

E.L.

Man Taking

Pottery

The oldest known 'pottery' emerged from excavations at Dolni Vestonice in Moravia, and dates from *c.* 25,000 BC: there fragments of fired clay were found which had been roughly moulded into the shapes of animals, many of them showing traces of stab marks. Sympathetic magic of this kind was clearly pottery's main function long before vessels were shaped; in the grain bins at Hacilar and Çatal Hüyük in the Near East modelled cult figures have been found, presumably to protect the grain, but at the former site excavators have also uncovered hand-modelled pots decorated with geometrical designs in red ochre and dated *c.* 5500 BC. In Japan pots have been found of even earlier date (*c.* 9000 BC). These are made of a spiral of clay rope in exactly the same manner as children today form their first pots in primary schools.

A primitive kind of potter's wheel – simply a turntable on a central pivot, worked by an assistant so that the potter had both hands free – first appears in Mesopotamia shortly after 3500 BC, but it is nowadays felt that the idea that the potter's wheel was an innovation of great social and economic significance is incorrect. A more pressing problem was the firing of the clay; primitive pots were dried in the sun or fired in a bonfire, but neither method was satisfactory, for slow firing in a kiln is what makes a pot permanently watertight. Vertical kilns, in which the fire is not in contact with the pots, existed in Mesopotamia and Persia before 4000 BC, and in Egypt about a millennium later. Some of these kilns have been preserved – while ordinary mud-wall houses have vanished – because constant firing has made them, as it were, large pieces of fired pottery.

A further step in making pots less porous is the technique of using a copper or lead glaze, which can also make them more attractive; an early formula for such a process is related in a tablet found at Tell 'Umar, Iraq, dating from the 17th century BC. Earlier methods of making pots impervious included burnishing the surface before it was fired or even dry, or dipping the pot in a smooth solution of clay – a 'slip' – after firing. Both techniques were in use in the Near East by 5000 BC. About a millennium later, we find a process which underlines pottery's role as an imitator of other materials: desiring to produce artificially the much-coveted deep blue of lapis lazuli, Mesopotamian potters covered the surface of pots with copper ores (principally azurite and malachite), and then fired them to produce a glazed surface.

But a lead glaze adhered better, and remained in use in Mesopotamia and later civilizations, to be joined by salt-glazing (for stoneware) which was developed in the Rhineland in the late 14th century. The use of tin glaze for producing an opaque white glaze on the second firing, also known in Mesopotamia, came into fashion in Europe at this time, via Islamic Spain, and painted well with metallic pigments. Its best-known type is the Italian majolica, which is tough enough to be used for outside decoration.

A combination of kaolin and felspar, fused at a very high temperature, gives porcelain; already being used in T'ang Dynasty China (AD 618–906), the process was successfully imitated in Europe only in the 18th century.

C. M. B. G.

Potter's wheel of a simple kind, painted on a 16th-century majolica plate.

Tanning

Tanning is basically the treatment of leather with vegetable, mineral or oil substances so that it will neither rot, let in water, nor go brittle. Before this treatment, the flesh remains and hair must be removed, often by protracted soaking in a mixture of dung and urine. It has been suggested that the process was first discovered when attempts were being made to dye leather with vegetable dyes also containing tannin: oak-galls, sumach, chestnut and oak-bark (the standard method in Europe until the 15th century AD). In the ancient Near East, tanning with mineral salts – usually a solution of alum and salt – was common: by Neolithic times, special knives and scrapers for the preparation of leather had been developed (these have also been found in Europe – for example at Swiss lake-dwelling sites).

The technique may be Egyptian in origin, for graves of the Tasian and Badarian periods (*c.* 3800 BC), the earliest known cultures of the Nile Valley, have been found with the bodies wrapped in garments of tanned leather. From about 500 years later we are well informed about the process in tomb-

Tanner at work, from a 16th-century Polish manuscript, the Behem Codex.

Blast furnace and bellows were essential for casting iron. This picture of a blacksmith's forge comes from Agricola's *De Re Metallica*, 1556.

Greek blacksmith, portrayed on an Attic black-figure vase. Weapons of wrought iron were fairly soft, and would not have taken a good cutting edge.

Bronze ritual wine vessel, in the shape of two rams, from Shang dynasty China.

paintings and inscriptions. Tanned leather had a wide variety of uses, including that of a writing medium (perhaps only when the information to be recorded was precious). The Egyptians had a thriving export trade in tanned skins and hides in both Hellenistic and Roman times.

In Mesopotamia, also, the process was well known: tanned skins have been found in the excavations at Ur, and we can see from the Standard of Ur (early 3rd millennium BC) that soldiers wore skirts of what are probably leather thongs (similar to the Roman soldier's sporran, or *cingulum militare*, which was looped over the belt and protected the stomach and groin), as well as leather cloaks and caps. The horses wore strips of stuff for protection, again presumably leather.

Tannery sites have survived from Roman times: one at Pompeii contains 15 pits for vegetable tanning and others for mineral tanning, as well as storage tanks for all the required liquids. Because of the suddenness of the disaster which overtook Pompeii in AD 63, we can see examples of tools as the tanners left them: they are similar to those in use today.

C.M.B.G.

Ironworking

Repeated heating and hammering is necessary for the production of iron. Copper and bronze could be formed in moulds, and hammered when cold for added hardness, but iron required a more intricate treatment. Starting with an iron-bearing ore, a furnace was needed, with charcoal as fuel, heated to a high temperature with blowpipe or bellows (the latter existed in Egypt by 2nd millennium BC); periodically the 'bloom' – a mass of iron and impurities – had to be taken from the furnace, hammered, and returned, until the impurities were reduced.

Such smelted iron has been found dating from *c.* 2700 BC at Tell Chagar Bazar (northern Syria), and a bronze hilt holding the remains of an iron blade was dug from Tell Asmar (*c.* 2400 BC). At Ur, a smelting site of *c.* 2000 BC has been unearthed. But at this period wrought iron would have been very expensive, and it was used for ornaments and ceremonial weapons rather than for everyday use: witness the ceremonial axehead from Ras Shamra, *c.* 1400 BC, or the few references Homer makes to iron as a precious metal on a par with gold. The 'Iron Age' of course starts at different times in different places, but the metal was scarce everywhere before *c.* 1000 BC.

Wrought-iron weapons are fairly soft, and will not take a lasting cutting edge; steeling was the answer, and this entailed the introduction of added carbon by hammering (some ores contain a high level of carbon, so anything made from them will be steeled in parts), and then hardening by sudden cooling and tempering to lessen brittleness. The process is complicated, and calls for considerable skill to achieve the required strength.

Small wonder, therefore, that fine swords were given names, that those smiths who knew how to control such processes were revered, and that rulers tried to maintain monopolies: the Hittite King Hattusilis III wrote the following reply to a request for iron goods (13th century BC): 'Good iron is not available in my seal-house at Kizzuwatna. . . . They will produce good iron, but as yet they will not have finished. When they have finished I will send it to you. Today now I am dispatching to you an iron dagger-blade. . . .' It has been surmised that the 'good iron' referred to is steel, or rather iron with steeled edges; the process was possible by this date although our knowledge of its spread is slight.

Metallurgical examinations have revealed swords from the Alpine La Tène culture with carburized edges, Roman saws which would blunt easily, and Saxon axes still without any sign of quenching and tempering. Such deviations are startling. However, there is no evidence of quench-hardening being known before *c.* 900 BC, and it is indeed only with the cultures of Hallstatt and La Tène (*c.* 750 BC to Roman date) that the use of iron really begins to outdistance bronze.

C.M.B.G.

Bronze

Before *c.* 3000 BC, copper had been the metal used for artefacts; but the discovery that a little tin when mixed with copper gave an alloy both harder and more easily worked and cast was to lead to more efficient tools and weapons. No one knows where the discovery was first made, but the view that it occurred near the sources of the metals, in Syria or eastern Turkey, to be subsequently bought by the rich Mesopotamians, appears reasonable. Again, it is difficult to say when tin was consciously smelted as an ingredient for real bronze; one metallurgist has affirmed that, prior to *c.* 2100 BC and perhaps later, the alloys called 'bronze' by archaeologists – found in artefacts from Sumer and Troy, probably originating from Anatolia – are too low in tin to be thus called, and that therefore tin-smelting can date only from between 1800 and 1600 BC, probably beginning in north-west Persia.

Use of the bronze alloy spread slowly; in Egypt, for example, little bronze was in use before the middle of the 2nd millennium; indeed, more of the furnishings of Tutankhamun's tomb (*c.* 1350 BC) are of copper than of bronze. In the Aegean, bronze began to supersede copper, particularly for tools and weapons in about 1700 BC, for the discovery was made that bronze could be cold-hammered, thus giving a much keener edge than could be achieved with copper or with an alloy very low in tin. By *c.* 1600 BC, alloys with about 10 per cent tin are found at Troy III, and the true Bronze Age is thus heralded: by *c.* 1400 BC the metal was predominant for tools, weapons and for casting objects such as the doors for the new temple at Egyptian Thebes (*c.* 1450 BC).

Because of the durability of the metal, we know a lot about bronze artefacts: they are to be found not only as grave-goods but also, because of the scrap value of the metal, in hoards perhaps collected and buried in times of trouble. Again, the moulds in which axeheads, spear-heads or ornaments were cast are frequently found. Because bronze flows easily, complex casting techniques are possible, and moulds have been found for use open, closed, and for the *cire perdu* process (which produces the best detail and, like the other types, is still in use today). Some moulds were massive, as the following description of casting figures to decorate Sennacherib's Palace at Nineveh (*c.* 600 BC) shows: 'Eight lions, open at the knee, advancing, constructed out of 11,400 talents of shining bronze . . . together with two colossal pillars the copper-work for which came to 6,000 talents. . . . I built a forme of clay and poured bronze into it, as in making half-shekel pieces, and finished their construction. . . .'

About 1500 BC, bronze-founding appears in China (without any preceding Copper Age, so it must have been introduced from the West); their craftsmen managed to cast even large cauldrons, using precisely detailed moulds, often in many sections.

C. M. B. G.

Pitch

Pitch is obtained from tar or turpentine and occurs in a mineral version as bitumen, or asphalt; viscous when heated, it is firm when cold, and was used in antiquity for building as well as for caulking. To produce it, wood was burned to charcoal, and wood-tar collected from the residue; the mineral variety occurs naturally in a more or less pure state in Mesopotamia and would have been collected from surface deposits. There were also deposits in the Holy Land.

The Sumerians used bitumen, as did the inhabitants of Palestine: in the excavations of Jericho a wall was found with the bricks cemented in place with bitumen, possibly dating from *c.* 2500–2100 BC. The main use of pitch was in fact as a mortar. It was the common cement for brickwork in Babylonia, and was used for this purpose at Mohenjo-daro in the Indus Valley (*c.* 2500–2200 BC); it was often mixed with bits of baked brick or vegetable fibres to give added strength. Again, it was used for insulation, or as a damp-course, as well as for the protection of brickwork: the piers of a bridge built by Nebuchadnezzar on the Euphrates were coated with bitumen below water-level to prevent rotting.

The other main use of the substance in antiquity was for waterproofing boats: the Babylonian Gilgamesh and his Biblical equivalent Noah both prudently coated their vessels inside and out with bitumen. Similarly, Moses arrived safely into the hands of Pharaoh's daughter because the cradle of bulrushes into which he was laid had been treated with bitumen before being placed on the river. In areas without natural bitumen, wood-tar was used for caulking: the Viking Gokstad ship, for example, was caulked with tarred rope, as was the Oseberg ship.

The main use of bitumen today is in road-building; it was perhaps first used for this purpose in the processional roads of Babylon, but then only as a mastic. Nowadays, under the name of tar macadam, roads are constructed of stone chippings bound in tar and pitch and well rolled. John McAdam, in spite of giving his name to the process, advocated the use of well-compressed broken stone, with no binding material, and it was in fact Thomas Telford (1757–1834) who developed pitch foundations.

C. M. B. G.

Glass

Pliny the Younger gives a romanticized account of how glass was first discovered: Phoenician merchants taking natron – a natural form of sodium carbonate – from Syria to Egypt camped on a beach of sand, and made a fire to cook on. Accidentally they placed a cooking pot on top of a block of natron, and the heat made the natron and sand fuse together to form glass.

But Pliny dated the invention too late: it appears to have originated in western Asia in the 3rd millennium BC, and was being made in Mesopotamia by 2000 BC; whereas in Egypt real vessels of glass – as opposed to beads and trinkets – date back only to *c.* 1500 BC. After the fragmentation of Aegean civilization in the 13th century BC centres of glass-making were established particularly in Mesopotamia and Syria, where they prospered from the 9th century BC onwards. By the 6th century there were manufactories on Rhodes and Cyprus, and Alexandria was an important centre of glass production from its foundation in 332 BC.

Several techniques for making glass vessels were in use concurrently, for the blowpipe was not invented, perhaps in Phoenicia, until the 1st century BC. The most usual method was to pick up the molten metal on a core of the correct contour, which was later removed. Vessels were also built up from glass

Above, *glass-making, from an early 15th-century manuscript of Mandeville's* Travels.

Right, *the first cast-iron bridge in the world, built across the Severn at Coalbrookdale, Shropshire, in 1779. This engraving was made three years after it was built.*

Opposite, *caulking the seams of a ship with pitch: French illuminated manuscript, mid-15th century.*

Right, *13th-century Syrian brass hand-warmer, pierced and inlaid with silver.*

Below, *a wall painting in the house of Julia Felix, Pompeii, shows a glass bowl of fruit.*

sections – called mosaic glass, and in use in Mesopotamia from *c.* 1500 BC – cut from a solid block of glass, or cast in moulds.

The Romans spread glass-making through their empire: there were factories in Cologne, for example, in the 1st century AD. But Egypt remained an important centre: the famous Portland Vase was blown in Egypt, from blue glass, around the time of the birth of Christ, and then cased in white glass, from which the figures were then cut.

From the 7th century AD the countries of Islam – Mesopotamia, Persia, Egypt and Syria – also had thriving industries, and some of their most impressive products are mosque-lamps in clear or tinted glass, painted and enamelled. But in Europe from *c.* AD 1000 until the 18th century, when George Ravenscroft invented lead-glass – heavier, clearer, and ideal for wheel-engraving – everyone got their quality wares from Venice, as a letter from a London merchant to a 'Senier Morelli' of Venice, dated 17 September 1669, shows: he draws outlines of glasses in the margin, and writes: '. . . let the plain glasses, as in the Margin, be first made in a readiness, for them we want most especially, and also we desire that no Looking Glasses be packed in the bottoms of the chests of drinking glasses; but we would have them very carefully packed in one or two strong cases. . . .'

Until well into the 18th century, however, drinking glasses and mirrors were luxury items.

C. M. B. G.

Cast iron

Cast iron was not well known in the Western world in antiquity; although it is easily cast, it is brittle and cannot be worked and reworked as can, for example, bronze. It did not, therefore, recommend itself for tools or weapons. A few ornaments were made in it: a hollow ring has been found at Býĉi Skála, of *c.* 600 BC; and the Romans sometimes used it for domestic pots. In China, however, cast iron appears to have been known and used with coal-fired furnaces from the 6th century BC – for cauldrons and agricultural implements as well as for gear-wheels.

The important event in the Western world for the large-scale production of cast iron was to be Abraham Darby's discovery in 1709 of a method of using coke – instead of the ever-diminishing supply of charcoal – for iron-smelting. Coupled with this was the development of the blast-furnace from the late 14th century (much earlier in China) which provided a much more efficient way of getting rid of impurities than having a smith beat out the 'bloom'. From the 16th century, cast iron was used in Europe for grave-slabs, fire-backs and, most importantly, ordnance: iron shot weighing 2 cwt. apiece were being cast in England by the late 15th century.

Using charcoal-fired furnaces, therefore, the European powers made guns as best they could during the 17th century, and anything in iron was better than the more expensive bronze; the English, however, were running short of timber and often bought their guns from Sweden from 1632 onwards, while in France Colbert fought a long battle to encourage the use of French iron ores.

But in the 17th century no one dared to use cast iron for anything structural (unless the water-pipes laid at Versailles in 1664 be counted as such). In 1777–79, however, Abraham Darby's company cast and erected the first cast-iron bridge in the world, at Ironbridge, Coalbrookdale, Shropshire. In 1772, cast-iron columns had already been used to support the galleries of St Anne's Church, Liverpool; and by 1800 its use, particularly for industrial buildings, for railway track, and for ornamental work, was in full swing. In conjunction with tile or brick, it was fire-resistant, and therefore recommended itself to Sir John Soane for the vaults of the Bank Stock Office at the Bank of England in 1794. Department stores in London and Paris were often iron-framed by the 1830s, and two decades later we have the Crystal Palace (1850–51) and the great London railway termini. And yet in China cast-iron beams had been used for pagodas in the 11th century, if on a much more restricted scale.

C. M. B. G.

Brass

An alloy of mainly copper and zinc, brass is versatile, and its properties depend on the proportions used. A mixture with more than 63 per cent of copper can be worked cold, and annealed, and is ductile; whereas a mixture with less copper and proportionately more zinc is worked hot, and has greater strength.

The use of brass in antiquity is obscure, because of the method used for smelting the zinc ore, or calamine. Zinc has a lower boiling point than copper, and when the heating of copper, charcoal and zinc ore takes place, it is difficult not to evaporate off the zinc. The Romans were possibly the first to use the process on a large scale, but it seems likely that brass was previously produced unwittingly by bronze-smelters, since the distinction between tin and zinc was not clear in early times. Only one example of true brass is known before the Roman period – from Gezer, in Palestine, of *c.* 1400–1200 BC. (We should beware of references to brass in the Bible, which are uniformly to bronze; and of the Roman coin called an *aes* – which means copper or bronze, but not brass – although, to complicate matters, they did use brass for coinage, and it was at first more expensive to produce than either copper or bronze.)

From medieval times, however, before it became associated with pots and pans, brass was a luxury product used in the production of monumental tomb slabs. They knew a European vogue from *c.* 1230 (the earliest known brass is to Bishop Yso Wilpe, died 1231) for about 300 years because they were much cheaper than large pieces of sculpture. They were made of ground zinc ore mixed with charcoal and bits of copper, and heated sufficiently to unite the two; with further heating the resultant alloy melted, and was poured into moulds. The first brasses in England were imported, often from the region of Tournai, whence clients would order their tombstones complete, already set in a handsome slab or Tournai marble; the technique was to cast the figure and frequently the surrounding canopy in silhouette, as it were, let it into a prepared stone slab, and then sketch in the detailing with incised lines. Sometimes alabaster or other inlays are used for hands and face. The fixing of the brass into its stone support was done with dowels set into lead plugs, the brass itself resting on a layer of bitumen; very large images would be cast in several sections and soldered together.

C. M. B. G.

Concrete

Concrete was much used by the Romans and led to important structural innovations. In and around Rome and Naples there are thick strata of *pozzolana*, a volcanic earth which, when mixed with lime and water, produces a mortar which will set even under water. Its cheapness and strength recommended it for use in the lower parts of the dome of the Pantheon, and it was also useful as a mortar, to bind together courses of brick or stone as in the strong brick exterior of Hagia Sophia, Constantinople (AD 532–37) – for the Byzantine Empire inherited Roman constructional techniques.

Later the art of making concrete appears to have been lost, only to reappear in 1568, when the French architect Philibert de l'Orme recommended a mixture of quicklime, river sand, pebble and gravel, plus water, for the foundations of bridges and harbours. However, true 'Roman cement' entailed the carriage of *pozzolana* from Italy, and it was not until 1756 that John Smeaton, after several experiments, built the third Eddystone Lighthouse in Britain on a foundation of concrete which would harden under water. In France, several houses were built of concrete before the French Revolution – including the Bagatelle in the Bois de Boulogne, of rubble and cement stucco, in 1779 – but the idea did not catch on.

The 19th century saw experiments to make concrete stronger by reinforcing it with iron bars. W.B. Wilkinson, an English plasterer from Newcastle, took out a patent for such a process in 1854, but it was the Frenchman François Hennebique, with patents in 1892 and 1893, who most energetically promoted his product. Anatole de Baudot's Gothic-looking church of St Jean de Montmartre in Paris (1894–99) established France's primacy in the medium for two decades. Hennebique himself built a wharf at Southampton in 1897; and his Swiss pupil, Robert Maillart, built bridges to prove the versatility of reinforced concrete.

Nowadays concrete is in universal use, for it is cheap, convenient and strong when reinforced. It can be cast *in situ*, or at the factory, with little skilled labour. It can be beautiful: the silhouettes of bridges built of reinforced and prestressed concrete often achieve an unrivalled finesse, but the merits of bare concrete walls – or walls still adorned with the marks of the planking in which they were cast – are more debatable. The clarity and modelling achieved by Le Corbusier at the monastery of La Tourette near Lyons (1953–59) is seldom captured by property developers.

C. M. B. G.

Paper

'In ancient times', says the Chinese Han Dynasty Chronicle, 'writing was generally on bamboo or on pieces of silk. . . . But silk being expensive and bamboo heavy these two materials were not convenient. Then Ts'ai Lun thought of using tree-bark, hemp, rags and fishnets . . . he made a report to the emperor on the process of paper-making and received high praise for his ability. From this time paper has been in use everywhere.' The date they give corresponds to AD 105 but Ts'ai Lun, who was a prominent courtier in charge of the furnishing of the palace, may have only made known officially an earlier invention of humbler people, for the oldest fragments of paper, found in a remote outpost of the Han Empire, go back to about AD 90. Nevertheless, he became a kind of patron saint of Chinese paper-makers, and his mortar was preserved as a relic. At first Chinese paper-makers used bast of the paper mulberry, but they soon went over to rags. Rag paper was so much more convenient than the bamboo strips, wood and bone that had been employed formerly that it soon became common for all but the most luxurious writings.

In the rest of the world, writing materials suffered from the same defects: they were either too cumbersome – like Babylonian clay tablets and Indian palm leaves – or too dear. Egyptian papyrus, a species of sedge, had been used since the 3rd millennium BC. The pith of the stalks was cut into strips, which were dried, and then laid across one another to form two sheets, which were glued together. This papyrus was successful enough to become a major

Above, *paper-making in 17th-century China. On the left, a sheet of wet pulp is placed under a wire screen for pressing. On the right, the pressed sheets of pulp are laid on an oven for drying.*

Right, *group of inlaid and painted papier-mâché furniture (c. 1845).*

Left, *one of the most beautiful of Robert Maillart's reinforced-concrete bridges – at Salginatobel, near Schiers, Switzerland.*

Above, Still Life with Pewter and Two Ming Bowls *by J. J. Treck, 1649.*

Egyptian export; it would seem that most of the literature of classical Greece was written on Egyptian papyrus. However, leather continued to be used there, especially after the perfection of a method of specially preparing it for writing, which is associated with Pergamon, whence the name 'parchment'. This involved a thorough cleansing of the skin, which was not tanned but rubbed with fine chalk and smoothed with pumice-stone. Unlike papyrus both sides of each sheet could be used, and erasure was not too difficult. So in Europe parchment, and in particular vellum (calf-skin parchment) superseded papyrus, which however survived in Egypt.

Meanwhile true paper had remained a Chinese speciality, although sometimes imported by her immediate neighbours. But in 751, the armies of advancing Islam came up against a Chinese expeditionary force, defeated it and captured, among others, a number of paper-makers who were brought to Samarkand. For over a century this city preserved a monopoly, sending paper to the rest of the Moslem world, but gradually the industry spread, to Baghdad (with Chinese craftsmen), to Damascus and Cairo, where it killed off the ancient native product; in one of the last surviving papyri, dated about AD 890, the author apologized, 'Pardon the papyrus'. But paper manufacture did not reach Europe for a while. In Moorish Spain, the paper of Jativa had won quite a reputation by 1150: in 1157, Jean Montgolfier, who had been a prisoner of war in Damascus, set up the first paperworks in Christendom, but the process was not properly established until the 13th century, first in Italy. The basic process has not changed much over the centuries. The rags are beaten and macerated to a pulp into which a wire screen in a wooden frame is dipped; the screen is brought up, lightly shaken to get rid of surplus water, so that a sheet forms over the wire, which is then placed on a piece of felt, to which the paper sticks. It can then be removed from the felt, sometimes after pressing, and hung up to dry. It could then be dipped again, this time in size, a solution that would give it a glaze. Originally gypsum was used, but quite early Chinese paper-makers turned to starch paste. The rags had always been pounded by hand in a mortar until the papermills at Fabriano pioneered the use of water-powered hammers, and of an animal-glue size.

A. K.

Pewter

This alloy of copper, with antimony and tin, is very easy to work: it does not go brittle with repeated beating, and it can be worked cold. But it is too soft to be used for heavy tools or weapons, and its use has always been confined to domestic utensils. The Romans made a lot of it, but after them the craft appears to have died out until the formation of the Craft of Pewterers in London, whose ordinances are dated 1348. These speak of 'all manner of pewter, as dishes saucers platters chargers pots square cruets square chrismatories and other things that they make square or cistils [i.e. with ribs] that they be made of fine pewter and the measure of Brass to the Tin as much as it will receive of its nature.' Regulations were also established governing the taking of apprentices and journeymen, and we soon find similar guilds in other cities of the realm.

For dishes and pots, pewter was as it were a substitute: it served the poorer aristocracy who could not afford gold or silver, and it served those on their way up in the world who could discard their wooden vessels. The distinctions are nicely drawn by William Harrison, the antiquary, in his *Description of England*, in 1577: 'As for drinke, it is usuallie filled in pots, gobblets, jugs, bols of silver in noblemens houses, also in fine Venice glasses of all formes, and for want of these, elsewhere, in pots of earth of sundrie colours and moulds, whereof manie are garnished with silver, or at the leastwise in pewter. . . .'

C. M. B. G.

Papier-mâché

Although its use is older in the East, and in Persia, papier-mâché only became fashionable in Europe in the middle of the 18th century, when it was used for small boxes, trays, ornaments, and mirror-frames. It was made of paper pulp mixed with glue, chalk and sand; the mixture was pressed by hand into oiled moulds and baked until set. Objects were then frequently 'japanned' – painted with a lacquer of Japanese origin. The lightness and cheapness of the substance recommended it for architectural decoration, and Robert Adam is recorded as commissioning sections of architrave from wooden moulds made to his own design; the alternative was, of course, expensive, time-consuming and uncontrollable by a busy architect – plaster had to be modelled *en place* by skilled craftsmen.

In 1772, Henry Clay of Birmingham patented a method of making the substance so hard that it could be treated as if it were wood; he described his invention as follows: 'My inventions of making paper panels for various purposes, are by pasting several papers upon boards or plates of regular thicknesses on each side the same, to prevent one side contracting or drawing with superior force to the other in drying; and when the same is rendered sufficiently strong for the purpose intended, it is then planed, or cut off at the edges until the board or plate appears . . . and put in a stove sufficiently hot to deprive them of their flexibility, and at the same time are rubbed over, or dipped in oil or varnish, which so immediately drenches into them as to secure them from damp etc. . . .' The product could indeed be treated like wood, and screwed or dovetailed; of course, the cotton rags he used for his paper pulp were much stronger than the wood-pulp used today.

Clay's patent expired in 1802, and his process was pirated by many firms in the Birmingham area; this led to improvements and extensions, for example Benjamin Cook's patents of 1842 and 1843 for casting metal with the stuff. Interest in the product declined after about the middle of the century, and in 1864, Jennings and Bettridge, successors to Henry Clay, closed down.

This might have been no bad thing, for the ease and cheapness of moulding the substance gave rise to some of the most outrageous productions of the Great Exhibition of 1851. It was used for chairs, couches, tables and cradles, but this was little more than a passing fad: perhaps, as one writer has suggested, the crinolines of the 1860s made it difficult for ladies to locate themselves on papier-mâché chairs with any steadiness and accuracy – or perhaps the growing use of plywood with its greater strength and simpler forms carried the day.

C. M. B. G.

Elastic

Elastic made from rubber was an invention of the 19th century, although the word was current earlier; thus an advertisement of 1812 for 'an elastic round hat' turns out to consist of a patented steel spring inside the crown which collapsed the fabric for carrying under the arm.

The existence of rubber articles, especially balls, had been known since the discovery of the Americas, but it was not until 1736 that samples of it were sent from Peru to Europe by Charles de la Condamine. People noticed how it would erase pencil-marks, and knew it came from the Indies – hence 'india-rubber'. (The great rubber estates in Ceylon and Singapore were founded much later, in 1876, from seeds grown at Kew.)

The substance appears in women's underclothing in 1830, as the following report from that year shows: 'Among the recent inventions at Paris an elastic stiffening of a vegetable substance has been seen, instead of that spiral brass wire now used for shoulder straps, glove tops, corsets etc; it is valuable because it neither cuts the cloth that covers it nor corrodes with verdigris; it is said to be made of Indian rubber.'

Some of the advertisements of the time sound positively dangerous, witness this one of 1835: 'Corsets 25/-. Patent caoutchouc instantaneous closing corsets; this novel application of India-rubber is by far the most extraordinary improvement that has ever been effected.'

It is to the chemist Charles Macintosh and the inventor Thomas Hancock that we owe the development of elastic: they were investigating solvents for rubber in about 1820, and the first manner in which Macintosh used its properties was by trapping it between two layers of fabric – creating the 'macintosh'. Hancock, in 1820, took out a patent for the first elastic – strips of rubber applied with glue to clothes and boots; the first patent for elastic-sided boots was taken out by James Dowie in 1837.

Since the Second World War many textiles – knitted, bonded or woven, man-made or natural – have been developed with a high degree of elasticity, which is partly due to the synthetic rubbers in their composition.

C. M. B. G.

Vulcanized rubber

Rubber was well known in ancient non-European civilizations: by the 13th century AD, rubber articles were in common use among the Mayas and the Aztecs. In Britain and the United States a rudimentary rubber industry began to develop in the early 19th century – Charles Macintosh made his first rubber 'macintoshes' in 1824. But raw rubber was an unsatisfactory material which readily lost its elasticity in cold weather and became sticky in hot. In the United States, Charles Goodyear (1800–60), a former hardware merchant from Philadelphia, set himself the task of producing a more usable form of rubber. In one experiment he accidentally overheated a mixture of rubber, sulphur and white lead, producing a substance which charred but did not melt. In 1841, he succeeded in making uniform, continuous sheets of 'vulcanized' rubber, elastic yet impervious to temperature changes, by passing the mixture through a heated cast-iron trough. Receiving little initial response in his native land, Goodyear tried to sell the idea to Macintosh in England. However, Macintosh's partner Thomas Hancock was able to work out for himself the process pioneered by Goodyear and secured a British patent a few days ahead of Goodyear in 1844. Two years later Alexander Parkes (1813–90) developed an alternative method of vulcanization by dipping the rubber in a weak solution of sulphur chloride, which was particularly good for producing very thin rubber, suitable for balloons, teats, etc.

The successful production of vulcanized rubber gave a firm push to the more traditional uses of rubber in clothes and footwear, but much more important was its use in the electrical and communications industries. Vulcanized rub-

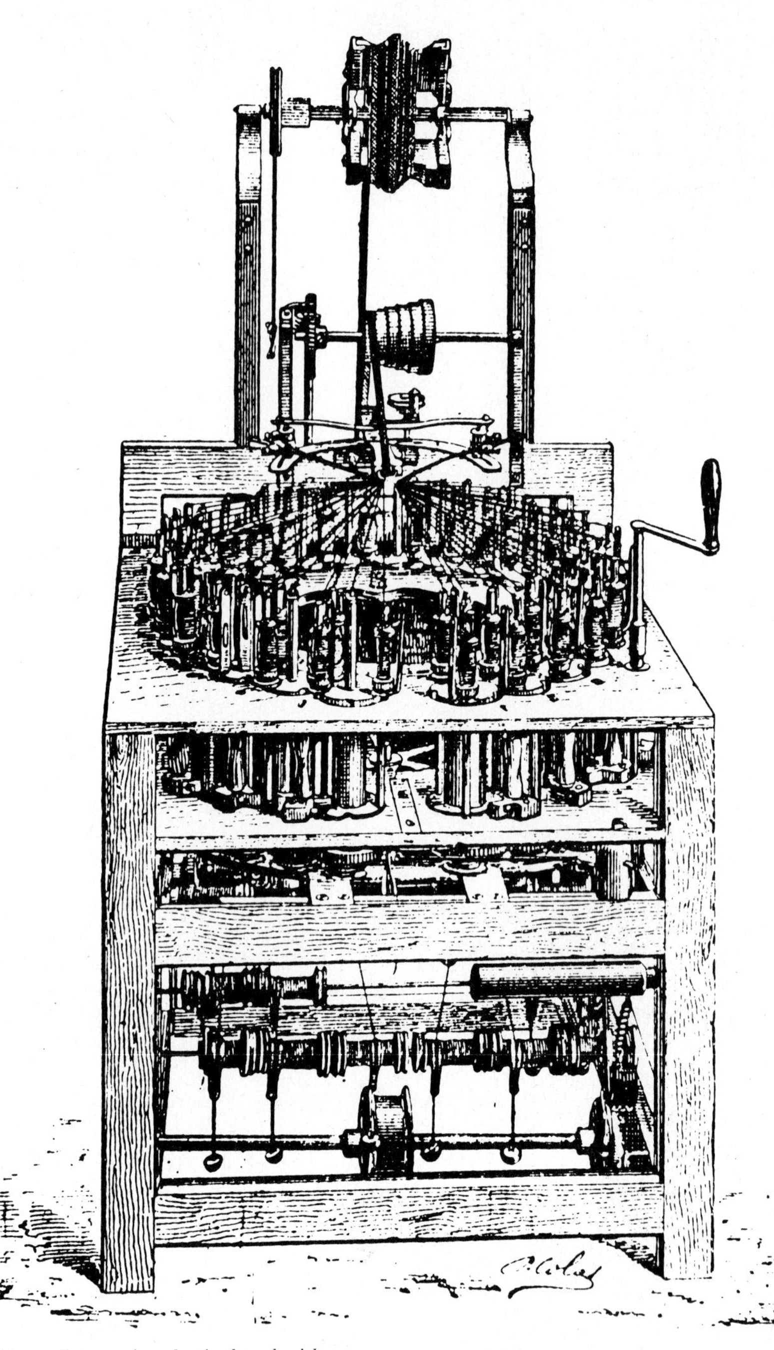

Frame for covering elastic thread with either cotton or silk (1892).

Hancock masticator, used for 'plasticizing' rubber in Britain before 1850.

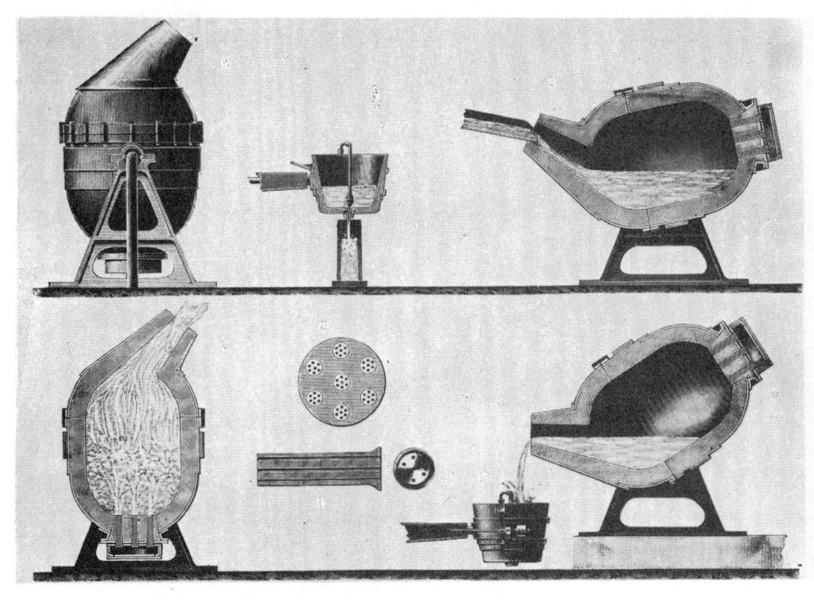

Bessemer's converter: a source of steel for the expanding industries of the mid-19th century.

ber was a perfect insulating material; on the railways it was used for shock-absorbers and cushions. Modern motor transport, as well as a century of relatively efficient contraception, has also depended upon vulcanized rubber.

A. M.

Bessemer steel

On 4 August 1856, Henry Bessemer (1812–98) read a paper to the British Association at Cheltenham. He called it 'The Manufacture of Malleable Iron and Steel Without Fuel'. One of the foremost authorities on iron and steel at the time, Dr John Percy, said later: 'I never witnessed any metallurgical process more startling or impressive.'

Iron was the basis of the Industrial Revolution. The inventions of James Watt, George Stephenson, Richard Trevithick and countless other engineers depended on abundant supplies of a cheap and reliable metal – iron.

Nowadays that useful metal takes many forms. But when Bessemer carried out his experiments, only two types of iron were the mainstay of manufacturing industry.

Cast iron, which could be melted and cast into many shapes, was used for engine cylinders, machine frames and a host of engineering products. Cast iron is brittle and has a low tensile strength. Much stronger and tougher was wrought iron, the more important of the two, which could be forged, rolled and shaped in various ways. This was the material of railway rails, the hulls of ships, bridges, and an endless variety of machine parts. Steel was made only by slow processes of limited scope. It was expensive, and restricted to cutting tools.

If molten iron is treated with oxygen the carbon (which makes cast iron brittle) will be burnt out. This was the principle used to make wrought iron in the puddling furnace, the source of oxygen being primarily rich oxides charged into the furnace with the iron. But it was slow and needed skilled manual labour. What Bessemer did was to blow cold air through molten iron to produce what he called wrought iron.

The chemistry of the process was simple and, after some fundamental problems had been resolved, it worked. Bessemer had actually produced, not wrought iron, but what we now call mild steel. It eventually superseded wrought iron altogether because it could be easily made in large quantities, and was for almost all purposes equally satisfactory.

The process was much faster than puddling and the product was more consistent and cheaper. So the world acquired a source of metal which fitted admirably with the expanding mechanization of the mid-19th century. The importance of its effect cannot be overstated. Although it has now given way to more modern methods of steelmaking – the last two Bessemer converters in Britain, at Workington, were scheduled to go out of production in 1972–73 – the Bessemer process has earned a secure place in history.

W. K. V. G.

Early plastic objects – from a fishing reel of 1862 to a powder box of 1937.

Plastics

About 1850 Alexander Parkes, professor of natural science at Birmingham, researching into colloids, experimented with nitrocellulose (or guncotton); mixed with camphor, it gave a hard but flexible, transparent material, which he called Parkesine. He teamed up with a manufacturer to produce it, but there was no call for it, and the firm went bankrupt. An American from New Jersey, John Wesley Hyatt, acquired the patent in 1868 with the idea of producing artificial ivory for billiard-balls. He improved the formula, worked out an efficient manufacturing process, and marketed the material, to make a few household articles, under a new name – celluloid. But the heyday of celluloid began only with the invention of cinematography: celluloid strips coated with a light-sensitive 'film' were found to be the ideal material for shooting and showing movie pictures. Even before the birth of the cinema, another American, George Eastman, had put his Kodak cameras, for use with celluloid photo film, on the market (1889).

Celluloid was the first plastic material and remained the only one until the turn of the century when two German chemists, Krische and Spitteler, discovered that the gas formaldehyde, which is soluble in water, could be combined with casein (the protein constituent of skimmed milk) to form a hard plastic which they called Galalith. It could be made in any colour and with many patterns, and it heralded the age of cheap, artificial substitutes for a whole range of natural and organic materials.

In 1904, a Belgian scientist and university lecturer working in America, Leo Hendrik Baekeland, succeeded in producing a synthetic shellac from formaldehyde and phenol. Called Bakelite, it was the first of the 'thermosetting' plastics – synthetic materials which, having once been subjected to heat and pressure, became extremely hard and resistant to high temperatures. He set up a company for marketing a moulding-powder as the raw material for shaping Bakelite products.

Jacques Edwin Brandenberger, a Swiss chemist, had been trying since 1900 to invent a waterproof protective covering for foodstuffs and other perishable goods; at last in 1912 he succeeded in producing a thin, supple film made from viscose, which he called Cellophane. After the First World War, the American chemical firm of Du Pont began to manufacture it under Brand-

Man Taking

enberger's licence. This firm had also developed the technique of polymerization, the chemical union of two or more molecules of the same compound to form larger molecules of a new compound with larger molecular weight; this resulted, in 1936, in the production of a fibre marketed as nylon.

Five years later, two scientists employed by the British Calico Printers' Association, J.R. Whinfield and J.T. Dickson, perfected the manufacturing process of a new polyester with a high melting point and excellent characteristics for textile production; they called it Terylene. The 'artificial glass' Perspex, called Plexiglass in the USA, was the brainchild of William Chalmers, a research student at McGill University, Montreal, about 1930, while a Belgian priest and chemist, Father Julius A. Nieuwland, was responsible for the development of a synthetic rubber based on an acetylene polymer, marketed by Du Pont and named Neoprene.

During the Second World War, when imported raw materials were scarce, British chemists produced a new group of plastic materials, the polyvinyl chlorides – PVC for short. Available as a white powder, they are now being made into a great number of articles, from light fittings and drainage pipes to long-playing records and handbags. After the war, fibre-glass became a popular industrial material, which can be woven and resin-bound to wood or metal. But today one of the most widely used plastic materials, whose early development began in the 1930s at the Alkali Division of Britain's Imperial Chemical Industries, is polyethylene, usually called polythene. Because of its electrical properties, it was first used during the war for the production of radar equipment.

The plastics industry is highly specialized, with dozens of different manufacturing techniques but few basic raw materials: oil, coal or carbon in combination with gases such as oxygen, nitrogen, hydrogen, chlorine or fluorine. The plastic products are made by moulding under high temperatures and pressure – compression moulding, injection moulding, or blow moulding – or by the extrusion technique, for thermoplastic materials such as pipes, films and sheeting: the plastic mass is extruded through shaped holes to form the desired profiles. The broad division is between thermosetting (heat-hardening) and thermoplastic (heat-softening) materials.

E.L.

Aluminium/Aluminum

Aluminium does not occur naturally as a metal and was first produced in its pure form by Hans Christian Oersted (1777–1851), the Danish chemist and physicist. Humphry Davy had, however, isolated its oxide in 1809, and it was the English inventor who suggested the name 'aluminum', in which form North America has retained it.

Aluminium is in fact the third most common element of the earth's crust and various gemstones – sapphire and ruby, for example – are compounds of it. Nevertheless the extraction process used by Oersted was extremely costly, and for the first years of its career aluminium was regarded as a semi-precious metal. The breakthrough came in 1886, when both Charles Martin Hall in the United States and L.T. Herault in France discovered, almost simultaneously, the economical process used to extract it today. This consisted of dissolving the purified oxide of aluminium in molten cryolite in a carbon-lined steel container; the latter acts as the cathode (negative pole) and carbon rods are dipped into the molten substance to act as the anode. When an electric current is passed through the apparatus the oxygen from the aluminium oxide gathers on the carbon anode to form carbon dioxide, and the resulting metal, being heavier than the molten cryolite, accumulates at the bottom of the container.

As soon as it could be obtained cheaply this most recently discovered of our common metals began to play an important part in a diversity of areas. It is used mostly in its alloyed forms, which give increased strength, and its uniquely valuable properties are lightness, high thermal and electrical conductivity, resistance to corrosion and non-toxicity. It is widely used in the transport industry, one of its earliest appearances on the field being in the German Zeppelin airships of the First World War; these were made of Duralumin, an alloy invented in 1906 which weighs only about one-third of the weight of mild steel.

However, aluminium is also employed in the building industry – the statue of Eros in Piccadilly Circus, London, was built of aluminium in 1893 – as well as in the manufacture of cars, aeroplanes and ships. It is also made into cooking utensils, carries high voltages of electricity and heals stomach ulcers, features in anti-perspirants and fixes colours in the dyeing industry. Most recently it has played a major part in aiding man's exploration of space.

S.F.

Mining technology

The development of mining machines has relieved the miner of a great deal of drudgery and danger in the pit while substantially cutting down the number of men working at the face. The first coal-cutting machine, nicknamed 'Willie Brown's Iron Man', was invented and introduced by an English mining engineer named William Brown about 1760: it aimed a pick at the coal

Obverse of a medal struck in June 1896 from the first batch of British aluminium.

Coal-miners of the 19th century being lowered down the pit shaft. Men rode in the basket, boys clung to the chain as best they might.

Right, *'corkscrewing' coal out of a thick seam above ground. This giant auger, 42 in. wide, bores up to 200 ft. into the seam.*

face, striking it harder and more frequently than a man could do it. The power was supplied by a horse. Circular steel discs for cutting coal were tried out at the beginning of the 19th century, but this and other early attempts at mechanizing underground labour were hampered by the absence of a prime mover in the pit – steam-engines would have produced too much smoke.

The introduction of the compressed-air engine in mining is generally credited to Lord Cochrane, a technically-minded British aristocrat, but its first important application was in the late 1850s on the boring of the Mont Cenis tunnel in Switzerland (a magazine article on this gave George Westinghouse the idea for his compressed-air brake for trains). The compressed-air engine is not in itself a prime mover like the steam-engine, but the compressed air which makes it a powerful mechanical tool can be produced above ground and conducted underground by tubes and hoses, or filled into cylinders which are transferred to the coal face. It is still a most important 'medium' for mining machinery, together with electricity. In the 1860s, when it was first used in the British and American mines for powering disc coal-cutters, electric engineering was only in its infancy. Subsequently, the development of mining machinery was mostly done in the US; in 1894, for instance, American engineers built the first efficient 'chain machine' which cut the coal with a continuous chain revolving around a long mechanical arm. Another early American coal-cutting machine worked with a circular steel bar with cutting-teeth.

Simultaneously with the mechanization of cutting coal and ore, transport in the mines was gradually improved during the second half of the 19th century. Until then, the inhuman conditions under which the miners had to work in many countries which called themselves Christian and civilized, also extended to the primitive methods by which they were conveyed between the surface and the galleries. Shaky windlasses with worn-out ropes which often broke, operated by horses, were in use for decades after the invention of the steam-engine; in some countries the miners were simply hauled up and down hanging in rope loops, in batches of half a dozen. Eventually, steam power was introduced, cages became larger, safety devices were incorporated – but again it took a long time until electric winding was adopted (in British mines only in the 1920s). Similarly, electric lighting remained the exception rather than the rule in the British mines before the First World War, while American mines introduced it widely before the turn of the century. The miners themselves used slightly modernized versions of Davy's safety lamps in European mines – or even highly dangerous naked lights – well into the 20th century; then the battery-powered miner's lamp took over.

Mechanization of coal transport underground was equally slow. As early as 1805, the British Colliery manager John Curr used a steam-engine for this purpose in his colliery near Sheffield: it was stationed at the shaft bottom and moved the loaded wagons along the 'tramway' (q.v.) to and fro on an endless rope. The idea did not catch on with the mine owners because it was considered too expensive: pit ponies, poor creatures which never saw the light of day, were cheaper. Later, compressed-air engines began to be used for underground transport, particularly in American mines, either stationary or in the form of small locomotives. Here, too, electrification came late, and diesel locomotives have also come to stay in some mines where ventilation is particularly good.

The general picture of mining technology shows rather slow progress until recent times, i.e. after the Second World War, when modernization was forced upon the mining industry in the western hemisphere by the competition with oil and, since the 1950s, nuclear energy. Many improvements have also been extracted from the managements – private or State-run – by the powerful miners' unions in the industrialized countries.

Today's coal miner is taken down to the working level, often a kilometre or more under the surface, in a high-speed cage; from the shaft bottom an electric train takes him to the distant coal face. The main galleries are lined with reinforced concrete, the side galleries have steel roof supports, and nearer the face, where there is the danger of rock fall, 'walking props' are used – hydraulic jacks which 'march' forward to take the strain of the roof.

In coal mining, the type of machinery employed at the face depends of course on the kind of seam. If it is high enough, a 'continuous miner' can be used, a very large machine with rotating cutters which rip the coal from the face – up to eight or more tons a minute. Some kinds of cutters, introduced in the 1960s, are fitted with electronic devices which enable them to follow the coal seam accurately, avoiding useless rock. In the wake of the cutting machines follow the conveyor belts – they may be 1,500 yards long – which take the coal to underground loading points, where it is tipped into trucks for the journey to the shaft; here it is hauled to the surface, washed and graded.

Where the deposit is soft and has a loose 'overburden', hydraulic mining – called 'hydrolicking' – can be used: powerful water jets are directed against the overburden, washing it away so that the deposit can be scooped up. This system is also widely applied in gold and tin mines; it needs an enormous supply of water, and large reservoirs.

Top, *automation in the mines: console of ROLF, the remotely operated long-wall face.* Above, *moving forward the hydraulically powered 'walking props'.*

Man Taking

In narrower seams continuous mining is often difficult; here, one miner drives a mechanical cutter mounted on wheels, cutting a slot along the bottom of the face, about six inches deep. This leaves the seam hanging from the overburden ready to be brought down by blasting. Following behind the cutter, a team uses pneumatic drills to bore the blasting holes into the face and pack them with explosives, which are detonated electrically after the miners have withdrawn to a safe distance. They return to scoop up the coal with 'crawling' loaders which place it on the conveyor belt.

Continuous mining, however, has been the aim of the mining engineers for a long time before it was developed into an effective technique in America in the late 1940s; Britain and Europe followed a few years later, while Russia had been experimenting with it since before the Second World War. It is the logical forerunner of more or less complete automation in the pits, of remote control in an industry where the human element ought to be eliminated as much as possible, for reasons of safety if not of efficiency. Automation underground, of course, cannot be introduced as a whole the way it was done in, say, the automobile industry where complete new factories can be built and taken into operation. A mine is almost an organism, and mechanization had to be introduced gradually: cutting machines at the face, mechanical transport, power loaders, conveyors came one by one within the space of decades. The pace quickened in Britain after the nationalization of the mines; from 1954 to 1960, for instance, mechanized output rose from 16 to 55 per cent. The next stage was the design and introduction of a number of remotely controlled machines at the coal face which can hew and load by themselves in one consecutive chain of operations.

A highly sophisticated automatic machine, first tried out in Leicestershire in 1960, worked with a control unit incorporating a 'sensing head' which emitted gamma rays from a radioactive isotope. The rays, reflected off the coal face, steered the cutting machine by means of an electro-hydraulic linkage system; this ensured that the machine remained in the mineral 'bed'. The next development was ROLF, the 'Remotely Operated Longwall Face', which came into operation at the Bevercoates colliery, Nottinghamshire, in 1963. From his control console some distance away from the face, on the roadway, an engineer supervises the automatic machine which cuts the coal and loads it on to the conveyor belt while the hydraulic jacks march forward, also controlled from the console.

There is a constant trend towards automation to relieve the miners of their most perilous and laborious jobs; an idea for utilizing coal without any underground work at all was the technique of 'underground gasification', tried out in the 1940s (but later abandoned) in the Soviet Union and Britain: shallow seams are made to burn continuously, and the resulting gas is conducted to the surface where it is processed for industrial and domestic use or made to power electric generators. The cost of such installations, however, has proved to be higher than the value of the product.

Open-cast mining requires a different technique altogether. This relatively modern process began in Illinois in the 1880s when huge power-driven shovels for the removal of soil and rock became available; it is an economical way of coal-mining up to depths of about 120 feet. Today, nearly a third of all coal produced in the USA is extracted by open-cast mining, which does not need the costly and elaborate pit-head installations of deep mining – there is just a vast trench from which coal is scooped up mechanically. In Western Germany, lignite mining is carried out on a large scale in this way; in Britain, open-cast coal-mining was begun only during the Second World War, but here only a few deposits lie so near the surface that they can be reached economically. 'Auger' mining, developed after the end of the war by American engineers, is an offspring of the open-cast technique: where open-cast seams continue in a hill, 'augers' – large boring machines – can follow them and extract coal cores, up to 5 ft. wide and 200 ft. long, in a corkscrew-like operation.

Iron-ore mining techniques are basically little different from coal-mining, but other raw materials from underground require production methods which have been largely developed during the last century – such as sulfur drilling. Though sulfur, most important in the chemical industry, is largely obtained in volcanic regions where it lies near the surface, huge deposits were found in Louisiana and Texas deep underground, covered by rock and sand layers. An American engineer, Herman Frasch, worked out a technique for extracting them around the turn of the century, based on the fact that the melting point of sulfur is only a little above the boiling point of water (100 °C). The Frasch process, therefore, consists of three operations: first, heating large quantities of water above its boiling point, then pumping it down a well into the sulfur deposit, and finally raising the molten sulfur up the well by hydraulic means.

Oil reserves can be located fairly accurately by the geologist with his modern sophisticated equipment, such as gravimeters and echo-sound receivers; however, the ancient Chinese found their oil wells, and exploited them, with very primitive tools as far back as the third century BC. The first oil well in modern times was drilled by a Colonel Drake near Titusville, Pennsylvania, in 1859, sparking off an 'oil rush' not unlike the Californian gold rush a decade earlier. Oil is usually trapped at high pressure below a layer of impervious rock – so as soon as that layer is pierced it will be forced up to the surface. The drillers bore a two-feet-wide hole and line it with a metal pipe, the drilling rig and overhead derrick are assembled, and the drilling starts. The drill has a rotating diamond bit which cuts through the rock; the process is a delicate one for the pressure put on the bit must be carefully watched so that the drill does not get stuck. Mud is forced down the pipe to wash away the rock waste and bring it back to the surface. More and more lengths of drill pipe are added, the bit has to be replaced when it is worn, and the hole has to be lined with more and more piping. Bore holes may be 10,000 or 20,000 ft. deep, and the drilling can take up to three months. And then the hole may turn out to be a dud, with all the expense and labour wasted!

Until a few decades ago, oil drilling was confined to the land, but geologists predicted great finds under the sea. It was only in the 1940s that underwater drilling was first begun in the Gulf of Mexico – and within ten years or so more than a thousand wells were gushing oil from under the sea in that region, some of them over 50 miles from the coast. For underwater drilling, completely new equipment had to be designed, and when the first extensive reservoirs of oil and natural gas were discovered in the North Sea in the late 1960s, drilling rigs or 'artificial islands' for exploiting that unexpected bonanza were already highly developed. Early platforms had been just wooden pile structures which were towed to their locations and then anchored; or converted barges were floated out and submerged. From 1949 onwards, special types of barge platforms were designed and constructed, with long telescopic legs which are lowered to the seabed; the platform which carries the rig is then jacked up until it is at a safe height above the sea. This type is suitable for drilling in depths of up to 250 ft of water. Where the water is too deep for bottom-supported platforms, the usual solution is a submerged but buoyant hull with the drilling deck supported above the water by thick buoyancy columns. The rig units are, as a rule, large enough to accommodate the crew in sleeping and living quarters; they 'commute' with the shore by speedboat, hydrofoil, hovercraft, or helicopter.

These are the alternative types of equipment used in North Sea drilling for oil and natural gas but at still greater depths drilling may have to be carried out directly on the ocean bed from submersible chambers. Their interior would be pressurized with a helium-oxygen mixture if human operators are at work, but the ultimate aim is robot drilling, remotely controlled from surface vessels or platforms.

Offshore oil rig in the North Sea.

American oil derrick, 82 ft. high, erected in 1891. Insets show the first derrick put up in America (1859, near Titusville, Pa.), 34 ft. in height, and Col. E. L. Drake who drilled the well.

New types of pipelines had to be developed to bring the oil and gas on land by way of the ocean bed, backed up by new coastal installations for processing them for industrial and domestic use. Thanks to these riches from the bottom of the North Sea, Britain and Western Europe will be largely independent of overseas energy supplies well before the end of the century.

E. L.

Synthetic minerals

Before the turn of the century only tiny crystals of emerald and ruby had been grown; nothing had been produced on a commercial scale. Then about 1902 the French scientist A. Verneuil perfected his 'flame-fusion' method of producing ruby and other corundums, which proved capable of growing unicrystalline 'stalagmites', weighing several hundred carats, in a matter of hours. Later this method was applied to the manufacture of spinel, rutile, strontium titanate and other gem materials. They were first developed for commercial purposes, as laser materials and semiconductors, for bearings and pivots in watches and meters, etc. Only 10 or 20 per cent of the millions of carats produced annually now is used as a substitute for natural precious stones, the majority being used commercially. As substitutes for natural gemstones, ruby and blue sapphire are the most favoured colours.

In the Verneuil 'flame-fusion' method of synthesizing minerals, finely powdered chemicals of the desired composition are sifted through the flame of an inverted oxy-hydrogen blowlamp, and fall as molten droplets on the stem of a pipeclay pedestal, enclosed in a small furnace chamber. Skilled control of the flame and powder concentrate the growth of the resulting mass into a single-crystal stalagmite. Finely powdered gamma-alumina is used for making corundum gems, and chromic acid is added for producing ruby, titanium and iron oxides for blue sapphires, nickel compounds for yellow corundums like topaz. A purple type to imitate alexandrite and a colourless corundum used as a substitute for diamond in cheap jewellery are also produced. The Verneuil method cannot be used with substances such as quartz or garnet.

There are other methods of producing synthetic gemstones, for example the Czochralski method of 'pulling' – a slow controlled withdrawal of a seed crystal dipped in a melt of the same composition – can result in the formation of a clear single crystal of indefinite length. Large crystals of scheelite, fluorite and rare-earth garnets have been made, also very pure rods of ruby for laser work.

Diamonds are more difficult to synthesize than other minerals, because of the structure of the crystals requiring very great heat and pressure to crystallize and stabilize graphite carbon. A team of scientists working for the General Electric Company in America, produced a synthetic diamond in February 1955, and the De Beers Research team in the Adamant Research Laboratory in Johannesburg, also succeeded in 1959.

Diamond being the hardest known mineral is much used in industry for tool-stones for truing and dressing abrasive wheels, for die-stones which are pierced and used as a die for wire-drawing, for drilling stones which are set into a drill bit or crown for drilling. The poorer qualities and smaller sizes are crushed to form diamond grit or powder, used as an abrasive or polisher.

T. C.

Stainless steel

Stainless steels are corrosion- and heat-resisting alloys composed chiefly of iron and chromium with further elements added to produce desirable modifications. Several inventors – Faraday among them – produced iron-chromium alloys in the 19th century but these were not true steels, and as late as our own century the English metallurgist Robert Hadfield had reached the erroneous conclusion that chromium actually *impaired* corrosion resistance. The story of stainless steel's invention is complex, therefore, not least because even when in the early 1900s various scientists (notably the Frenchmen Leon Guillet and A. M. Portevin) did produce alloys in the stainless steel group they failed to recognize the outstanding property of what they had produced, that is, its power to resist corrosion.

This quality was first recognized by the Germans P. Monnartz and W. Borchers, of whom the former obtained a German patent on his steel in 1911. But it was the Englishman Harry Brearley, a self-taught metallurgist who was head of a research laboratory run jointly by the steel firms of John Brown and Thomas Firth and Sons, who really deserves to be regarded as the 'inventor' of stainless steel. In 1912 Brearley discovered the important martensitic alloy and developed a tough, magnetic, anti-corrosive steel to be used for the manufacture of naval guns. The military authorities were unenthusiastic, however, so Brearley suggested it might be useful for cutlery. But his employers, apparently without consulting him, had some knives made from his alloy and pronounced them worthless.

Brearley was undeterred, however, and had some more knives made himself which were far more successful. Production of the martensitic type began in 1914. But by now Brearley had taken his knowledge elsewhere, obtaining an American patent in 1915 since Firth's

Synthetically produced crystals of boron nitride.

had not thought it worth their while. In 1920 his new employers, Brown Bayley's, introduced commercially the ferritic alloys, a development of the martensitic which can be worked hot or cold, are non-hardenable and are especially useful for architectural use and such things as motor-car trimmings.

Many other inventors are associated with the beginnings of stainless steel, however, and it would be incorrect to give the impression that Brearley was a lonely and misunderstood genius. In the United States F.M. Becket played an important role, as did Elwood Haynes, an Indiana inventor who had devised a process for producing tungsten chrome steel as far back as 1884 and was also a pioneer in the automobile industry (he built a gasoline engine in 1893). A third group of stainless steel alloys, the heat- and shock-resisting austenitics, now widely used in the food industry, for chemical equipment and in combustion chambers, was developed by Guillet and W. Giesen, but again its corrosion resistance was imperfectly realized and the credit for this group should more fairly be given to Edward Maurer and Benno Strauss of the research department of the German firm Krupps, where it was first produced in 1912.

Over a hundred different stainless steels are produced today, and stainless steel is used in a wide variety of products from spacecraft to jewellery.

J.G.

Float glass

The first glass was blown into shape as it cooled, then various additives such as calcium and lead were found to improve the brilliance and purity. Highly complicated processes were evolved to produce high-quality flat glass. One method involved drawing a ribbon vertically from a furnace up the side of a tower (much as a sweet-maker draws up a flat ribbon of hot toffee). This resulted in a brilliant glass acceptable for, say, domestic windows, but the larger the size the more the characteristic wavy distortions could be seen.

The high-quality glass needed for shop and car windows and mirrors was supplied by polished plate. Molten glass from a furnace was rolled into a continuous ribbon, but though the surfaces were parallel the roller inevitably left marks. The ribbon had to be ground and polished on both sides, which meant a lot of glass wastage and cost a lot of money.

The ultimate achievement was to combine the best features of both processes – a brilliant surface of drawn sheet glass and the parallel surfaces of rolled polished glass. As with most revolutionary ideas the concept appears obvious in retrospect: why not float the glass on the naturally flat surface of a liquid? Alistair Pilkington conceived the basic idea in 1952 but it took seven years and £7 million to bring it to fruition, during which time it had to be kept a closely guarded industrial secret. His studies interrupted by the war, Alistair Pilkington returned as a married man to Cambridge in 1945 to complete the Mechanical Sciences Tripos, and joined Pilkington Brothers, the firm of glass-makers, in 1947. After heading the team that developed the invention, he was knighted in 1970 and in 1973 the ultimate prize became his when he became chairman of Pilkingtons – the fairy-story quality of his achievement enhanced by the fact that although the name is the same, he had no family connection with the famous firm.

This is how float glass is made: a continuous ribbon of glass is drawn out of a melting-furnace and floated on to the surface of a bath of molten tin. The molecular structure of the tin is denser than that of the glass and so the tin supports it at a high enough temperature for a long enough time for the irregularities to melt out and the surfaces to become flat and parallel. Natural forces determine a glass thickness of 6 mm., which by lucky chance meets half the market demand, but refinements can now vary this thickness. The ribbon is then gradually cooled while still floating, until it solidifies sufficiently to be rolled out of the bath without marking and finished off in annealing chambers. The result is top-quality glass, even in thickness, and with a brilliant surface that needs no grinding or mechanical polishing. As this can be produced for the same price as the old, cheaper flat glass, it is no wonder that manufacturing licences have been granted to 22 manufacturers in 12 countries; in 10 years float glass has virtually replaced all other makes.

S.M.

Above, *the research laboratory run jointly by two British steel firms, where Brearley invented stainless steel.*

Below, *continuous ribbon of float glass, 4 mm. thick. The reflection suggests the perfect flatness produced by this method.*

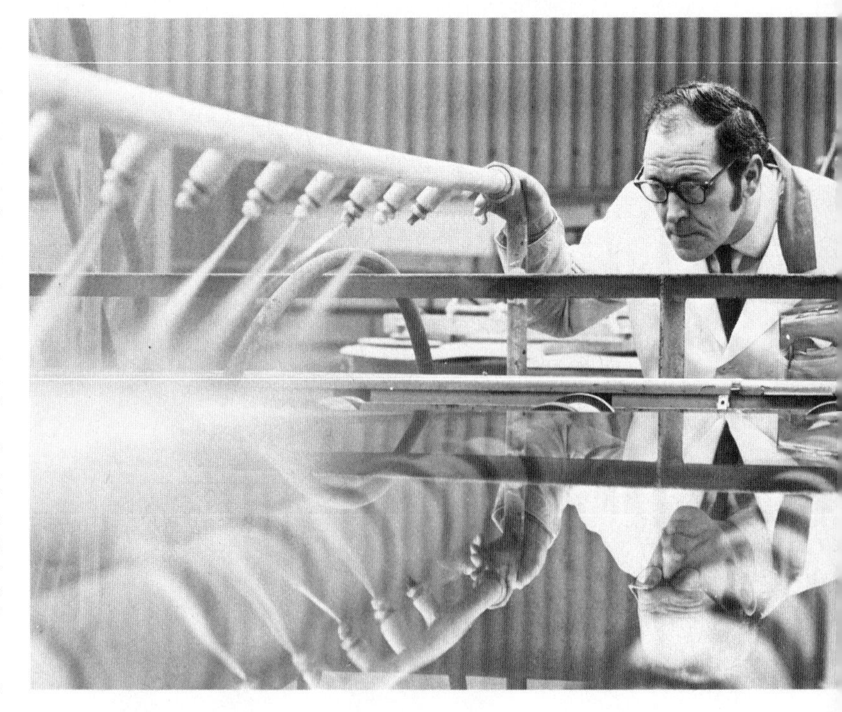

MAN LIVING

Food and agriculture

Clothing

Building

Domestic

Health

Protection

Organization

MAN LIVING

Food and agriculture

It seems paradoxical that the pressure for new ideas and inventions has never been very great in the area of food and agriculture. This is because, until quite recently, the established methods of obtaining food were on the whole adequate to cope with the established demand for food in a relatively stable population. Food and agriculture are based on natural processes and these continue to function with or without man's ideas.

We know that Jethro Tull, a musician and lawyer turned farmer, invented in the early 18th century the seed drill which allowed farmers to plant more effectively and in straight rows. But we have no idea who invented the plough or the exact circumstances in which it came to be invented. It probably developed gradually over a long period and as a result of modifications introduced by many different people.

At first, most food was obtained by hunting. We know that people were eating mammoths about 200,000 BC but have no information on how they caught these huge creatures. Unless they waited for the animals to drop dead or be killed by other creatures, they must have used traps or weapons of some sort. Over the ages a variety of traps have been developed and many of the most basic forms are still in use. Weapons such as spears, harpoons, arrows, blow-guns were also used in hunting. For catching fish, men used spears and women fish-hooks. The first fish-hooks, probably in use about 8000 BC, were made from bone or shell. A simple slip of bone tied in the centre and sharpened at each end would be swallowed by the fish; tension on the line would then cause the slip to turn crossways and jam in the gullet. The curved fish-hook and the barbed fish-hook came later. Like many inventions, they had to await metal technology for their perfection.

When hunting was replaced by agriculture as the major source of food in many cultures, a whole new area for ideas opened up. There was the need for irrigation, for mechanical devices both to work the soil and to harvest the crop, for methods of using the crops, for new crops, for fertilizers, weed-killers, pesticides. It has been suggested that the great civilizations of the ancient world developed in areas requiring deliberate irrigation because this need created a variety of inventions. Irrigation required land survey and measurement, an organized central bureaucracy and administration, mechanical devices (such as pulley, screw, gear, piston, etc.) for raising the water, and perhaps the need for an army to defend scarce water supplies. The ultimate in irrigation is the artificial creation of rain. To some extent this has been achieved by the seeding of rain-bearing clouds so that they discharge their content of rain. A practical method, using ice crystals, was first suggested by the French meteorologist Bergeron just before the Second World War. Since then there have been many different ideas, including the use of silver iodide smoke suggested by B. Vonnegut in 1947. This provides a nucleus on which water vapour can condense. It is a far cry from sacrifices to the ancient rain gods to do-it-yourself rain-making.

Simple mechanical devices have played a large part in agriculture. The simple scratch plough was adequate in light Mediterranean soils but useless for heavier, wetter soils in northern countries. Agriculture in such areas had to await the development of the hea[vy] plough or carruca which actually turned the soil instead of ju[st] scratching it. The use of animal power and, later, steam power [to] pull the plough increased the area of land a single man could ploug[h] but at the same time may have made some sort of local organizati[on] necessary (larger fields, and the capital to buy the equipment [or] share it). When the land was prepared, the next stage was to se[ed] it. Scattering seed by hand was always adequate, but it is surprisi[ng] that the seed drill was so late developing. When the crop was rea[dy] it had to be harvested. The Romans experimented with a mechanic[al] reaper, but interest in this device revived only at the beginning [of] the 19th century. An inventor who also happened to be an act[or] demonstrated a mechanical reaper on stage in 1814. Twelve yea[rs] later Bell developed a different machine, but refused to patent i[t]. The result of this humanitarian gesture was that there was litt[le] money for its development. But even so, there may have be[en] meagre interest in mechanical reapers because agriculture was a wa[y] of life for so many people that labour-saving as such was not [a] particular goal. For instance, the invention of a threshing machi[ne] led to riots, because threshing was something done during the wint[er] by people who had no other work and could not work on the la[nd] at that time of year. In fact, the mechanical harvester came into i[ts] own only when the American Civil War made agricultural labo[ur] scarce. Once the path of development had been set, it could b[e] followed right up to the combine harvester we know today.

It is interesting that the Romans (and the Chinese) had qui[te] advanced agricultural ideas which were then lost for about 1,0[00] years in the European Dark Ages and their aftermath. For instanc[e] crop rotation was practised by the Romans but only rediscover[ed] 1,000 years later by 'Turnip' Townshend and others. It was n[ot] known that certain plants returned to the soil the nitrogen extract[ed] by other plants, but on an empirical basis it did seem to work. [As] soon as science had worked out why plants needed nitrogen the[re] was a rush to artificial sources that could be fed into the soil. Sodiu[m] nitrate occurred naturally in Chile, but it seemed that technolog[y] ought to be able to make agriculture less dependent on a sing[le] source, and one so far away. Technology *was* able – in the person [of] Fritz Haber, who found out how to make the first artificial fertiliz[er] by chemical synthesis.

Growing crops are attacked by insects. Homer mentions t[he] burning of sulphur to protect crops and the Chinese evidently us[ed] arsenic sulphide. Today a variety of insecticides are availab[le] including the famous DDT, which was discovered by one man, Pa[ul] Muller. DDT is an example of an invention which came just at t[he] moment when everyone was ready to use it – towards the end of t[he] Second World War when insect-borne diseases such as typhus we[re] liable to become rampant with mass movements of population a[nd] the breakdown of hygiene. DDT was so effective that its use so[on] spread into agriculture, where it was dominant for a long time un[til] it was realized that it could constitute a serious pollution hazard.

The development of weed-killers followed the classic path [of] chance discovery. A disease among the Bordeaux vines made cessary to spray them with a mixture of copper sulphate and lime a fungicide. A grower called Bonnet noticed that this mixture led off some weeds growing near the vines. From this came the velopment of chemicals that were selective in their action and uld be used to kill weeds and leave the crop intact. The develop-nt of these is now so sophisticated that one variety actually makes of the weed's own hormones to make it grow in so unnatural a nner that it kills itself.

The ultimate area for man's inventiveness is the crops themselves. lective livestock breeding in cows, pigs, sheep and poultry has netimes shown remarkable results in terms of yield or resistance disease. 'Miracle' rice is an example of a deliberate technological ort that succeeded in doubling the yield of rice from a given area der cultivation. This effort was carried out by the International ce Research Institute in the Philippines. Since rice is the staple d of 60 per cent of the world's population, an invention such as s can have profound effects.

The preservation of food is an area that has exercised man's entiveness over the ages. Drying, smoking, salting, pickling, oling, freezing, pasteurizing, canning, gamma radiation have all en used. Many of the processes provide no clear history of inven-n, but certain ones do. Jacob Perkins first worked out the idea of rigeration as a deliberate method – although the use of ice d snow for preserving was known from ancient times. The first ctical refrigeration devices were, however, made by a printer, nes Harrison, who had emigrated to Australia. The need to carry d from Australia to Britain over a long sea voyage must have ovided a considerable impetus to the development of the inven-n. Deep freezing was also the idea of one man, Clarence Birdseye, o noted that the speed of freezing was more important than the v temperature eventually reached.

The interesting things about food canning are that it was invented g before the reason for its effectiveness was known, and also t it is an invention that resulted from the offer of a prize. During Napoleonic wars there was a great need for food preserving in ler to feed the ever-mobile armies. Nicolas Appert won the prize 1809 for a food-preserving method with a sealed bottle into which ated food had been introduced. The metal can came later, and er still the can opener.

It was Pasteur who showed, more than fifty years later, that heat uld kill micro-organisms that made food rot and that if such ganisms were kept out (as they would be in a sealed vessel) the d would be preserved. The heating part of the process we recog-e today in the 'pasteurization' of milk, which is carried out as ch to kill dangerous organisms, such as the tuberculosis or un-lant fever bacteria, as to preserve.

But what about man's inventiveness as applied to actual food lf? This has been rather limited. There are processes, like the king of wine, beer, and spirits, which have been known from tiquity but which have been improved with new techniques, thods and apparatus. Then there are methods of processing food, h as the making of pasta and breakfast cereals. Items like instant fee could be included under this heading. Finally there are totally ificial foods, of which the main example to date is margarine. some cases, such as breakfast cereals, which started as health ds, there has been an individual inventor. This was true also of tant coffee, the success of which was guaranteed by the First orld War which caused the US government to buy vast quantities t.

In general, the history of invention in agriculture is rather more dued than in other areas. There is, for instance, nothing to com-e with the invention of the steam engine, printing, radio or the Bessemer steel process. Because people have always eaten, there have always been ways of getting food by agriculture or hunting. New inventions may have made things easier, less labour-intensive or more suitable to different terrains, but at the moment of the invention the world could probably have done without them. In the long run, of course, the cumulative developments have been most necessary to feed the ever-growing world population.

Clothing

With regard to clothing, the history of man's inventiveness has been remarkably dull. That is to say, there have been very few ideas which have affected the final product. On the other hand, the process of making the material for the clothing has shown an extraordinary series of improvements, to the extent that the number of man-hours involved has decreased a thousandfold over the last two hundred years.

Man's clothing has been made from leather and animal skins, felt and woven (or knitted) cloth. That is all, unless one includes armour or the rubber wet-suits used for skin-diving today. In the case of felt the animal hairs are pressed together under conditions of heat and moisture, and form a pad. This is much easier than weaving and seems to have been extensively used in Central Asia from very early times. Today there is a good deal of development work directed at going back to the felting principle by melting layers of artificial fibres without weaving them. The process would be very much faster.

Spinning to produce the yarn and then weaving it into cloth are basic processes that have been practised for a long time. Beginning in the 18th century, there was a series of inventions which took the processes out of the cottage and into the factory: Kay's flying shuttle, Hargreaves's spinning jenny, Crompton's mule. Each of these mech-anized and speeded up processes that had otherwise to be done rather slowly by hand. (Like several other inventions, Hargreaves's spinning jenny was the result of a competition.) At first, these inven-tions were strongly resented by the workers who would be put out of a job (just as the threshers resented threshing machines). But as soon as the work was organized on a factory basis with power supplies, and some entrepreneur to take the profits, the interest in new devices increased if they could be shown to have a definite advantage (as did Crompton's mule, which made much finer fabric). The import-ance of the entrepreneur is shown by the story of the knitting frame invented by the Reverend William Lee. He took his idea to the court of Queen Elizabeth I but found no response. His efforts on the Conti-nent were no more successful. After he was dead, his brother was lucky enough to meet a merchant who was interested, and from then on the idea prospered. Without it, there would be no stockings or tights today.

The invention of the sewing machine, too, showed both the opposition of the threatened workers and lack of interest elsewhere. It is interesting that the workers actually showed a keener apprecia-tion of the potential of such ideas than the entrepreneurs of the time. The first sewing machine, made by a French tailor to speed up the making of army clothing, was wrecked by a mob. The next develop-ments occurred in the US, and seem to have been effected by different people in parallel. As no one showed much confidence in the invention, Elias Howe travelled to England to try to find more interest. Eventually he returned to the US to learn that by then several manufacturers were actually making his machine – without bothering to ask his permission.

The value of quite a simple device is illustrated by the cotton gin devised by a young lawyer, Eli Whitney. At one stroke this simple device turned the Southern states of the US into competitive cotton producers. Before that, the variety of cotton they could grow required too much hand cleaning to compete with other sources.

Cotton, wool, silk and linen were the basic fibres until man succeeded in inventing his own. The story of this invention is a particularly curious one, because several people in turn invented the process for making artificial fibres but no one perceived its potentialities. At one time the process was re-invented in a search for a suitable filament for electric light bulbs. Eventually Count Hilaire de Chardonnet produced some artificial silk and displayed it at the Paris Exhibition of 1889 – after that it was easy. Once artificial fibres were in use the search for better ones accelerated. In 1938, after 10 years' research, Dr Wallace Carothers discovered nylon, which still remains the most-used artificial fibre.

Bleaching and dyeing have always been an important part of the preparation of material for clothing. Many methods of bleaching had been developed by trial and error and tradition. The discovery of chlorine in 1774 by Carl Wilhelm Scheele led to the production eleven years later of a chlorine-based bleaching fluid by Claude Berthollet, who passed on the information to James Watt, who in turn passed it on to Charles Tennant of Glasgow. As so often happens with inventions, Charles Tennant improved the process to provide the basis of modern bleaching powders. This whole development was an unusually swift instance of the passage from a scientific discovery to a practical invention.

An 18-year-old student working in a shed at the bottom of the garden discovered aniline dyes. The fact that the mauve colour he first produced was taken up by fashionable ladies assured his success. From this discovery came all the other aniline dyes and derivatives; almost overnight the whole business of colouring cloth had changed.

The history of clothing fasteners concerns pins, buttons and the zip. Pins were in use from ancient times; even the safety pin itself is old, but the modern version of it was invented, so it is said, one afternoon by Walter Hunt, who was twisting a piece of wire into various shapes. The story of buttons includes various types and methods of manufacture, which gradually reduced their price to the point where they were no longer the prerogative of the rich. The zip fastener, in contrast, is a single, specific invention. The idea was suggested in principle by Whitcomb L. Judson but his idea was not quite perfected; the successful zip was eventually designed by a Swede, Dr Gideon Sundback.

The aesthetic side of clothing also exercised man's ingenuity, sometimes in odd directions. The corset, which appears first in Minoan Crete, is a formidable apparatus with a long history and is designed to give to women the shape they wished they had. A breakaway from the corset was the brassière designed (or re-designed) by a New York debutante – one of the very few examples of a patented invention made by a woman.

Cosmetics are included in this section because they are a sort of aesthetic clothing. They have a very long history indeed. The principles have remained much the same, though the actual materials used may have changed.

Building

About 40,000 BC, man was making tents with mammoth bones as tent-poles. He probably used wood as well, but bones last and wood does not. The tents in use today by certain nomadic tribes and by the ultra-modern holiday camper may not differ much in style althou the materials have certainly changed.

As a building device, the column is related to the tree, the te pole, and the modern skeletal steel structure. It performs the v simple function of support. From a tree-trunk column to one of sto or marble is a small step. The Greeks were so delighted with t structural and aesthetic values of columns that they did not chan their style of building for over a thousand years, even though th knew about the arch.

The arch can be looked at as two bent columns meeting at t top. Its advantage is that there is no need for a lintel to lie across t columns and support weight. The strength of this lintel has alwa been the weakness of column construction. As a device, the ar solved this problem instantly and perfectly by leading the support weight back into the columns. The Assyrians, the Babylonians, t Greeks and the Egyptians understood the principle of the arch b did not need it. The Romans, who were much more engineerin minded, made great use of the arch. It was much cheaper than a f load-bearing wall and involved less transport of materials. (T Romans were probably more interested in such practical matte than in aesthetic ones.) The Romans used the round arch, whi needs a keystone, yet in many ways the pointed arch, as used Gothic and Muslim architects, is much more efficient since it requi less massive support and the tendency to splay the supporti columns is less as the supported weight is led more directly to t ground.

The vault, like the arch, uses the same compression-type buildi materials to cover spaces. Again the Romans could only constru rather heavy barrel vaulting, whereas the Gothic builders used t pointed vault and also developed the ribbed vault and fan vau These were genuine developments because they meant that the weight was taken by the ribs and the space between could light in-fill, unlike the solidity of a barrel vault. In the fan vault the was the ingenious device of using a pendant weight to hold the ar together – in other words a column that worked by hanging dow wards instead of pressing upwards.

The dome, which first appears in 2500 BC, is a sort of rotated ar and again serves to cover space without any need for 'tensio building materials. In this area we find a unique example of inve tion in Brunelleschi's building of the dome of Florence Cathedr He invented the external tie or hoop which, by preventing an ou ward bulge, made it possible to support the dome on a simple dru

Scaffolding used for constructing buildings, and especially arch and vaults, was probably not much different from that used toda except that it is now made of metal. The other essential device f building was the crane. Again the many early variations have su vived to this day with only slight modifications. Pulleys and winch were a necessary part of the crane and are described elsewhere.

From the crane as a lifting device used in the course of constructi to the lift as a device used in the function of the building is a sma step but in its effects a big one. The safety lift was invented by o man, Elisha Otis. Before his invention, buildings could not be mo than five stories high – the limits of stairway use. In a way, the devi is as basic as the arch or the dome. It is true that high buildings cou not have been constructed without steel framework constructi techniques or possibly concrete technology, but nevertheless t lift remains the key device.

As a building material concrete is described elsewhere (see pa 86). Stone and wood are obvious building materials, in which ve little change has taken place. One of the few inventions in this fie was that of plywood, to provide a thin but strong wooden materi The theory and practice of plywood were first worked out ichael Thonet in about 1830, and patents for the process were ing taken out in the US from 1865. The principle of using thin eets glued together with the grain at right angles was an excellent .e; the result was to create a 'tension' material.

The other basic building material has been bricks, which go back at least 6000 BC. At first they were hand-formed and sun-dried, it later they were moulded and fired in a kiln. The smaller size of ick, like the one we know today, was probably first used about 3500 : because it was easier to fire (not because it was easier to carry). ie Romans used bricks widely but, like so many other things, the ocess seems to have been neglected for many hundreds of years itil it emerged again at the time Henry VIII was reigning in England. the interval, important buildings were made of stone and unim-rtant ones of wood (with lath and plaster).

Though there have been steady developments in building, not any of them have been significant changes. Once something was equate there seemed little reason to change it. The situation is tally unlike war, when a weapon may be seen to be obsolete in a v moments and the urge to obtain an advantage is very great. .cept for skyscrapers, and recent developments of such ideas as flatable buildings, our ordinary buildings are probably similar the ones built four or five thousand years ago.

omestic

ie most striking thing about man's domestic innovations (at least the Western world) is the way progress went backwards. At the ight of the elegance, luxury and civilized living of Versailles, there re no bathrooms or toilets in the palace, yet the Romans, many ndreds of years before, had developed bathing to a high standard. itil the 19th century most cities in Europe had no proper drainage tem and yet the Hindus at Mohenjo-daro had drains and latrines 2500 BC and the Romans again had well-developed drainage tems. The Romans had a central-heating system which used hot from a furnace, conducted through channels in the walls. The ne principle was rediscovered only in 1908. The water closet, too, a relatively recent invention (Bramah in 1748 and Cummings in 75). The idea, however, must have existed before then because ieen Elizabeth's godson had told her about them and she had one ilt for herself.

Soap is probably the most ancient chemical invention, and seems have been used by the Sumerians about 3000 BC. It was obtained boiling animal fat with alkali such as that obtained from wood or crude soda. It was not until the 1950s that soap was displaced some of its functions by synthetic detergents. The history of mestic lighting shows a smoother curve of development in so far there were several innovations following each other: the candle oil lamp provided basic illumination for thousands of years; rious improvements were made, including one suggested by onardo da Vinci; the hollow wick and chimney, invented in 84, reduced smokiness. The most effective invention was made by tson in 1885 when he introduced the mantle lamp, which burned porized oil to heat a mantle into incandescence. The basic nciple is still that of an oil lamp but the increase in light is enor-ous. By then, of course, there were other types of lighting. Gas hting had been tried experimentally ever since 1760 but was only en up in a practical commercial way after Lebon's demonstra-n in Paris in 1801. Even so, the use of the idea awaited a com-rcial company with the money to invest in piping, etc. The gas ntle was an important innovation (by Auer von Welsbach) which made it possible for ordinary coal gas to give a much brighter light. But by this time the electric filament lamp had been discovered (Swan and Edison). In its turn, the filament lamp was superseded for many purposes by the fluorescent lamp developed by 1934.

If we examine individual domestic appliances, we can see in them instructive examples of the different paths of innovation. With the washing machine it was a matter of applying the convenient electric motor to a method that had already been developed. Agitation within a box by use of a 'dolly' arrangement or by moving the box itself with a handle were already in use, and the addition of a motor was obvious. The non-stick frying pan may be obvious with hind-sight but there was a gap of seventeen years between the discovery of PTFE and the notion of using it on frying pans. The development of the vacuum cleaner is similar to that of the washing machine in so far as there were old vacuum cleaners worked by bellows, but this time there was a crucial invention. This was made by Hubert Booth in 1901 when he introduced the idea of filtering the sucked air through cloth to remove the dust. The pressure cooker is an example of a device which was fully developed more than two centuries before it was re-invented. In 1679 Denis Papin carried out several demonstrations showing that cooking under pressure was faster than ordinary cooking and retained more flavour. A good example of deliberate invention is that of the safety razor by King Gillette. He was resolved to invent something and one day the idea of the dis-posable razor came to him as he shaved. It took him nine years of determined effort to get his invention under way and make it a huge commercial success. The fluorescent lamp is a case not so much of determination as of development. Fluorescence had been known for a long time and neon lights were using the principle for adver-tising. But it still took years of research before the right colour of light could be produced. The vacuum flask is an example of a household idea taken straight from the science laboratory. The first Dewar flask was made in 1892, but it was in 1904 that Reinhold Berger saw the commercial possibilities of the idea. Another example of an invention taken from the area of science and engineering is microwave cooking. The cavity magnetron which made this possible had been developed to provide a way of generating radar waves and was a vital device during the Second World War.

The carpet sweeper was developed by a Mr Bissell, who kept a china shop but seems to have been allergic to the straw in which his wares were packed. This personal reason for invention contrasts strongly with Gillette's quest for a commercial invention. Wallpaper became cheap enough for general use only when someone thought of using a calico-printing machine to print it in continuous rolls instead of hand-blocking each piece.

Domestic appliances offer a fertile area for invention because there is a definite market, the items are relatively small to make, and the coming of advertising has made it possible to broadcast claimed advantages more rapidly than ever before.

Health

In this area there is no dividing line between science and technology or between discovery and invention. Invention usually implies putting together in a new way knowledge that has been available before; discovery implies finding something new. But in the field of health and medicine a discovery may follow an inventive idea that looks at something in a new way, or it may involve following up an old idea and using it in a new way. The examples included in this section could equally be regarded as scientific discoveries. On the other hand, many other examples would have as much right to be included here as the examples actually used. The examples do, however, illustrate many of the processes of innovation.

Chance has probably played a more important part in medicine than in any other field of science. This is hardly surprising in view of the extreme complexity of the body; until there is complete understanding of it many discoveries do have to come from chance observations. The very basic principle of inoculation was discovered by chance by Louis Pasteur. It involves giving a weakened dose of an infective agent in order to protect the body against serious infection with the same agent. The principle is still at the core of preventive medicine (anti-typhoid, anti-cholera, anti-rabies, anti-polio inoculations). It was only much later that science caught up and showed that the inoculating dose excited the antibody system to prepare defences against the infecting agent.

Antibiotics, which as a group are probably the greatest invention in the field of medicine, were also the result of chance. It is well known that the first antibiotic, penicillin, was discovered by Alexander Fleming when an airborne mould contaminated a culture plate. The incredible X-rays which allow a doctor to look through living skin and flesh at the bones beneath were discovered by chance by Röntgen in 1895 (in fact their greatest use has been in detecting tuberculosis rather than in looking at bones). Even the humble stethoscope was the result of a chance observation by Laënnec, who noticed some children playing with a piece of wood through which sounds were conducted.

Another source of innovation has been the pursuit of folk medicine or old wives' tales. Aspirin, the most used of any medicine, had long been a folk remedy, as it occurs naturally in the bark of certain trees. It was synthesized in 1899 by Hermann Dreser, but even today no one knows exactly why it works. Digitalin, one of the main drugs used in heart disease, was also part of folk medicine. The incredibly effective principle of vaccination, which at once removed the scourge of smallpox, was the result of an old wives' tale listened to by Edward Jenner. But it took him twenty years to summon up the great courage needed to vaccinate an eight-year-old boy with cow-pox and then deliberately to infect the boy with smallpox.

Transference has played a part in many medical inventions. For instance, Sir Humphry Davy had suggested in 1799 that nitrous oxide, used to give people a 'high' at fashionable parties, could also be used as an anaesthetic. Eventually ether, which was played with in a similar manner, was the first anaesthetic used for a major operation (by William Morton at the Massachusetts General Hospital in Boston).

Medicine has been largely governed by ideas. Usually these have been disastrous (for instance bleeding, or the water treatment of the insane), but occasionally a good idea is taken up ahead of the scientific proof for it. This happened in the case of the germ theory of disease, which became current about 1860. Lister, who took up the idea, applied carbolic dressings to wounds and devised his carbolic spray for disinfecting the air. Although this spray was probably useless it served to symbolize in a practical form the antiseptic attitude which eventually led to the aseptic form of surgery we know today.

Deliberate search has, of course, also played a part in medicine. The search by Paul Ehrlich for chemicals that could kill germs resulted in Salvarsan for treating syphilis. The search for further chemicals led to Prontosil, the first of the sulphonamides (which were actually much more 'scientific' than the more powerful antibiotics). Then there was the search by Gregory Pinus and his associates for a hormone that would have a contraceptive effect – with the outcome we know as 'the Pill'.

Alongside these biochemical innovations there have bee mechanical ones, which are somewhat less vital in their functio For instance, the invention of the iron lung cannot compare wit that of the polio vaccine. Ambroise Paré first started the developme of artificial limbs because as a military surgeon he had to deal wit so many amputees. The artificial kidney was invented by Willia Kolff in 1944 but was made more practical when Scribner showe how tubes could be permanently inserted in artery and vein f periodic attachment to the machine. George Washington's dentis John Greenwood, pioneered the dental drill by adapting his mother spinning wheel. Einthoven was able to perfect the electr cardiograph, which measures the electrical performance of th heart (and is the major diagnostic tool in most heart disease), b cause he had developed the string galvanometer which was sensitiv enough to record the tiny amount of electricity produced by th beating heart.

We take spectacles for granted and yet it could be argued th they are the most important health aid of all time. A vast number people wear spectacles and would be at a very great disadvanta without them, since the sense of sight is so predominant in man. Y we do not know who actually invented the first spectacles thoug it probably happened in about 1280. On the other hand, we d know who invented the ophthalmoscope, which allows a docto to look right into the eye, and the story of this invention illustrat three basic principles of innovation. The first inventor seems t have been Charles Babbage in 1847 – the same man who is regarde as the father of the computer. So *the first principle is that a person wit an inventive frame of mind often invents things in widely different field* Babbage gave the device to a surgeon friend but it was neglecte In 1851 Hermann Helmholtz independently invented the ophtha moscope and this time it was taken up by the medical professio *The second principle is that independent inventors often invent the same similar devices at about the same time* (as in the case of the intern combustion engine and the sewing machine). *The third principle that it is not enough for an invention to be good. It may still be totally neglect unless the inventor or someone else has the determination to carry it throu until the world sees its value.*

Protection

Man has found the devising of weapons with which to kill his fello men an area remarkably suitable for the exercise of inventiv genius. It is a far cry from a lump of rock hurled by one caveman another to an inter-continental ballistic missile travelling 7,00 miles at 15,000 m.p.h. and carrying a hydrogen-bomb warhead but the principle is the same. Often man has claimed, sometim with sincerity, that he needed to develop good weapons to prote himself from others who were likely to attack him. But wheth protection or aggression is the aim the result is the same: the inve tion and development of new weapons.

When Leonardo da Vinci wrote to apply for a job with Lodovic Sforza, ruler of Milan, he offered his services not as a painter but a military engineer and inventor. He realized that weapons had measurable value whereas paintings did not. The invention an development of weapons has always been an ideal situation for a inventor, for a large number of reasons. The need for new weapo is continuous. It is not a matter of solving a problem, like conqueri cholera or designing a water-closet or printing at speed. The proble is open-ended because there is always a better weapon that cou ll more people at a greater distance. Perhaps the hydrogen bomb the ultimate weapon, but that is quite recent. The situation is an calating one, with each side trying to leapfrog over the other's novations. Swords and spears are countered with shields, armour id castles with gunpowder and guns, gas with gas masks, mines with inesweepers, missiles with ABMS.

In weaponry the direction of invention is clear because, unlike ıy other situation, the competition is not with nature but with her men. The inventor knows that he is striving for an advantage oadly defined as more destruction or better protection. Sometimes e advantage can be very great: in the First World War German bmarines sank 12 million tons of shipping and in the Second 'orld War 20 million tons. This sort of advantage was almost cisive until it was countered by the convoy and by better submarine tection. The advantage of nuclear weapons ended the war in the icific.

Where weapons are concerned, an advantage is usually clear – rhaps not always at once, but at any rate in potential. For instance, e first cannon were probably less effective than the existing tapults, but the potential was obvious.

Even greater than the need for an advantage has been the fear being left behind. Methods of fighting, and weapons, can quickly ecome obsolete. The invention of the rifle made cavalry and right uniforms obsolete. The invention of the breech-loading fle made the muzzle-loading gun obsolete, not because the new ın was more powerful but because it could be loaded from a prone)sition. The development of mobile divisions and airborne troops ndered fixed fortifications obsolete, as did the development of the nk for trench warfare. Since it is impossible to be sure of keeping ıce exactly, the usual aim is to be ahead at all times.

Another great advantage of weaponry for the exercise of inventive- ss is that the market structure is simple. Instead of having to rsuade the general public to buy something you just go straight the king (in bygone days) or to the military today. This does not ean that these gentlemen were progressive buyers. Sometimes ey were and sometimes they were not. The tank, which proved so fective a weapon in the Second World War, had to overcome gorous resistance before it was accepted. It required the unlikely -operation of Winston Churchill and the Royal Naval Air Service get it tried out. Even between the wars there was a body of)inion that was firmly against tanks; one of the few officers in vour of them was the young de Gaulle. In the beginning, aero- anes had just as hard a time and at an early stage in their develop- ent the US Congress actually forbade the Army to waste its money 'flying machines'. Nevertheless it is still true that, prejudice)art, the market acceptance of a new weapon is rather easier than other things. Above all, there are no cost restrictions, which can hinder the emergence of an invention in other fields.

It is often claimed that war technology gives a general spurt to ventiveness and that there is considerable spin-off which is useful other areas. This is undoubtedly true, but just how much spin- f is less clear, and there would have to be a tremendous amount to cuse the destructiveness of the technology. In its early days metal chnology was probably accelerated by the need for better weapons. our own days aviation technology, and especially the jet engine, rive directly from war technology. So, probably, does nuclear wer used for peaceful purposes.

Apart from weaponry man has had to protect himself in other ıys. For instance, Benjamin Franklin invented the first lightning nductor but was fiercely opposed by the clergy, who thought it rong to protect oneself against divine wrath. In a more practical ıme of mind insurance companies set up the first fire engines and fire brigades to protect themselves against fire losses. The story of locks and keys is a very old one going back (like so many devices) to the Chinese about 4,000 years ago. A beautiful tale of invention is the miner's safety lamp developed by Sir Humphry Davy. The problem was a definite one: he worked carefully on it and came up with a very simple and elegant solution. I suspect that if the same problem had to be tackled today the solutions would be very different indeed and would range from complex ventilation systems to infra-red gas detectors. Davy, however, had to exhibit true inventiveness. not just the application of technology.

Organization

The most general idea of inventions concerns mechanical or electrical devices which carry out some useful function. But the mechanical device is only the expression of an idea. A new idea or invention is a putting together in a new way of known effects in order to produce a new and worthwhile effect. So a new idea can apply to organization as well as to mechanical devices. Banking, insurance, hire purchase, building societies, stock market, are all organization ideas. Some of them have developed gradually, whereas others may have been the idea of a single man even if we do not know who he was.

We do know that the supermarket was invented by Michael Cullen in the US in 1930. We do not know who invented chess in about AD 500 in India. It seems that the very basic invention of coins was made by the kings of Lydia in about 800 BC. Later came the invention of paper money. Coinage was a brilliant invention because it provided units of fixed value around which all sorts of rules could be made. It provided the storage, transport and delay element in a barter system. In a way, ordinary playing cards provide a similar system. They are units of recognizable value around which a structure of rules and achievement can be built. The history of playing cards, like so many organizational ideas, seems to have been one of modification and change rather than instant invention.

The slot machine is an invention that lies halfway between a mechanical idea and an organization idea. The organization idea involves the removal of a person at the exchange point so that coins can be immediately exchanged into goods. The mechanical idea has then to be developed to carry out the organizational idea. With the one-armed-bandit type of slot machine the actual mechanical device itself creates the organization idea by providing a new type of gambling.

The assembly line used to construct motor-cars was an organizational idea which had probably been used at various times by various people (e.g. Eli Whitney, for making guns) but it is usually associated with Henry Ford. Automation, made possible by the technological developments of feedback control and computer handling, takes the process a step further by making human intervention unnecessary. This is another example of the way technical developments open up organizational possibilities. The credit card instead of money is a further example.

These few examples of organizational ideas are included here simply to illustrate the fact that new ideas and inventions are not restricted to the mechanical or chemical fields. In fact, in the future the most important new ideas are likely to be in this organizational area rather than in those areas which have shown so much inventiveness in the last hundred years.

Trap

Trapping is probably as old a method of catching animals and fish as the spear itself, but we have no evidence from the earliest times, and must rely on the hazy pictures in prehistoric art. By inference, we know that traps must have been set at least in the Middle Pleistocene (*c.* 200,000 BC), for at this date the men at Dolni Věstonice, in Czechoslovakia, were catching and eating mammoths, and presumably they could not have caught them with spears alone (or, indeed, have relied on the odd animal getting caught in a crevasse like the deep-frozen mammoths of Siberia).

But prehistoric art gives us no definite and agreed representations of traps: at La Pasiega there is what might be a hind caught in a trap; at Lascaux a cow in front of an enclosure; at Niaux a bison in front of a ring of spikes; but even if these sets of lines *are* traps, and not just magic symbols of unknown meaning – which are common throughout prehistoric art – there is no means of reconstructing them. Not until the Late Bronze Age do we have any actual traps. One found in Drumacaladerry Bog, County Donegal, Ireland, is a tread-trap, and examples similar to it have been found, often in groups, all over northern Europe, showing that trapping was methodically practised at water-holes, etc.: this was at first thought to be a musical instrument, or even a machine for making peat bricks, until a stone relief of 10th century AD was noticed which showed how such a trap might be used. An oblong piece of wood has a hinged trapdoor in it, with a bendy branch inside forcing it closed and flush with the wooden surface; when the animal steps on it, its weight partly opens the trapdoor, but the springy wood, forcing the trap closed, imprisons the animal's foot. Similar devices were used into this century for trapping deer and bears in Poland.

We can turn to primitive peoples of today to see the kinds of traps which must have been used by their ancestors. Birds were trapped by their own weight: Australian aborigines smear bird-lime on tree branches so that birds landing there cannot extricate their feet; the same tribes embed spikes along animal tracks so that wallabies and the like impale their feet. They also dig pits with concealed surfaces, sometimes lining the bottom with spikes. American Indians catch marmots with spring-poles and nooses, and the Athabascan Indians of Alaska combine the noose with a delicately set log attached to it for trapping caribou (this is called the 'deadfall', and is popular in South America and Africa as well).

As for fishing, the weir is used all over the world; the Nootka Indians supplement this with a wooden fence in tidal waters which leaves salmon stranded when the tide retreats. Further upstream, they use basket traps which rely on the current to drive the fish into them.

The history of the mousetrap probably goes back to the Romans who, however, preferred to keep weasels to hunt their mice for them. Several designs were current in the Middle Ages, and there are illustrations of them in a Hebrew book of parables completed in AD 1281. A German MS. version of this, possibly from a Spanish original, shows a metal contraption with a large spring and jaws, rather like a mantrap; also a wooden one with jaws, the upper of which has clamped down on to the mouse which, perhaps, has moved a stick inside. But an Italian MS. shows a cage of wicker and wood: the mouse enters and takes the bait, but this is attached to a lever which lets down the 'portcullis' entrance.

Traps were of professional interest as well. Italian Jewish loan-bankers of the 15th century were obliged by the terms of their contracts to do everything in their power (including the keeping of cats) to curb the mice which might damage the pledges deposited with them.

C. M. B. G.

Fish-hook

It seems likely that the fish-hook is an innovation of the Neolithic period, at least in central Europe. At Lepinski Vir, for example, between the Black Sea and the Adriatic, there is a riverside settlement; and archaeologists excavating the site have come across large spoil-heaps of kitchen rubbish (called middens), which contain many fish-bones, particularly of carp. In the Mesolithic layer here (probably *c.* 5500–4500 BC), there are no fish-hooks, so presumably the fish were caught by hand, stunned with rocks, or speared; but the Neolithic level (4500 BC onwards) has revealed

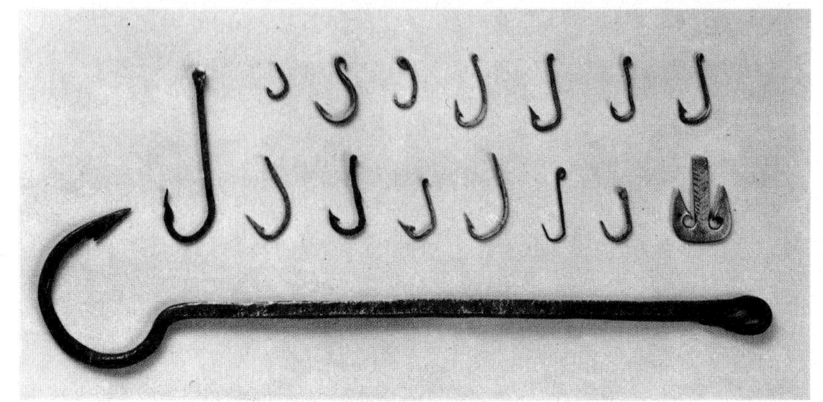

Left, *eel trap in a mill stream, from the Luttrell Psalter,* c. *1340.*

Left, centre, *plaited fish-trap of Stone Age date from Holbaek, Jutland.*

Left, below, *barbed fish-hooks from Egypt, 3rd millennium BC.*

Below, *camels at the wells at Wajir, in Kenya's Northern Frontier Province.*

hooks. It has been surmised that in some cultures the use of fish-hooks rather than harpoons was regarded as effeminate: the Stone Age cemetery of Västerbjerg, on Gotland, has graves in which harpoons for sealing are buried with the men – and fish-hooks with the women. Barbed spears, or harpoons, are a feature of the Magdalenians (*c.* 13,000 BC, Upper Palaeolithic), for example, and it may be that such contraptions are the ancestors of the hook.

In northern Europe, on the other hand, the hook is of Mesolithic and not of Neolithic times: the Maglemosians fished in dug-out canoes, with paddles, and perhaps made quite long journeys. Certainly, Neolithic middens in Denmark contain the bones of deep-sea cod and other deep-water fish.

In Palestine, perhaps as early as 8000 BC, the Natufian culture was using hooks of bent bone; while at Chaheinab in the Sudan the shell of the Nile oyster was employed. This last practice was widespread: Australian aborigines, for example, fashioned hooks out of shell, as did the Chinese.

As we might expect, with the coming of metals hooks were made from what was available: the Boian culture of the Lower Danube (*c.* 3800 BC) fashioned hooks from strips of copper but, like all that had gone before, these were unbarbed. Barbed hooks – much more efficient because the hook stays in the jaws, even under pressure – appear in different areas at different times: they have been found at Vinča-Pločnik (west Balkans, *c.* 3000 BC), but are introduced much later as one gets further west and north – metal hooks appear in northern Europe only in the Late Bronze Age, and difficulties were surely found in making barbs in strip metal.

How were unbarbed hooks baited? The first 'hooks' were no more than small doubled-ended pointed toggles tied along the bait's flank, which, it was hoped, would lodge in the prey's throat when it took the bait; proper hooks may have been attached in the same way. Such simple gorges are still in use today: Eskimo gorges are formed from a waisted piece of bone, with a pointed cone to either side. The Nootka Indians on the north-west coast of America fish for salmon with hooks made with a wooden shank, to a groove in which a bone barb has been lashed.

Up to the 17th century, when the fishing reel was invented (somewhat earlier in China), fish would have been caught by trailing a line in the water, or trolling behind a boat, perhaps with a piece of attracting shell attached to the hook. Nets were, of course, a much less time-consuming way of fishing; they were used equally for sea-fishing and for trapping the bred fish in the artificial fish-ponds which were a feature of medieval Europe.

C. M. B. G.

Well

From the first, men have tried to place their homes close to the water on which they are so dependent. To protect their springs from contamination and to make it easier to draw water, they began to reinforce them with stone walls or wooden linings, to form 'tites' or enclosed springs. Most of the so-called 'holy wells' of Europe are really springs with a surround of this kind; a very elaborate timber tite was built at St Moritz in Switzerland, dating from the Bronze Age.

If the water-source was small, it would be natural to try and enlarge it, digging round it or digging deeper. In time of drought, attempts might be made to dig in dried-up river-beds, or in other places where even if there was no water above ground there might be some not too far down. By classical times an extensive lore had grown up about the indications of subsurface water: excessive condensation, vapour rising from the ground in the early morning, the presence of frogs, of plants that demand plenty of moisture. Wherever rainfall was slight and rivers few, or dry in summer, wells were an absolute necessity of human settlement, valued, purchased at a price, often fought over, since the rights to use particular wells were the normal source of conflict in all desert lands.

In the earliest wells there would have been no lining but the naked rock, but to prevent falls, 'steining' or lining with dressed stone, timber or brick came into use. Perhaps the oldest wells now known, those at Chandu-daro and Mohenjo-daro in Pakistan (the former *c.* 3000 BC) have a circular lining of well-designed wedge-shaped bricks, complete with a brick coping and pavements on which to stand water-jars. The Indus civilization may have been the first to reinforce wells like this, but the practice spread westward and by the 2nd millennium BC, wells over 200 ft. deep in Egypt and Canaan were being excavated in the Middle East: Jacob's Well at Cairo was nearly 300 ft.

These wells would have a parapet to prevent anything, or anybody, from falling in. But where water was relatively close to the surface, and in country districts, there might be no more than a low coping, like the well in the Jordan Valley where two messengers were concealed from King David's rebel son Absalom, by simply spreading a sheet over the mouth of the well and covering it with corn as if to dry in the sun. These copings are usually deeply scarred by the grooves made by the ropes hauled up by so many generations of villagers on their daily visits to the well. But if the shaft was wide and too deep to haul up a bucket, steps were often provided right down to the water. This practice survived at the great well of San Patrizio at Orvieto, begun in 1527, which had a double spiral staircase, big enough for donkeys to be driven right to the bottom.

All these wells were dug, not bored. In China the salt wells of Szechuan were drilled, down to 300 ft., and the exploitation of this natural brine goes back to the Former Han Dynasty (2nd – 1st centuries BC). But the Chinese did not use this method for ordinary wells, and boring does not seem to have become established much before the 17th century, perhaps first in Italy, and with hand-augers: even then it was after first digging to a considerable depth: they would then bore to test if it was worth while going on.

A. K.

Irrigation

Drainage is a necessity for any settled community, whether living in a rainy or temperate climate, or in desert conditions; in a dry climate, the crops must be irrigated, and in a wet one they – and indeed the houses and streets – must be kept drained.

Irrigation in early cultures was mostly natural: the great centres of population were by the banks of rivers – the Tigris, Euphrates, Nile, or Indus – which flooded regularly, washing silt over the fields, the process being controlled by dykes. Better crops, particularly where the plough was used, would result, and thus higher populations. Partly for these reasons the first great civilizations were in the river valleys named above.

Dams, as well as canals, would be needed for full control, and large forces of men and good government organization were required for such operations: it appears, indeed, that the city states of Sumer often fought over irrigation disputes. Later, *c.* 1700 BC, Hammurabi's Code of Laws provides for such questions as the upkeep of canals, the protection of tenants against flooding, and the guarantee of sufficient water for one's needs.

In ancient Egypt great respect was accorded to gods of irrigation, of whom Osiris was one. It is not difficult to imagine how the arts of land measurement, and of water-storage and lifting, were stimulated in an attempt to make irrigation and drainage efficient. During the Greek period, these arts declined, partly because of climatic differences and the resulting differences in crops from those grown in the river valleys. But as early as the 3rd century BC the Romans were active in drainage, embanking the Tiber and draining lakes. The Pontine Marshes, to the south of Rome, were a constant source of malaria, and attempts were made to drain them from 160 BC onwards, but without success until the efforts of the Dutchmen van der Pollen and Meyer in the 17th century.

C. M. B. G.

Plough

Most historians agree that the first farmers grew their seed in plots that had been prepared by burning the vegetation. They covered the seed with the soft ash that remained and learned that all soil must first be broken up before it will make a good seed-bed. At first they used a simple digging stick that was dragged along the surface, or a bent stick that served as a hoe.

It was in the fertile valleys of the Middle East that the plough was developed from these beginnings. If a hoe is hauled through the ground instead of being used to chop at it, it makes a rudimentary furrow. The idea of the plough could have stemmed from this. Two men could haul a lengthened hoe-handle along the surface while a third guided the implement with a stick fixed to the foot of the hoe. But even if primitive man developed the plough, not from a hoe but from a single, suitably branching piece of natural timber as used by the early Romans, either device would suggest the use of animals to haul it. The simple hand-tool had to become a machine; and the application of animal power to the breaking up of the soil was a revolutionary step in man's progress. By ploughing more land he could grow more corn, thus making possible a greater concentration of people.

The primitive scratch plough, the 'ard', was suitable for the dry, friable soil of the Mediterranean and the Middle East where it was developed. It is still in use today in those areas, just as developments of the primitive digging stick, the *caschrom* or foot-plough, and a similar device, the breast-plough, were in use in remote parts of Britain until after the last war. The ard did not normally turn the soil over: it parted it, leaving a wedge of undisturbed soil between each furrow. Cross-ploughing was therefore needed, and this resulted in fields being squarish in shape whereever the ard was used.

But the scratch plough was not suitable to much of northern Europe, where rainfall is more abundant than further south, and where soils are generally stiffer. Here a heavier plough was necessary to make a true furrow along which surface water could be led away. This heavy plough, the *carruca*, emerged in the 1st century AD in the foothills of the Italian Alps, and was fully developed by the 6th century in central and southern Germany. The social effects of the use of the *carruca* can be seen in the denser population of northern Europe in this period, followed by the 'bleeding off' of surplus people to less-developed areas. Thus, there could well be a direct connection between the emergence of the *carruca* and the movement west and south of the Anglo-Saxons and Danes. With its heavy share, its 'coulter' or knife and mouldboard, which turned over the furrow-slice, the *carruca* could make a fruitful onslaught on the stiff soils as long

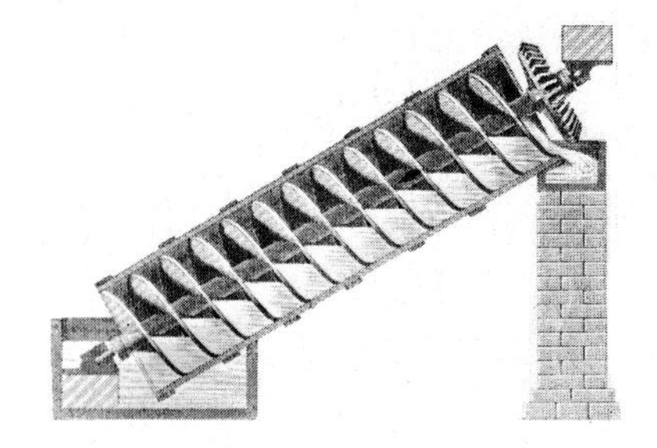

Screw conveyor invented by Archimedes that irrigated the fields of Egypt.

Below, *plan of an irrigation survey from China (1247).*

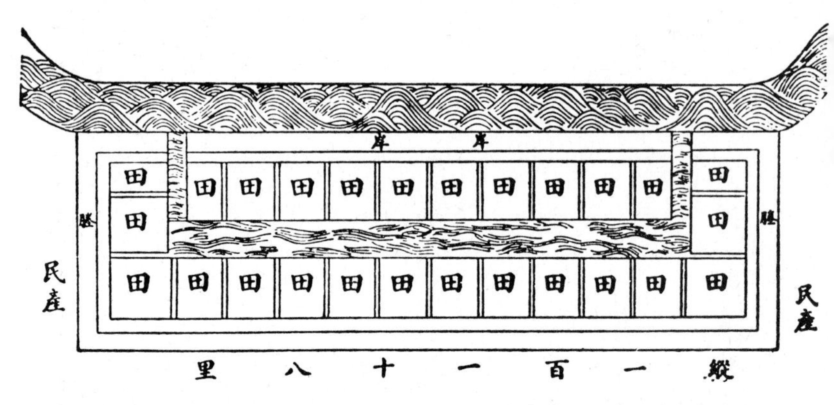

Left, *Egyptian scratch plough, pulled by oxen because horses were needed by the army.*

Below, *northern plough and harrow, from a 16th-century Book of Hours.*

The Last of England *by Ford Madox Brown. Vegetables were kept fresh in 19th-century emigrant ships by being slung in nets along the bulwarks.*

A large brewhouse of the 1760s. Malt and water are being stirred in the tuns, before being drained off into the fermentation vats below.

as there was enough power to draw it. The power came from oxen yoked in pairs. But this meant that it was now impossible for a peasant farmer to own and operate a *carruca*. He was compelled to co-operate with his neighbours and to farm in long strips in huge open fields. And it was around this basic framework that the medieval manor evolved. A similar process is going on now that the multi-furrow tractor-plough and the machines linked with it are developing beyond the resources of today's small farmer.

G. E. E.

Food-preservation

The problem of how to preserve his food, to do something to it that would lock up its energy-giving qualities until he chose to eat it, has been with man from the first day that he ceased to be a nomadic food-gatherer, moving with the changing seasons to where the food was. Settled communities required food all the year round, for both men and animals, and agricultural goods were one of the earliest commodities with which trade could be conducted.

Drying is the oldest method of preservation, and cereal grains dried by the sun and the air man's earliest preserved food. The Old Testament tells how Joseph rose to become second only to the Egyptian Pharaoh in power as a result of his precautions in storing up grain during the seven years of plenty in preparation for the seven years of famine. Besides wheat, figs and raisins were dried in ancient Palestine and Mesopotamia, as were some herbs and plants from which drugs were made. Small quantities of dried meat, too, would be found, though only in the houses of the rich.

In the less hot and dry regions of Europe prehistoric man applied artificial heat to his stored food by making fires in his shelter, and thus learnt the advantages of smoking meat and fish. Smoking not only dries food but also changes its taste by decomposing fat, avoiding the need for spices. It was developed into quite an art by the German tribes and by the Romans later on. We have a record of a warning from the 8th-century Pope Zacharias to Boniface, for example, which abjures him to stick to boiled or smoked bacon and pork during his travels among the 'heathen tribes'. Even in relatively recent times, however, smoking was not practised in the Near East owing to the shortage of fuel. Instead they relied on the sun and the wind, burying their fruit and vegetables in the hot desert sand.

In India, on the other hand, spices were early used as preserving agents. If cooked meat was curried, people discovered, food not eaten at one meal could be kept for the next day. As a result of this important discovery the spice trade rapidly came to assume great international importance, later occasioning numerous wars and the founding of European colonies from South Africa to Indonesia.

The discovery of salting is ascribed to the Median tribes, and salted, dried and pickled fish, for example, was a staple food of the ancient Greeks. In the classical period various Greek writers gave detailed instructions, as indeed Chinese authors had done earlier, for the pickling and salting processes: they recommended that an egg (or lotus seed in China) be floated on the surface of the brine as a final test of readiness. In the Roman world, where pickled wares were under the control of *salsamentarii*, the use of salting increased enormously, and not only established civilizations but also barbarian tribes such as the 'lotus-eaters' of Africa are known to have salted their fish.

Even the chemical additives so extensively used in modern preserving plants have their forebears in remote history. Benzoic acid, for example, occurs naturally in prunes and cranberries, as does tartaric acid in tamarind juice, and the acid in vinegar was used to make sauerkraut in very early times by the German tribes. Wine and vinegar, of course, are themselves preserved grape juice, the unstable fruit-sugar being converted to a stable acid by fermentation, a preserving process used as much as 2,500 years ago when Indian yogis laid down health rules which recommended the use of sour milk. Animal milk and its by-products had been a part of man's life even when his way of life was still nomadic, and fermentation, which normally results in food decay, was used instead to preserve it by encouraging the growth of micro-organisms and then controlling them when they had completed that part of their work which suited man. Curdled milk not only kept for longer periods than fresh but could also be converted into cheese.

Cold as a method of preservation was not widely used by ancient civilizations, although it was certainly practised by the Chinese and traces of ice-pits have been found around the Mediterranean and the Black Sea. The Peruvians in pre-Columbian times used cold to produce the first instant mashed potato, spreading potato tubers on the ground at high altitudes to catch the frost and become brittle; after this they were beaten to a pulp and ground to a blackish-grey powder which could be reconstituted by adding liquid as and when required.

S. M.

Brewing

Some of the earliest records and inscriptions in the world contain descriptions of brewing, and it is evident that beer, of which several types were already known, was a popular drink in Egypt and Babylon between five and six thousand years ago. The initial steps in brewing – grinding barley or some other cereal between stones, adding water to make a dough and baking the resulting loaves at low heat – were identical with those in baking bread and the two crafts were usually practised together. To produce beer the loaves were broken up, macerated with water and left to ferment by exposure to warmth until the resulting liquid could be strained off through a woven filter.

Although the production of fermented drinks was common in all the early empires of the Near East, the Greeks and Romans thought beer a barbarous concoction, suitable only for Celts and Germans. The centres of brewing in Europe thus tended to be in the north, where until the 12th century brewing was mainly practised as a household chore done by the women. It was the Germans who, by the 13th century, introduced the modern type of beer flavoured and preserved with hops (which were not grown in England until about 1400). Two hundred years later it was again the Germans who first produced light, 'bottom-fermented' beer, which soon spread from Bavaria throughout Europe.

Brewing remained a cottage industry for centuries. Powerful brewers' guilds developed in Britain and Germany in the Middle Ages, however, and many monasteries became major producers, serving whole neighbourhoods in addition to their own needs. It was not until the 19th century, which also saw the development of the first large brewing plants, that the art changed significantly. Pasteur's studies of yeast fermentation in the 1860s and a succession of investigations into the brewing process which rapidly followed led to a series of improvements, including the addition of cereal adjuncts, carbonization (which increases the sparkle of beer) and the stabilization of beer colloids (which made beer chill-proof), not to mention numerous advances in brewery equipment, refrigeration, bottling, pasteurization and so on.

Other fermented drinks, such as cider and perry, have a history almost as long as that of beer; but porter, which rapidly became popular because it kept for longer than ordinary beer and so could be more widely distributed, first appeared as late as the 17th century.

J. G.

Wine press

The first solution to the problem of expressing grape-juice was simply to trample the fruit underfoot. The ancient Egyptians improved on this by wrapping the trodden grapes in a linen cloth which was then tied at each end to a stout pole. Two operators pulled the linen bag tight and so squeezed the must. An advance on this was a wringing device on the kitchen tammy-cloth principle.

More progress was made by the Greeks, who invented the 'basket' – a vertical cylindrical vessel into which trodden grapes were put; a plate like a lid fitting inside the cylinder was then forced down by hammering in wooden wedges. Next, the principle of the lever was used to force down the plate on top of the must.

The screw was first applied to the wine press in the 2nd century BC. Cato the Elder describes such a press in detail in *De re rustica*, and this type was used at the vineyards in the region of the present Châteauneuf-du-Pape.

In the following centuries evolution followed that of all presses, with the blacksmith's work replacing the carpenter's work, until the application of machine power.

E. H.

Two ways of pressing the juice from grapes: by a wine-press (background) and by leg-work.

Pasta

The world thinks of pasta and thinks of Italy. But in very ancient times it was prepared in China from rice and bean flour, and traditionally it was Marco Polo who introduced pasta into Italy when he returned from his travels in China bringing with him an appetite for noodles and recipes for their preparation. We do not know when its consumption became general in Italy, but it was firmly established in the 14th century as Boccaccio's *Decameron* makes clear: 'In a region called Bengodi where they tied the vines with sausages and where one can buy a goose for a farthing and a gosling included, there is a mountain made of grated parmesan cheese on which men worked all day making pasta and ravioli, cooking them in capon's sauce and then rolling them down, and who grabs most eats most. . . .'

In fact the farinaceous foods that we call pasta are by no means exclusively Italian: similar products can be found in peasant cultures all over the world. To make it, the floury starch extract of cereal grains (not necessarily wheat) is processed into one of a number of shapes – ribbons, tubes, even 'bow ties' – and boiled in water until it softens. The distinctive ingredient of pasta is the gluten it contains, which prevents the dough from dissolving when boiled.

Rolling and cutting pasta by hand is hard work. The earliest travellers to Italy from the New World were amazed by the sturdy arms of the peasant women who rolled out the pasta in the southern regions famous for it. For many centuries pasta was made by hand, and in fact it was such hard work that it was by no means a common dish. It was first made on a large scale in 1800 in Naples, where workers operated crude wooden screw presses and hung long strings of spaghetti

Pasta is not an Italian monopoly: macaroni factory in the Soviet Union, 1923.

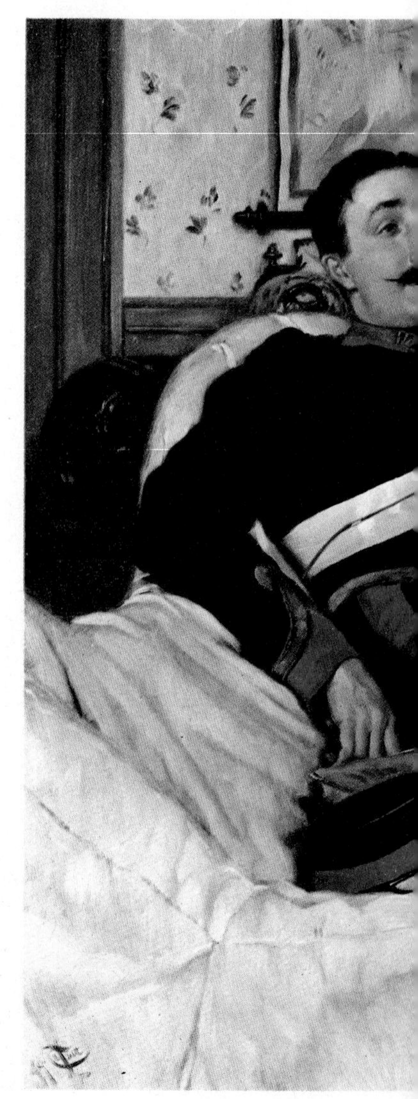

By the 1870s the humble cigarette was gaining an increasing share of the market, even in the best regiments. Portrait of Frederick Gustavus Burnaby by J. G. Tissot.

Early 20th-century ice-cream seller in a South London street.

out to dry in the sun. The dough still had to be kneaded by hand, however, until in 1830, again in Naples, a mechanical kneading trough was invented which was soon widely adopted in Italy and initiated the era of mass production. Innumerable other improvements in pressing, kneading and drying equipment followed.

Today there are some thousands of pasta factories in Italy, producing about half a million tons a year. Nevertheless the tourist in southern Italy will still occasionally see what he takes to be fishing-nets and lines strung out on trestles on the beach and realize that they are in fact lengths of home-made spaghetti drying in the sun.

S.M.

Ice-cream

Courts and kings figure largely in the story of ice-cream, for to produce the great luxury of a chilled dish in a warm climate signified great power. So Pharaohs gave their guests fruit juices kept cold by snow packed inside double-walled silver goblets, Alexander the Great learnt these party pleasures from the Persians when he conquered them in the 4th century BC, and as early as 250 BC the Chinese were systematically using ice to cool food. Somewhere along the line fruit pulp itself was frozen by being mixed with snow – perhaps accidentally during Nero's reign, when ice and snow were brought from the Alps to Rome and used to pack fruit for transportation.

While the Goths were sacking Rome in the 5th century, across the world in China a thousand blocks of ice were being stored each winter in vaults at the imperial palace. Sugar had been known in China (but not in Europe) since 100 BC but was not in common use; a special dish for the emperor was a frozen mixture of rice and milk flavoured with camphor. Other dishes with a milk base remained popular in China and frozen milk delicacies were sold from handcarts in the streets of Peking at the time of Marco Polo's visit. It is probable that Polo brought back recipes on his return to Venice in 1295, though others believe ice-cream was independently invented by a certain Bernardo Buontalenti of Tuscany in the early 1300s.

When Catherine de' Medici married the future King Henry II of France, in 1533, the Venetians used the opportunity to show the rest of western Europe what true elegance was. During the 34-day wedding celebrations a different water-ice was served each day, including lemon, orange and wild strawberry flavours. There was also a semi-frozen dessert made with cream, for which the recipe was kept a closely guarded secret. Charles II of England is said to have acquired the taste for ice-cream during his exile on the Continent, where, in the year of his Restoration, an enterprising southern Italian, Procopio Coltelli, had opened the famous Café Procope in Paris which introduced the Neapolitan ice-cream to France.

Creamy mixtures became fully freezable in 1550, when Blasius Villafranca, a Spanish doctor living in Rome, discovered that freezing point could be rapidly reached if saltpetre (and, later, common salt) was added to snow. Using this knowledge the Florentines were the first to produce true ice-cream on a large scale.

In 1870 an ice-making factory was set up in London and shortly afterwards a flood of Italian immigrants brought ice-cream within the reach of almost everyone. Victorian children called the Italian ice-cream sellers 'Hokey Fokey men' from their street cry 'Ecco un poco' ('Here's a bit').

The American contribution to this story (apart from chocolate brought back by the Spanish conquistadors from Mexico) was the cornet, said to have been invented by the girl friend of an ice-cream salesman at the Louisiana Purchase Exposition of 1904 who rolled a wafer to stop her ice-cream dripping after she had similarly rolled another to hold some flowers he had given her. Our modern 'soft-serve' ice-cream was also an American innovation, following on the homogenizer invented by the French scientist Auguste Saulin.

S.M.

Cigarette

In most places in the world today the offer of a cigarette would be considered a polite gesture, but this attitude stands in sharp contrast to the prevailing sentiments of less than 200 years ago. Prior to the late 18th century, in fact, no one with any pretension to social standing would have deigned to accept such a paltry substitute for the cigar, which itself had only been introduced to Europe around 1600 by Spanish conquistadors returning from expeditions in the New World. The *cigarro*, which derived its name from *sik'ar*, the Mayan word for smoking, was for many years a symbol of conspicuous wealth among Spaniards and later among other Europeans. The paper-wrapped cigarette, on the other hand, was an improvisation of the beggars of Seville, who used tobacco from discarded cigar butts and referred to their product as a *cigarrillo* or *papelete*. By the time of the economic crisis of 1873, however, enough tobacco addicts had found their financial resources sufficiently reduced to give the cheaper cigarette a larger share of the market, even among the upper middle classes.

From Spain, the cigarette found its way to Italy and Portugal and thence was carried by traders to Russia and the Levant. The French and the British first became acquainted with the 'light smokes' while fighting on Spanish soil during the Napoleonic campaign of 1808–14, and it was in France, some few years later, that the term 'cigarette' was first used. In 1853, a cigarette factory was established in Havana, Cuba, but among English-speaking people this 'weak, feminine article' did not become really popular until the Crimean War, during which British soldiers were introduced to the mild Turkish variety. Shortly thereafter, Britons opted for straight flue-cured Virginia tobacco, while American tastes favoured Turkish and other blends.

In both Europe and the United States, the first manufactured cigarettes were either hand-rolled in the factory or sold as 'roll-your-owns'. Today, hand-rolled cigarettes are practically unobtainable, while 'roll-ups' are generally considered to be a smoke of the working class. Machinery was first introduced to the industry in 1860, when W.H. Pease patented a shredding and cutting machine.

By the 1920s, cigarette sales exceeded those of cigars, and the demand has grown steadily ever since. Recent research linking cigarette smoking with lung cancer, heart disease and numerous other ailments has scarcely affected the cigarette market, and in the United States alone, annual taxation on cigarette sales still exceeds $2,000 million.

E.W.

Seed-drill

Jethro Tull's claim to have invented the seed-drill was contested during his lifetime and long after. But there is no doubt that his original cast of mind enabled him to make the first machine to drill seed effectively. He was a musician and lawyer, and went into farming out of necessity. His servants and neighbours wrote him off as 'a gentleman farmer', which in this context meant a man who could raise little except his hat. He tried the experiment of growing clover on his farm, but his men jeered at him and finally went on strike: they thought he was a Notionist who had 'too many bees in his Bonetcase'.

But, as often in farming, it was the intelligent outsider who brought in a truly seminal idea. Tull saw that if he could invent a device to supplant broadcast sowing he could put his theories about farming into practice and at the same time be less dependent on his men. Up to this time the principle of drilling seemed simple enough: the only snag was that when it was tried out it did not work. You had a wheeled vehicle which held the seed in a container; and as the wheels turned the seed ran down through metal tubes or hollowed coulters underneath. The front of each coulter made a small furrow in the soil and the seed ran into it. A bush-harrow drawn behind the drill restored the soil and covered the seed. Earlier machines had failed because they could not effectively control the flow of seed from container to soil. Tull, the musician, solved the problem by adapting the mechanism on the sounding-board of an organ to the drill he was making. He controlled the flow of the grain by means of a brass cover and adjustable spring, copied from the tongue in the organ mechanism.

Tull saw that the drill, along with his system of horse-hoeing, could revolutionize farming: by sowing in rows instead of broadcast, the farmer could weed and also aerate the soil round the growing plants. But Tull's drill did not travel far beyond his farm, although the Scots were enthusiastic about his whole system. It was not until the 1800s that drilling became widely practised in England. Two Suffolk drill-makers, Smyth of Peasenhall and Garrett of Leiston, helped to popularize it, Smyth by sending round travellers who offered to drill seed at half a crown an acre. Tull was a neglected man whose ideas were either scorned or plagiarized; but he never lost his conviction that his system would one day become the bedrock of British farming.

G.E.E.

Factory farming

In 1771 Arthur Young, in one of his exhaustive, first-hand surveys of English agriculture, described a system for fattening oxen, practised by a farmer named Moody, of Retford in Nottinghamshire. Moody believed that the hotter the conditions in which the cattle were kept, the faster they would fatten, and 'accordingly keeps them shut up and, for some time, does not so much as let in any air through the holes in the doors; the breath of so many, with the natural heat of their bodies, brings them soon to sweating prodigiously, and, when this is at its height, they fatten the best and the quickest.'

This would seem to be an early start to intensive beef-rearing, but no one from that period would recognize the methods used today. Poultry have produced the biggest changes because of the shorter maturation period of birds, and their larger number of offspring. In the USA, both layers (for egg production) and broilers (for eating) were housed in intensive units as early as 1920, but Britain is now the world's recognized leader. In 1952, after studying American methods, Geoffrey Sykes pioneered the broiler system in his native Wiltshire. A year later, animal foodstuffs came off the ration, and this, combined with the introduction of antibiotics without which disease would have decimated the confined birds, led to a tremendous upsurge in battery chicken-farming.

Above, *drawing of a seed-drill in Jethro Tull's* Horse Hoeing Husbandry *(1733).*

Left, *precursor of the seed-drill: through regularly spaced holes in the setting board a dibber made holes in the earth.*

Below, *'sweatbox' piggery.*

Beehives, from a medieval manuscript of Virgil's Georgics.

A field of barley, half sprayed with weedkiller. In the untouched half, charlock still flourishes.

Intensive methods up to a high degree are now used in practically all types of stock-farming. As an offshoot of its vast dairy industry, Holland pioneered the factory farming of veal calves. Originally they were kept in total darkness (to 'blanch' them, like celery), and also periodically bled while alive – both done to produce the 'white veal' that was in such popular demand. A more scientific approach separates the calves at birth from their mothers and keeps them for 12 weeks in dimmed isolation pens (as a precaution against cross-infection, and mutilation by sucking) while they are fattened on synthetic 'milk' containing the legal minimum of iron. This inhibits the formation of red blood cells and so results in white flesh.

Pigs have been successfully fattened in the 'sweat-box' devised by James Gordon of County Down. In this system, widely used in Northern Ireland, the animals are closely packed into large unheated pens, which they warm entirely by their own body heat. Pigs lack sweat glands, but through a combination of a high level of liquid intake, the moisture they breathe out, and the evaporation of urine, a 'Turkish bath' effect results. The high temperature, and a humidity reaching 90 per cent, appear to minimize respiratory ailments, although inspection is difficult and heat stroke can occur.

Sheep have adapted less well to intensive farming, but in any case they graze on land unsuitable for intensive cultivation.

S. M.

Beehive

Long before bees were kept conveniently in hives of bark, straw or pottery, wild honey would have been collected by nomadic peoples, for honey was the main source of sugar in the ancient world. In Egypt, honey was much prized; it served from early times as an offering to the gods and the bee featured, often in its hieroglyphic form, as a symbol of the Kingdom of Lower Egypt. Honey appears to have been an important element of trade and tribute: a Sixth-Dynasty official (2420–2258 BC) took honey to barter with when he went south to visit the Negroes, and it was common for Egyptian kings to return from campaigns in Palestine and Syria with a tribute of jars of honey. Honey had several uses: for religious observances, for medicinal purposes (e.g. as unguents), as an alcoholic drink in its fermented form, and as a constituent of honey cakes – a mixture of honey and flour – both for everyday eating and again as an offering to the gods (a practice which continued into the classical world).

The Hittites also kept bees, and their cultivation, as well as playing an important part in Hittite mythology, was of sufficient practical importance to be the subject of a law: 'If anyone steals bees in a swarm, formerly they used to give one mina of silver, but now he shall give five shekels of silver. . . .' Apiculture was well established in Greece in Homer's day and remained popular throughout the classical period.

It would be inaccurate to claim the beehive as the invention of man rather than of the bees themselves. All men did until the end of the 18th century was to reproduce in pot or straw the form of the hive the bees constructed. But in 1789 the Swiss naturalist François Huber built the first movable-frame hive, the 'leaves' of which opened like a book: each frame would be 'primed' with a small piece of comb, to which the bees would add. The type was further developed by the Reverend L. L. Langstroth in 1851. In 1843 experimenters developed synthetic comb-foundations on which the bees could build their honeycomb cells; extractors to remove the honey from the comb by centrifugal force were developed at about the same date.

C. M. B. G.

Weedkillers and pesticides

Toxic chemicals to combat pests were in use in Homer's time (he mentions the fumigant value of burning sulphur), and Pliny advocated the insecticidal use of arsenic. At least 300 years ago the first natural insecticide – the nicotine extract of tobacco – was being used against the plum curculio and the lace bug. By 1828 a second natural insecticide was provided by pyrethrum, the dried and powdered flower-heads of a variety of chrysanthemum. During the mid-19th century – in a return to the Greeks – sulphur was advocated as a fungicide on peach trees: 'take tobacco one pound, sulphur two pounds, unslaked lime one peck . . . pour on the above ingredients ten gallons of boiling water.'

The control of weeds (as distinct from insects and fungi) was a later development, and can be said to have originated from fungus control. For centuries, farmers had had to rely on hand-weeding, harrowing, burning off and other such methods that were either too laborious or too unselective. But in the 1880s and 1890s, when the Bordeaux vineyards were being severely affected by downy mildew, the vines were sprayed with a mixture of copper sulphate and lime (still in use today under the name of 'Bordeaux mixture'). In 1896 a vine-grower named Bonnet noticed that this fungicide blackened the leaves of yellow charlock which was growing near his vines, and killed it off. Shortly afterwards, it was observed that iron sulphate, when sprayed on to a mixture of cereal and dicotyledonous weeds, killed the weeds but ran off the leaves of the cereals and left them undamaged. This was the breakthrough.

and many other simple inorganic substances were soon found to have selective properties when applied at suitable concentrations.

During the inter-war years, research produced organic substances which mimicked plant growth hormones; selectively applied, these had the effect of killing off weeds by an exhausting and unnatural spurt of growth.

The story of weed and pest control is incomplete without mention of Paul Müller's development of DDT in Switzerland during the Second World War. Though it brought incalculable benefits to a war-torn and devastated world, and earned its discoverer a Nobel Prize, it has since proved to be one of the worst, and most persistent, sources of environmental pollution, and its use has been banned in some countries and restricted in many others.

S.M.

Corn-reaper

The Elder Pliny mentioned a mechanical reaper in the first century AD. This was a two-wheeled, cart-like vehicle which the horse or ox *pushed* into the standing crop. Iron teeth at the front of the machine beheaded the corn, and the ears fell into the cart.

Dozens of patents for reapers were taken out in Scotland and England early in the last century; in 1814 an actor-inventor demonstrated one on stage where he had first 'planted' a crop. But it was a Scot, Patrick Bell, who made the first workable machine in 1826.

He made a number of triangular knives which he joined to two horizontal bars at the front of the reaper: the lower bar was fixed and the upper had a reciprocating motion through being geared to the ground wheels. Revolving sails held the corn to the clipping knives, and a canvas drum laid it aside in a swath. For humanitarian reasons, Bell refused to patent his machine in spite of being out of pocket through his invention. Because of his inability to pay, indifferent material and workmanship went into Bell's machine and it was only moderately successful. Moreover, no big firm would take it up, probably because of opposition then to the threshing machine in England.

Some of Bell's machines reached America, and Bell himself went there in 1832, giving advice freely to anyone who was interested. Conditions there were different: a huge continent was waiting to be exploited. Two Americans, McCormick and Hussey, brought their reaper to England in 1851. They got the orders, but it was later proved that Bell's machine, when properly made, was equal to theirs.

Later, the American Civil War saw a disastrous drop in farm labour, and, whatever their misgivings, farmers were forced to try using machinery. This gave McCormick and others the impetus to develop the reaper into a machine which would not only cut and deliver the corn in swaths, but also tie it into neat sheaves with string.

G.E.E.

Food-canning

Canning was a direct response to the needs of the French Revolution and of Napoleon's war machine. In 1809 Nicolas Appert, a Paris pastrycook and chef, won a prize of twelve thousand francs offered for the discovery of a practical method of food preservation using glass jars with corks.

The idea of the tin canister was introduced by a London merchant, Peter Durand. He sold his patent to a firm called Donkin, Hall & Gamble, who set up a 'Preservatory' in Blue Anchor Road, Bermondsey, in London. Canned food reached the shops in England by 1830.

M.C.

Lawn-mower

The first lawn-mower, Thomas Plucknett's 'machine for mowing corn, grass, etc.', patented in 1805, consisted simply of a pair of shaft-steered carriage wheels geared to revolve a large circular blade parallel with the ground and was soon superseded. In 1830 Edwin Budding of Stroud, Gloucestershire, patented his 'machine for cropping or shearing the vegetable surface of lawns, grass-plots and pleasure grounds', which represented a major advance on Plucknett's model. Budding's nineteen-inch roller-mower, in fact, looked basically like today's machine, except that the collecting-box resembled a large seed-tray. It used the principle of rotating cutters operating against a fixed one – an adaptation of the method of shearing nap from cloth employed in the Lister textile factory where Budding worked as an engineer. The inventor stressed that his machine would cut dry grass (for scything, the mowing method then most commonly in use, it had to be damp), but admitted that a second run might be necessary against tall grass. He added that 'country gentlemen may find in using my machine themselves an amusing, useful and healthy exercise'. In fact few country gentlemen accepted his invitation: those with strong waist muscles preferred to display their skill with a scythe.

In 1832 Ransome of Ipswich obtained a licence to manufacture Budding's mower; others did the same, but the Ransome model won through. In 1869 side wheels were introduced and so costs were cut. Horse-drawn mowers were often used at first, but hooves tended to

Top, *McCormick and Hussey's reaper, brought to England for the first time for the Great Exhibition of 1851.*

Above, *roast veal, canned by Donkin, Hall & Gamble in 1824.*

Right, *canned Christmas pudding made in 1900 by E. Lazenby & Son.*

Edwin Budding's lawn-mower, the forerunner of the modern machine, patented about 1830.

Mid-19th-century threshing-machine.

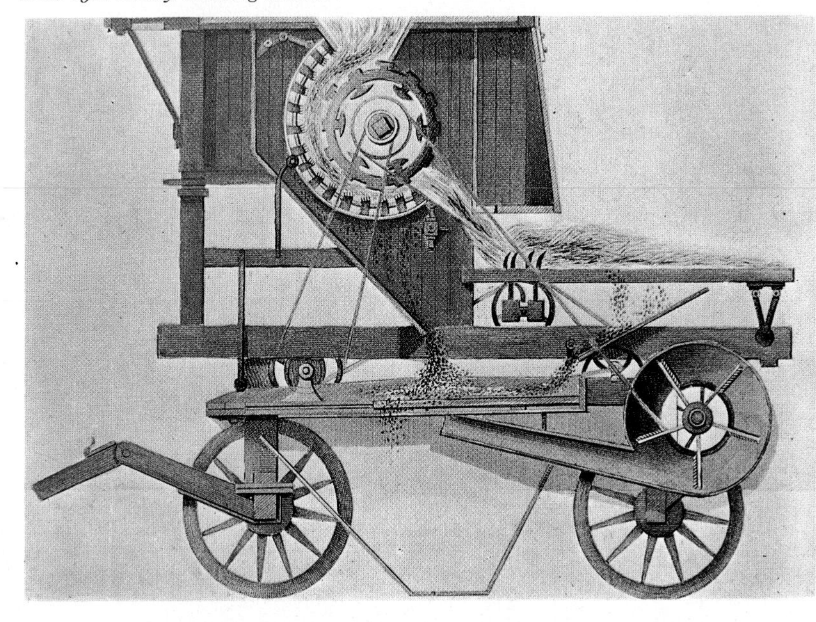

'Turnip' Townshend, eloquent exponent of crop rotation.

scar the lawn and droppings had to be swept up. Mechanization began in 1893 with a huge Leyland steam-mower for large estates; about 1899 Ransome began to adapt the internal-combustion engine to mowing, introducing their self-propelled forty-two-inch mower (with a saddle for the driver) for sports grounds in 1902. The gang mower – a group of machines linked in formation – was developed in America in 1914 and in 1926 the first electrically driven model was introduced. Today we have hover-mowers and models operated by remote control from deck-chairs: amusing, useful, but offering no exercise.

E.S.T.

Threshing-machine

The Scot Andrew Meikle, son of a millwright, was the inventor of the first successful threshing machine. But Meikle's corn thresher was the culmination of a long social process to which many ingenious men had contributed. Simple threshing machines had existed centuries before this. The Chinese had used rollers driven round over the corn: the Romans used a heavy wooden sledge with flints or iron teeth fixed in its underside.

The Scottish inventors adapted the principle of rotary motion to the beating action of the flail. A similar idea had already been exploited with the winnowing machine. James Meikle, Andrew's father, brought the first successful 'fanner' to Scotland. He built a stout wooden frame which turned on a roller. To the frame he fixed strips of strong cloth or leather: as the roller turned, an artificial wind blew the ears and chaff away from the grain. The son undoubtedly learned from his father; and in making his thresher he took up the principle of scutchers 'acting on the sheaves by their velocity and beating out the grains of corn'. Horses drove the machine, but later on steam and water took their place.

However, the threshing machine was not welcomed in England (there was to be more scope for it in America), and was one of the reasons for the great rural unrest in the 1830s. Threshing with the flail in the barn had meant constant winter work for the farm men when nothing could be done on the land. It saved them from asking for parish relief. Now the machine threatened to change all this. It was the main target for the 'Captain Swing' riots which broke out in Kent in August 1830, soon spreading to the corn counties. To prevent the burning down of their barns many farmers moved out their machines into the yard for the men to destroy.

In spite of the ruthless suppression of the revolt, the spread of the machine slowed up; and many farmers in East Anglia were still using the flail well into the second half of the nineteenth century. It was called the 'poverty stick' in Cambridgeshire because farmers compounded with parish officers to give the unemployed work flailing in the barn. During the latter part of the century most threshing was done by the 'threshing-drum', driven at first by a portable steam engine but later by a traction-engine, which also towed drum and tackle round the farms when needed.

G.E.E.

Crop rotation

Crop rotation depends on the recognition of a crop's influence upon the soil and thereby upon subsequent crops. Men became aware of this interaction in various parts of the world long before the Christian era. Very simple systems, first two-year, then extending over three, four, five and even seven years evolved; in Central Africa even a 35-year system was practised, the same land being cleared for cultivation by burning trees and scrub every 35 years.

When crop rotation was first practised it is hard to say. The earliest organized farming has been traced to Jericho, settled by men of the later Stone Age because of its good water-supply, but there is no evidence that the idea occurred to them. Likewise another of the earliest centres of civilization, Egypt, seems to have been unfamiliar with the practice despite the other advances made in agriculture there, because the yearly flooding of the Nile brought down rich silt to feed the land and effective irrigation was the farmers' main preoccupation. It was, however, known in ancient Greece, and Roman writers such as Cato described crop rotation in some detail around 200 BC.

After the fall of the Roman Empire, however, the secret was lost and Europe was not to rediscover efficient crop rotation until the introduction, 1,000 years later, of the Norfolk four-course system which can be regarded as crop rotation at its most developed. The feudal system was the midwife of crop rotation in Europe. The descent of the Dark Ages had reduced trading and forced manorial units to become self-sufficient. When feudalism disappeared crop rotation survived and grew into the Norfolk four-course system. This brought productivity to a peak: three fields out of four would be sown in rotation with wheat, turnips and barley and the fourth sown with clover and allowed to rest. The advantage of clover is that nitrogen-fixing bacteria live in nodules in its root system, which convert the nitrogen in falling rain into a form that plant life can assimilate. At the end of the season the clover is ploughed under and both organic matter and nitrogen replenished in the formerly depleted soil. Cattle can be grazed on the fallow field, thus keeping down unwanted herbage and contributing manure while feeding themselves at the same time. The alternation of crops further ensured that weeds and insect pests stood less chance of establishing themselves with a compatible crop.

In Britain the crop rotation movement threw up many eloquent exponents, such as 'Turnip' Townshend and Jethro Tull. Their promotion of the system, aided by landowners such as Coke and the 5th Duke of Bedford, led through the writings of Hamel du Monceau in the mid-18th century to the French government's appointing officials to compile reports on the British system and effect improvements in their own agriculture. The first experimental station for crop rotation was set up in Illinois in 1876.

Chemical fertilizers and weedkillers have meant that crop rotation is less practised today. Nevertheless an increasing number of groups, most notably the Soil Association, consider it to be the most logical and ecologically sound use of Nature's resources and are arguing that a return to methods which preserve a balanced life-cycle may be the only answer to the problems of today's polluted world.

S.M.

Artesian well

A well is called 'artesian' when the water it taps is contained in a layer of pervious rock trapped between two impervious beds sufficiently inclined for hydrostatic pressure to build up over the lower part of the pervious stratum. When a well boring penetrates it, this pressure finds an outlet and forces the water up to the surface. The name records what is traditionally the oldest such well, at Lillers in Artois, sunk in 1126. But while the ancient Egyptians were exploring the oases of the Western Desert and trying to find local sources of underground water, they came upon the same phenomenon, first probably at Dakhla, where the permeable bed was only about 100 ft. down, and later at Kharga.

Here, as in Artois and long after wherever such wells might be found, they were quite accidental – the diggers might well be killed by the upsurge of water. If well water rose of its own accord, that was just a lucky strike, to be enjoyed but not to be explained. There were some famous artesian wells around Modena in northern Italy, and the first attempts to derive a scientific explanation from the recently discovered concept of hydraulic pressure were made here in the last years of the 17th century, not by hydraulic engineers but by Cassini, an astronomer, and Ramazzini, a professor of medicine. According to an English account, 'with a new digged well they make two cylindrical walls concentric to one another', filling the space between with clay, 'which done they sink the well deeper into the ground and continue the inner wall so low till the earth underneath seems to swell by the force of the water rising up; and lastly they bore this earth with a long wimble [i.e. an auger] whereupon the water breaks through the hole with a great force so that it doth not only fill the well, but overflows so as to provide a constant supply for local irrigation.'

If their physics was on the right lines, their geology was less good; they clung to the old belief, still maintained by some of the most careful scientists of the time, that rainwater was not enough to maintain all the springs, rivers and wells of the world, so that there must be huge reservoirs or subterranean lakes deep in the heart of the mountains. Most springs were supposed to be connected to them somehow, and they could also provide both the water and the hydrostatic pressure needed to push water up the artesian well. Those of Emilia were presumably drawing on such a reservoir in the Apennines. Acting on these principles, Cassini sunk new wells, one of which 'cast up water five foot high above the level of the ground'. But the real expansion of artesian wells only began after scientific stratigraphy was sufficiently precise to enable seekers after water to recognize which rocks were pervious and which not, and allow them to estimate the underground course of the beds, so as to know where to bore and how deep they could expect to go. More powerful methods of deep drilling were equally necessary, so in practice it was only in the 1830s and 1840s that hydraulic engineers had recourse to them as a regular practice.

A.K.

Refrigeration

Refrigeration as known today was invented by Jacob Perkins, an American who spent most of his long and active life in England. In his British Patent Specification of 1834 he described the vapour compression cycle, in which cooling was produced by the evaporation of volatile fluids, 'yet at the same time constantly condensing such volatile fluids and bringing them again and again into operation without waste'. The story goes that one summer's evening a working model succeeded in making a small amount of ice, which the excited mechanics wrapped in a blanket and carried by cab across London to Perkins's lodgings.

Perkins did not develop his invention, however, which he made late in life. Possibly independently James Harrison, a Scottish printer who had emigrated to Australia in 1837, was working on similar lines. He is said to have noticed the cooling effect of ether while using it to wash type. Machines to his design

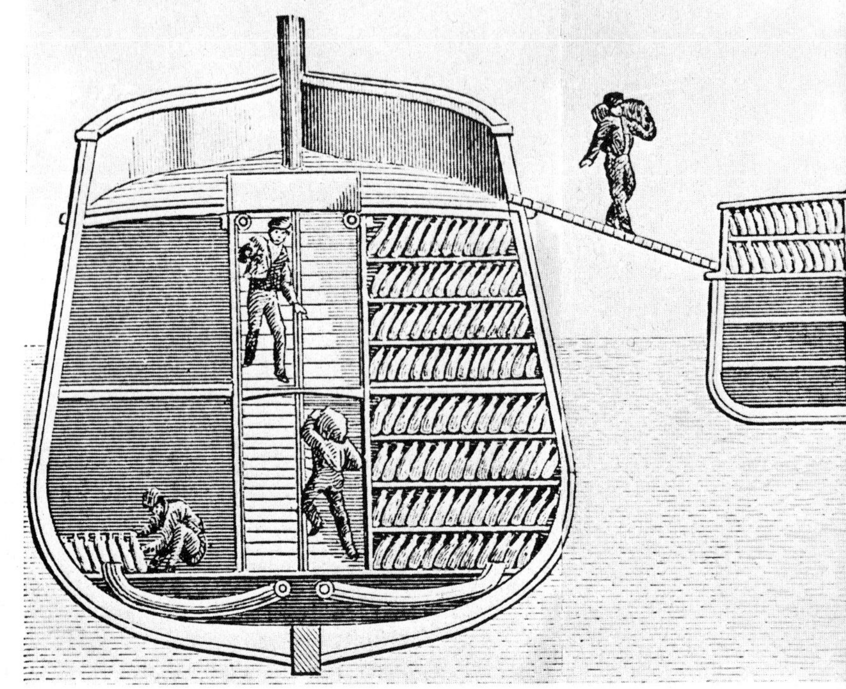

Opposite, above, *cross-section of a 19th-century artesian well at Passy, near Paris.*

Opposite, below, *early refrigerator ship on the Australia run,* c. *1880.*

Below, *domestic refrigerator made by Electrolux in 1927.*

The bull's head can-opener, possibly the earliest type made.

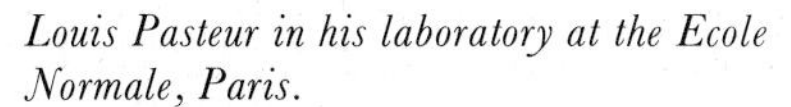

Louis Pasteur in his laboratory at the Ecole Normale, Paris.

were made by Daniel Siebe and shown at the International Exhibition of 1862. These were the first refrigerators to be marketed.

Harrison had been inspired by the need to bring Australian meat to Britain, but his attempt to ship meat in 1873 was a disastrous failure. The need remained pressing, however. In 1867 *The Times* contrasted the superabundance of food in Australia with the starvation at home, and called it a 'cruel reproach to modern science'. The reproach was removed in 1880 when the *Strathleven* arrived in London with a small cargo of sound meat. It had not in fact been refrigerated by Harrison's method, but the first link in the 'cold chain' which now extends round the world had been forged.

By the turn of the century refrigeration was being freely applied for preserving food and also for a host of industrial processes. Soon the American Willis H. Carrier was showing how it could be applied to give comfortable conditions in buildings by air-conditioning.

The social effects of refrigeration have been profound, and are likely to be even more so in future. It has made possible the growth of large urban conglomerations, and has divorced the consumer of food from the producer. Freezing of food enables surplus produce to be stored against times of famine. Air-conditioning can provide comfortable living conditions in any part of the earth. Refrigeration enters at some stage or other into almost everything we buy: plastics, drugs, clothes, gas, tobacco, films, colour magazines, to name only a few. Artificial cooling of the human body, or hypothermia, has opened up entirely new possibilities in surgery. The preservation of sperm by cold makes possible the birth of an offspring years after the father has died. And if the problem of keeping bodies alive under refrigeration is solved, man's life span may be lengthened at will. These accomplishments may be expected to raise social and ethical problems even more difficult than those already faced in transplant surgery.

W.B.G.

Can-opener

Having discovered how to seal food *into* cans, the inventor rested on his laurels without bothering too much about how to get it *out* again. An 1824 can of veal, taken on one of Parry's Arctic expeditions, carried the instructions: 'Cut round on the top with a chisel and hammer.' The can was made of iron and weighed more than 1 lb when empty. Only when thinner steel cans, with a rim round the top, came into general use (in the 1860s) could the can-opener possibly have been invented, probably in America. Its first appearance in that omnium gatherum, the Army and Navy Stores catalogue, was in 1885. It is believed that the first can-openers may have been elaborate mechanisms with which the shopkeeper opened each can before it was taken away. Possibly the earliest domestic can-opener was the bull's-head type for bully beef which still turns up occasionally in old kitchen drawers and has recently been revived by trendy kitchen boutiques.

E.G.

Pasteurization

Pasteurization, which takes its name from the 19th-century French chemist, Louis Pasteur, was originally developed not for the treatment of milk but in order to retard the abnormal fermentation of wine and beer. In launching his successful attempt to save France's wine and beer industry from certain ruin, Pasteur was undoubtedly familiar with earlier experiments of a similar nature, especially those of Lazzaro Spallanzani, the Italian biologist who in 1768 demonstrated the value of heat in conserving foodstuffs. Still, all previous theories concerning fermentation prior to Pasteur's work in the field during the early 1860s had been advanced without experimental support and corroboration. For this reason, the scientific investigations of the Frenchman also had the effect of disproving the 'spontaneous generation of life' theory, since Pasteur showed that the organisms that cause abnormal fermentation do not spring to life of themselves, but come from similar organisms which are ever-present in the atmosphere. In the same vein, Pasteur's more definitive work on milk, done between 1857 and 1862, revealed that souring is caused by the multiplication of bacteria found in the air.

The actual pasteurization process, which subjects liquid food or beverages to a mild heat treatment for a specified period of time, is based on the resisting capacities of *Mycobacterium tuberculosis*, one of the most heat-resistant non-spore-forming micro-organisms capable of causing disease in human beings. Wine and beer are heated to temperatures of approximately 57·2 °C., while with milk every particle must be kept for at least 30 minutes at 61·7 °C. to be effective. Other approaches, such as the ultra-heat treatment method and sterilization, further necessitate the process of homogenization, by which the cream is evenly distributed throughout the milk. Solid foods can also be heat-preserved, but the specifications are less uniform.

The primary purpose of pasteurization is twofold: to enhance the keeping qualities of the treated beverage and to destroy pathogenic micro-organisms. In fact, the process only partially achieves the desired results, since not all the pathogens are destroyed and of those that are, the reproducing endospores remain unaffected. Subsequent refrigeration will, however, prevent certain spore-forming pathogens from growing.

Today, many researchers and proponents of natural foods are attacking pasteurization for its adverse effects on the food values of treated liquids, especially milk. Both vitamins B_1 and C are greatly reduced, acidity is increased, and when the high temperatures necessary for the elimination of all harmful bacteria are applied, valuable proteins and enzymes also begin to break down. In place of such costly and inadequate laboratory treatments, critics of pasteurization and sterilization emphasize the need for a continued reduction of bovine diseases and for more sanitary methods of handling fresh milk. But there are numerous businesses with a vested interest in the pasteurization process, and they are unlikely to admit to its deficiencies.

E. W.

Artificial foods

Today the use of synthetic foods is increasingly arousing public discussion, but how many people realize that men have been happily eating an artificial food for over 100 years? For margarine was the world's first synthetic food, and so important has it since become that the export of vegetable oils for margarine plays a major role in the economies of numerous developing countries.

During the Industrial Revolution in Europe large numbers of people gave up working on the land. Population figures rose fast, however, and soon there was scarcely enough animal fat to go round. In response to this situation a French chemist named Hippolyte Mège-Mouriez was commissioned by Napoleon III to create a cheap butter to help 'increase the national larder'. The result, patented in 1869, seems more likely to have increased the national lard, for it looked and tasted a good deal like that unappetizing commodity. Nevertheless Mège-Mouriez felt more romantically about his product; he thought it resembled a pearl in colour and named it from the Greek word *margarites* (pearl).

Mège-Mouriez did his research at the imperial farm at Vincennes, where he made an exhaustive study of cows, reasoning that what the cow did naturally he must be able to do by mechanical means. His experiments hinged on the butterfat content of milk, and he decided that fat mixed with milk, some water and a special ingredient should result in artificial butter. His original method sounds gruesome: he processed beef tallow to produce a soft fatty raw material called *oleo*; this was then liquefied and mixed in equal quantities with milk and water and his 'special ingredient' – cow's udder chopped up fine.

The outcome of his final experiments was patented in France and England in 1869. Poor Mège-Mouriez, however, was taken prisoner in the Franco-Prussian War a year later. He retired to England after this, dying there, a bankrupt, in 1883 – the year his patent expired. However he had divulged his process to the brothers Jan and Henri Jurgens, butter merchants in the village of Oss in Holland, and soon the Jurgens and a rival firm in the same village, the Van den Berghs, were making large quantities of the new fat. Business prospered and the product, sold as 'Butterine' in England until the description 'margarine' became compulsory in 1887, was launched on its successful career.

As the demand for margarine increased, a shortage of beef and other animal fats led to a major change in the raw materials used. Vegetable fats from tropical oil-bearing plants (such as palm fruit, coconut and groundnuts), from sunflower seed and soya beans, and marine oil from the unfortunate whale began to be substituted. Initial difficulties caused by the use of these softer fats were solved in 1910, when hydrogenation – a process that hardens oil by adding hydrogen – was introduced, and a generation later vitamins A and D were added to give a more perfect resemblance to butter.

Mège-Mouriez's pioneering efforts have today led to a wide range of animal-meat and animal-milk substitutes, all of which follow his reasoning in converting vegetation directly into nutritious, digestible food for humans, instead of waiting for an animal to convert it. Today we can eat 'kesp', a 'knitted steak' spun out of soya-bean protein to resemble meat fibres. There are cream substitutes which contain absolutely no dairy product, and even a totally vegetable milk called 'Planil'. Imitation milk powders made from cheap skim-milk and vegetable fats are common, as is vegetable-fat ice-cream. All of these commodities are cheaper than the real thing because animals are being outstripped by the growing human population of our planet.

S. M.

Cream separator

Gustav de Laval was born in Blasenberg and came of a family of soldiers. A student of Uppsala, he graduated in engineering, with honours in every subject and later studied metallurgy which may have given him the idea, in 1876, of developing the centrifuge for cream-separating. Shown at the Royal Show at Kilburn in 1879, it achieved an immediate success, and mechanical cream separation completely displaced the skimming of cream by hand.

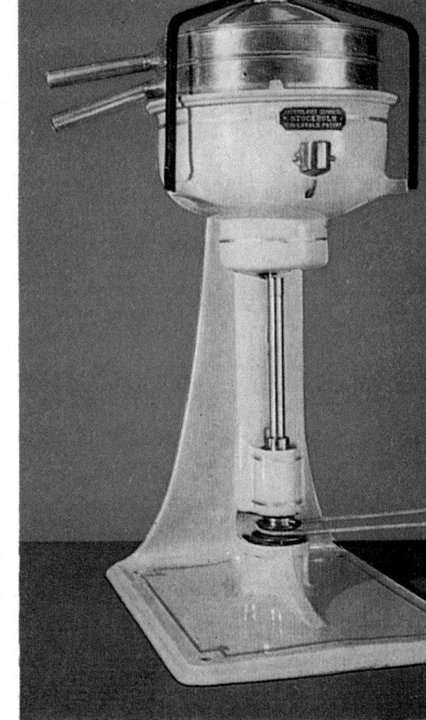

De Laval's cream separator (1879).

Calendar-holder advertising margarine, c. 1890.

Tasting pastries made with margarine, at a Milan exhibition in 1921.

An international tractor-drawn combine harvester made in the late 1930s.

Bulk delivery to Battle Creek food stores, 1913.

An early advertisement for breakfast cereals.

One of his earliest machines was driven by a whirling-arm steam turbine – similar to a lawn sprinkler – mounted in the base, while others were driven by belts, by hand-cranks through step-up gearing, and finally by the impulse steam turbine invented by de Laval in 1888, for which he devised the convergent-divergent nozzle used in steam turbines today. The machines were operated at the then remarkable speed of 6,000 r.p.m.

The centrifugal cream separators used today are substantially the same as the machine invented by de Laval.

A. BU.

Combine harvester

An American wrote at the end of the last century that he had seen a combine working in California as early as 1887, propelled from behind by sixteen mules. Shortly afterwards, mules and horses were replaced by steam-power. The combine uses the essential features of the reaper and weds them to a travelling thresher. It has cutter-bar, elevator, threshing drum and shaker, and cuts and threshes in one operation. To carry the corn from the cutter to the drum, early combines used canvases, as did the reaper; but most modern machines use an Archimedean screw or auger instead. For many years the tractor was used to pull the combine, but present-day machines are self-propelled. This makes a machine much more manœuvrable; and, at the busiest time of the year, releases the tractor for other work. The tractor-drawn combine usually carries an auxiliary engine for the cutting and threshing.

The earliest combines were evolved and used in America and Australia, where the scale of farming gave full scope for the development of new machines. But these did not prove very satisfactory when introduced into Britain. Like the ancient *vallus*, they were 'headers' and left the greater part of the straw standing, and this did not suit the British farmer. Again, little corn threshed in British fields is dry enough to go straight into storage, and drying machines had therefore to be developed to supplement the use of the combines. Finally, crops there were heavier than in the new countries, and longer in the straw; and they demanded machines that were modified to suit them. The latest combines have introduced no new principles into the cutting and threshing of crops. There have, however, been big improvements, especially through the use of hydraulic power.

The name of Cyrus H. McCormick is widely associated with the combine harvester, but apart from his development of the reaper, his contribution was as the founder of the giant International Harvester company.

G. E. E.

Breakfast cereals

The two great names in the history of breakfast cereals are Kellogg's Corn Flakes and Post Toasties; both of them originated as health foods. The Kellogg brothers were vegetarians and Seventh Day Adventists, descended from an emigré to America from Debden in Essex. One of the brothers, John Harvey, was a Fellow of the Royal College of Surgeons in London and a member of the American College of Surgeons. As a medical student, in an age of stuffy dress and stodgy food, Harvey found he was able to think more clearly on a light breakfast of apples and wholemeal crackers. When he started a 'sanitarium' on food-reform lines in Battle Creek, Michigan, in the 1880s, someone objected, 'There is no such word as "sanitarium".' He retorted, 'But there will be soon,' and *Webster's Dictionary* offers it as an alternative to 'sanatorium'. His brother Will, a born organizer, was soon producing foods of his own invention for the guests.

Charles William Post, an irrepressibly energetic and inventive man of 36, came to the Kellogg sanitarium in 1890 to recuperate after an illness. It is not recorded whether, or to what extent, Post's ideas were influenced by the Kellogg health foods, but both parties were evidently thinking along the same lines, for in 1892 Post bought farmland and a building on the outskirts of Battle Creek and opened La Vita Inn for 'the treatment of persons afflicted with nervous prostration and overwork'. In a little white barn he experimented for over two years with different combinations of grain and emerged with Postum Cereal, the Health Beverage. He formed a company and went into production with only $70 investment, which included a second-hand two-burner stove, a small peanut roaster, a coffee-grinder and several mixers.

The business grew rapidly, but the sales of Postum, a hot drink, fell off during the summer; to counteract this and spread sales and production more evenly, Post set about devising a cold cereal product, one he had been making for some years for his own use. Made of wheat and malted barley and baked in the form of bread-sticks, it was marketed in 1897 under the name of Grape Nuts because he believed that grape-sugar was formed during the baking, and because the cereal had a nutty flavour.

Grape Nuts was one of the first ready-to-eat cereals on the market, but not the first. This distinction goes by a few months to the Kellogg brothers' Corn Flakes, which had been one of the products served at the sanitarium. Patients who returned to their homes created such a demand for the Kellogg health foods that Will Kellogg went into business. From these small beginnings two very large enterprises grew.

S. M.

Man Living

Instant coffee

It all began, so the most prominent legend goes, sometime around AD 850 with an Abyssinian goatherd named Kaldi. Curious about the antics of his goats, Kaldi decided to try for himself a few berries of the evergreen bush at which the animals had been nibbling. Much stimulated, he dashed off to proclaim his find to the world, whereupon he met an old Moslem monk who was depressed by his tendency to doze off while at prayers. Kaldi introduced the good ascetic to his eye-opening discovery and coffee was on its way to becoming one of man's favourite beverages.

The organic ingredient which produced such an invigorating effect in goat and goatherd alike was caffeine, an alkaloid which in small amounts functions as a stimulant and diuretic by acting on the central nervous system, the heart, blood-vessels and kidneys.

Because of its stimulating effect on people, coffee became not only a very popular drink within a few hundred years of its initial 'export' from Abyssinia to the Yemen and the rest of the Arab world, but a highly controversial one as well; and there are numerous instances of its consumption being banned. Nevertheless, anything the people want badly enough they will eventually get, and by the time the first European coffee-house was opened in London in 1652 another of man's many habits had taken root. Less than three hundred years later, with the development of instant or soluble coffee, popular opinion had the opportunity to outwit the conservative minority. In this case the sages were the expert coffee-testers, who predicted that the public would never accept such an inferior substitute, no matter how economical. Today, however, roughly a third of all coffee prepared in Western households is of the instant variety and its use in other parts of the world and in restaurants everywhere is steadily increasing.

Serious consideration was first given to a coffee extract, in a liquid form which has never proved practical, in 1838, when the United States Congress began substituting coffee for rum in the rations of American soldiers and sailors. Powdered instant coffee was probably invented by Satori Kato, a Japanese chemist living in Chicago, who in 1901 sold his product at the Pan American Exposition held in Buffalo, New York. Five years later, an American chemist by the name of G. Washington developed a 'refined soluble coffee' which was subsequently marketed on a broad scale. During the First World War, the US War Department gave the instant-coffee business an enormous boost by purchasing its entire domestic output, while the next big war saw the American Government buying up some 260,000,000 pounds of it – all for the use of its military forces. Obviously, it is not only the armaments manufacturers who derive financial benefits from war.

Until the mid-1950s, two kinds of instant were generally available, the product type, to which carbohydrates were added, and the all-coffee type. It is this latter variety which has survived. As with regular ground coffee, the soluble concentrate is produced by first blending, roasting and grinding the green *Coffea* bean. Extraction is then effected by a process of evaporation, after which the bubble-like particles of powdered instant are created by spray-drying. In the 1960s, a new method of drying was introduced whereby the extract is frozen and then placed in a vacuum chamber which sufficiently sublimes the moisture content to leave a solid mass. This mass is in turn reduced to readily soluble granules. The success of freeze-dried instant coffee is attributed to its excellent flavour, which more closely approximates the 'real thing' than its powdered counterpart.

E.W.

Synthetic fertilizer

World production of wheat and rice, man's basic food, is about 550,000,000/600,000,000 tons a year. Grain crops require phosphatic, potassic and nitrogenous fertilizers, without which crop yields would fail and there would not be enough food and fodder.

The importance of fertilizers and the key role of nitrogen were already recognized a hundred years ago. Nitrogen was then commercially available in the form of sodium nitrate from deposits found only in Chile, and sulphate of ammonia, a by-product of coal gas. As demand was rising fast there were fears that the former would soon be exhausted and the latter insufficient. Thus the fixation of nitrogen, which makes up almost four-fifths of the atmosphere, attracted much scientific attention.

Fritz Haber (1868–1934) found the most successful solution in 1908–09 by reacting hydrogen and nitrogen in the presence of a catalyst at high pressure and temperature. His synthetic ammonia process incorporated many novel concepts and is still in use, though the operating conditions have been altered. In less than four years a team at Badische Anilin & Soda-Fabrik under Carl Bosch developed a cheap and robust catalyst, designed the big reactors, compressors and special control instruments, and provided for the bulk supply of hydrogen and nitrogen. The first small unit, producing nine thousand tons of ammonia a year, was started in Germany in 1913. Though enlarged it could not meet more than a fraction of the country's nitrogen requirements when war broke out. The situation soon became critical, and it is likely that Germany would have collapsed before 1918 but for the nitrogen-fixation plants built in 1914–16.

Many synthetic ammonia works were built between the wars. They were very expensive, but governments considered the Haber process so important that they helped finance construction and often operated their own works. Tariffs were usually imposed on competing imports. By 1930, it was possible to produce 2,500,000 tons of synthetic nitrogen a year. During the Depression supply exceeded demand, but when agriculture recovered the growth of nitrogenous fertilizer consumption resumed.

High-pressure techniques were extended first to the synthesis of methanol, then to petroleum refining and during the 1930s to the production of oil from coal. Catalytic processes characterize the modern chemical industry and are often applied on the scale of several hundred thousand tons a year per plant. Yields and production costs depend critically on high and sustained outputs; markets must therefore be found and developed. Thus the technological momentum imparted by Haber's discovery continues to influence the evolution of the industry.

L.F.H.

Deep freezing

Since very early days men have realized the value of cold in slowing down the decomposition of food. An eminent martyr in this field of research was Francis Bacon, who died from a chill caught when trying to freeze a chicken by stuffing it with snow.

It seems to have been Clarence Birdseye who, travelling in Labrador soon after the First World War, first observed that the key to the preservation of meat and vegetables lay in the speed of freezing rather than in achieving extraordinary extremes of cold.

Back from Labrador in 1923, Birdseye experimented on rabbit meat and fresh fillets in his own kitchen and then in a refrigeration plant in New Jersey. He developed a process whereby the cartons of food were pressed between refrigerated plates. This system is still in use, but has been largely replaced by 'blast-freezing' in a sort of wind tunnel. The first retail sales of packaged frozen foods were made at Springfield, Mass., in 1930. Birdseye died in 1956, aged seventy, with three hundred patents to his credit.

In the wake of quick-freezing came the home-freezer, enabling people to preserve their own food as well as commercially frozen goods.

M.C.

Opposite, *instant coffee: filling and sealing the cans.*

Opposite, below, *Fritz Haber, father of the synthetic fertilizer industry.*

These four pictures show the dramatic effects that can be achieved by cloud-seeding. Within minutes, a huge head of cumulus builds up.

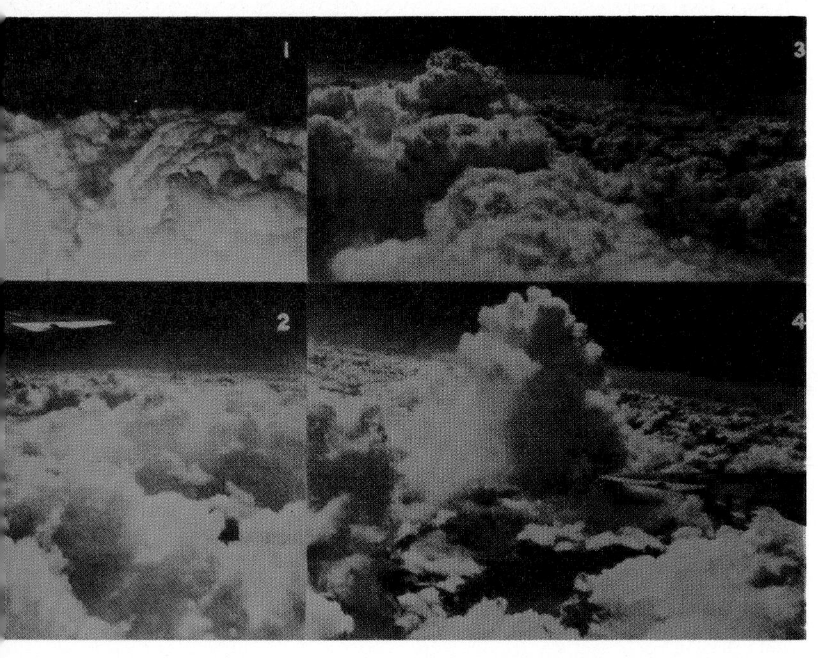

Treating stored grain with insecticide.

Cloud-seeding

For thousands of years men have tried to influence the weather – by prayers and sacrifices to the weather gods, by magic and even by shooting cannon at the clouds to make them release their moisture. Yet droughts, particularly in Africa and Asia, still kill off hundreds of thousands of people every few years.

The most effective technique of 'making rain', developed only during the last few decades, is cloud-seeding – the only scientific method at present at our disposal. The French meteorologist Bergeron invented his 'precipitation initiation process' shortly before the Second World War. It works, however, only with clouds which contain ice-crystals as well as water droplets; the ice-crystals attract the droplets, and the result is often snowfall. By dropping more ice-crystals into the cloud from high-flying aircraft, the cloud is 'super-saturated' with ice and 'subsaturated' with water; the ice drops through the cloud and turns to rain in the higher temperature of the lower regions.

More research was carried out in the late 1940s at the General Electric Research Laboratory in Schenectady, New York. Here, V.J. Schaefer discovered that very small particles of 'dry ice' – carbon dioxide in its solid state – would do the trick better than the Bergeron process. In the first experiment the dry ice was mechanically released from an aircraft flying through a large altostratus cloud at an altitude of 15,000 ft. Within a short while, rain began to fall.

Another technique, working with 'coalescence', uses clouds which do not contain ice but only suspended water droplets; these can be made to 'coalesce', forming larger drops which then fall to earth. The process can be started by seeding the cloud with salt, which attracts water; or water can be sprayed into the cloud from an aircraft to stimulate the formation of larger drops.

Perhaps the most economical method of cloud-seeding was developed by another General Electric research worker, B. Vonnegut, in 1947. He discovered that in a very cold cloud, silver iodide smoke can produce snowflakes; the silver iodide particles serve as freezing nuclei, since their crystal structure is similar to that of ice-particles. The seeding is also carried out from aircraft, and at lower levels the snowflakes turn to rain. This technique, and cloud-seeding in general, has been found very effective in cumulus clouds. Still, the practice of cloud-seeding needs a good deal of further development; apart from the United States, Australia is now in the forefront of rain-making research.

E.L.

Insecticides

To our knowledge, the first people to attempt chemical control of insects were the Chinese, who used naturally derived arsenic sulphide against crop pests, while in the West the ancient Greeks and Romans are known to have employed sulphurs and other substances for the same purpose. By the late 19th century, in Europe, lead arsenate had come into vogue among farmers, but it was not until 1939, in Switzerland, that DDT – the first of the big-time insecticides – was put into action. The Swiss used this long-lasting and very toxic poison to fight the Colorado potato beetle. The Second World War gave the insecticide business a tremendous boost, as numerous chemical compounds – with DDT in the forefront – were developed to protect the fighting troops against typhus, malaria and other insect-carried diseases. Today, a very large number of insecticides are in common use all over the world, ranging from household aerosols to ultra-sophisticated industrial sprays, often applied by aircraft or helicopter over large tracts of farmland.

'Pest poisons' fall into three categories: stomach poisons, which must be ingested by the insect in order to disrupt its metabolic process and cause death; contact poisons, which penetrate the integument or cuticle; and fumigants, attacking the creature via its respiratory tubes. The most modern and lethal insecticides incorporate all three methods and form a combination of synthetic and natural chemicals. In addition, the industry has developed a variety of repellants and chemosterilants, the latter seeking to inhibit the reproduction of certain insects rather than kill them. Clearly, the prime motivation for pesticide development and use is economic: in one country alone, the United States, some 500 species of insects annually cause up to $8,000 million worth of agricultural damage.

In the end, however, the insects may have the last laugh, even if it is a posthumous one. The pollution of the environment caused by large-scale spraying operations, the degeneration of the quality of vegetable life, as well as the direct poisoning of man through the ingestion of treated vegetation present an obvious and immediate threat to the survival of the human race. DDT, for example, is known to contain cancer-inducing chemicals. Furthermore, most insects tend to build up an immunity to the various insecticides within a few years of their introduction, while the non-selective nature of most chemicals used often results not in the destruction of unwanted pests but in their proliferation, since their natural enemies are also destroyed. Perhaps in time man will find a middle course in the matter of pest control somewhere between his present extreme and that of the fanatical

Buddhist who will not allow the destruction of insects by any means.

E.W.

Fishing technology

The tremendous postwar expansion of the world's marine fisheries has been made possible by the application of increasingly sophisticated technology, particularly that concerned with detecting fish and handling and preserving the catch. Much of this technology has been brought in from outside rather than devised within the fishing industry itself: it has evolved within the fisheries environment rather than originated there. In essence, fishing practice has changed very little, but technical innovation and the advent of new materials have increased dramatically the catching power and range of modern fishing boats. As a result, many stocks of fish are now either over-exploited or are in danger of becoming so. The tuna, one of the most popular sources of food, is seriously threatened in several sea areas. Stocks of cod, haddock, halibut and salmon have diminished. The blue whale has been hunted to the point of extinction.

Catching fish must rank among man's oldest ways of exploiting the oceans, but to deploy modern fishing vessels economically, large quantities of fish have to be found within range of the fishing gear. Most modern vessels therefore carry some kind of fish detection system, usually sonar. Sonar systems all work on the same basic principle: an acoustic signal is transmitted towards the seabed or ahead of the boat and the echoes returning from obstacles in its path are recorded and displayed. The time-lapse between the initial signal and the echo gives the distance to the obstacles, in this case the fish. Sonar systems are now being devised for many fishery applications, from the discovery and assessment of new fish stocks to the active pursuit of large shoals of pelagic (surface-dwelling) fish. These developments, based originally on the echo sounder and wartime submarine detection systems, are taking place in many countries – Denmark, Germany, Japan, Norway, the Soviet Union, Sweden, the United Kingdom and the United States. Now, instead of wrinkling his nose and watching for familiar signs of fish, the modern skipper is likely to display his skill by the way in which he interprets his sonar and other instrument recordings.

The electronics revolution in the fisheries is not limited to the deployment of sonar and the like. Large fishing fleets can also benefit from the rapid processing of fisheries data and meteorological information by computers. Japan, for example, provides its fleets with forecasts and information on conditions known to favour the presence of fish. These forecasts are based on large quantities of information relayed to fishery centres by research vessels and the fishing fleets themselves. Thanks to modern navigational systems, productive areas can be pinpointed even in the open ocean. In fact, the vast array of navigational and communications equipment on some fishing vessels, particularly those that operate far from their country of origin, have led to many suspicions that they are involved in some work other than catching fish. Certainly, these suspicions are sometimes well-founded: a large fishing fleet with vessels bristling with antennae and navigational aids provides excellent camouflage for the military 'spy ship'.

The actual techniques used by the commercial fisheries, the electronics revolution notwithstanding, have changed very little in many respects for centuries and certainly for decades. One of the oldest organized fisheries is that for herring – records of a British herring fishery go back to Saxon times. The principal gear used in this fishery is a gill net known as a drift net. The net, made up of sections, can extend for three miles or so, hanging down like some giant tennis net from the surface. Once it has deployed the net, the fishing boat drifts along with the wind and awaits nightfall when the herring rise to the surface waters to feed. The net does not alarm the fish. On meeting the barrier they attempt to swim through it. The mesh is so gauged, however, that only small fish can pass through. The larger fish are retained, ensnared by their gill covers as they attempt to retreat out of the mesh. At first light, the net is hauled in and the captured fish, possibly several thousand or more, are shaken free onto the deck.

Seine fishing, too, has changed little in principle over the years. In this technique the fisherman tries to corral the shoaling fish rather than wait passively for them. One end of the net is attached to a buoy before the fishing boat encircles the fish, paying out the net as it goes. On arriving back at the buoy, the two ends of the net are brought together. A warp, which is threaded through rings along the base of the net, is drawn in, an action known as 'pursing' the seine. The shoal is now trapped in the basin-like net, which is hauled in until the catch is a seething mass alongside the boat. The fish are then pumped out of the sea and into the holds or, if they are large, scooped out with nets. The reasons for this can be better appreciated if one bears in mind that some seine nets are large enough to enclose St Paul's Cathedral and a single haul might yield as much as 400 tons of fish.

The final stage in fish hunting comes with the pursuit of the prey typified by trawl fishing. The first trawls were cumbersome affairs with a heavy beam along the top of the net to keep it open. The frame of the mouth was completed

Above, *the* Vostok, *a Russian factory ship.*

Left, *fishing for tuna with rod and line.*

Opposite, above, *bringing the catch aboard.*

Opposite, below, *hauling in a stern trawl, full of haddock.*

Below, *lowering an otter board.*

by sidepieces with metal runners, like those of a toboggan, to help the heavy trawl skid over the seabed. The breakthrough in trawl design came in the 19th century with the introduction of the so-called otter boards. When towed through the water the boards, one on each side of the net, sheer away from each other like kites, keeping the mouth of the net extended. The beam was no longer needed.

The precise origin of the otter board is uncertain. An Irishman by the name of Musgrave is believed to have experimented with them between 1860 and 1885. Certainly the use of these boards spread first from Irish and British fishermen to Europe and then to more distant parts of the world. By 1905 Japanese fishermen were using them.

At first the otter boards were attached directly to webbing on the sides of the net. This webbing had originally helped to increase the spread of the trawl by funnelling fish towards its entrance. With the wider gape made possible by the otter boards, the webbing could be reduced. In fact, the boards did not need to be attached directly to the net at all, as was proved by Vigneron-Dahl in the 1920s. The V-D trawl has long wire bridles between the net and the boards, giving a very wide gape to the mouth of the trawl. This principle has since been drawn upon by many net designers. Today numerous designs of otter trawl exist and new trawls continue to be developed, particularly for fishing in the middle and surface waters of the sea rather than along the seabed. Trawls that operate away from the seabed usually have a square mouth. In some, the bottom of the net protrudes, to prevent fish from diving beneath it; others have elaborate floats and hydrofoils to keep them in position, and the majority have specially designed otter boards to hold the mouth of the net open.

When nets cannot be used, the commercial fisherman falls back on gear more often associated with leisure fishing – hooks and lines. In at least one type of tuna fishing, the technique is not so very different either. When a shoal of tuna has been spotted breaking the surface, the tuna boat approaches cautiously and releases live bait from special compartments. At the same time, the boat begins to churn the water to conceal its presence and to attract the fish to the bait. The feeding tuna snap at anything that flashes in the swirling water, including unbaited shiny hooks. Teams of fishermen working from platforms lowered over the side of the fishing vessel cast their lines as fast as they can. They use unbarbed hooks. As soon as the tuna bites it is jerked out of the water and onto the boat where it falls away from the hook, which is returned immediately to the water.

Hooks and lines are also used in some demersal (deep-water) fisheries. Long lines with as many as five hundred branches are lowered in the water and each branch carries a barbed hook baited with a morsel guaranteed to attract a hungry cod or haddock. Once the line has been shot, the upper end is attached to a marker buoy and the fishing boat moves on to lay down more long lines. Baiting the lines as they shoot out into the water can be a hazardous job. Some long liners are now equipped with large rotating drums from which the branches hang, making it easier to bait the hooks as the line is slowly paid out.

Apart from the traditional fisheries, other techniques, some based on existing ones, are being developed. For example, the United States National Marine Fisheries Service is experimenting with an electric shrimp trawl. The trawl delivers short pulses of electricity to the seabed, sufficient to force the shrimps out of their mud burrows. The American scientists have even worked out what strength and frequency of electrical stimulation is needed to make the crustaceans move to a suitable height for a particular speed of trawling. In other US fishery experiments, underwater light lures are being combined with suction pumps. Initially the technique was used to collect small crustacea needed to feed the larvae of pelagic fish being reared in the laboratory. On a totally different scale are Russian experiments to harvest the krill, a large shrimp-like crustacean, that swarms in the waters of the Antarctic Ocean. A research ship was equipped with a large funnel-shaped net to drag through the krill shoals. To enable continuous harvesting a pump was attached to the end of the net to allow the catch to be drawn continuously aboard the fishing vessel. Finally, there is all the new technology of fish farming: this ranges from techniques of hatching and cultivation to those of constructing suitable enclosures or marine fields.

Recent progress in nets and other fishing gear would have been impossible without corresponding improvements aboard fishing vessels. In some cases, the traditional fishing methods have been completely revolutionized. A good example is the power block invented in 1955 by an American, Mario Puretic, who joined forces with the Marine Construction and Design Company of Seattle to produce the unit for the local salmon purse seine fleet. The Puretic power block is a large hydraulically-powered V-shaped sheave through which the entire net, rather than just the warps, is pulled. It has had remarkable success. Soon after its development the power block was being used by the American tuna fleets. Within 12 years over 7,000 power blocks had been installed on fishing vessels throughout the world. It has been argued that the power block was instrumental in making possible the fantastic growth of the Peruvian fisheries, where the catch increased five-fold in the decade from 1953. The power block has also influenced some European fisheries. Norway has built over 500 purse seiners, all fitted with power blocks, and is now the top fishing nation in Europe.

Mechanical handling, on-board processing of the catch and automation are all becoming part of everyday operations in modern commercial fisheries. They reach a peak in large trawlers and have influenced one of the most important advances in fishing vessel design – the stern trawler. As the name suggests, the fishing gear is handled over the vessel's stern and not over the side as in conventional trawlers. The new concept resulted directly from the whaling activities of a British company, Christian Salvesen of Leith. Faced with a decline in whaling the company decided to explore the possibilities of transferring to the fishing industry experience gained in handling and processing the giant mammals. Their first stern trawler, the *Fairtry*, went into service in 1954. The vessel was equipped with a stern ramp similar to that of the whaling 'factory' ship and like the factory ship it also carried a considerable amount of processing machinery. The *Fairtry* was to be the first of many stern trawlers built by fishing nations all over the world. Not all of them have been of the factory type. Since the early sixties a smaller vessel with a swinging gantry instead of the stern ramp has also appeared on the fishing scene.

The large factory trawlers are part of a new dimension in commercial fisheries where roving fleets operate on fishing grounds far away from their home ports. One of the best examples is the Soviet fishing fleet. Russian trawlers operate on fishing grounds off both North and South America, Africa and many parts of Asia. Both side and stern trawlers are used, but by far the most impressive vessels are the factory trawlers. These stern trawlers, about 3,200 gross tons and nearly 300 feet long, are capable of processing the catch down to the last scrap. Below decks they have complete automated production lines. Fish are filleted, packed and frozen. The inedible parts – skin, bones and viscera – are reduced to fish meal and oil. Nothing is wasted. The vessels return to their home bases only occasionally. Support ships ferry out crews and supplies, returning to base loaded with frozen fish packs, bags of meal and barrels of fish oil. Finally the Soviet fleet has floating fish factories. These may act as a flagship for as many as 150 fishing vessels, including up to 10 factory ships.

The tremendous catching power typified in the roving fleets of eastern bloc countries and Japan mean that advances in fishing technology must be considered against a background of conservation and controls on exploitation. Without adequate management, progress in fisheries technology will lead to a decline in the very industry it is designed to support. The present-day fishing industry is big business. Few people, perhaps, realize that the value of the world catch exceeds that of the oil tapped from off-shore fields. On the other hand, the days when expensive ships and fishing gear can be moved to new grounds once existing ones become less productive or unprofitable are fast coming to a close. Symptoms of this realization that fish are a limited, finite resource are seen in the increasing number of claims by coastal states to large expanses of sea for their own exclusive use. These states are no longer willing to allow valuable and accessible stocks of fish to be exploited by the fishing vessels of other nations. Increasingly, new technology will be needed to bring fish populations in deep waters and remote from the land within reach of the fisherman. New methods of processing and presentation will be needed for these new species. The story has by no means come to an end.

T. L.

'Miracle' rice

'A miracle', wrote Bernard Shaw, 'is anything that creates faith.' Just how much faith IR-8, BPI-76, C-18 and other strains of so-called 'miracle' rice have created among Asians is still a moot question. Consumers dislike it because it tastes flat and does not cook fluffy (in Thailand people refer to it as 'sticky rice'). Poorer farmers are down on it because it requires modern techniques of cultivation. Nutritionists decry miracle rice by claiming that it has less food value than traditional fine-grain varieties, while health-food partisans are put off by the amount of artificial fertilizer such hybrid types need to grow properly. The United Nations Food and Agriculture Organization (FAO), on the other hand, thinks short-stemmed high-yield rice is the best thing that has happened to Asia in years, and many experts – especially economists – agree.

Miracle rice, the cornerstone of Asia's much-heralded 'green revolution', was developed at the International Rice Research Institute of Los Banos, in the Republic of the Philippines. The IRRI, which was established in 1962 as a joint venture of the Rockefeller and Ford Foundations, in co-operation with the Philippine government, was motivated by a dire world shortage of rice. This shortage, the institute realized, could not be solved by the age-old method of opening up new land areas to rice cultivation whenever a higher domestic yield was desired. Agronomists therefore set about breeding a new strain of the plant that represents the staple diet of 60 per cent of the earth's population. They did this by mating a short-stemmed *Indica* from Taiwan with a tall Indonesian *Indica* and carefully nursing the resultant offspring through five generations over a four-year period. Out of this process, during which the plants were often exposed to disease and harsh climatic conditions to develop their resistance, came the basic stock of today's miracle rice, a 36- to 42-in. dwarf known as IR-8. This rice variety, which has a huge appetite for nitrogen, matures quickly and has a per-acre yield more than twice that of local types. And its low height makes it resistant to 'lodging', the tilting of plants that occurs when fertilizer is applied.

Among the more recent problems of miracle rice, which is very similar to the *Japonica* breed found in temperate-zone countries, is that its luxuriant growth also favours the rapid multiplication of disease organisms and insect pests. In Pakistan, one of the 80 nations now working with the new crop, farmers wryly comment that 'miracle rice has produced a miracle locust to eat it'. Nevertheless, this agricultural child of American technology has, if modestly, begun to attack the root cause of a major food crisis. New strains, such as the 'tastier' IR-24, are constantly being experimented with, and it is hoped that future miracle rice crops will be more nutritious than their predecessors. Perhaps, too, benighted consumers will eventually realize that it is their own buying habits, which consistently favour polished rice over healthier brown rice, that reduces their daily intake of valuable vitamins, fats, proteins and minerals.

E. W.

Appliqué felt hanging for the wall of a tent, found in a tomb at Pazyryk, E. Siberia. Date, c. *400 BC.*

From Danish peat bogs come some of the oldest surviving garments, made from textiles in the Bronze Age.

Felt

Animal hairs – not woven, but compressed into felt – were much used in antiquity, particularly in the cold climates of central Asia. The date of this process's invention cannot be guessed at, but it seems likely that it preceded weaving. To make felt, heat, moisture and pressure are applied to hairs which, once pressed together, do not subsequently disintegrate; the fabric has no grain and will not fray, but it holds its shape badly unless quilted or stitched in some way.

The range of uses of felt can be assessed by an examination of the Iron-Age (5th century BC?) Pazyryk burials in eastern Siberia; for there high-born noble families were interred in all their finery, and the barrows decorated to resemble their normal habitation. Thus in Barrow 2 the floor was covered by a thick black felt, and the walls draped with the same – turning the whole into a tent. In Barrow 1 there was coloured felt on the walls, with festoons of red, white, yellow and blue, and cut-out lions' heads stitched carefully to the material. Saddle-cloths were found which used the same technique, with felt and leather silhouettes of animals attacking their foes stitched or pasted to the body of the work. Stockings of very thin, fine felt (2–3 mm.) were also found here, as well as bonnets of the same material.

Because of its simplicity of manufacture, felt is found in every community where wool-bearing animals are domesticated. Today it is much used for insulation, for vibration isolation, and for packaging – and the Greeks and Romans lined their helmets and shields with it for much the same reasons. A fanciful story of the invention of felt demonstrates these qualities: some merchants became footsore because their shoes rubbed them, so they plucked soft hairs from their camels and pushed them inside their shoes – the heat, sweat and pressure produced felt.

Much more economical than wool-felt has been the recent development of bonded fibre fabrics, usually with man-made rather than natural fibres, particularly polyamide, polyester and acrylic (the process involves the use of an adhesive – they do not naturally felt). Use of such materials for clothing is restricted to interlining and the backing of other materials (also by bonding) because it has the same disadvantages as true felt.

C. M. B. G.

Textiles

Textiles, like pottery vessels, were an essential part of the Neolithic Revolution. The idea of spinning yarns and making them into textiles probably derives from basketry, of which the earliest known example is from the Proto-Neolithic level of the Shanidar Cave, in northern Iraq. (Possibly the same is true for pottery, for early pots are often decorated in imitation basket-weave.) The oldest textiles we have are from Çatal Hüyük in Anatolia – bits of wool cloth adhering to human bones, and dated *c.* 6500 BC. Early textiles would have been made from fibres spun into thread between the palms; later a spindle was introduced to add momentum.

Wool was the first common material, sheep having been domesticated from Neolithic times; its use was widespread in the eastern Mediterranean and in Europe (although the earliest woollen garments here come from coffin burials of *c.* 1300–1100 BC). In Egypt, on the other hand, flax was the staple, and gave linen cloth of varying standards from Neolithic times onwards; wool, in any case, was considered ritually unclean. In Ur of the Chaldees, however, woollen garments clothed the people, and the finest linen was reserved for the statues of gods.

Cotton was being cultivated at Mohenjo-daro in the Indus Valley by *c.* 2500 BC, for traces of cotton textile have been found inside a silver jar, attached to its patina. Spindle-whorls which are much too light to have been used in spinning wool have been found at the same site. Although the use of the plants for textile spread east only slowly, reaching China in the 13th century AD, it came west more quickly; King Sennacherib of Assyria was planting cotton trees by *c.* 700 BC; after describing his palace, he describes its great park 'wherein all kinds of herbs such as grow on the mountains and in Chaldea as well as trees bearing wool were set out . . . the wool-bearing trees they sheared and wove into garments.'

Silk is eastern, and its production was restricted to China and India for most of antiquity (although there were reputedly silk-moths on the island of Cos, and in Assyria, in Greek times); but trade brought it west, and there are traces of it at Hallstatt sites of the late 7th century BC in Switzerland.

C.M.B.G.

Cosmetics

The painting of face and body had religious significance long before it became a social convention reserved exclusively – or almost – for the female sex. That significance remains in some primitive tribes today, when initiates are sometimes tattooed.

The techniques of make-up have not changed over the centuries, but the materials used have. Cosmetics were probably first used in China, but the earliest records of their use come from early pre-dynastic Egypt (about 3750 BC onwards). We can imagine the Egyptian lady of the second millennium lining her eyes with galena and green malachite, rouging the cheeks with red ochre, on a foundation cream tinted with yellow ochre, which would be applied to face, neck and arms; in addition the eyebrows would be plucked, and long false ones painted in. Thuthu's toilet-box (*c.* 1300 BC) contains slippers, pumice-stone, eye-pencils, and pots which would have contained creams and ointments such as wild castor plant oil for cleansing and softening the skin. Such boxes often contain elaborately decorated bronze mirrors. Female cosmetics probably reached their height at the time of Cleopatra (about 50 BC).

Much earlier, a tomb at Ur has revealed a lipstick of *c.* 4000 BC, and a cosmetics case of gold, with instruments for manicure. These may well have belonged to a man; Xenophon tells us that Astyages, King of the Medes, was 'adorned with pencillings beneath his eyes, with rouge rubbed on his face, and with a wig of false hair – the common Median fashion.' The splendid beards worn by Assyrian kings were often false – symbols of royalty and status, rather like wigs in the 17th and 18th centuries.

We can learn much about cosmetics fashions from the Romans: Ovid wrote a book about them, and the epigrammatist Martial, of the late 1st century AD, writes to a male friend, 'I would not have you curl your hair. . . . Your beard should be neither that of an effeminate Asiatic nor that of a condemned person.' But Martial reserves his most biting attacks for his girl-friends: 'While you, Galla, are at home, your hair is being dressed at the hairdressers; though you lay aside your teeth at night with your silk garments, and lie stowed away in a hundred boxes; though even your face does not sleep with you, and you ogle me from under eyebrows which are brought to you in the morning . . . nevertheless, you offer me delights.'

Has anything changed since then? Poppea, the wife of Nero, made herself face masks from bread-crumbs and asses' milk, and left them on all night; and another of Martial's girls, with a body-odour problem, overcame it with a paste of chalk and vinegar.

After the Dark Ages cosmetics and perfume (mainly imported from the East) began to be used again in Western Europe, and the rich and powerful were able to buy or concoct an assortment of outlandish beauty preparations. If a queen bathed in wine at least her young ladies in waiting could bathe in milk. By the 18th century 'painting' was taken for granted among the upper classes, despite the opposition it encountered from such individuals as the English Member of Parliament who actually introduced a Bill to ban 'artificial beauty aids as used in the seduction and deception of the opposite sex'. In France in particular the manufacture of perfumes and beauty products was established on a scientific basis by the last century, but it was not until comparatively recently that the majority of women of all classes came to accept and use cosmetics regularly.

C.M.B.G., M.H.

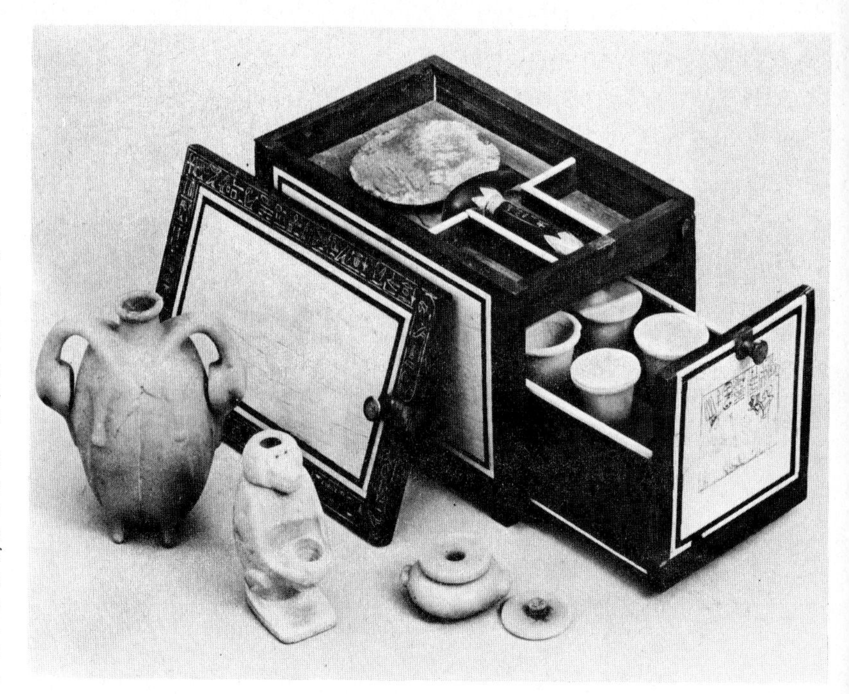

Egyptian make-up box of wood and ivory, and 'blue marble' (anhydrite) cosmetic jars.

Button

The button was well established by the beginning of the 3rd millennium BC: at Mohenjo-daro in the Indus Valley a carved shell amulet was found, pierced with two holes, and it might possibly have served as a button. Stuart Piggott has pointed out jet buttons found at sites in northern England and Scotland, dating from the early 2nd millennium BC; perhaps that is why Britain was slow to adopt pinned garments which were the fashion around the Mediterranean until well after the birth of Christ.

From the late Middle Ages, the button grew in popularity both as a fastening and as a decoration – and it was probably aesthetic before it was practical. The etymology of our word will elucidate how buttons were made: from the French *bouton*, from the verb *bouter* – 'to raise into relief'. The word was used to describe decorations on book-bindings, or on metalwork or embroidery, before it designated a fastening. By the 13th century in England, clothes were ornamented with them, usually in metal, either silver or gilt: see the accounts of Edward II, of 1321: 'To Robert le Fermor, bootmaker of Flete Street, per six pair of boots with tassels of silk and drops of silver-gilt, price of each pair, five shillings, bought for the King's use. Westminster, 24 May. . . .'

The fashion for button fixings grew in the 14th century: of precious metal, or

Detail from a sculpture by Giovanni Pisano, c. 1300. The buttons on the sleeve would be more for decoration than for use.

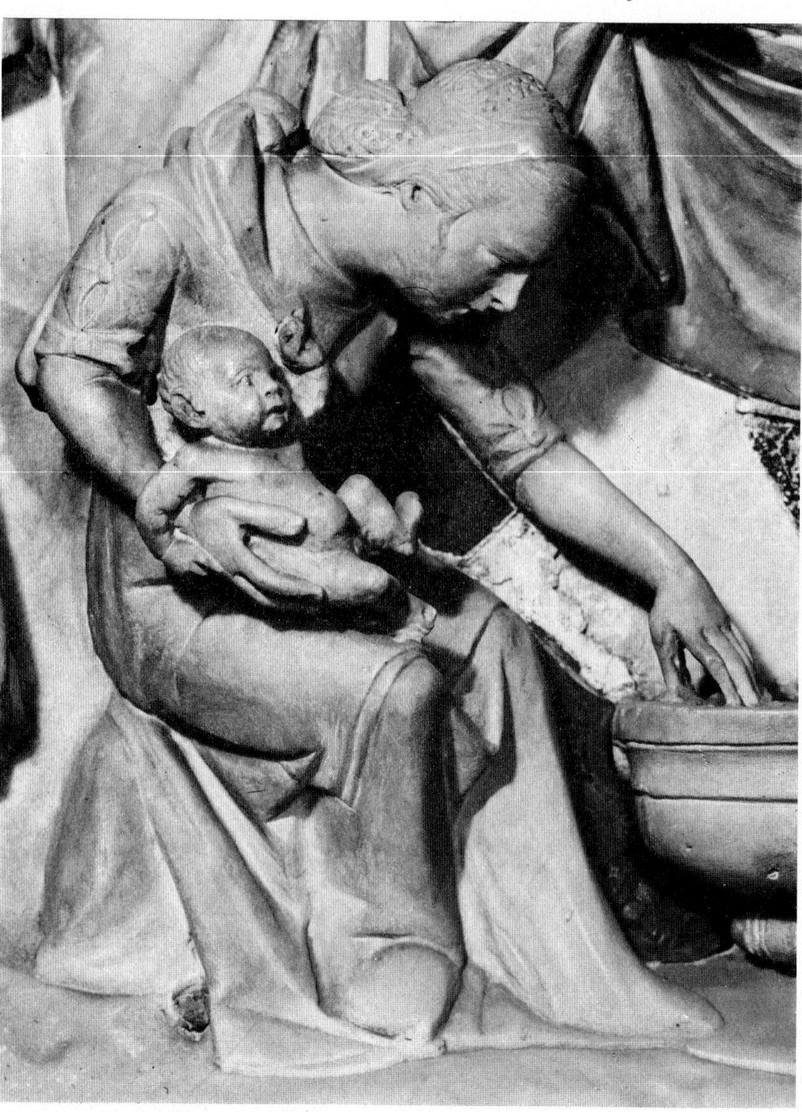

Detail from Fouquet's Virgin and Child with Angels*: corsets, from the 14th century on, were often worn as outer garments.*

copper, crystal, glass or fabric-covered; they appear particularly on the sleeves of women's gowns, buttoned from the elbow. Tomb monuments, particularly brasses, form the best documentation we have for the fashionable classes – the ones most likely to use such delicate and expensive devices, all of which were formed by hand. They were, indeed, an unofficial badge of status, just as the nature of the button on the mandarin's hat – sapphire, shell, gold, silver, etc. – indicated his grade within the hierarchy of officialdom.

Buttons could not become popular until they became cheap: Birmingham became the British manufacturing centre in the 18th century, and the product gradually got cheaper as the quantities produced got larger and the mechanization more efficient. Most buttons were made from sheet metal with punch and die, and with the shanks (if needed) soldered on; the imitation marcasite buttons, very popular from the late 18th century, were made of steel, with the facets riveted on, and were very expensive, at least until the process was imitated with a simple stamp: Matthew Boulton was selling them from his Soho works in 1767 for 140 guineas a gross. In 1807 a Dane called Bertel Sanders perfected a method of locking together by machinery two discs of metal which could trap in their fixing any required material; the metal shanks he used were replaced by his son with flexible canvas shanks (in 1825). Such covered buttons competed with the more expensive gilt article: the latter were 'as cheap as Brummagem buttons' (counterfeit coins, which they resembled), but linen-covered ones, for underwear and working clothes, were to be cheaper still.

Metal buttons were a strong feature of uniforms during the 19th century; their use perhaps goes back to the livery in which the nobility clothed their retainers: in 1757, for example, the Duke of Bedford paid nearly £5 – enough to pay a skilled workman for a month – for an orange suit of livery, decorated with about a hundred gilt buttons. Materials for button-making were legion during the 19th century, but two of the most important innovations were a much cheaper horn button made from softened cattle-hoof, and the even cheaper vegetable ivory button, made from the kernel of the corozo nut.

C. M. B. G.

Below, *button-making in a Birmingham factory, late 19th century.*

Safety pins: below, *French, 500 BC;* right, *Greek, 8th century BC.*

Safety-pin

It is not known where the safety-pin was invented – Mycenaean Greece, Italy or Sicily – but it is known when: in the Bronze Age, in the 2nd millennium BC. The straight pin had been in use at least since Sumerian times, but it was unsatisfactory, with its sharp, unguarded point. Some inventive genius therefore bent it double and made the point fit into a slot or hook (though it was not until modern times that the point was completely concealed).

To prevent the pin from slipping, something else was needed: a spring of some kind to hold it firmly in place. The device that was adopted, and which is still in use today, was to give the pin, at the bend, a circular twist, which acted as a coil-spring. This form is said to have been invented by the prolific American inventor Walter Hunt in three hours one afternoon in the 1940s, when a draughtsman to whom Hunt owed a small sum of money offered to pay him $400 for all rights to the various shapes into which Hunt twisted an old piece of wire.

J. L.

Corset

The idea of a corset designed to modify and alter the contours of the body may well originate from Minoan Crete, where a statue of the 'Serpent Goddess' (18th century BC) illustrates what was probably the earliest form: it is formed out of a framework of copper plates, which made the skirt lie flat on the hips and accentuated the slimness of the waist. But most women in antiquity tended to dress in loose-fitting garments, and, although corsets were worn (usually as outer garments) in the courts of France and Burgundy from the 14th century, it is not until the arrival of Spanish Court fashion (*c.* 1550–1620) that we find tight bodices of material reinforced with wood or iron, or even in some cases made entirely of iron. Metal frames fitting tightly over the chest (for a flat chest was one element of the fashion) were needed to support the wide skirts, as can be seen in the paintings of Velazquez. A little later we are told that Louis XIV's queen made such strenous efforts to follow the fashion that she actually achieved a waist measurement of 13 in.

Corsets returned with a vengeance with the tight bodices of the First Empire in France – and these were used by men as well as women. In 1811 we find an advertisement for the 'pregnant Stay . . . to compress and reduce to the shape desired the natural prominence of the female figure in a state of fruitfulness', and there are painful accounts of the struggles undergone by fashionable ladies attempting to get into corsets which were basically a rigid shell; one of their more unfortunate effects was that they tended to make the right shoulder larger than the left.

After a period around 1850 when little whalebone was used, corsets became more dominant as crinolines got smaller, and about 1860 the desirable waist measurement was 17–21 in. Some 15 years later the bustle went out, and long corsets came in; they were worn by all but 'aesthetic' ladies, who favoured a looser look. The straight-front corset was introduced in about 1900, and was supposed to flatten the stomach; later, corsets became even longer, occasioning such ditties as 'I do not need a country seat / Because I can't sit down.'

In about 1912 the 'girdle' arrived, and freedom with it; if we follow naughty postcards from the Habsburg Empire to the Paris of the 1920s, we can see how various have been the means of keeping stockings up. The introduction of tights from about 1958 onwards (not to mention Women's Lib a few years later) spelt the death-knell of the corset for all but the excessively figure-conscious and the middle-aged.

C.M.B.G., M.H.

Shoe

Foot protection appears to be a very recent innovation, introduced much later than body clothing. We may surmise that men in colder climates wrapped skins round their feet, but no shoes have survived from earlier than the middle of the 2nd millennium BC to give us a clear idea of their appearance. We can follow their introduction by looking at Egyptian frescoes and sculptures: in the First Dynasty (3200 BC onwards) the kings are shown clothed, but barefoot, and sandals of leather or plaited fibre only appear about the Eighteenth Dynasty (*c.* 1500 BC).

The same applies in the military world of the Near East: the Egyptians, Sumerians and (probably, though ornamental pottery boots have been found) the Hittites went into battle barefoot. The Assyrians (1380–612 BC) at first sent their infantry barefoot, but reliefs at Nineveh show sandals, and also calf-length laced boots, possibly reinforced with metal. The Persians (550–331 BC) wore slippers of felt, but we can see from their decorated pottery that the Greeks of the heroic age fought barefoot – though their shins were protected by greaves; Achilles' end came, it will be recalled, because his heel was unprotected.

Civilian footwear in ancient Greece was varied; most people went about the house and often the city with nothing on their feet, but sometimes a sandal was worn (made to size by the cobbler by tracing round the person's foot directly on to the leather). For journeys or hunting the *limbas* was worn – an ankle-high boot, laced in front. Women wore slippers, a favourite type being the *persikà*, the name of which gives away its origin.

The Hittites were renowned as shoemakers; and one story of their god of agriculture, Telipinu, tells of him getting in a bad temper, and 'putting his right boot on his left foot and his left boot on his right foot'. But it is probably to the Assyrians that we owe the ancestor of today's army boot, with a thick leather sole – very different from the Roman military issue of sandals, with hobnails for added grip and wear.

We are less well informed about shoes from northern lands, however. Bogs at Guldhoy and Jels in Denmark have yielded shoes and sandals of leather, as well as shoes of spun wool; primitive 'stockings' – to stop the foot chafing – were merely strips of cloth wrapped round the foot. These finds are of Early Bronze Age date. And waterlogged frozen barrows of approximately the same date show how men were shod in the Noin-Ula Mountains of Scythia near Lake Baikal – with heavy felt boots (worn to this day from the Baltic to China, and from Siberia to Tibet) luxuriously embroidered with silk.

Little except style changed in the manufacture of footwear until the introduction of the sewing machine by Elias Howe in 1846. In this century man-made fibres and artificial substitutes for leather which are yet able to 'breathe' have been developed, partly because of the growing shortage of natural leather.

C.M.B.G.

Loom

The loom, in its simplest form, was merely a device for holding in tension a number of parallel threads, known as a warp; through these and at right angles to them the weaver interlaced other threads, the weft, so as to form cloth. Even on the most primitive looms, the weaver could produce cloth of every kind and quality; but the process was laborious.

The history of the loom consists of a long series of improvements designed to increase productivity while reducing labour. Already in ancient times systems of rods and cords were introduced which enabled the weaver to control the warp threads in groups instead of singly; originally operated by hand, from the Middle Ages onwards these systems were more efficiently operated by means of treadles. Other systems of rods and cords allowed patterns to be woven into the cloth and repeated as often as desired.

Looms with such systems, comprising thousands of moving parts, were the most complex machines used in the medieval and Renaissance periods. Leonardo da Vinci and others endeavoured to increase the speed with which the weft threads were passed through the warp, but this problem was not satisfactorily solved until the 18th century. The application of mechanical power to the loom, replacing the weaver's muscle-power, was achieved in the late 18th and early 19th century, and at the same time J.-M. Jacquard perfected his punched-card system of automatic pattern-weaving. Since then

most cloths have been woven on automatic power-looms, with human operatives intervening only to prepare and tend them.

Without the loom, and the cloth which it has provided, civilized life is hardly conceivable. For thousands of years we have worn clothes, not only as a protection against cold, heat or wet, but also to beautify ourselves, to conceal or emphasize parts of the body for erotic purposes, to reassure ourselves and influence others by creating in our dress and household furnishings personal works of art which display or disguise our temperament and tastes, our place or pretensions in society. Cloth coverings for furniture, floors, walls and windows have much enhanced the comfort and beauty of the home.

Cloth has always been a major element in the industry and commerce of the world. In antiquity and the Middle Ages, thanks to a consistently high demand for the product and the heavy cost of raw materials, looms and finishing processes, cloth production was already organized on modern industrial lines, with a small number of wealthy capitalists employing a large number of indigent workers. As a result, weavers have always been prominent in social protest and revolution and it is true to say that the loom has played some part in polarizing the doctrines of capitalism and Communism.

D.K.

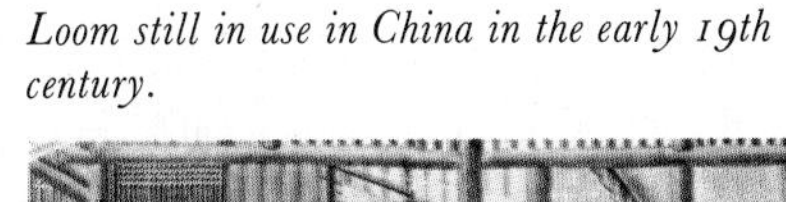

Loom still in use in China in the early 19th century.

This detail from Pinturicchio's Penelope and the Suitors *shows a 15th-century Italian loom. The treadles move the warp up and down while the shuttle is moved from side to side by hand.*

Left, *Roman cobbler mending a shoe, from a relief in the Ostia Museum.*

The Luttrell Psalter (c. 1340) has this picture of a spinning-wheel – an introduction, probably, from Italy.

Spinning-wheel

It is women who are most likely to have been the inventors (possibly as far back as 6500 BC) of the first spinning devices: the distaff or rock and the spindle. The distaff was a cleft stick, about 3 ft. long, on which the wool was loosely wound by hand; it was held under the arm or stuck into the spinner's girdle so as to leave her hands free. A continuous lock of wool was drawn from the distaff, usually through the fingers of the left hand while the thumb and forefinger of the right hand twisted it and wound it by means of a hanging spindle. This was a thin wooden rod, 8 to 12 in. long, with an incision at the top for attaching the thread. The spindle, of course, had to revolve, and to this end its lower part was inserted into a 'whorl', a wooden disc acting as a flywheel to keep the spindle turning at a uniform pace.

This assembly, with some regional modifications, was in use for thousands of years in the Middle East and in Europe; the oldest examples in the British Isles were found in Pictish settlements in Scotland, dating back to Roman times. Distaff and spindle were still being used in many British villages, although a 'modern' invention, the spinning-wheel, had arrived in England already in the 14th century, probably from Italy.

At first a hand-wheel was used to move the spindle more easily and simply, by the aid of its momentum. The spindle was fixed in a frame and turned by a belt whose other end passed over the wheel. A book published in the early 19th century describes the operation as follows: 'In spinning with the hand-wheel the roving was taken fast hold of betwixt the left forefinger and thumb at six inches distance from the spindle. The wheel, which by a band gave motion to the spindle, was then turned with the right hand, and at the same time the left hand . . . was drawn back about half a yard. The roving was thus drawn out into weft [thread], the necessary twist was then given by a few turns of the wheel, and finally the weft was wound upon the spindle.' That description was given in the past tense because by that time the spinning-wheel was generally made for operation by foot, leaving both hands free for manipulating the thread. The foot-operated spinning-wheel, invented by an unknown benefactor of spinning womanhood, had come into use around the middle of the 18th century. At that period, however, a whole spate of new machines for spinning and weaving had already begun to introduce the modern age of textile manufacturing, thus sparking off England's Industrial Revolution.

E.L.

Knitting-frame

The Elizabethan age was a difficult time for the clerics of the Reformed Church. Many were suspicious of the new faith, which seemed to them an evil, frightening thing, foisted upon the people by the unscrupulous seekers of power who surrounded the throne.

It was against this background that the Reverend William Lee lived. He was a misfit, and he might have lived and died as such had it not been for the tormenting clicking of his wife's knitting needles.

Suddenly he had an idea. As he watched his wife's hands rapidly knitting the stitches from one needle to another he found himself wondering why there should not be hundreds of small needles instead of one large one, and a series of hooks to lift the loops over the wool and produce a whole row of stitches at a time. Why not an automatic knitting-frame?

There was nothing particularly new in Lee's idea. The nomads of North Africa had used knitting-frames and hooks in the centuries before Christ, and the carpet weavers used a frame technique very similar to the one envisaged by Lee. What was new was the concept of a frame and a row of hooks that would lift the loops or stitches over the wool with one simple action.

Once Lee thought he had found his way to fame and fortune, his calling was

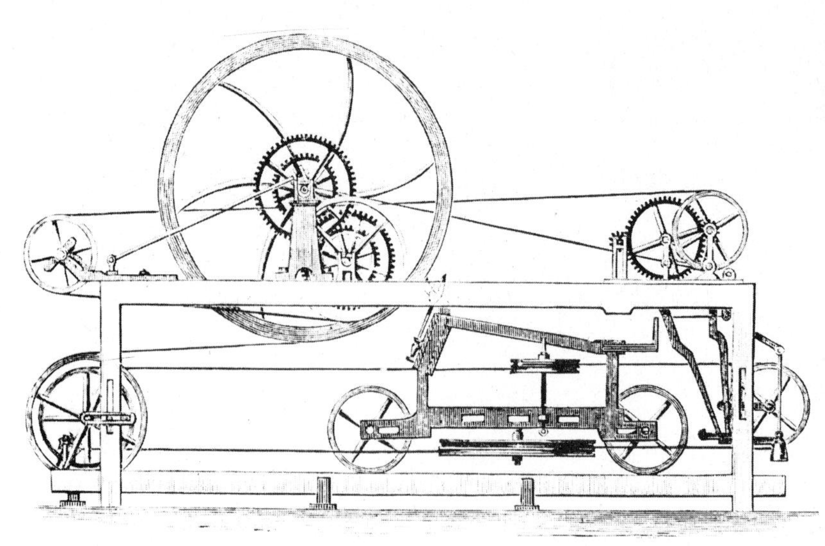

Crompton's mule (1772–8), which combined Hargreaves's movable carriage with Arkwright's use of rollers to produce the finest thread on the market.

forgotten. He stacked his books away and went to present himself and his machine at the Court of Queen Elizabeth. He desired two things. First, the queen's patronage in the form of a grant towards the cost of exploiting his invention; secondly, monopoly and patent rights to protect his interest and to bring him a steady income. But the queen sent Lee packing, and he fared no better when he sought financial aid in the City. No one had any faith in his new-fangled ideas and before long he boarded a ship for the Continent, where he hoped to find readier ears. Together with his brother, he tried to persuade financiers to open a frame-machine knitting industry. But all his efforts were in vain, and William Lee died unknown and obscure in Paris in 1610.

His brother, equally disheartened, returned to England with the machine. Suddenly the tide turned. He met a merchant from Nottingham who showed interest, and together they established the first frame-knitting factory, in the North Midlands. The venture was so successful that a century later the hand-knitters of Leicester petitioned the mayor and aldermen to protect their interests by refusing to allow any more knitting-frames to be set up in the county. Thanks to Lee's invention, machine-knitted stockings were being produced much more cheaply than hand-knitted ones and were no longer a luxury.

J.N.

Flying shuttle

Three inventions in the 18th-century British cotton industry are known at least by name to every schoolboy: Kay's flying shuttle, Hargreaves's spinning-jenny and Crompton's mule. Together they made possible the great expansion of cotton production which characterized the world's first Industrial Revolution.

In 1733 John Kay, 12th child of a Lancashire yeoman farmer, invented the flying shuttle. For centuries handloom weaving had been carried out laboriously on the basis of the shuttle bearing the yarn being passed slowly and awkwardly from one hand to the other. Kay introduced shuttle boxes at each side of the loom connected by a long board, known as a shuttle race. Inside each box was a horizontal metal rod or spindle and free to slide along each spindle was a leather driver or 'picker'. A loose cord was attached to each picker and these were joined at the centre of the loom by a stick or picking peg which was held by the weaver. By jerking the picking peg from side to side one single weaver, using one hand, could cause each picker in turn to slide, taking with it the shuttle and throwing it across the loom to the opposite shuttle box. Kay also introduced wheels to the shuttles to reduce friction along the shuttle race.

One weaver could now produce at far greater speed than ever before cloths of different widths. Yet Kay became extremely unpopular with the handloom weavers, some of whom refused to pay his charges for using the new devices. Eventually in 1747 he left England for France. He died in poverty.

Nevertheless, the simplicity of the new invention guaranteed its success, and the fact that weaving output greatly increased in consequence put pressure on businessmen to find a quicker way of spinning yarns also.

A. BR.

Spinning-jenny

In 1761 the yarn shortage after Kay's invention of the flying shuttle led the Royal Society of Arts to offer prizes for the invention of a machine that would 'spin six threads of wool, flax, hemp or cotton at a time, and that will require but one person to work and attend it'.

James Hargreaves was one of the inventors who took up the challenge. His spinning-jenny, which reproduced mechanically the actions of the hand spinner, is said to have been invented in 1764, but it was not patented until 1770. Hargreaves had already devised other machines and, like Kay, he suffered from opposition both from other inventors and from working people. Yet he did not die in poverty.

One of his rivals said that the name 'jenny' was taken from the name of one of his own daughters, but most likely it was a term derived more prosaically from the word 'engine'.

The principle of his 'jenny' was to be quickly copied by others. Bobbins filled with roving (textile fibre) were placed at the bottom of a frame carrying several spindles and a sliver from each bobbin was attached to a spindle, passing on the way between two rails forming a bar which slid backwards and forwards on the frame. The spinner drew out the roving by moving the bar back to a certain distance. The rails were then pressed together to hold the sliver fast while the backward movement of the bar and the turning of the wheel which moved the spindles was continued. When enough twist had been given, the bar was once more moved forward and the spindles were slowly turned to wind the yarn. In the meantime the spinner pulled a lever which depressed a wire to push down the thread into a position where it could be wound. The processes patented by Hargreaves were improved by other inventors: they were also exploited by Richard Arkwright, whose claims to have actually invented anything himself were in dispute. Nevertheless, Arkwright was one of the pioneers of cotton-spinning in factories and his

The membership card issued to a Leicester frame-work knitter shows William Lee's original machine.

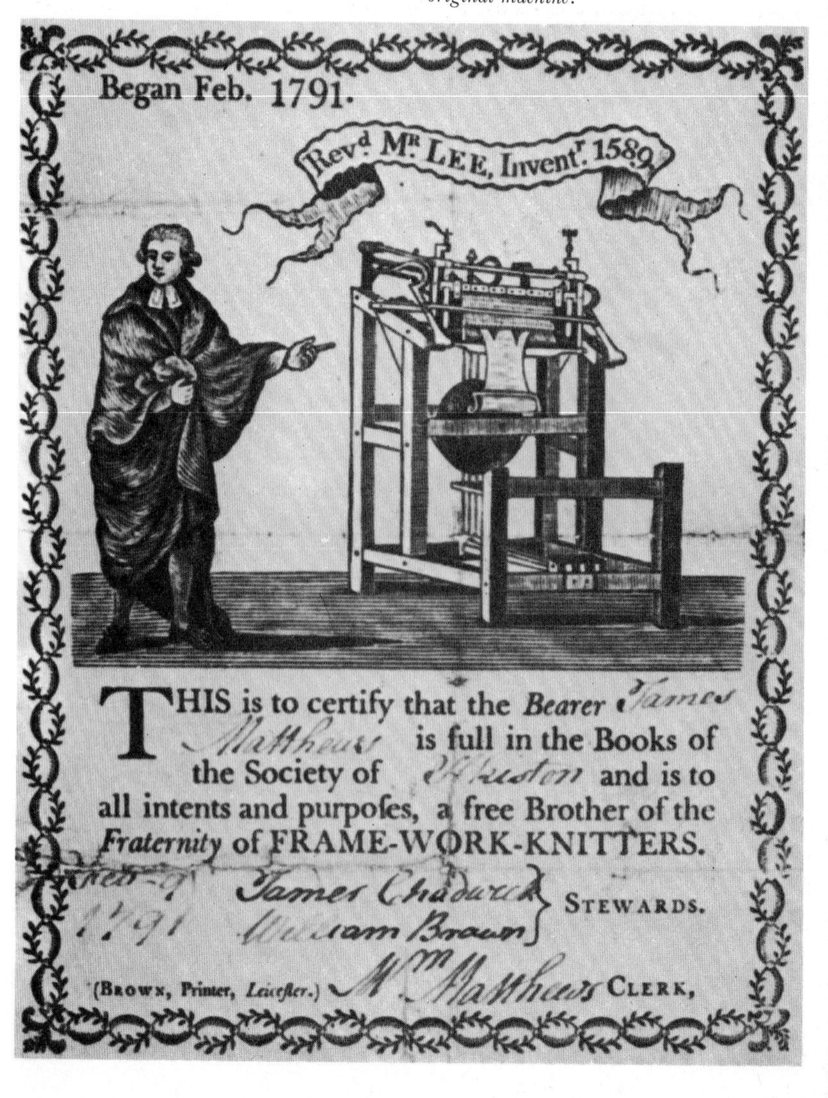

Began Feb. 1791.

Revd Mr LEE, Invent r 1589

THIS is to certify that the Bearer James Matthews is full in the Books of the Society of Ilkeston and is to all intents and purpoſes, a free Brother of the Fraternity of FRAME-WORK-KNITTERS.

Nov 9 1791

James Chadwick
William Brown } STEWARDS.

(Brown, Printer, Leiceſter.) Wm Matthews CLERK,

'water-frames' were envied throughout the textile districts.

A. BR.

were working up to 18 hours a day at the roaring machines.

T. J.

Rioting weavers attacking the house of John Kay, inventor of the flying shuttle.

Crompton's mule

Crompton's mule was the culmination of a snowballing series of inventions. Kay's shuttle had meant that the weaver's family had to work overtime on their spinning-wheels to make enough thread to keep his loom busy, a problem James Hargreaves went some way to solve with his spinning-jenny. Then in 1769 Richard Arkwright invented – or more likely pirated – the water-frame, which could spin coarse but strong threads and be driven by water power, and used it to found one of the first great industrial empires based on the factory system.

But both the spinning-jenny and the water-frame had serious limitations, and in 1774, when he was 21, Samuel Crompton set out to try and make something better. He was a shy, introspective man and he worked alone in the attics of 'Hall i' th' Woods,' a rambling half-timbered manor-house now swallowed up in the suburbs of Bolton in Lancashire where his domineering widowed mother was a caretaker-tenant. It took him five years, working in secrecy, but in the end he achieved what he was striving for, and helped by his new wife started spinning and selling cotton yarn of a fineness and quality no one else could get anywhere near equalling.

Crompton soon found he was virtually besieged, with rivals climbing up ladders to the attic windows and boring holes through walls to discover the secret of the wonderful new machine. Since he could not afford to patent the idea he was forced to reveal it in return for a public subscription which raised a miserable £60. What he had done was to make a hybrid – hence the name 'mule' – of Hargreaves's and Arkwright's ideas with the vital addition of a carriage which carried the rotating spindles of newly spun yarn, and moved back and forth to keep the tension on the delicate threads light and even all the time.

Crompton envisaged small hand-powered versions of his machine being used in weavers' cottages in the traditional way, but instead it turned out to be perfectly adapted to being driven by steam or water power and it was seized on by the budding capitalists of the era as a means of solving the chronic shortage of yarn for weaving.

By the time he was awarded a belated £5,000 by Parliament in 1812 in recognition of his invention, 4 million spindles were producing yarn on Crompton mules in Britain (the equivalent of 4 million women with 4 million spinning-wheels, only faster), huge fortunes had been made and thousands of children

Right, *the cotton-gin – a simple invention that brought prosperity to the American South.*

Below, *Arkwright's 'water-frame' – a powered spinning-wheel.*

Cotton gin

Never has six months' work by one man so changed the course of a great nation's history as the solitary winter Eli Whitney spent developing the cotton gin. The son of a struggling New England farmer, Whitney graduated in law from Yale in 1792. Worried about his debts, the only job he could find was as a tutor in South Carolina.

But on his way to take up the tutorship he stayed at Mulberry Grove, a run-down plantation near Savannah, Georgia, belonging to Mrs Catherine Greene, the 40-year-old widow of a Revolutionary War hero.

Whitney seems to have nurtured a lifelong but unspoken passion for the gay and charming Mrs Greene, and when the problem of cleaning freshly picked cotton bolls came up, he seized on it as a chance to extend his stay.

The problem the Southern planters were complaining of was that although the Industrial Revolution was opening up a huge new market for cotton across the Atlantic, the only variety of cotton which could be grown in most of the South was the upland type, which has short fibres and tightly clinging green seeds. The primitive cotton gins of the time could not cope with this variety, so the seeds had to be separated from the wool by hand, and it took a whole day for a slave to clean just 1 lb. of cotton in this way. If only an economical way of cleaning upland cotton could be found, it could open up huge new markets for the South. Whitney took up the challenge and, drawing on his tremendous manual dexterity, he produced the first model of his gin within 10 days.

Essentially the gin consisted of two rollers. One, covered with wire spikes, tore the cotton away from the seeds, and the other, covered with bristles, brushed the cotton off the first so that it did not get clogged up. With this simple machine a slave could easily clean 50 lb. of cotton in a day, and the crop became an economic proposition in the South overnight. The region was quickly transformed from one of backward farming communities to a prosperous society based on plantations with plenty of slave labour to pick and gin the cotton – and the historic aftermath of Whitney's stay at Mulberry Grove is still with us today. As for Whitney, he almost wore himself out trying to enforce his rights to the invention, with very little success; he went on to pioneer mass production and some of the earliest machine tools.

T. J.

Bleaching powder

The art of bleaching textiles such as linen and wool is of great antiquity and seems to have been familiar to all the ancient civilizations. The Egyptians, Phoenicians, Greeks and Romans are all known to have produced white linen goods, but little is known of the methods they used. After the Crusades the Dutch emerged as the leading exponents of the craft, retaining a near-monopoly of the industry down to the 18th century: in Britain, for example, though bleaching grounds are known to have existed near Manchester as early as 1322 and a major bleaching works operated at Southwark in the mid-17th century, most of the brown linen manufactured was sent to Holland for bleaching. The Dutch method was to steep the linen in alkaline lyes for several days, then to wash it clean and spread it on the ground for some weeks. These processes of 'bucking' and 'crofting' were repeated alternately five or six times, after which the cloth was soaked in sour milk or buttermilk for some days then crofted again. The process was far from being ideal, since it required a large amount of space, and could take several months, especially in northern countries.

In 1736 an Act was passed in England which permitted the addition of cotton to the composition of cloth. This provided a great stimulus to the bleaching industry, and in 1756 Dr Francis Home suggested the substitution of dilute sulphuric acid for buttermilk, thus drastically reducing the length of time required by the traditional process. Then in 1774 the Swedish chemist Karl Wilhelm Scheele (1724–86) discovered chlorine (he followed this by discovering glycerine in 1779 and, it is thought, had already discovered oxygen the year before). A French chemist, Count Claude Louis Berthollet, who had been Napoleon's scientific adviser during his Egyptian campaign, was the first to recognize the power of this new gas to bleach the natural colour of linen, and in 1785 he introduced a chemical bleaching process by means of *eau de Javel,* a solution made by passing chlorine over lime in a solution of potash and water until effervescence began. Berthollet described the process to James Watt, who passed on the information to Charles Tennant of Glasgow. Soon Berthollet's process was being used in Scotland, but it was inconvenient (the bleacher had to make his own chlorine), disagreeable and damaging to the health of the men who worked with it, and in 1799 Tennant introduced an innovation of his own, which can be said to have been the origin of bleaching powder as we know it today.

Tennant used a variant of the *eau de Javel* process to produce a solid reagent by passing chlorine over lime. The introduction of the bleaching powder thus produced, which was offered at £140 a ton at first but by 1830 at £80 a ton, was an extremely important event without which the cotton industry could never have achieved its enormous expansion; chemical bleaching was also valuable in the manufacture of paper. By 1830 annual production of bleaching powder in Britain was in the region of 1,500 tons. Subsequent attempts to replace bleaching powder have met with little success, except where cheap electricity is available to make sodium hypochloride from common salt.

J.G.

Sewing-machine

The invention of the eyed needle was one of the greatest technological advances in human history, comparable in importance with the invention of the wheel and the discovery of fire. The needle appears astonishingly early, for large numbers, made of mammoth ivory, reindeer bone and the tusks of the walrus, have been found in Palaeolithic caves, where they were deposited 40,000 years ago.

Had the two men who turned their attention to the problem of making a mechanical aid for sewing been able to combine their efforts, the sewing-machine might have been invented before the end of the 18th century. In 1755 an Englishman, Charles F. Weisenhall, patented a double-pointed needle with the eye in the centre, and in 1790 Thomas Saint patented an apparatus which had many of the features of the modern chain-stitch machine, but without the eye-pointed needle.

It was not until 1830 that Barthélemy Thimmonier, a poor tailor at Saint-Etienne in France, devised a real working machine – albeit a rather clumsy one, made chiefly of wood. It was used for making army clothing in the 1840s, but Thimmonier's workshop was wrecked by a mob and he only narrowly escaped with his life. Determined to persevere, he improved his machine and in 1848 took out patents both in England and America, but he never managed to market his invention and died in poverty in 1857.

It was in America that the next step forward was taken; several inventors working independently, unaware of one another's existence – Walter Hunt of New York, Elias Howe of Spencer, Massachusetts, and Allen B. Wilson – all devised working models. Howe came to England and sold his patent rights, but poverty forced him to return to America in 1849. He found several people manufacturing sewing-machines there, but he managed, after much litigation, to vindicate his patent rights and get a royalty from all subsequent makers. The most notable of these was Isaac Merritt Singer, and soon there was hardly a middle-class household which did not possess a sewing-machine.

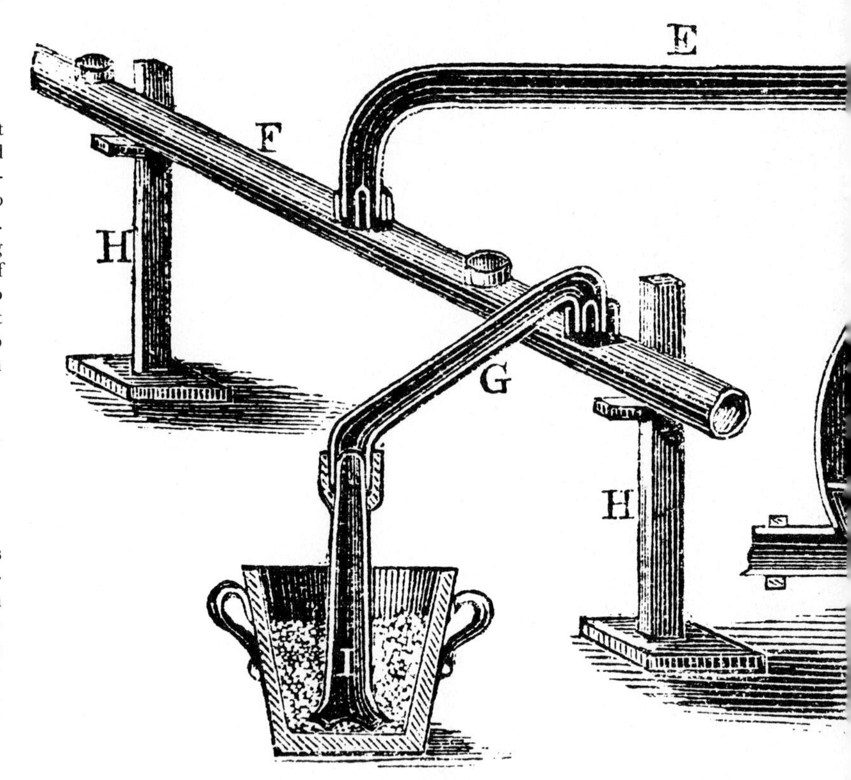

Mid-19th-century apparatus for making bleaching powder by combining chlorine from the still with hydrate of lime.

Barthélemy Thimmonier using the sewing-machine he patented in 1830.

Coloured fashion-plate of 1870 accompanying paper patterns.

The new invention, which did so much to lighten the work of the seamstress, had a notable effect on the fashions of the late 1860s and early 1870s. Clothes became extremely complicated and elaborate. Since then design improvements have been introduced and electric motors added, but the basic nature of the sewing-machine remains unchanged.

J. L.

Paper patterns

Paper patterns, in the sense of full-scale models of the pieces in which cloth was to be cut, do not appear until the early 19th century. They were, however, preceded by small-scale illustrations in books, one of the first of which was Juoan de Alega's *Libro de Geometrica y Traca*, published in 1589. There is little, if anything, of the kind in the 17th century but in the 18th we find quite elaborate diagrams in such works as *Description des Arts et Métiers*, issued by M. Garsault in Paris in 1769; this was a general encyclopedia. Works devoted entirely to cutting soon began to appear: *The Taylor's Complete Guide* in 1796 and *The Tailor's Friendly Instructor* in 1822.

The new ladies' magazines, with coloured fashion-plates, which began to be issued in increasing numbers in the early 19th century often included patterns for embroidery, lace caps, etc., but it was not until the 1840s that the full-scale paper pattern, issued with the magazine, became usual. The first number of *Bekleidungskunst für Damen – Allgemeine Muster-Zeitung* (Dressmaking for Ladies – Universal Pattern Journal) appeared in Dresden in 1844, and in the following decade *The World of Fashion* was publishing full-scale patterns for bodices, mantles, underwear, etc. These were drawn on a single sheet of paper which was folded and inserted into the magazine.

Other magazines followed suit and ran a regular paper-pattern service. It should be noted that paper patterns (described as 'models make-up in paper') for professional use rather than for the home dressmaker, were being sold 'in "sets" at ten shillings (comprising four articles) packed for any part of the United Kingdom at three shillings extra' as early as 1834. Butterick's Patterns, intended for both professionals and amateurs, began to be issued in 1863; Weldon's in 1879. McCall's Patterns began to appear in America in 1870.

There was a rapid development from 1890 onwards, for the home dressmaker now relied almost entirely upon paper patterns, issued with each number of the women's magazines. In Britain *The Lady* ran a pattern service from its beginning in 1885 until 1935. *Vogue,* which first appeared in America in 1892, issued patterns, and continued when most of the magazines had given up the practice, with such success that a *Vogue Pattern Book* was issued separately and attained a high reputation. The English *Vogue Pattern Book* first appeared in 1927. Paper patterns continue to be issued, but it is probable that their heyday is over, owing to the increasing efficiency of the mass production of ready-made garments.

J. L.

Aniline dyes

Among the many acts by which Queen Victoria's husband Prince Albert demonstrated his devotion to the advancement of science was his invitation to A. W. Hofmann (1818–92) to be the first Superintendent of the Royal College of Chemistry. Hofmann was one of the first to appreciate the importance of coal tar (a by-product of the expanding gas industry) as a source for many different substances of great chemical interest – particularly benzene, aniline (formerly prepared naturally from indigo, called *anil* by the Portuguese), and allyl-toluidine, which in chemical composition is a little like the important anti-malarial drug, quinine.

In 1856 Hofmann had as a research student an enthusiastic 18-year-old, William Henry (later Sir William) Perkin (1838–1907). Late into the evening, and all through the holidays, Perkin laboured in a home-made laboratory erected in a shed at the foot of his garden. With commendable pertinacity, he set out to prepare quinine, by, as he hoped, oxidizing allyl-toluidine through the addition of sulphuric acid and potassium dichromate (which was supposed to yield the necessary oxygen). Unfortunately this procedure left out of account the thus far unknown concept of chemical structure, and all Perkin got was a dirty reddish-brown precipitate.

Despite Hofmann's strong disapproval of this diversion from pure science, Perkin now tried the same process with impure aniline. This time he got a black sludge; but when boiled with water the sludge – thanks to the impurities – produced purple crystals which Perkin found to be capable of dyeing silk. Receiving a favourable report from the famous dyeing firm of Pullar's of Perth, Perkin persuaded his father to put up the money for a dye factory near Harrow. Shortly afterwards one of the Pullar family wrote to Perkin: 'I am glad to hear that a rage for your colour has set in among that *all-powerful* class of the community – *the ladies*. If they once take a mania for it and you can supply the demand, your fame and fortune are secure.' Perhaps more important, the colour, mauve, was a favourite with Queen Victoria; it was also used for the penny postage stamp.

The first synthetic dyes: some samples of the aniline dyes discovered by Perkin.

By the time he was 35, Perkin had made so much money from his dye that he was able to retire from business and return to chemical research. He had opened up a fascinating and lucrative area of investigation and other workers quickly moved in: even the disapproving Hofmann was responsible for developing the rosanilines, or 'Hofmann's violet'. In the race to develop an artificial process for manufacturing the red dye alizarin, previously derived naturally from the plant madder, Perkin himself was just beaten by Heinrich Caro (1834–1910). Britain's initial debt to Germany was now more than repaid, and dyestuffs became a central element in the massive German chemicals industry, which on the eve of the First World War completely overshadowed the British industry. But everywhere the dyestuffs industry, and therefore production of finished textiles, had been revolutionized by the development of synthetic aniline dyes.

A. M.

Man-made fibres

In 1664 Dr Hooke, Curator of Experiments to the Royal Society, wrote in his *Micrographia* that he had often thought that there might be a way 'found to make an artificial glutinous composition much resembling, if not fully as good as – nay better than – that excrement or whatever substance it be which the silk worm wire draws its clews. This hint may give some ingenious inquisitive person an occasion for making some trials which if successful . . . I suppose he will have no occasion to be displeased.' He was right; the giant firms of today, Courtaulds, ICI, Du Pont and others have absolutely no reason to be displeased.

The 18th-century French scientist Réaumur also foresaw the possibility of an artificial fibre, but it was not until the 19th century that experiments with dissolving and re-precipitating wood-pulp and other cellulosic materials were successful and a patent for the manufacture of artificial silk was taken out by a Swiss chemist, George Andeman. Once again the idea was allowed to lapse, but later in the century both Edison and Swan were looking for suitable filaments for their electric lamps. A method of producing fibres by forcing nitrocellulose in acetic acid through a series of holes was successful, and was patented by Joseph Swan in 1883.

Yet again there was a failure to grasp at once the full implications of this man-made thread, but some years later Comte Hilaire de Chardonnet, looking into the causes of disease in silkworms, decided to set up a factory in France to manufacture artificial silk under the name of rayon. Chardonnet silk was displayed at the Paris Exhibition of 1889, and at last the world realized what an important process had been invented.

A patent to manufacture cuprammonium silk followed in Germany where the product was called *Glanzstoff*, and the third breakthrough came in England with the viscose rayon process arrived at by three chemists, Cross, Bevan and Beadle.

From this 'one-man-and-his-bunsen-burner' type of process came an overwhelming stream of new miracle fabrics. In 1910 the first rayon knitted stockings were made by the German firm of Bemberg, using viscose rayon. Although rayon was at first mainly used as a cheap substitute for silk, in 1924 Dr Leon Lillienfeld patented his high-tenacity viscose process and in the 1930s the first high-tenacity viscose rayon was delivered to automobile tyre manufacturers for use as tyre cord.

In 1938, after 10 years of research, the American scientist Dr Wallace Carothers announced the discovery of nylon 6.6. The basis of this is a very long chain of molecules, and nylon is known as a polyamide fibre. It was first marketed by Du Pont in the USA, and was the first real synthetic yarn – that is to say that the raw material was chemically produced, or synthesized, as opposed to 'man-made' fibres which are produced from a raw material such as wood-pulp (although both types come under the general heading of 'man-made'). The chemical substances used to make nylon are benzene, oxygen, nitrogen and hydrogen.

Nylon has so far been the most important discovery among the man-made fibres. Apart from nylon stockings and nylon knitted underwear, which have had a revolutionary effect on women's lives, it can be used for almost anything from clothing to bed linen, carpets to safety-belts, fire hoses to artificial fur. Being synthetic it can be moulded and welded into shape. Garments made of nylon respond to 'memory' pressing, so that once 'baked' they will always return to their original shape.

Terylene, a synthetic polyester fibre that is chemically different from nylon, was discovered in 1941. It is made from ethylene glycol and terephthalic acid derived from petroleum. Its uses, too, are endless and it blends particularly well with natural fibres such as wool. Like nylon it is thermoplastic so that it can be permanently pleated and shaped.

Other synthetic fibres in daily use include elastomeric fibres (stretchable), acrylic fibres (mainly produced as fleecy, knitted or tufted fabrics including carpets) glass fibre (upholstery and curtains), metallic fibres (non-tarnishable glitter fabrics for garments and upholstery), polyolefin fibres (mainly for industrial purposes such as cordage, fishing-nets and webbings), bonded fibres (for paper-like disposable clothing

One of the earliest fashion illustrations for rayon yarn, from a book published by Courtaulds in the 1920s – The Art of Artificial Silk.

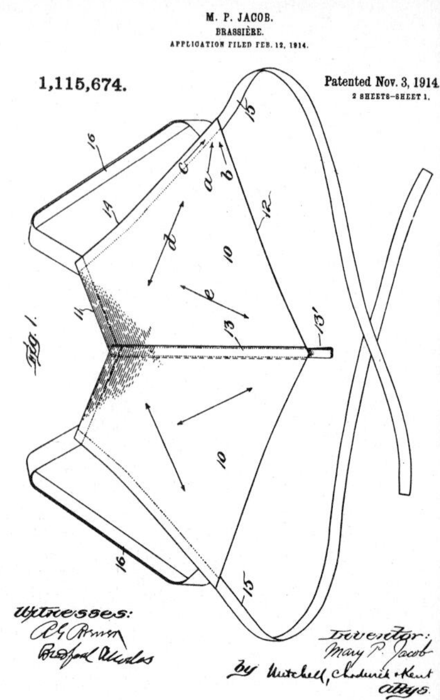

The original patent for the brassière, granted in 1914.

and bed linen) and polyvinyl chloride (PVC, a leather-like substance originally destined for domestic and industrial uses but now used for clothing as well).

It is doubtful whether even the learned and inventive Dr Hooke could have imagined such developments from his idea of 'an artificial glutinous substance fully as good as – nay better than – the excrement' of the silkworm.

J.I.

The first zip-fastener. This drawing gives an impression of the relative crudity of the mechanism in the early days.

Zipper

Today's nylon and microscopically machined brass zips – tolerances fined down to 0·0002 in., or one-thirtieth of a hair's thickness – are relatively jam-safe. It was not thus in a flash. A century ago sporadic experiments were going on around central Europe to wipe out the age of buttons and bows with a rapid-fire one-movement substitute for laces and hooks and eyes. It was in America that the zip began to look like a workaday appliance. In 1893 Whitcomb L. Judson, a Chicago engineer, filed the first patent for a 'slide fastener' as a blow against toiling with a buttonhook at the rows of couplings on high boots. It didn't catch on, mainly because the juvenile zip tended to catch on things – or to spring open at disquieting moments in alarming places. Judson had found the solution in principle, but it was a Swede, Dr Gideon Sundback (who died in 1954 aged 74) who, in 1913, refined the crude brainwave to a dependable commercial fitment. His brilliant manoeuvre was to attach the matching metal locks to a flexible backing. That is how a zip works, in case you have brooded, foxed, over one: each tooth is a tiny hook which engages with an eye under the adjoining hook on the opposite tape. You cannot push a zip together, because each head is too wide to slot between the adjoining teeth on the other side – until the slider splays out the teeth as it runs and leaves them meshed in its wake.

Yet the mass-production industry remained nervous of them. The first bulk order came from the American Army for troops' clothing in the First World War. Even after that it did not infiltrate civilian garments. It acquired its present name in 1926 when the novelist Gilbert Frankau, at a promotion luncheon for yet another model, is reported to have exclaimed: 'Zip! It's open! Zip! It's closed!' It took Schiaparelli in 1930 to dictate that the zip was, at last, in: she decided that it was correct for women's dresses, so overcoming the lingering worry among women that there was an improper alacrity about the zip. Schiaparelli's answer to the Slump was to make a gown with a dramatic zip from neckline to hem.

Now the zip is all-purpose. The master patents on slide fasteners in general form expired in 1931 and they are made throughout the world and for most imaginable purposes. One firm supplied hundreds of zip bootees for sheep in a foot-and-mouth zone; an Austrian surgeon sewed a zip into a man's stomach so that it was instantly accessible for internal dressings.

K.A.

Brassière

There is a mosaic in a villa at Piazza Armerina, Sicily, which shows a female Roman athlete wearing bikini pants and a bra. British newspaper advertisements in 1902 offered 'patent bust improvers – flesh coloured 7*s.* 6*d.*' similar to the modern bra, and in *The History of Ladies Underwear* Cecil Saint Laurent claims that the first brassière was invented in 1912. So who actually originated the bra is open to argument. But the first person to patent the idea, and produce the design that eventually killed off the corset, was Mary Phelps Jacob, in November 1914 (later better known as Caresse Crosby, a glittering, almost fictional American heiress).

She was a descendant of Robert Fulton, who devised an early steamboat. In her book *The Passionate Years* she wrote: 'I believe that my ardour for invention springs from his loins – I can't say that the brassière will ever take as great a place in history as the steamboat, but I did invent it, and perpetual motion has always been just around the corner.'

She got the idea when she was a New York débutante. Fashionable women in those days wore a sort of box-like armour of whalebone and pink cordage underneath, and these corsets prevented women from moving with ease. Caresse Crosby finally rebelled against wearing a corset, and before a dance one night she and her French maid, Marie, devised the prototype of the bra from two pocket handkerchiefs, some pink ribbon, and thread. Friends liked it, and persuaded her to make bras for them. When a total stranger wrote asking for a sample of her 'contraption', enclosing a dollar, she decided to exploit her invention.

An expert designer produced a series of drawings for her for which he asked $50; she gave him $5 and christened her brainwave the Backless Brassière.

The patent was granted in the following year, and she and her maid produced a few hundred samples, but they did not sell. She lost interest, until shortly after her first marriage when she bumped into Johnny Field, who was working for the Warner Brothers Corset Company; she asked if his firm would be interested; Warners offered $15,000 which she accepted: her patent has since been estimated to have been worth $15 million.

M.MCC.

Man Living

Tent

Perhaps the oldest tent site is of the Middle Palaeolithic (*c.* 40,000 BC) and was found at Molodova, southern Russia: the site is an oval of 4–5 m. in diameter, and surrounded by mammoth-bones, the longest of which were probably used as the tent-poles, and then covered with skins. Similar sites, with mammoth-bones or stones for holding the covering or guy-ropes, are found as late as the 6th millennium BC in France, Germany and the Iberian Peninsula – after the end of the last glacial period.

People who live a nomadic life, seeking water and grazing for their animals from place to place, must take their houses with them; although tent-dwelling nomadic tribes still exist today, a watershed in mankind's development was the change from a hunting and food-gathering economy to one of food production during the Neolithic period (*c.* 8000 BC in western Asia, later further west) when men began to live in permanent dwellings of mud, brick, rushes or stone.

The tombs at Pazyryk, in the Altai Mountains between Mongolia and China (5th century BC) have yielded hints of how a nomadic horse-herding people made their tents; larch- and birch-bark sheets, pre-boiled to make them pliable, have been found stitched in sections, which were probably used to make tents or huts (sometimes on wagons). Herodotus has described an Eastern tribe called the Argippaei as living 'under a tree which in the winter they cover over with thick white felt, and in the summer leave open'; if we take the notion of living under trees rather loosely, this accords well with the excavator's report of Pazyryk, which mentions 'conical, albeit miniature, hexapod felt covers found in all the large barrows, used in smoking hemp. Their rods were like the frame of the light Kazakh shepherds' *kos*, with the upper end lashed with a thong, which could be assembled and covered with felt in a minute. We have seen such a *kos* still in use among Adaev Kazakhs as recently as 1926.' The same tombs have also revealed sections of felt, decorated with horsemen, with swans and with monsters – perhaps for use as tent hangings.

Armies on campaign have always used tents: we can see on reliefs from Nineveh tents similar to those in use in the same region today, and we may imagine that the tent in which Achilles sulked was not too different from those erected by Lord Raglan and his forces in the Crimea, except that the former was probably of leather, instead of canvas or cotton duck.

C.M.B.G.

Bricks

Prior to about 6000 BC bricks seem to have been formed by hand, not in moulds. Such hand-modelled bricks are to be found before this date on the Iranian Plateau, and examples dating from 5000 BC survive at Jericho, where they look rather like short French loaves.

Brick is traditionally the oldest substantial building material: 'Here we can make bricks, they said to one another, baked with fire; and they built, not in stone, but in brick, with pitch for their mortar. It would be well, they said, to build ourselves a city, and a tower in it with a top that reaches to heaven . . .' (Genesis 11: 3–4). But if the Tower of Babel was not exactly a success, the Babylonian ziggurats – from which the legend of the Tower of Babel perhaps derives – were more lasting, and they, too, were faced with brick, as were the city buildings of the Babylonian Empire. The plains of the Tigris and Euphrates are alluvial, so stone and marble were scarce while clay was plentiful. The use of columns was prevented by the lack of stone, and the resultant architecture is massive and simple, with much use of buttressing.

The building techniques of the Babylonians (*c.* 3000–1250 BC) were transmitted to the succeeding Assyrians (to 612 BC) and we find a similar use of brick for domestic construction among the Egyptians of the early dynasties (3200 BC onwards). The more durable stone – of which they had a good supply – was reserved for their monumental structures, the survival of which was imperative to their religion.

Bricks seem to have been fired in a furnace, rather than just put out in the sun to dry, by about 3500 BC in Mesopotamia; previously bricks had been large, but they were now reduced to about the size of the modern house-brick, because the smaller size proved easier to fire successfully. Fired bricks were more durable than sun-dried ones, and would have been reserved for those parts of the building where greater than

Above, *this drawing of the building of Rennie's London Bridge (in the 1830s) shows how the arch is built up over a wooden supporting structure.*

Opposite, above, *tent of an Assyrian officer in the 7th century BC: alabaster relief fron Nineveh.*

Opposite, below, *brickmaking, from a Flemish manuscript of the mid-15th century.*

Right, *column design by Sir William Chambers, to whom the classical was the only true style of architecture.*

normal strength was needed, or wear expected. Often other expedients had to be resorted to, for fired bricks were expensive to produce in great quantities: thus the great Ziggurat of Ur (*c.* 2000 BC) is mainly of sun-dried brick, the necessary strength being provided by intermediate layers of reed matting and fired bricks being reserved for the cladding. Even more durable finishes could be obtained by glazing, and we find experiments in this direction in Mesopotamia *c.* 1000 BC.

The Romans were highly skilled in brick-making, and it is possible they learned their expertise in Egypt and Greece; but after the fall of their empire the art died out in Europe and, although there are brick buildings of the 13th century in England, it was not until the reign of Henry VIII that they were widely used, and then only for prestige buildings such as Hampton Court Palace. The increase in their popularity was due partly to their cheapness – brick-fields were to be found on the outskirts of any town with suitable clays – but also to their fire-resistant properties, a lesson put into effect with the rebuilding of London in brick and stone after the Great Fire.

C. M. B. G.

Column

The origin of this structural device is obscure, for any wooden structure, once rotted, leaves only post-holes; and while this gives archaeologists the plan of the building (if the soil is suitable) they can only speculate about its true elevation. The column was probably first of all a simple tree-trunk with the branches lopped off, supporting a beam on which would lie the rafters of the building, also of wood. The Greeks surely first built their temples of wood, and then retained the simplicity of the technique when they switched to marble.

Such ideas on the origins of architecture are as old as Vitruvius, an architect of the time of Hadrian, and they were still current in the time of Sir William Chambers, to whom we owe many of the pavilions at Kew. The column thus holds a crucial place in the classical tradition of architecture, which men like Sir William saw as the only true style, sanctified by the usage of the Greeks and the Romans.

Classicism can be explained in terms of the simplicity of the column and lintel arrangement, a system which has few structural potentialities or excitements – for there are limits to the space one can span with a stone lintel, and the covering used was generally either flat joists or a simple pitched roof. For about a millennium the Greeks used this same building technique for their temples; and the growing sophistication of these from the rugged strength of the structures at Paestum or Selinus (6th century BC) to the calculated beauty of the Parthenon (447–432 BC in its present form) is aesthetic and not structural. Indeed, the great Temple of Diana at Ephesus (356–334 BC) was placed among the Seven Wonders of the Ancient World precisely because of its structural daring: at the entrance front, the central intercolumniation was extremely wide, and the whole temple was built up on rafts over marshy foundations.

Chambers's view of the origins of classical architecture (compare the view of the Romantics who thought the Gothic cathedrals had their spiritual and formal origins in the gently inclining tree-trunks of the leafy forests of northern Europe) begins too late, however, for the Egyptians were using true columns by the Third Dynasty (2780–2680 BC), as the entrance-hall of the Zoser Step Pyramid at Saqqara shows; moreover, columns of square section attached to walls (i.e. pilasters) had been employed even earlier, in the First Dynasty brick-tombs at Saqqara (3200–2780 BC).

C. M. B. G.

Arch

An arch may be defined as a structure of wedge-shaped bricks or stones arranged in a curve, each of which holds the others together by bearing upon it; the thrust of the materials is led down to the ground through columns or pillars. If any element of the structure is removed, the arch falls. Its strength and sophistication as a construction method derive, the one from its inherent stability, the other from the need for some kind of centring to help in its erection.

The arch was known to the Assyrians and the Babylonians: an early example is the 6th-century BC Ishtar Gate of Babylon made of sun-dried brick. Reliefs from Nimrud, of the 9th century BC, also show arches, and the Egyptians and Greeks knew of the technique, though they did not make much use of it.

The arch came into its own in the Roman period, when its lightness and airiness, and its cheapness when compared with a solid wall, combined with its great load-bearing potential to make very tall constructions such as aqueducts and amphitheatres possible. The Romans favoured the round-headed arch (and Renaissance Europe was to copy this preference), but the architects of the Gothic period tried to develop types of pointed arch which made the churches they erected very different from those of the Romanesque period: the Romanesque architects, as the name implies, followed Roman building practice, and their churches tend to be massive and gloomy, with long barrel-vaults and small apertures for windows. The more adventurous architects of the Gothic period sought a combination of column and pointed arch which, put together to make vaulting, ensured a structure which was lighter in weight because greater spans were possible with a pointed as opposed to a round arch and also because of the growing use of stone-faced pottery vaults (a technique derived from the Romans). Again, pointed arches, which led thrust down through the columns to the ground much more efficiently than did a round-headed arch, gave a much lighter interior because the larger spans permitted windows where previously there had been walls.

C. M. B. G.

Vaulting

Formed by a series of arches, a vault is a device for spanning a larger area of ground than would be possible with a column and lintel arrangement. Although, like the arch, vaulting was known in Babylonia, it began its proper career under the Romans, when 'barrel'-vaults (of semicircular section) were extensively used in drainage systems.

Barrel-vaults, usually of stone, are heavy and cumbersome, because the thrust of the blocks is not all channelled down to the ground through the supporting walls: the keystone – the central uppermost stone of the arch – obviously bears directly downward, forcing the supporting walls outward. Hence the need for supporting buttresses.

A pointed vault can, on the other hand, be much lighter because, since it has no keystone and its sides incline more steeply, the thrust is more easily led down the side walls or arcades to the ground. Through a process of trial and error, the rib-vault was developed during the Gothic period: here the basic structure of any vault became a framework of ribs which carried the thrust, and the spaces in between – serving little structural purpose – were filled in with lighter materials.

The apogee of virtuosity in vaulting is to be seen in late Gothic work, as at King Henry VII's Chapel, Westminster (1503), or King's College Chapel, Cambridge (vaulted 1508). This is fan-vaulting: as well as ribs, an elaborate system of panelling is used, and the 'fans' themselves hang from the transverse ribs. Although they are highly decorated, they are not superfluous to the construction of the vault for, were their downward thrust removed, the remarkably large span of vault would collapse, by breaking outwards and upwards at the points where they are attached: the fans act as anchors or, to use an analogy from the barrel-vault, as pendant keystones.

Thomas Fuller, the historian, wrote as follows of the interior of King's College Chapel, in 1655: 'the stonework . . . carrieth away the credit (as being a Stonehenge indeed) so geometrically contrived, that voluminous stones mutually support themselves in the arched roof, as if Art had made them to forget Nature, and weaned them from their fondness to descend to their center. . . .'

C. M. B. G.

Dome

A dome is a kind of vault erected over a square, circular or partile space in a building. It can best be envisaged as a series of arches, doubly curved – curved around the dome's perimeter, and also from the base of that perimeter up to the apex. It is much more stable than just one arch, because each 'arch' of stones is braced against its fellows by the built-up 'rings' of masonry which form the layers of the dome's construction.

The technique is perhaps first found in the tombs of the Mesara Plain in Crete (*c.* 2500–2000 BC). The so-called 'Treasury of Atreus' at Mycenae (a beehive-shaped tomb, in fact) was built *c.* 1500 BC. There are examples of domes in Mesopotamia of the 2nd millennium BC, but little use was made of the idea until the Romans.

The Pantheon at Rome (*c.* AD 112) is the first important example of a dome, and it provided the inspiration for a whole series of circular churches, large and small, with domes or vaults, from the Renaissance onwards; in it the main problem of any dome – weight – is controlled by a careful grading of materials and wall thickness. The dome has a diameter of 142 ft. 6 in. and it contains a central 'oculus' 27 ft. in diameter; the upper parts of the structure are of pumice and tufa, for lightness, and are without decoration. The topmost portions are 4 ft. thick, but the construction is more massive lower down; nevertheless, the weight is kept down by a system of coffering which, although it might appear at first sight to be merely decorative, in fact adds to the strength of the structure. This, the main structural part, is in concrete.

The dome of San Vitale, Ravenna (AD 526–47) is even lighter, being constructed of small interlocking pots topped by a timber roof. This rendered unnecessary the complicated series of lower barrel-vaults and semi-domes without which Hagia Sophia, Constantinople (AD 532–37) would collapse.

There are two devices enabling architects to place a dome over a square space: that used at Hagia Sophia is the pendentive – a spherical triangular section formed between the supporting arches at right angles to each other, and providing a useful surface for painters to decorate.

The other device is the squinch: this is an arch built obliquely across the right angle at the top of the square section, on which the dome sits. It appears to be of Eastern origin – an example is the palace at Sarvistan, near Persepolis (*c.* AD 350).

Right, *the east end of Bourges Cathedral, c. 1192–1225. The ribs of the vaulting carry the main thrust.*

Opposite, above, *looking west from the Forum, along the Via Nova: a row of typical Roman arches. The cement holding them together has lasted 1,800 years.*

Opposite, below, *timber scaffolding lashed together with hemp ropes, from a 15th-century Flemish Book of Hours.*

Below, *the Pantheon, Rome – the largest dome built in antiquity.*

The most famous dome of modern times was that of Florence Cathedral, built by Filippo Brunelleschi in 1420–34: it was the wonder of the age, and looked upon as a great technical feat. It covers an octagon 138 ft. 6 in. in diameter, and is raised on an octagonal drum, each side of which has a window; of pointed form, it consists of 24 ribs and is the first structure to use a hoop of tie-bars on the outside to stop the dome bursting outwards. Michelangelo followed this practice in his dome for St Peter's, as did Christopher Wren at St Paul's, London. The reason for using ties (timber and iron hoop for Brunelleschi, iron chains for Michelangelo and Wren) is aesthetic and practical at the same time: no buttresses are needed to stop the dome breaking outwards, and the dome can therefore be raised on a simple drum, giving a clean silhouette.

C.M.B.G.

Cement

In the ancient Near East, pitch was the substance most often used for binding together courses of brick or stone: such is the case at Sumer (5000 BC) as in Babylonia two millennia later. Such bitumen mortars included an amount of vegetable fibres for added strength, or other expedients: in Nebuchadnezzar's Palace at Babylon (*c.* 605–563 BC), the bitumen was mixed with chalk, and vegetable fibres are absent – but appear in the form of matting every five courses for reinforcement. In other cases, the substance is mixed with earth or cinders.

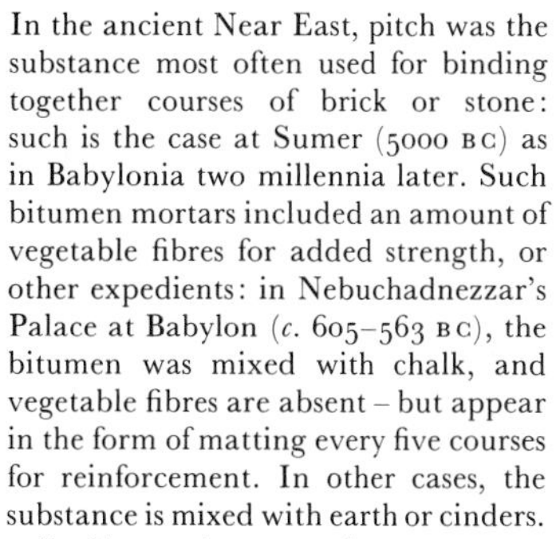

In Egypt, however, burnt gypsum mortar makes its appearance at the Pyramids of Gizeh (*c.* 2500 BC); and by the time of the Persians, gypsum and lime had replaced pitch over a wide area.

For the highest-quality work, of course, cement was not used: the Greeks clipped the blocks of stone on the entablatures of their temples with bronze or lead cramps, and the Etruscans and early Romans continued the practice, or used dowels, or sometimes mortar. At Mycenae (*c.* 1300 BC) the great foundation walls of the citadel are of carefully matched blocks of stone, with only a mud mortar – this probably being the oldest binding material of all.

It is important to distinguish between cement and concrete: cement mortar was being used by the Romans as early as *c.* 300 BC at Pompeii, and it consisted of lime and sand. But concrete – a mixture of lime and *pozzolana* – is a very different matter, and first occurs in the theatre at Pompeii, of 55 BC.

Similarly, Portland cement is not concrete, although it can be an element in it. Smeaton, while experimenting prior to building the third Eddystone Lighthouse in 1756, remarked that the concrete he had invented 'would equal the best merchantable Portland stone in solidarity and durability'. But what we know as Portland cement was patented by Joseph Aspdin, a Leeds bricklayer, in 1824. This is essentially a mixture of chalk and clay, heated to a high temperature; the resultant material sets hard when mixed with water.

C.M.B.G.

Scaffolding

Scaffolding must be at least as old as the pulley, the device which allows large weights to be hauled up vertically from the ground. The alternative, when labour was cheap, time plentiful and construction techniques of the simplest, was to erect a ramp of packed earth up which men pulled the building blocks to the working face. The Egyptians under the Old and New Kingdoms built like this, and the monoliths of Stonehenge, including the horizontal members, were probably positioned by the same method.

But more advanced construction techniques, and somewhat less expendable slaves, needed scaffolding in timber (never in metal until recently): centring was needed for arch or vault, for such constructions relied on a forme which could be moved when the mortar had set. The Romans, great builders, used scaffolding a lot, and we often find recessed bricks in their walls to take horizontal scaffolding members; their name of 'putlog holes' is self-explanatory.

Medieval account books and contracts give much information about scaffolding, which probably had not changed much since Roman times: the higher the scaffolding the thicker the poles used, and these would be lashed together with hemp ropes or withies, and the set-up then tightened by driving in wooden wedges. Platforms would have been of woven wicker, for planks were not used until the introduction of mechanical sawing in the 15th century. For repair work, masons would be let down in a cradle, or scaffolding bracketed out from the wall high off the ground, to save the expense of full scaffolding.

In high buildings, particularly vaulted ones, scaffolding and centring would be very expensive: while the former could be quite flimsy (and often caused serious accidents), the latter had to be rigid, of planed planks trussed together and held firm by boards, the whole being supported from ground level by an equally firm structure – otherwise the vault might set unevenly. Both are frequently specified in contracts, for example, the one for King's College Chapel, Cambridge (1512–15): 'And the said John Wastell and Herry Semerk shall provide and fynde at their costs and charges . . . lyme, sand, scaffolding, cynctours [i.e. centrings] moldes . . .

and euery thyng concernyng the same vawtying. . . . Except that the said Mr provost and scolers with thassent of the said surveyour graunten . . . for the great cost and charge that they shalbe at in remevying the great scaffold there to haue therfore in recompence at the end . . . the tymber of ij seuereys of the said grete scaffold by them remeved to their own use and profight. . . .' These 'ij seuereys' – two bays of centrings – would have given them a lot of valuable timber.

While the Gothic masons discovered the rib-vault as a kind of permanent centring, it was not until Brunelleschi's great dome for Florence Cathedral (1420 onward) that a much reduced level of scaffolding became possible. Brunelleschi showed that, whereas all previous domes had required scaffolding inside and out at huge expense, his workmen needed only a light rig which was easily moved up the internal face of the dome as they worked – and probably not attached to the ground, although its exact design is a matter for conjecture. Such a device had another advantage: it obviated the delicate maintenance needed to keep ordinary scaffolding and centring in good order because of the expansion and contraction of both wood and stone.

C.M.B.G.

Crane

Where the pulley that carried a rope from a winch could not be secured to some part of the building under construction, or where it was to be used to unload a cargo ship, something else had to carry it over the load to be raised. The obvious solution was to hoist it to the top of a mast as if it were on board ship; and so the crane was born. The exact date, even the right century, is anybody's guess. The first descriptions are those of Vitruvius and Hero, contemporary with a magnificent Roman carving of a crane at work on the façade of a tomb. With one mast the pulley could be swung freely round to wherever it was needed, but it was obviously unstable and hard to manage. In consequence two masts were used, linked together at the head, and joined at a convenient height by a winch. Even so, the relief shows well the mass of guy ropes that was required. With particularly heavy loads a secondary capstan could turn a wider drum mounted on the winch. Or else – evidently a quicker method and one that avoided the danger of slipping while employing the whole weight of the workers all the time – a tread-wheel could be used, the famous 'squirrel cage'.

Cranes seem to have been the normal way of raising materials in medieval churches – once you got them up to the roof. Sometimes a third mast was added, forming a sheerlegs, or even four for exceptional lifts, but in that case the crane had to straddle the work, and could scarcely be used to get up to any height in a building. With the lofty towers and pinnacles of Gothic Europe, the units of stone might be lighter, but the heights to which they had to be raised were far greater than in the past. This may be the reason for the invention of the jib-crane, with a projecting horizontal arm, called the 'hawke' in medieval English building records, which speak of its use at Winchester as early as 1257, and later in the Tower of London, where two iron rods were used, presumably as struts. The beaks of these jibs often appear in late medieval manuscripts, always at the highest point the building has reached, so they dominate the scene. Like the treadmills, they were often left on the site afterwards in case they were needed for restoration work; one installed in Cologne Cathedral early in the 16th century is supposed to have lasted until 1819.

Similar cranes, with jibs that were hinged so they could slew freely through as much as a semicircle, began to appear on the quaysides of Europe in the 14th century: a treaty of 1446 specified that English ships at Antwerp were to discharge 'where is the machine called die Crane', and many miniatures of that period feature the harbour cranes. Some are in effect just mast and treadwheel, enclosed in heavy wooden housing, and rotating on a pivot supported by quarterbars – like the contemporary postmills, on which they were doubtless modelled. But others appear in townscapes, in roundhouses with conical roofs, surmounted by secondary caps: the intervening space presumably allowed for a jib which could swivel right round the mast.

A.K.

Chimney

The chimney is a comparatively recent invention. For centuries after man had learnt to make a fire, even for centuries after he had learnt to build a house for himself, he was content to let the smoke from his fire escape into the air through a simple hole in the roof. It seems not to have been much before the 14th century that somebody had the idea of extending the hole, so to speak, above the level of the rest of the roof, thereby increasing the protection from the elements and at the same time reducing the risk of smoke re-entering the house on gusty days.

Chimneys do, however, begin to appear in wealthy houses in Europe from about the 14th century onwards, and in England, where brick was used more extensively for building than in many other countries owing to a shortage of native timber, they became com-

mon fairly quickly. An English writer of 1577 notes that, whereas at the beginning of the century 'there were not above two or three chimneys, if so many, the religious houses and manor places of their lords always excepted, but each made his fire against a reredos in the hall, where he dined and dressed his meat', now chimneys are common in villages.

Now it looks as though domestic chimneys are on the way out. Modern fuels do not require them, and open fires in the home, however desirable in theory, are a rarity in Western society.

Plywood

Perhaps plywood was first used by Michael Thonet, a cabinet-maker of Boppard-am-Rhein, who set up shop in 1819; he pioneered the steam-bent wooden chair, and designed a chair with beech frame and plywood seat in 1855. By 1830, he had made chairs from strips of veneer glued together under pressure in a wooden mould. Bentwood chairs similar in conception to those by Thonet were being made at Reval (Estonia) by 1844. Thonet's designs must have been good, for they are still being produced in Czechoslovakia today.

Americans were patenting processes for using plywood from 1885; one such patent, dated 1874, reads: 'If three layers of veneer are used, the grain of the middle one runs crosswise to that of the outer layers . . . the elasticity and strength of the veneers is thereby increased considerably. The chair is made of three parts . . . and the front part . . . extends up over the upper part of the back. . . .'

All early plywoods – and, of course, the veneers used for furniture manufacture in the 17th and 18th centuries – would have been cut by saw, and would have been rather thicker and more expensive than those used today. A process perhaps introduced at the end of the 19th century was the placing of a de-barked and steam-softened log in a large lathe, which turned against a knife: the tree could then be 'peeled' almost to nothing, and the veneer cut into pieces with a guillotine.

A big boost to research into plywood was given by the construction of aeroplanes during the First World War; high-quality plies were needed, particularly as a covering for the wings, and for various ribs and spars, and resin glues for dry bonding were developed for this purpose.

C.M.B.G.

Elevator

'All safe, gentlemen, all safe.' The speaker was Elisha Otis (1811–61), the occasion the Crystal Palace 1853 Exposition in New York. Here every night Mr Otis, who like many successful inventors was a bit of a showman, demonstrated his invention – an elevator with a safety device – by mounting its platform, being hauled aloft, ordering the rope to be cut, then remaining, to the astonishment of the crowd, suspended safely in mid-air waving his triumphant top hat.

It was not his first elevator – he had designed one, a year or so before, to carry freight in the bedstead factory where he worked. Nor was it, of course, the first elevator ever devised. Primitive hoists had been known in Roman and medieval times; a Frenchman, Velayer, invented a counterweight lift in the 17th century; hydraulic elevators were fitted in several European factories by 1830; and two Englishmen, Frost and Strutt, had built in 1835 a hoisting-sheave system called the 'Teagle'. Most of these early lifts were hydraulically powered, the cage, partly balanced by a counterweight, being mounted on a plunger working in and out of a cylinder. They were safe but hopelessly slow. To hang the cage on a rope meant greater speeds, but was too dangerous for passengers. This was the problem. Elisha Otis provided the answer with his safety device, and made passenger lifts, and thus tall buildings, practicable. Until then five storeys, the maximum height tolerable for staircase access, had been the limit.

The principle of his invention was a wagon-spring-activated ratchet-and-pawl device which operated only if the tension in the rope failed. His first safety lift – driven by steam – was installed in the New York store of E.G. Haughwout & Co. in 1857, and commercial use of his electrically operated elevators began in 1889. In 1904 the Otis Company pioneered the gearless traction elevator which through its speed, simplicity and low cost made the skyscraper possible – the 41-storey Singer building in 1907, the Empire State in 1932. True, none of this would have been possible but for the unknown inventor of the pulley and for those many ingenious men, famous and unknown, who developed everything from electric motors to automatic signals. But it was the old wagon-spring used by Elisha Otis, the master mechanic who died comparatively poor and obscure in 1861, that made Manhattan possible.

A minor but disastrously conspicuous spin-off from the lift has been the roof-top eruption housing the override and motor room – an opportunity for an architectural flourish splendidly seized upon by the designers of the earliest skyscrapers and, with the exception of the post-war work of a number of Japanese architects, almost totally neglected since.

H.C.

Above, *building the Tower of Babel, from the Bedford Book of Hours, 1425. On an upper storey a crane can be seen – the 'sheerlegs' of classical times.*

Opposite, above, *in a modern plywood factory. Slicing the veneer off a rotating log is a method introduced in the 1890s.*

Opposite, below, *detail from Carpaccio's painting* Healing of the Madman *(c. 1495). By this date chimneys, found only in wealthy houses from the 14th century onwards, were becoming more common.*

Right, *with this advertisement, in 1881, the Otis Company proudly announced their contract to install the first elevators in the White House.*

Paint

Not to be confused with dye, paint is usually formed from a dry pigment mixed with a medium. From *c.* 23,000 BC, cave-paintings appear in which the pigments used are iron or manganese oxides (giving yellow, red and black; white was not possible), with hot fat or perhaps urine as a medium. There were several ways of conveying the paint to the wall: it could be applied by hand, or by using bits of fur or chewed stick for a brush or (as at Lascaux, *c.* 15,000 BC) blown through a tube – an early aerosol spray! Scapulas to mix the paint, and palettes to do it on, are common finds; the beautiful outfit from the tomb of Tutankhamun (*c.* 1350 BC), for example, includes palettes in wood and ivory, a brush-holder, and one ivory palette still with some blocks of dry colour in place.

In the Middle East, from Badarian times in Egypt (pre-4000 BC), naturally occurring ochres were the usual pigments, giving red, yellow and brown; for green, powdered malachite (copper carbonate) was used; for blue, real lapis lazuli or a substitute made by heating silica, malachite, calcium carbonate and natron (from *c.* 2500 BC); white was obtained from chalk or gypsum, or white lead.

Grounds for painting on included gypsum plaster, pottery, ivory, stone and wood, a gypsum layer being applied to the last two before painting. The medium used would be water and some glue – from size, gum or white of egg, or casein (from milk curds). 'Frescoes' were often varnished with wax, or with a true varnish, to preserve them. But true fresco – water-paint suspended in lime and applied to wet lime plaster – dates back no further than *c.* 2500 BC (in Crete); it was to be the method by which the Italian Renaissance measured true excellence in painting, and was used by Michelangelo for the Sistine Chapel ceiling. In classical times, too, true fresco and *fresco secco* were in use: the latter was easier to use, for the artist could work slowly and correct mistakes, whereas true fresco required quick and accurate work – but also produced an image which sank into the plaster, and was therefore more than surface-deep.

From the 5th century BC encaustic painting – in which the medium was hot wax – was used for decorating battleships as well as for painting pictures; it allowed richer colouring than was the case with fresco or tempera on panel (the latter using egg as medium).

But true oil-painting, using linseed oil, although its properties had been known since the 6th century AD, only became current in the 15th century, first in Flanders, then Italy. The use of oil for a colourless or a coloured glaze enabled a depth and subtlety of hue not possible in fresco or tempera, where only the crudest shading could be obtained; thus a figure with a blue robe could be overpainted with glazes of differing colours so that the base colour shimmered and sparkled.

Other techniques, such as watercolour (colour ground up in water-soluble gums), pastel (powdered colour with enough gum to make it sticky, and made up into sticks) and gouache (opaque watercolour, or poster-paint) have recently been joined by a host of new materials, including acrylic (acrylic vinyl polymer emulsion) and others.

C.M.B.G.

Oil lamp

Wood fires and blazing torches provided man's earliest form of indoor illumination. It seems to have been about 50,000 years ago that the earliest lamps were made out of hollowed stone in which a piece of fat or grease would be placed and set alight. Early examples have been found in France and Mesopotamia, Egypt and Persia but we cannot be sure whether the device was used generally or merely in areas, such as those inhabited by the Eskimos, where wood was scarce. There is no evidence, certainly, that wicks, which made for a clearer flame and less smoke, existed much before 1000 BC or were widely distributed for a further 500 years after that. Gradually, however, the familiar ancient oil lamp evolved, consisting of a wick, usually made of a reed or other vegetable fibre, set in oil or fat in a saucer-shaped vessel, sometimes with a handle. In the classical world oil lamps became quite sophisticated. Made at first of pottery, but later sometimes of iron or bronze which might be elaborately chased, Greek and Roman versions occasionally had holes or spouts for holding two or more wicks, a hole for pouring oil into the vessel and a handle; often, too, they were made in fanciful shapes.

Drawing of bison and a horse in the 'salon noir' at Niaux, France – a very early domestic use of paint.

Rock-cut cistern at Masada, near the Dead Sea.

Boat-shaped oil lamp from Puteoli, Italy, with multiple holes for wicks.

Boiling soap: illustration from The Book of English Trades, *1821.*

In this form the oil lamp continued to be used for centuries all over the world. A few innovations were introduced to its design, however, one of them due to Leonardo da Vinci, who added to the basic lamp a water-filled glass tube through which ran a cylindrical glass chimney; the flame, shielded from draughts, thus burnt more steadily, while the water acted as a lens to magnify the light emitted. A further improvement came in 1784 when the Swiss physicist Aimé Argand added a hollow chimney in conjunction with a hollow wick through which a current of air passed upwards, providing the inside of the flame with oxygen; this did away with yet more of the smoke, smell and flickering that still accompanied oil lighting.

Although oil was first used in street lighting (in London) as long ago as 1681, it remained an expensive fuel for domestic lighting in Europe until the discovery of petroleum in the mid-19th century. In 1885 the 'Kitson lamp' was invented in which the wick was supplanted by an incandescent mantle fuelled with oil vaporized under pressure.

By allowing mankind to 'burn the midnight oil' the humble oil lamp has given centuries of service; even in the second half of the 20th century it continues to have its uses – for instance in tents, on board sailing craft, and during power cuts.

S.F.

Cistern

While excavating Megiddo and Gezer in Israel, archaeologists discovered tunnels which led deep down beneath the ancient towns to caverns where they found springs which had provided the Bronze Age Canaanites with a secret water-supply, safe in time of siege. The earliest cisterns would have been natural caves where there was no spring, but into which the local inhabitants had diverted some outside source. In the arid conditions of most parts of the Middle East and the Mediterranean not a drop of rainwater could be wasted; in a region where the roofs of houses were usually flat and surrounded by a parapet, each house had its own miniature catchment area, whose run-off could be directed into a cellar and stored there like food, perhaps originally to keep it cool, but it was soon realized that there would also be less evaporation underground. At Knossos in Crete the careful Minoans ensured that the rain which fell on their great stone staircases was properly conserved in this way. And in the mountains of the Levant natural sinks and fissures in the rock provided ready-made cisterns for the storage of water; it was easy to enlarge them then and perhaps cut steps down to the water-level; there are many of these 'improved' natural cisterns still to be seen. Even where other sources of water were adequate, people often preferred rainwater because of its purity, and kept it, sometimes in cisterns lined with timber or plaster. Sometimes what look like wells prove to be cut in impermeable rock, so they must have served as cisterns. All these are underground, or at least roofed over.

But even in relatively humid regions it was often thought worth while to store rainwater, for instance in India and Ceylon. Here, as evaporation is less rapid, the cisterns could be left open to the sky: this was the original meaning of the word 'tank'. Sometimes where really large quantities were needed, big open-air tanks were dug in quite arid places, despite the inevitable losses, as at Carthage, or the ancient 'Pools of Solomon' south of Jerusalem, or the medieval Pool of Mamilla to the west of that city. However, as the art of construction advanced, underground cisterns could also achieve a respectable size. Among the largest are those under Istanbul, such as the Hall of a Thousand and One Columns (actually about 420), the work of Byzantines in the 6th century AD: James Bond was chased through it in *From Russia with Love*. As soon as pumps were used to raise substantial amounts of water, that, too, had to be stored in cisterns from which it could be dispensed – our water-towers are cisterns on stilts. And aqueducts fed cisterns or reservoirs; England's oldest was the Round Pond at Islington, which held the water of the New River cut in 1609–13 to take water from Hertfordshire springs to quench the thirst of a growing London.

A.K.

Soap

Soap is such an everyday and commonplace commodity that it is difficult to think of it as having been invented in the same way as, say, the zip-fastener and the ballpoint pen. It has been around for a long time; in fact it was one of the first great chemical inventions.

Soap was made by the Sumerians in 3000 BC; they boiled together alkalis and used the residue on their bodies and equipment. The Phoenicians who settled at the mouth of the Rhône in 600 BC also used a salve made in a similar manner. Soap was, however, unknown to the Romans, until they apparently relearned about it from the Gauls. Pliny the Elder in his *Historia Naturalis* mentions a salve made of beech ash and goat fat which the Gauls apparently used as an ointment and hair dye. Galen mentions soap used specifically for cleaning in the 2nd century, and by the 8th century it was known in most of southern Europe. It was chiefly produced at Marseilles, Genoa, Venice and Savana, because of

Above, *an 18th-century French candle-maker's shop.*

Right, *the solitary candle in Jan van Eyck's* Arnolfini Wedding *(1434) symbolizes the all-seeing eye of God.*

Below, *the great Roman bath, at Bath.*

the availability there of olive oil and crude soda – and it is from the name Savana that the French and Italian words for soap are said to derive.

In general use in the Middle Ages, soap was made in London by 1524, though not until the 17th century was the first soap patent issued in England. About this time its use also appears in America.

T.C.

Drains

Drainage for domestic cleanliness was an early development: while the houses at Çatal Hüyük in Anatolia (pre-6000 BC) have only holes filled with sand or gravel for drainage, and those at Haçilar not even that, the city of Mohenjo-daro in the Indus Valley (*c.* 2500 BC) had drains, and latrines too, in most houses.

True drainage also appears in the Western world in the 2nd millennium BC: the Cretan Palace of Mallia (16th century BC) has drainage pipes, and the Palace of Knossos boasts not only piped drinking-water and bathrooms, but also water-closets of a kind. Less sophisticated techniques were known to the Mycenaean Greeks: large drains were built into the citadel of Mycenae III (13th century BC) and cisterns as well, and the Palace of Tiryns (*c.* 1450 BC) had a bathroom – but probably no piped water.

After the fall of the Roman Empire, standards in all the public services declined, and we must imagine the streets of any small town or big city, until at least 1800, as being foul and noisome with frequent open sewers and cesspools. Cities like Paris and Bordeaux, where the mid-18th century saw projects for more civilized sanitation, were exceptional.

C.M.B.G.

Bath

Today we consider bodily cleanliness one mark of civilization, but such beliefs were not widespread in the ancient world – or, indeed, until very recently, in Europe.

Bathing is a religious imperative for the Hindu, and this may have been the main reason for a large bath and a group of smaller baths at Mohenjo-daro in the Indus Valley (*c.* 2500 BC), though there are private bathrooms in some of the houses as well. The Great Bath has stairs into it, and is surrounded by a paved walk; it has a drain-hole, and a drain with inspection-cover, though it is not known where this led to. A well served the eight small baths, and perhaps the Great Bath as well. All the baths are of burnt brick, but this is backed by a layer of bitumen to prevent waterlogging.

The Babylonians bathed in much the same kind of bath in the Palace of Mari (*c.* 1800 BC) – or at least the king did, probably for reasons of purification, not cleanliness. The Egyptians, on the other hand, apparently had no baths before the Romans introduced them. But it is the Greeks who introduced a modern attitude to bathing: they considered bathing toned up the body, and baths are sometimes to be found in gymnasia dating from the 5th century BC. While their ornamented sarcophagi were sometimes used as bath-tubs by the nobility of medieval Byzantium, the first purpose-built tub which we would recognize comes from the South-east Bathroom at Knossos, and is shaped like a hip-bath. This, like others found on Crete, would have been filled and baled out by hand.

The Romans turned bathing into a social occasion and built massive bath complexes. The Baths of Caracalla, for example (AD 211–17), had hot, warm and cold baths, sweating-rooms, shampooing and manicure shops and a gymnasium; they also held libraries, lecture-halls and works of art; lead pipes brought water from the Marcian aqueduct via a reservoir, and furnaces heated it as it was needed.

In post-Roman times, most people in Europe had a bath when they were baptized by immersion, and infrequently thereafter; the rich used perfume, and the poor stank. The Knights of the Bath did not embody a love of cleanliness in the Middle Ages; they were merely a group of notables chosen to help the king take a purificatory bath prior to his coronation.

Although there were bath-houses in Germany in Dürer's time, they were rare even in exalted circles in that most civilized of all countries, France: the Château of Fontainebleau had one (*c.* 1540), but at Versailles (despite the complicated fountain-systems) toilets and bathrooms were absent during the *grand siècle*. In England, public baths are the creation of the Baths and Washhouses Act of 1846.

C.M.B.G.

Candle

Artificial lighting is at least as old as Palaeolithic times, when it was discovered that a wick of fibrous material fed by oil or fat would keep burning. Stone Age lamps, of triangular shape with a saucer-like depression (the wick lying in the saucer with its point sticking out) are the obvious antecedents of the candle, but we have no evidence for dating candles – which are, after all, self-destroying articles – before the Bronze Age, from which time we have candlesticks which served to collect the precious oils and fats for re-use.

For the fats, oils and waxes used for candles and tapers *were* valuable, be-

Plan of the Benedictine Monastery of Christchurch, Canterbury, showing the projected water-supply and drainage system, c. *1160.*

Detail from Virgin and Child *by Massys,* c. *1490 – one of many such paintings that show a carpet under the Virgin's throne as a symbol of rarity and beauty.*

cause they were edible and often, as with fish-oil, nutritious. As late as the beginning of the 19th century it is reported that the Elders of Trinity House were concerned over the high consumption of tallow candles used for their lighthouses – and discovered that the keepers were supplementing their diet with them.

Torches and tapers would have been placed in special containers to lessen the risk of fire; from these developed the 'pricket', or spiked holder for torch or candle – considered by the ancients to be older than the socket holder. Because of the vague terminology adopted by ancient authors, for whom a single word could mean torch, taper or candle, no precise date for the candle's discovery can be given. Experts consider it Etruscan, pointing to a representation in an Etruscan tomb near Orvieto in Italy; it was certainly much used by the Romans, and the Christians adopted it and made it a symbol very early – for example, its snuffing and overturning featured in the act of excommunication, while the solitary lighted candle in Jan van Eyck's *Arnolfini Wedding* symbolizes the all-seeing eye of God.

From the Middle Ages, the craft of candle-making was institutionalized; the Wax Chandlers of London were flourishing in the 13th century, although they were only incorporated by Royal Charter in 1484; part of their Ordinances, of 1488, reads: 'Also, that no manr of person nor persons, occupying the crafte of Waxchaundlers within the Cite of London . . . make no do to be made any torches, quarerres [large square candles for funerals], pricketts, sises [small court and church candles], chamber morters, tapers, candelles, nor ymagery, but of goode and able wax . . . that is to say, not puttying thereto any rosyn, code, turpentyn nor tallough, wherein the Kyng's liege people may be deceyved . . . and who so be taken and found gilty in this, to paye for his fyne, the first tyme, xiijs 4d, the ijde tyme, xxvjs viiid; and the iij tyme, four marks. . . .'

Certainly there were plenty of incentives to adulterate candles: ordinary ones were made of tallow, wax being reserved for church and court use. Thus they were a perk of officers of state: in 1135 the Lord Chancellor was getting 5*s.* a day, plus his keep, plus 'i fat wax candle and xl pieces of candle' – which he presumably sold off. Until the 14th century, when moulding was introduced, they were made either by dipping the wick through molten fat or wax, or by pouring the substance down the suspended wicks, or they were modelled by hand.

A technical advance was the introduction by Cambacères in 1825 of the plaited wick (which did not need trimming: hitherto the charred end had hindered burning, eventually making the candle go out) and another the patent taken out in the same year by Chevreuil and Gay-Lussac for making stearine candles from crude stearic acid. Machines, producing up to 500 candles an hour at first, were introduced in 1834.

C.M.B.G.

Carpet

People were weaving baskets from reeds long before 5000 BC, and we may assume that the carpet was a development from rush matting; but it is not known when man started weaving carpets as opposed to cloth, for although pictures of looms have been found of *c.* 3500 BC we do not know precisely what they were used for. Tapestry (a flat carpet, or hanging, with interwoven threads, not tufts) presents the same problem; in the *Odyssey*, for example, mention is made of *tapetia*, but these may have been woven or piled.

One of the oldest carpets found to date was discovered in 1947–49 in a 5th-century BC frozen tomb in the Altai Mountains, between Mongolia and China. The finder, S. I. Rudenko, estimated the number of knots per square inch to be 2,700, and the carpet to be of Iranian origin – indicating a trade in luxury goods. From other graves in the area came carpets of felt, and also of cut-out felt fragments stitched on to a canvas backing.

In modern times, while carpets were produced in Egypt and Syria in the 12th century, the finest carpets were produced in Persia under Shah Abbas (1586–1628), in India about the same time (using imported Persian labour), and in regions of Asia Minor, notably the Caucasus and Turkey.

In western Europe such products were prized; perhaps the rugs which Eleanor of Castile brought with her when she came to marry Edward I of England in 1225 were of Eastern inspiration woven by the Arabs of Spain; certainly, 'turkey' carpets were known in the 13th century, for we find Giotto painting them. From the 15th century onwards they were highly prized items, and we see them as such – perhaps as hangings or carpets for the Virgin's throne – in paintings by Memling, Giovanni Bellini and Hans Holbein the Younger.

Pile carpets were being made in England in the 16th century, possibly by Flemings, and about 1620 Pierre Dupont and Simon Lourdet started a manufactory in a former soap works near Paris – the origin of Savonnerie carpets. In 1701, William III granted a protective charter to carpet-makers in Axminster and Wilton.

Greater carpet production came in about 1801 with the perfection by J. M. Jacquard of the loom which bears his name; here, a continuous loop of perforated card controlled the woven pattern. Such machines were worked by steam by the middle of the 19th century.

C.M.B.G.

Central heating

The first evidence of a central heating system dates from the beginning of the Christian era in the Roman Empire. At that time, furnaces were constructed either beneath the floors of buildings or adjacent to them, and the heat-giving combustion product was conveyed through flue panels in the walls and finally discharged to the atmosphere. Radiation was the main heating principle involved, and it was not until 1908 that the technique was independently rediscovered by A.H. Barker, an Englishman who noted the 'comfortable effect obtained from walls warmed by chimneys'. Barker patented several arrangements for heating floors, walls and ceilings and these 'radiant systems' later gained wide popularity in both Great Britain and the United States. A few original hypocaust systems, relics of the Roman occupation of Britain, were uncovered in the rubble of post-war London.

The next major attempt, in the West, at building grand heating contrivances came during the 11th and 12th centuries, the heyday of feudalism. For the most part it was a poor show. Huge fireplaces were installed in the great halls of castles, and these – since they allowed about 80 per cent of the heat to escape out of the chimney – are notoriously inadequate. The saving grace of this method was the simultaneous construction, directly opposite the fireplace, of a thick wall with sufficient thermal capacity to absorb radiation and then re-radiate its heat even after the hearth fire was very low. But this sensible idea did not reappear until the early 19th century, in the United States. As for the ordinary rooms of the feudal castles, they were badly heated by small coal or wood braziers.

In 1624, Louis Savot developed a fireplace for the Louvre in which air was drawn through passages under the hearth and behind the fire grates, then discharged back into the room through a grille in the mantelpiece. Not until 100 years later did someone figure out a way to draw air in from outside a building. Stoves, which are far superior to fireplaces, have been around since the late 15th century, but it was Benjamin Franklin, that great American polymath who in 1744 established the basic principles of modern stove design.

Central heating, which has only recently begun to replace more primitive heating methods throughout Europe, began to re-emerge from the Dark Ages with the development of hot-water heating in 18th-century France. In 1792, Sir John Stone contrived such a system for the Bank of England, while around the same time James Watt and others were investigating the practical possibilities of steam heat. Since then numerous innovators have applied themselves to the development of sophisticated fuelling systems, which include the use of gas, oil and electricity, and more elaborate means of conveying heat through the rooms of larger and larger buildings. Nevertheless the days of the small-space heater are far from being over.

E.W.

Smoke-jack

Montaigne was very impressed with his first smoke-jacks, seen in Switzerland in 1581 – 'broad light sails of pine that they place in the funnel of their chimneys, which turn with great speed in the draught caused by the smoke and the steam of the fire: and they turn the roast slowly'. He did complain 'they dry out the meat a little too much', but one can hardly expect a Frenchman to admire foreign cooking without reservations.

The first hints we have of this device are to be found in Italian sketches, so it is not certain whether they were first constructed north or south of the Alps. A famous Renaissance cookery book shows that the Papal kitchens had one. Essentially the smoke-jack was just an upturned windmill, with the updraught of hot air heated by the fire rotating a wheel of vanes supported in the chimney-flue on a spindle, which was linked by a train of gears to the spits. No great changes were subsequently introduced, except for the use of bevel-gears (some great houses still have their 19th-century smoke-jacks).

This was the oldest machine to translate heat energy directly into mechanical power, modest as it was, and the earliest, too, in which the motive force was artificially supplied.

Benjamin Franklin tried to turn his jack by electricity, using the discharge of the Leyden bottle to revolve a disc, with a fowl spitted on its shaft. In April 1749 he held a famous electrical picnic on the banks of the Schuylkill, at which a turkey was 'killed for our dinner by the *electrical shock,* and roasted by the *electrical jack* before a fire kindled by the *electrical bottle*', and the healths of all his colleagues were 'drunk in *electrified bumpers* under the discharge of guns from the electrical battery'. This device provides a neat link between the Renaissance smoke-jack and the electric-powered automatic spits of 20th-century grills.

A.K.

Wallpaper

Wallpaper dates only from the early 16th century AD, the earliest fragment known being of 1509; it was clearly a substitute for embroidered work at this date, and sought to imitate it in design. Such cloth hangings would have been, in their turn, a substitute for tapestries:

Above, *the smoke-jack, from a book of 'new devices' published in 1662.* Below, left, *diagram of a Roman hypocaust or underfloor heating system.* Below, right, *a fragment of early 16th-century wallpaper, showing the royal crest.*

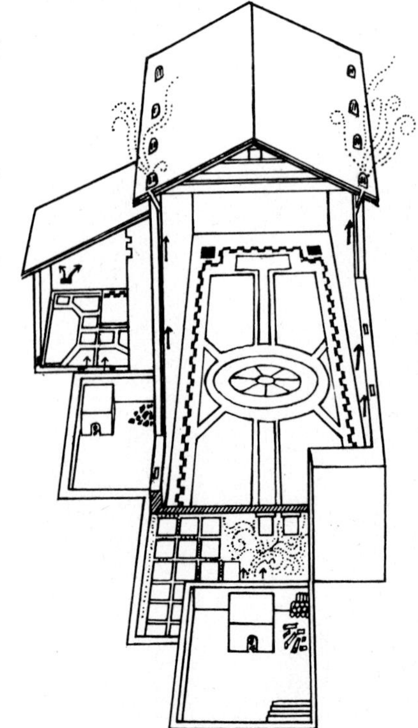

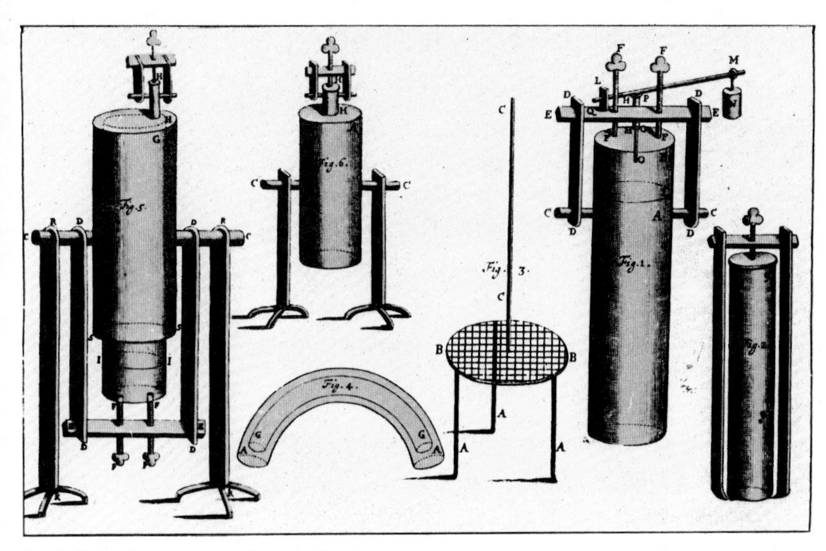

Denis Papin's pressure cooker: explanatory diagrams from his booklet A New Digester.

Early 18th-century advertisement for 'figured paper hangings'.

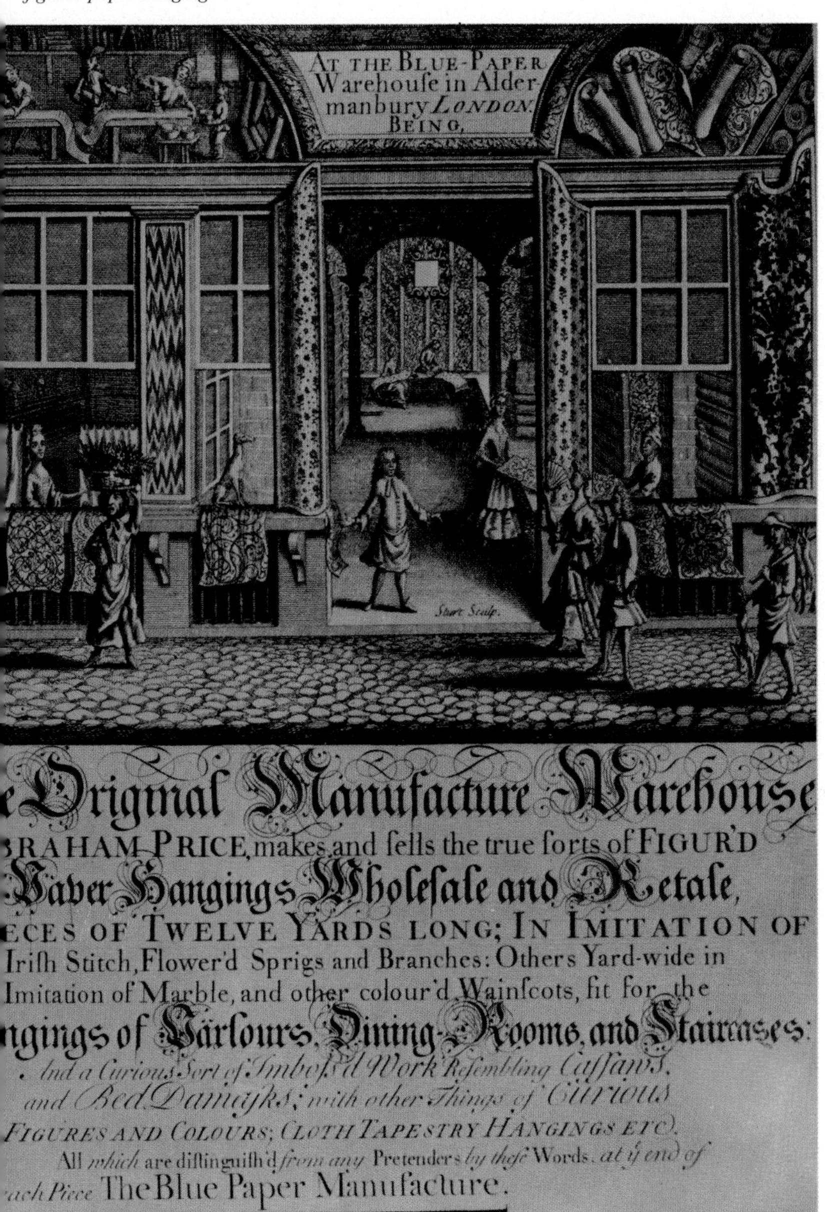

although used by the Roman nobility, these reached their apogee in both quality and quantity in the 15th century, and were costly enough to make French monarchs trundle sets from castle to castle on their peregrinations, rather than keep some in each house they owned.

A cheaper substitute for tapestry had been leather, usually embossed and gilded; this was first introduced in the 11th century by the Arabs, and made in Spain by the 16th century, then all over Europe a century later. Embossed wallpapers, which have always been popular, are a direct descendant.

'India' papers – so called by John Evelyn in 1693, when he admired some in the Queen's collection – were in fact imported from China by the trading companies from about the middle of the 17th century. Decorated with birds or flowers on a brightly coloured ground, their popularity with the upper end of the market lasted until well into the 19th century, though they were by no means cheap: they were sold in sets of about 25 rolls, at 3 gns to 5 gns a roll.

The production of English-made wallpapers was at first in the hands of the stationers, until specialists, like leather-embossers, took over – but the product remained expensive at about 1*s.* to 4*s.* per roll. The types available are listed in this advertisement of 1690: 'At the old Knave of Clubs, at the Bric ge foot in Southwark liveth Edward Butting who maketh and selleth all sorts of Hangings for Rooms in Lengths or in Sheets, Frosted or Plain. Also a sort of Paper in imitation of Irish Stitch, of the newest Fashion, and several other sorts, viz, Flock Work, Wainscot, Marble, Damask, Turkey Work, etc.'

The work was done by hand, or with the aid of a press to print the wood-blocks, until a firm of calico printers, C.H. & E. Potter, of Darwen, Lancashire, adapted a calico-printing machine to the purpose, which was patented in 1839; this used rollers with the designs in relief, rather like a continuous wood-block (a machine for producing the paper by the roll, rather than by hand and by the single sheet, had been developed at the beginning of the century). By the mid-century, wallpaper was a popular and cheap commodity. As the *Economic Library* remarked in 1851, 'it is almost as little cost to paper a room as to whitewash or colour it. A papered room has a comfortable look which no ordinary material can impart, and the hanging of the paper is not a tedious operation nor unpleasant. . . .'

C. M. B. G.

Pressure cooker

The inventor of pressure cooking, Denis Papin, was also one of the very early pioneers of the steam engine. Born at Blois, France, in 1647, he eventually came to London, where he was assistant to Robert Boyle and moved in Royal Society circles. He demonstrated his 'steam digester' in 1679 – a container with a tightly fitting lid, which increased the pressure inside and considerably raised the boiling point of water. The lid was fitted with a safety valve, also invented by Papin.

At a meeting of the Council of the Royal Society on 8 December 1680 Christopher Wren ordered the publication of a booklet 'intituled a New Digester . . .'. Papin, in his dedication to the Royal Society, says modestly that there is nothing in the work 'worth offering to such sagacious Wits'. In the preface, addressed to the general public, however, he claims to be offering a considerable improvement in the ancient art of cookery – 'seeing that by the Engine here treated of, the oldest and hardest Cow-Beef may be made as tender and savoury as young and choice meat'.

The booklet contains a thoroughly explained steel engraving which shows the construction of the 'Engine' (including the safety-valve), and chapters which describe in detail experiments in the cooking of mutton, beef, lamb, rabbit, pigeon, mackerel, pike, eel, beans, green peas, cherries, gooseberries and plums. Papin more than once makes the point that by cooking in this way you can preserve flavour and nutritive qualities that would otherwise be lost. Other chapters describe the value of the invention to the ship's cook, the confectioner, and in making drinks.

Some years later, Papin became the Royal Society's 'temporary curator of experiments'. He died in London about 1712. The principle of the 'steam digester' was more used in industry – as the autoclave – than in cooking until this century, when the modern pressure cooker was developed and became popular first of all in America. Many manufacturers entered the field after World War 2, stressing the savings to be achieved in both time and fuel.

The main sales resistance has always been due to the fear of accidents caused by the high pressure, but manufacturing standards are very strict and leave a margin of safety which even the most careless housewife would find it difficult to cross.

M. C.

Water-closet

In the 1950s an elderly lady staying at a guest-house asked for a chamber-pot in her bedroom. The manageress gave her one but said: 'We don't put them in the rooms now, the staff don't like emptying them.'

The staff never did like it. In 1731 Swift, in his sarcastic *Directions to Ser-*

vants, tells the chambermaid how to treat a mistress who uses the commode in her dressing-room instead of going out to the privy in the garden: 'Convey away the utensil openly, down the great stairs and in the presence of the footmen, and if anybody knocks, open the street door while you have the vessel, filled, in your hands. This, if anything can, will make your lady take the pains of evacuating her person in the proper place.'

Two types of water-closet had been patented by the end of the 18th century – Cummings's in 1775 and Bramah's in 1778. As the inconvenience and nastiness had been for so long a burden on the entire human race, mankind had always made jokes about it: elegant (and now very valuable) chamber-pots with comic pictures and rude remarks were found behind screens in the dining-rooms of mansions. But sanitary plumbing in houses, with its comfort, cleanness and privacy, and the provision of public lavatories out of doors took away the strain and distress and therefore the impulse to make jokes, though these still survive in unsophisticated circles. The improved conditions also produced a reticence and shamefacedness about the mere idea of urination and defecation unknown in previous centuries.

Before the establishment of the water-closet and its drainage system, the contents of commodes and privies were collected early in the morning by the 'night-soil men' who carried it away in carts and tipped it into cesspools. When these were full they were covered over and new ones dug. The problem of disposal became steadily worse with the increase of population in the 19th century and the challenge was met by the tremendous feats of sanitary engineering in the Victorian age.

Though the drainage of London, engineered by Bazalgette, was not in operation until the 1860s, the idea of the water-closet was known as early as 1597. In that year, Queen Elizabeth's godson Sir John Harington published a tract, *Ajax* (a pun on 'jakes', the 16th-century word for privy). In it he described, with diagrams, the working of a valve water-closet, which he had picked up from an Italian. Elizabeth, whose sensitive nose was a torment to her, had the invention tried in her palace at Richmond, where it proved a delightful success.

E.J.

Gas lighting

In the 18th century both coal gas and natural gas were called, indiscriminately, 'inflammable air', because of the clear, bright flame they gave when burned at a nozzle. George Nixon lit a room with coal gas as early as 1760. Carlisle Spedding lit his office with natural gas in 1765, and offered to use it to light the streets of Whitehaven. In 1782, Archibald Cochrane, 9th Earl of Dundonald, collected gas from his coke ovens in a container like a large tea urn and burnt it in Culross Abbey to amuse his friends. From this time onwards similar experiments and demonstrations were made all over Europe. Two of the experimenters, Jan Pieter Minckelers in Louvain and William Murdock in Cornwall, have each been competitively but unjustly awarded the title of 'inventor' of gas lighting.

If the title of inventor of gas lighting can be conferred on anyone, it must be awarded to Philippe Lebon, who began experimenting with gas at the age of 25 at about the same time as Murdock.

Lebon exhibited his Thermolamp in a house in Paris in 1801. Though the demonstration was a technical and aesthetic success, the French government was unwilling to finance Lebon's plans for gas lighting on a grand scale. In 1804 he was stabbed to death in the Champs-Elysées for reasons that have never been explained.

Philippe Lebon's idea caught on, not in France, but in England. An eccentric German promoter who described himself as a 'professor of commerce' and anglicized his name to Frederick Albert Winsor built himself a Thermolamp and brought it to England early in 1802. Meanwhile Boulton and Watt, alarmed by the news of Lebon's Paris demonstration, urged William Murdock to pursue his experiments.

From the beginning, Winsor appreciated that gas must be generated centrally and conveyed to the consumers along a web of mains under the streets. For this, money would be needed on a scale beyond the resources of any single industrialist. So Winsor set about cajoling Parliament into sanctioning a charter to float a limited liability company.

Equally, Boulton and Watt were determined to establish gas lighting within their own financial resources. So their solution had to be a piecemeal one – a gas-generating plant in every factory and every private house. This worked after a fashion in the cotton mills, where it was mainly applied, and in very big houses, but it proved impossible to install a gas plant in the basement of every house.

In 1804 Winsor started to demonstrate gas lighting at the Lyceum Theatre, at the same time mounting an advertising campaign promising a colossal fortune to anyone bold enough to subscribe. He became one of the jokes of London, but soon found himself supported by a powerful and influential committee, and even managed to enlist the interest of the Prince of Wales.

In the face of the bitterest opposition, the National Light and Heat Company was founded in 1812. The first street lighting began in 1814.

At first, methods of gas production were haphazard. Escapes of gas were frequent, and in 1813 there was a

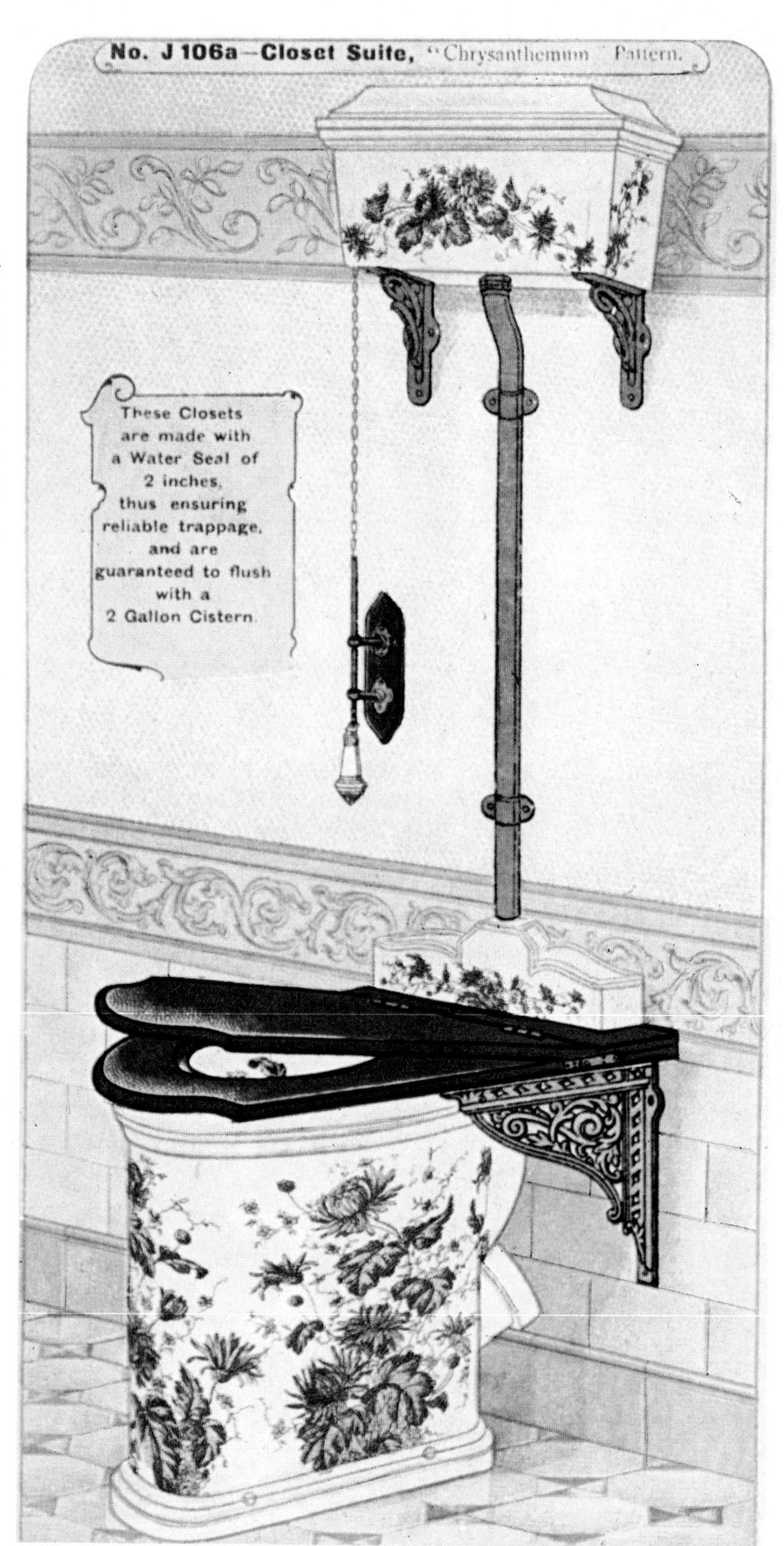

'Chrysanthemum pattern' closet suite, from an 1890 catalogue.

Right, *the Bissell 'Grand Rapids' carpet sweeper, and* (opposite) *an early Ewbank model.*

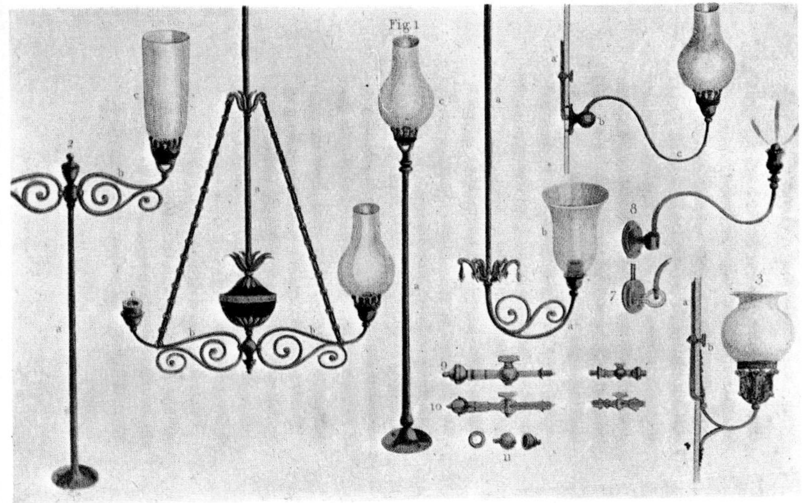
Early 19th-century gas lamps and fittings.

Printed pattern on linoleum, 1895.

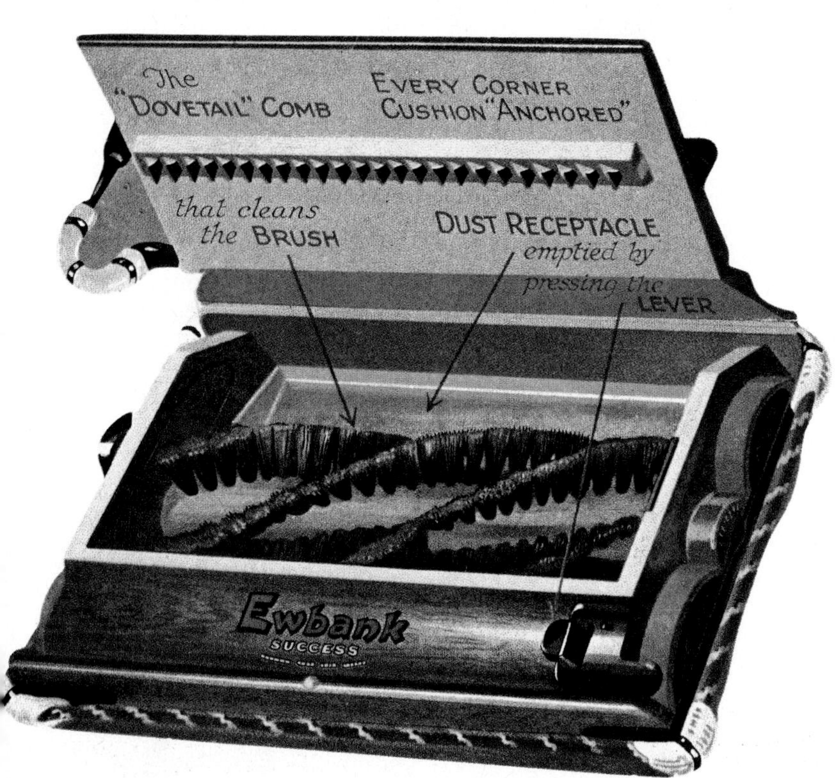

violent explosion in the company's first gasworks in Peter Street. However, when Sir Joseph Banks led a party to Peter Street to investigate the dangers of gas, the chief engineer sent for a pickaxe, drove a hole in the side of a holder and ignited the gas with a candle. He made his point. From that time, nothing could stop the progress of gas. By 1819, 288 miles of mains had been laid in London to supply 51,000 burners.

As for Winsor, to his great mortification a ruthless and unsympathetic management forced him off the board of what by then had become the Gas Light and Coke Company – which went on to serve Londoners for nearly 150 years until nationalization.

A. E.

Linoleum

Today we tend to take floor coverings for granted; even in the poorest home, a carpet is a point of pride. But this state of affairs is very recent and, if we go back to the middle of the 18th century and examine painted portrait groups, we can see how rare the carpet is: except in the most wealthy *milieux* most people walked on bare boards.

Then in 1860 Frederick Walton patented a process for oxidizing linseed oil mixed with resin and cork-dust on to a cotton or flax backing. The flax (*linum*) and the oil (*oleum*) gave the product its name – linoleum. It had forerunners: at Kirkcaldy, Fife, where the first linoleum factory in Scotland was to be set up in 1877, Michael Nairn had been experimenting with a mixture of oil paint and cork fibre as early as 1847; such experiments seemed natural in an area traditionally associated with the production of sail-cloth and canvas. Walton's process was a tediously repetitive one, requiring much space and time: the oil had to be boiled down slowly, and built up on the material in layers, each of which had to dry thoroughly before the next was applied – and this entailed large sheds for the drying process. Only in about 1900 did he develop a method of oxidizing the oil in kettles, mixing it with a filler, calendering it through rollers, and then applying the backing, after which artificial heat would be used to harden the finished linoleum.

Linoleum is in a sense a substitute for rubber, a material which obviously had great possibilities as a floor covering. Elijah Galloway was experimenting with it in the 1850s, heating it and mixing it with cork-dust; the product, called 'kamptulicon', was then glued in place, or affixed to a fabric backing. It was used by Sir Charles Barry for certain less important parts of the Houses of Parliament, but was too expensive ever to be popular, especially when competing against the cheaper and harder-wearing linoleum.

The production of linoleum reached large proportions in the 20th century, Britain alone producing 126 million sq. yds. in 1937; but in the last 20 years other synthetic coverings have begun to take its place.

C. M. B. G.

Carpet-sweeper

Though the first attempts at mechanical sweeping had begun in 1811 (with an English patent issued to James Hume) it was decades before any device proved really practical: dust was still stirred up as if by a broom, and not collected. In 1842, Sir Joseph Whitworth invented a wheeled road-sweeper with a revolving brush, and a carpet-sweeper on the same principle followed in 1859. This was soon succeeded by another in which the wheels activated a fan intended to blow the dust into a box (there was no brushing of the carpet surface), an idea which became really feasible only with the development of the small electric motor.

But none of these captured the housewife's loyalty, and in 1875 the advice still being given in the *Manual of Domestic Economy Suited to Families Spending from £150 to £1500 a Year* was: 'To dust carpets and floors, sprinkle tea-leaves on them, then sweep them carefully. Fine carpets should be gently done, on the knees. Those parts that are most soiled may be scrubbed with a small hand brush when it is not considered necessary to undertake a general washing of the whole; always adding a little gall to the water to preserve the colours.'

Liberation from all this came in 1876 when Grand Rapids (Michigan) gave its name to a sweeper patented by M. R. Bissell. Bissell, who had a china shop, suffered from allergic headaches caused by the dusty straw in which the china was packed. To solve his problem, he invented a sweeper with a box to contain the dust, and – crucial feature – a knob adjusting the brushes to the variations in floor surfaces. Most of the components were made by women working at home, and assembled by Bissell and his wife. Soon the 10*s*. sweepers were in use at Queen Victoria's court, in Arab palaces – and on Scottish putting-greens.

The timing was right. Pasteur's recent germ theories had made the public as hygiene-conscious as Florence Nightingale had already become when, at Scutari, she had insisted on damp-sweeping to reduce airborne dust, saying 'air can be soiled just like water'. The new invention caught on so widely that housewives spoke of 'Bisselling the carpet' rather as we now speak of 'Hoovering' it.

E. G.

Early filament lamp made in America by Swan's great rival, Thomas Edison.

Filament lamp

Joseph Wilson Swan (1824–1914) said his thoughts were first drawn to the subject of electric lighting by a lecture he went to at the Sunderland Athenaeum. In his extensive reading, he found a reference to an apparatus which produced electric light by means of continuous metallic and carbon conductors in a vacuum.

Swan soon saw that there were three important factors to be considered in producing an electric lamp with a fine filament. The first was to provide a suitable source of electricity. The second was the need to find a substance from which to construct the filament, which had to withstand the conditions encountered in manufacture and the high temperatures in use. Thirdly, as the filament would be destroyed if there were any air in the lamp, the existing vacuum pump had to be improved.

In about 1848, he used a selection of papers cut into strips, sometimes saturated with treacle, tar or other liquids, which he packed in a mass of powdered charcoal and baked to a high temperature in a pottery kiln. These experiments lasted over several years and on 18 December 1878 he was able to demonstrate, at a meeting of the Newcastle-on-Tyne Chemical Society, a reliable lamp with a thin carbon filament in a vacuum.

He was not, however, able to get his lamp into quantity production until 1881, and in the meantime the American inventor Thomas Alva Edison had been working on the same problem and along much the same lines. Edison is usually held to be the inventor of the filament lamp but, strictly speaking, priority must be given to Swan. Edison's carbon filament lamp – using a scorched cotton thread as the filament – was lit on 21 October 1879 and burned for 40 hours continuously. With his usual flair for publicity, Edison triumphantly lit the main street of his home town, Menlo Park, New Jersey, with the new incandescent lamps, took out a patent and began to produce it in quantity. Eventually the rival companies amalgamated as the Edison and Swan United Electric Light Co. Ltd.

P. D.

Gas mantle

An Austro-Hungarian aristocrat, Freiherr Carl Auer von Welsbach, invented the incandescent gas mantle, that fragile object which converted the blue flame of burning coal gas into a softly romantic yellow glow.

The reign of the mantle was brief. It was a late Victorian invention – Auer von Welsbach took out his patent in 1885 – and the mantle was not widely used until the next decade. By that time electric light was clearly on the way – and in 1900 Auer von Welsbach himself made a substantial contribution to the success of the electric light bulb with his Osmium Glühlampe, which has come down to us as the 'Osram' trade mark.

Gas lighting was mostly confined to outdoor use until the 1860s. Faulty connections in pipes, with the risk of explosion, were too common. This was remedied, and the search for a better way of illuminating by gas began.

One way was to add to the candle-power of the gas by enriching it with a gas containing more carbon particles, which would let out more light when burning in the flame. Benzol vapour was added after the retort stage, and oil gas was sometimes employed.

But chemists knew that if the gas could be made hot enough, certain metals and rare earths would glow incandescently in the flame. The first breakthrough came with Professor Robert Wilhelm von Bunsen (1811–99). The famous burner attributed to him works on the principle of mixing air with coal gas before combustion, and the resulting mixture produces more heat than the gas burning alone.

Auer von Welsbach was working under Bunsen, and his special field was the rare earths. These glowed brilliantly in the hot gas flame. He had the idea of soaking fabric in a solution of metal salts, and then burning away the fabric, leaving a shell of incandescent material. The first mantle was exceedingly fragile, but a combination of 99 per cent thorium and 1 per cent ceria proved stronger.

In commercial production, a cylinder of cotton net was soaked in the thorium-ceria solution until the cotton took up no more. It was squeezed and dried. The little cylinder was sewn at the top with asbestos thread, and then the cotton was burnt away. During the burning process, the nitrate salt of thorium became an oxide, and an enormous expansion of the material took place: the mantle filled with sponge-like holes, helping the conduction of heat. For additional strength the mantle was soaked in collodion.

G. N.

Auer von Welsbach, who had the idea of soaking fabric in a solution of salts to make a gas mantle.

The 'Baby Daisy' – one of the most successful of the early vacuum cleaners.

Vacuum cleaner

Blowing or sucking as a method of extracting dust from carpets had inspired a number of inventors before Hubert Booth made the idea really work in 1901.

Pre-electric cleaners had used bellows as a method of sucking. Some were operated by hand, some by foot, and one by the wheels of the sweeper. This last contained water to filter the dirt out, unlike others which conspicuously failed to retain all the dust they had sucked up. The filtering problem had defeated most inventors.

The invention of the vacuum cleaner was soon followed by the invention of the mobile office cleaner, operating from the street and extracting the dust through hose-lines.

Early American safety razor. The American Safety Razor Company was formed in 1901 (above, right, *Gillette's patent of December 1901*).

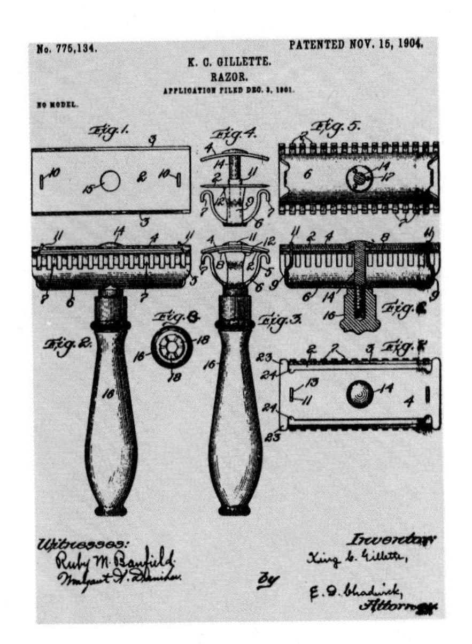

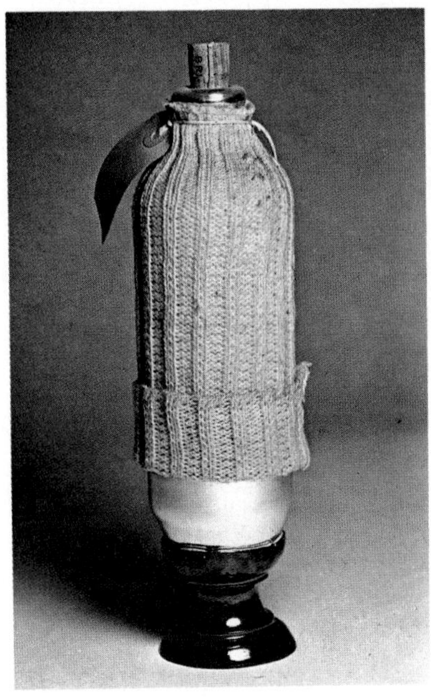

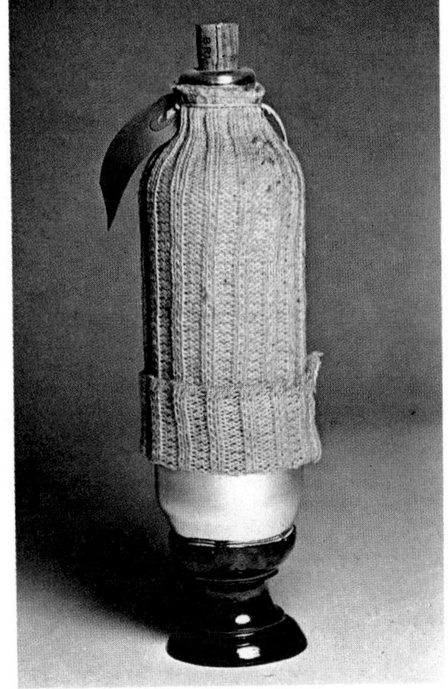

Vacuum flask made by Sir James Dewar in 1902 for his baby son. His mother-in-law, doubting its effectiveness, knitted a woolly cosy.

Many of these cumbersome devices required the services of at least two housemaids to operate them: one to crank or pedal the bellows, the other to apply the nozzle to the floor. Their main virtue seems to have been durability: some are still in use in conservative homes.

The coming of the small electric motor paved the way for advance.

Hubert Booth was mainly a bridge-builder and designer of such big wheels as the one at Blackpool and the one at Vienna that was featured in *The Third Man*. It was by chance that he solved the problem of filtering dust from air and thus became the inventor of the first viable vacuum cleaner. He had attended a demonstration of a new cleaner at St Pancras Station in London: it was designed for use in railway carriages and was based on the principle of blowing dust out. When he suggested that sucking might be more effective, he was told it had been tried but didn't work. Back home, Booth lay on the floor and, with a handkerchief over his mouth, sucked hard. The dirt thus trapped in his handkerchief gave him the clue which led to cloth filters.

His first version was a large horse-drawn contraption which, at spring-cleaning time, stood in the road and sucked the dirt from a house through an 800 ft. hose. He had trouble with the police because this noisy serpent caused cab-horses to bolt.

Basically, all cylinder vacuum cleaners since have relied upon Booth's principle.

E. G.

Safety razor

'Why don't you invent something which will be used once and thrown away? Then the customer will come back for more.' This advice, from his boss William Painter, inventor of a disposable bottle-cap, haunted the 40-year-old salesman, King Camp Gillette, from Fond du Lac, Wisconsin. In 1895 the vision splendid came as Gillette stood before a mirror honing his cut-throat razor. The only part of it which did any work, and not much at that, was the edge. Why forge and temper a hunk of steel just as a backing to an edge? Why not market a sliver of steel to be used in a clamp and thrown away? 'I stood before that mirror in a trance of joy,' he wrote to his wife. 'I have got it. Our future is made.'

The idea of promoting shaving safety, of staunching a nation's blood, was far from his mind. Others had thought of that. In 1771 Jean-Jacques Perret, cutler-author of *Pogonotomy, or The Art of Learning to Shave Oneself* (with a special appendix on bleeding) had invented a plane-type razor designed so that only the edge could scrape the skin. Sheffield offered a razor with a guard along the edge in 1828. The Duke of Wellington, a great student of advertisements, sent for a safety razor but refused to endorse it. In 1880 Oliver Wendell Holmes gave a testimonial to a hoe-shaped razor with which it was 'impossible to cut oneself'. Obstinately, however, men clung to their open razors. Gillette, the fame-hungry habit-changer, struggled for six years to market his dream. His trance of joy evaporated when experts said no rolling mills could produce the steel thin enough, or hard enough, or flat enough, or sharp enough, or cheap enough. His friends jeered, but he said, 'I believe in the gold at the foot of the rainbow.'

In 1901 William Nickerson, an inventive mechanic who had devised a push-button for lifts, was induced to tackle the problem on a half-time basis and the American Safety Razor Company was formed. By 1902 Nickerson had overcome most of the technical difficulties. In 1903 sorely tried backers groaned at the year's sales figures: 51 razors and 168 blades. But in 1904 the amazing shaving revolution was under way: 90,000 razors and 12,400,000 blades sold. Gillette had to resign himself to buyers using the blades more than once. Other inventors devised ways of stropping and preserving what was meant to be thrown away; then, in 1930, Gillette merged with the firm AutoStrop.

The next shaving revolution was led by a retired American Army officer, Lt-Col. Jacob Schick, who patented his electric razor (with reciprocating blades under a shearing head) in 1928 and marketed it in 1931. Gillette died in 1932, a year when lumbermen in Oregon were reported to be giving each other perfect shaves with honed axes.

E. S. T.

Vacuum flask

A vacuum flask is a glass container with double walls completely sealed at the top, with the air between these walls pumped out to reduce heat-transference. By silvering the inner walls, the transfer of heat by radiation is diminished.

Although the vacuum theory had been established in 1643 by Torricelli, the Italian inventor of the mercury barometer, it was only in 1892 that a vacuum flask was made (it is still preserved at the Royal Institution in London), purely for laboratory purposes. Named the 'Dewar vessel' after its inventor, this type of flask is still used to contain liquid gases used in low-temperature research.

In 1904, a German, Reinhold Burger, spotted its potential for domestic use, and offered a prize for the best name to give this new convenience. 'Thermos', Greek for hot, was the winning entry; it is now a protected brand name. At first it took a skilled craftsman to make only about eight flasks a day.

Thermos flasks went with Shackleton to the Antarctic and with Hillary up Everest – the latter used them to keep scientific samples at an even temperature. Vaccines, serums and even tropical fish are often transported in them.

E. G.

Detergents

Strictly speaking, a detergent (from the Latin *tergere,* to wipe) is anything which cleans, and that includes soap. For many cleansing processes, ordinary soap – made from seed oils or animal fats and caustic alkalis – is still the best agent, but synthetic, soapless detergents can be developed 'to measure', i.e. for special cleansing jobs. Although their recent rise in popularity has run parallel with that of man-made fibres, which require special cleansing treatment, the story of detergents goes back to the end of the last century.

Around 1890, a German research chemist, Krafft, discovered the soap-like properties of some non-soapy substances, and an American inventor, Twitchell, continued this line of investigation by experimenting with fat-splitting catalysts. A Belgian chemist, Reychler, found in 1913 that the long-chain alkane sulphonates would make good detergents and were, in fact, more stable than soap under acid conditions. Just before the outbreak of the First World War, two British scientists, Martin and McBain, investigated the washing qualities of these alkane sulphonates and of other synthetic substances with long hydrocarbon molecular chains; they found that such detergents would be free of some deficiencies of ordinary soap, but too expensive for commercial production.

In 1917, when Germany was extremely short of natural fats owing to the Allied blockade, two chemists of the Badische Anilin- und Soda-Fabrik, Gunther and Hetzer, invented the first commercial detergent, which they called 'Nekal', as an *ersatz* product for soap. It was not very satisfactory, but IG Farben, the German chemical concern, marketed it after the war as a wetting agent for the textile industry. Further research and development was carried out in the late 1920s by various groups of German scientists; IG Farben marketed their 'Igepon' detergent, invented by Platz and Daimler, and supplied it also to Unilever for certain industrial uses. The next step towards a general-purpose detergent was the discovery by a Swiss chemist, C.A. Agthe, that the addition of complex phosphates greatly increased the cleansing powers of synthetic detergents. Other research teams found in the 1930s that detergents made from petroleum-based raw materials were cheap to produce and had most satisfactory properties; as a result, the oil companies played from then on a leading part in the development of detergents.

Perhaps the decisive factor which appealed to the general public was the addition of fluorescent brightening agents, the 'whiter-than-white' factor. A German analytical chemist, Krais, had the idea in 1929: minute quantities of fluorescent substances – which convert the ultra-violet rays of the sun into visible bluish rays – make the fabrics *look* brighter, although they do not make them any cleaner. IG Farben took out the first patents for this innovation during the Second World War, and after 1945 detergents of various types grew rapidly popular, helped by enormous publicity, in the industrialized countries.

We do not yet understand fully the chemical action of detergents, but the main point is that their long molecules contain two groups of atoms: one which dissolves in oil but not in water, and one which dissolves in water but not in oil. The detergent, by lowering the surface tension of the water, wets and penetrates the soiled fabric, coats its surface and the dirt particles and thus gives them a similar electric charge so that they no longer attract each other. Lather provides 'mechanical' assistance in the removal of dirt from the fabric, and the dirt particles are kept in a fine suspension during the washing process so that they do not settle again on the fabric; oil particles are emulsified.

One major advantage as compared to soap is that detergents overcome the handicap of hard water by dispersing the scum. As every housewife knows, there are several distinct types of detergents – powdered ones for washing wool, silk, and man-made fabrics; powdered and liquid ones for washing and dish-washing machines; heavy-duty liquid ones for general purposes. Most of them also kill germs. Detergents have by no means ousted soap from the field of personal hygiene, but in industrialized countries consumption of detergents is now three times as high as soap consumption.

E. L.

Washing machine

For centuries washing was done by putting clothes in a tub and agitating them with a 'dolly' – a device which looked like a milkmaid's stool turned upside down on the end of a stick. At sea, clothes had been towed behind the ship, the principle being to get the water *through* them. But around 1860 the idea of putting the clothes into a wooden box and tumbling them over by handle was beginning to catch on. By the 1880s the famous firm of Thomas

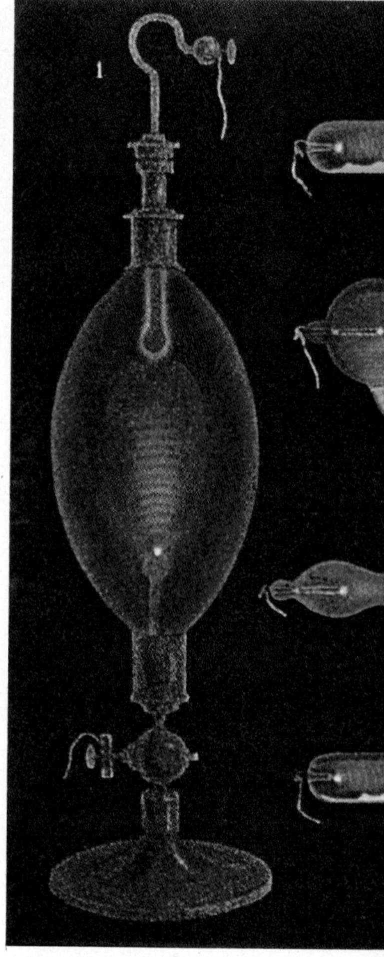

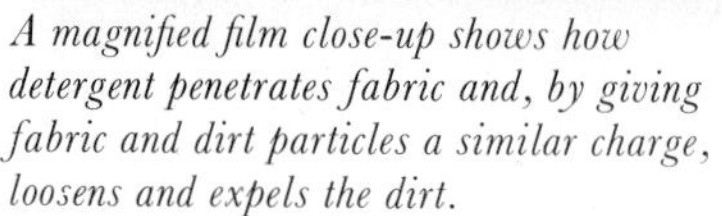

A magnified film close-up shows how detergent penetrates fabric and, by giving fabric and dirt particles a similar charge, loosens and expels the dirt.

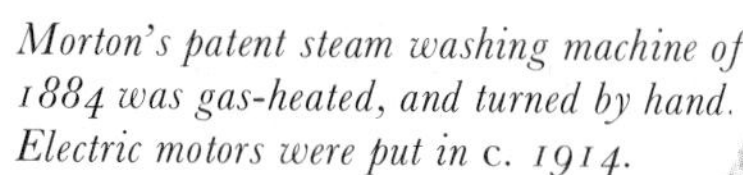

Morton's patent steam washing machine of 1884 was gas-heated, and turned by hand. Electric motors were put in c. *1914.*

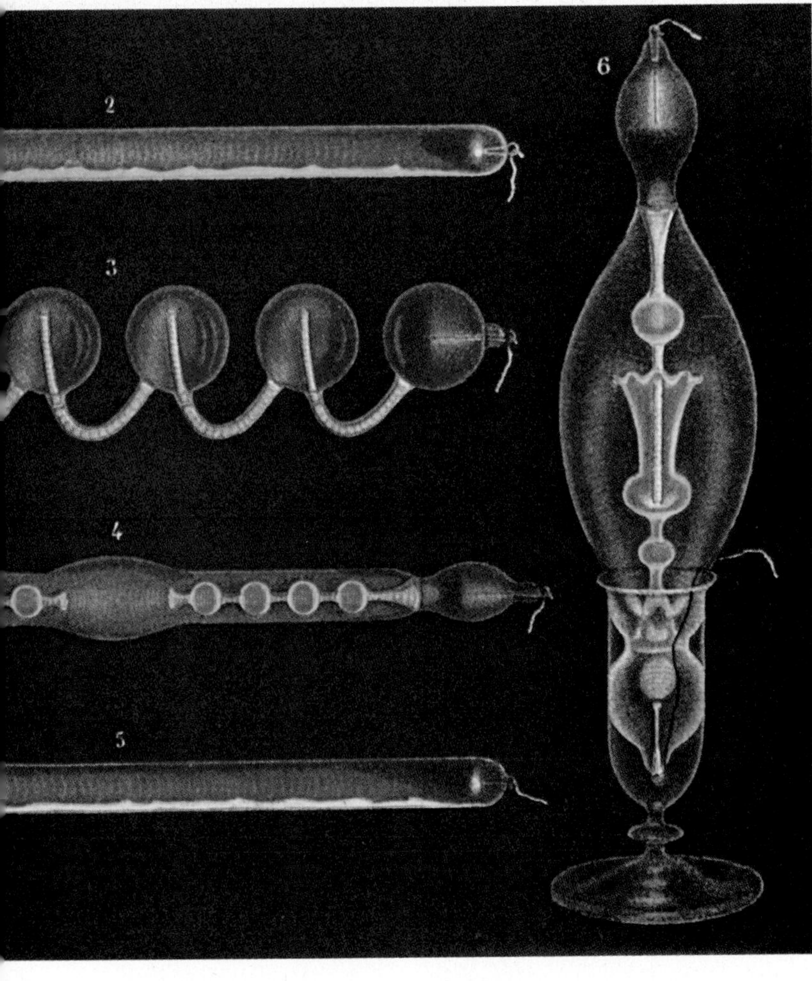

Above, *Geissler tubes, showing electrical discharge in a rarefied gas.*

Below, *diagram of the wiring of an electric blanket, 1904.*

Bradford was making home washing machines consisting of an ordinary mangle with a wooden octagonal box underneath, in which the clothes were placed. A handle rotated the box and this system of turning clothes by hand persisted until about 1914, when electric motors were introduced.

At first, these motors were fitted under the tub and were of the open variety; water often dripped into them and gave paralysing shocks to the operators.

In the 1920s, mechanical tubs were introduced and washing as we know it today was born. However, as early as 1677 the scientist Robert Hooke had recorded an early washing-machine invention: 'Sir John Hoskins way of Rinsing fine linnen in a whip cord bag, fastened at one end and strained by a wheel & cylinder at the other. NB whereby the finest linnen is washt wrung and not hurt.'

Another housewife's friend, the spin drier, was made, albeit primitively, by Bradfords in the 1880s. But the spin drier as we know it today was introduced in America in 1924, forgotten until the 1960s, and then seized upon by the automated age.

Advertising copy in the washing-machine industry has not changed much. An early example, written about Morton's patent steam washing machine, dates back to 1884: 'The Machine is so easy to work that a child can wash six sheets in 15 minutes and the clothes are washed whiter by this Machine than by any other, and they will also *last twice as long*.'

G. B. L. W.

Fluorescent lamp

After nearly 60 years of undisputed rule, the incandescent lamp, developed and first marketed by Edison and Swan in 1879, found its great rival in the strip light or fluorescent lamp.

The first attempt at producing fluorescence was made by the French scientist Henri Becquerel, who discovered radioactivity (but left its investigation to his pupil, Marie Curie); in 1859 he built a kind of fluorescent lamp with a Geissler tube (used in laboratories to demonstrate the glow caused by an electrical discharge in a rarefied gas). He coated the inside of the tube with a chemical (a phosphor) which had the property of fluorescing under the influence of the glow. Other scientists of many countries began to work on the same lines and soon found hundreds of minerals which fluoresced when exposed to radiation of the right wavelength. One by-product of this activity was the neon lamp (invented by the French chemist Georges Claude in 1910), which shines red when an electric charge is sent through it, and which became extremely popular for advertising purposes in the 1920s.

The production of ordinary light for homes, factories and offices by fluorescence was eventually found to be possible by coating the inside of a glass tube with some chemical fluorescent substance. An American, Dr Arthur H. Compton of General Electric, developed the first effective lamp of this kind, with the right colour for general use, around 1934. Operating with low-voltage current it was found to be much more economical than incandescent light, which needs high voltages.

The modern fluorescent lamp, with light in five colours to choose from, was introduced at the New York World's Fair in 1939, and 15 years later the strip light overtook the incandescent lamp as the chief source of electric lighting in the USA. Its usual form is that of a long, gas-filled tube, coated inside with fluorescent powder; this lights up when excited by the invisible ultra-violet rays of an arc passing between the two electrodes at either end of the tube. The reason why this lamp needs only a low-voltage current is that its light is 'cold': that is, very little energy is turned into waste heat as in incandescent lamps. A modern strip light is roughly 50 times more effective than the first carbon-filament lamps, and it is also more adaptable than most incandescent types of bulbs because by varying the composition of the fluorescent coating one can produce light in many shades and colours (there are, for example, at least half a dozen 'whites', from 'daylight' to 'super de luxe cool white').

Related, yet very different in their effect, are the 'gas-discharge' lamps now widely used in street lighting; they can give only one kind of colour, either yellow (sodium-vapour lamp) or greenish-blue (mercury-vapour lamp). Developed from the neon light in the early 1930s, they use metallic vapour at low pressure as a 'conductor': the electric current raises the vapour to incandescence. The light of these 'monochrome' lamps is anything but flattering to human faces, but it makes the roads safer at night: it is free from glare, it makes it easier for the motorist to distinguish objects on the road, and it is less scattered by fog or mist.

E. L.

Electric blanket

An American, S. I. Russell, invented the electric heating pad in 1912 as a medical aid for tubercular patients sleeping outdoors: it was a small square of fabric with electrically heated tapes running through it. Each was made to order to suit the patient's requirements and cost \$150. Some customers started stitching four or more squares together and using the result as a bed-warmer, but industry only grasped the idea of heating the bed rather than the body in

1937, when electric overblankets first appeared in America. Parallel with this development the British began producing the electric underblanket, which monopolized the British market at least until the 1960s.

The British involvements in fact dated from 1926, when the Ex-Services Mental Welfare Society, which had been seeking a product which could be manufactured by its patient-members in an industrial colony, was given the patent for an electrically heated pad by a sister organization in Germany. This led to the Thermega underblanket, which was rapidly copied by several commercial companies. Most subsequent refinements of the idea have simply been in the direction of greater safety: one, the use of vinyl plastic round the elements, was developed from wartime research into electrically heated flying suits. In 1970, however, Thermega pioneered a completely new principle, by which the heat is conveyed through the blanket by hot water in tubes, rendering the use of an electric element quite unnecessary.

E. G.

Microwave cooking

Microwave cooking can fairly be described as the first absolutely new method of cooking since prehistoric man invented the making of fire. The claim is justified by the fact that there is no application of fire to the food, direct or indirect. The food in the microwave cooker is bombarded with electromagnetic waves only 12 cm. in length and with a frequency of 2450 Mc/s. This creates molecular activity in the food and therefore heat, thereby cooking it.

The electronic tube that produces the microwave energy – magnetron – was the brainchild, in 1940, of Sir John Randall and Dr H.A.H. Boot. The development was carried out at Birmingham University. However, the thoughts of the team were not on gastronomy, but on how to cook the Nazi goose. The 'cavity magnetron' was needed by Britain's radar defences in the Second World War.

The first microwave cooker was produced in the late 1940s by the American firm of Raytheon Inc.; since then its use has proliferated, especially in America and Europe. It is more usually used to reheat food, for unless an oven that makes use of microwave energy and heated recirculating air together is employed, the appearance of food cooked from the raw state by microwaves is not the same as that cooked by conventional methods.

An oven has been designed and patented by a British firm to overcome this problem, but it has not met with an enthusiastic reception.

M. C.

Microwave oven, using re-circulating air and microwave energy together.

Non-stick pans

The non-stickiness of non-stick pans depends on the plastic PTFE (polytetrafluorethylene) discovered almost accidentally in 1938 by Dr Roy Plunkett of the Du Pont company while working on refrigerants. PTFE has some remarkable properties. It is a fine electrical insulator, it is unaffected by heat or cold over an enormous temperature range, and it is very slippery.

There are innumerable industrial applications for PTFE, and non-stick pans account for only a small part of the output. It is marketed by Du Ponts as 'Teflon', by ICI as 'Fluon', and there is a German variant called 'Hostaflon'. The idea of using it on cooking utensils appears to have occurred independently to a number of people, all at about the same time in the mid-1950s. Mark Grégoire, now living in retirement in Paris, founded the Tefal Company in 1955. A keen fisherman as well as an engineer, he used PTFE to prevent his line sticking. Mme Grégoire thought it was unfair and there was nothing to prevent things sticking to her cooking pans, and he agreed.

In 1956 the same idea was put into the head of the late Philip Harben by a viewer who saw an egg stick to his frying-pan in one of his television shows. The Harbenware brand was a pioneer in Britain and is still one of the leaders. Early production of non-stick hollowware was everywhere tentative and experimental, and Du Pont date the start of large-scale production in the United States as late as 1962. By now three out of four American housewives are reckoned to possess at least one non-stick pan and sales in the rest of the world are rising rapidly.

M. C.

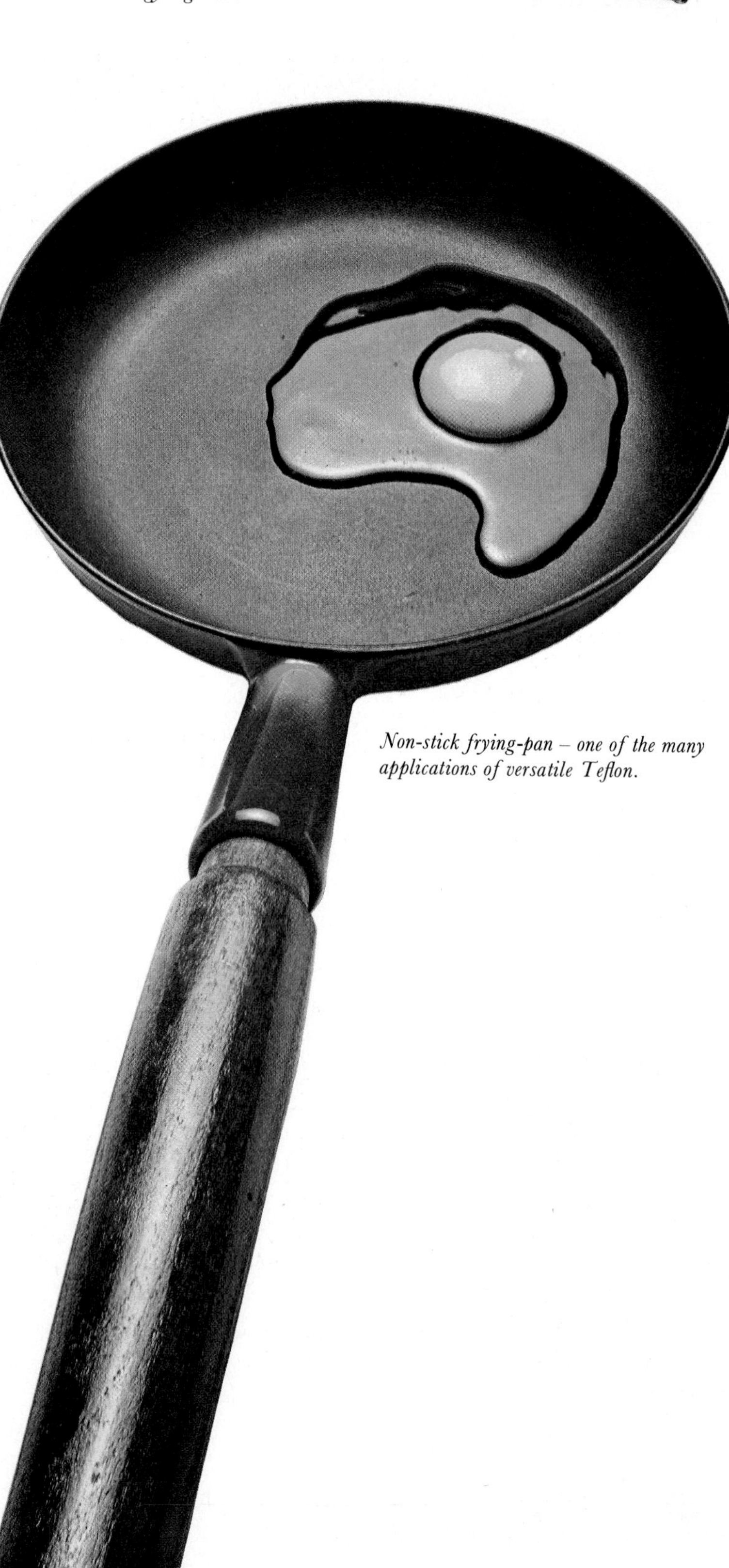

Non-stick frying-pan – one of the many applications of versatile Teflon.

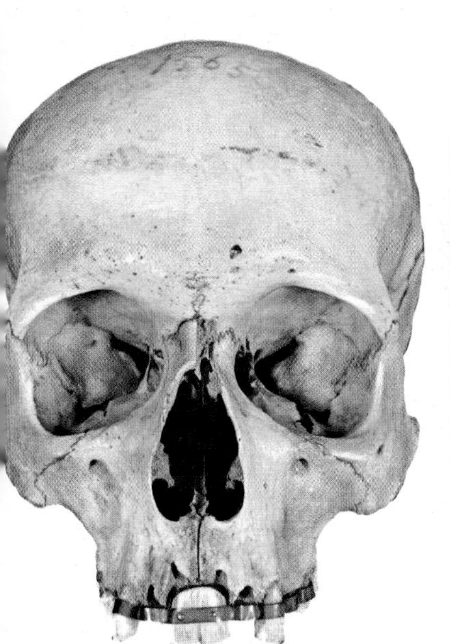

Partial denture with gold bridgework: Etruscan, 700 BC.

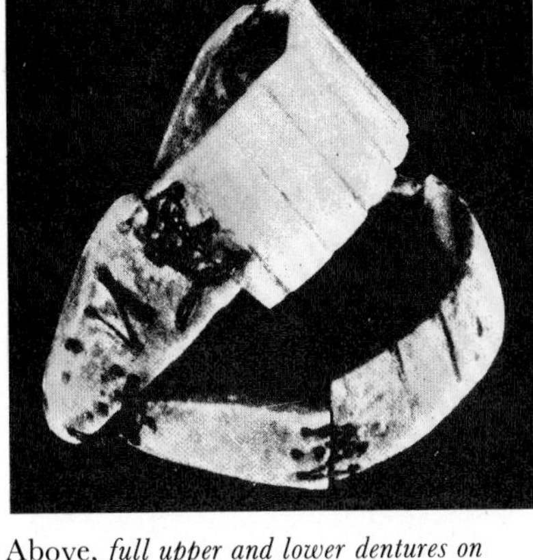

Above, *full upper and lower dentures on spring: Swiss*, c. *1500.*

Below, *carved ivory teeth: early 19th century.*

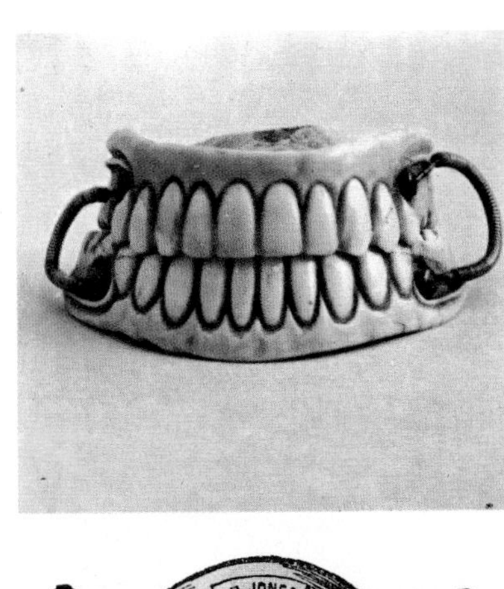

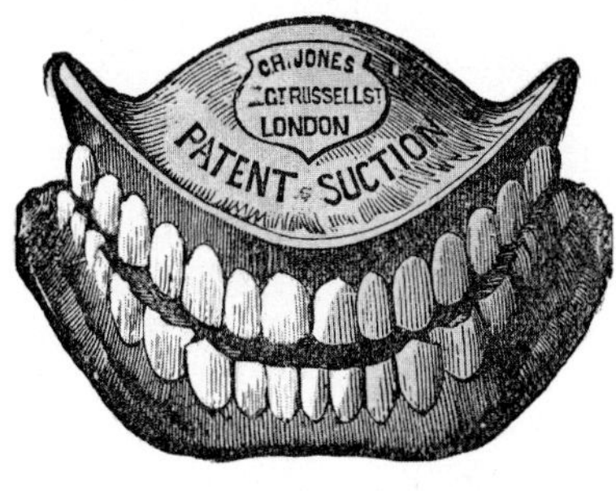

Advertisement, from the 1880s, for an 'improved' form of false teeth.

Left, *Chinese 18th-century acupuncture chart.*

Acupuncture

When modern practitioners are confronted with an ancient technique or philosophy for which they can find no precise physical explanation, they tend to look upon it as a joke. Should the joke actually work, bewilderment and confusion is the result. Sometimes there is repression, as well.

In 1972, with the dramatic thaw in relations between China and the United States, American journalists and other professionals were given their first opportunity to visit the Chinese mainland in over 20 years. Among the stories they brought back with them were descriptions of complicated surgical operations being performed on conscious patients without the use of anaesthetics. Instead, one or more pins were merely inserted at various points of the body, a technique also used to relieve internal congestion in the treatment of such disorders as arthritis, headaches, convulsions, lethargy and colic. Thus was 20th-century America informed of acupuncture, which had been developed in China around 2700 BC.

One of the first official reactions was to ban the practice until further studies could be made. Hundreds of Chinese doctors resident in the Chinatowns of cities like New York and San Francisco, who had been quietly treating their communities with acupuncture for years, became outlaws. True, acupuncture had been partially accepted in some European countries, particularly in France and Germany, after the 1930s; but the American doctors simply could not believe their ears nor, later, even their eyes.

The techniques of acupuncture were first recorded 2,300 years ago in a brilliant treatise of medical philosophy entitled *The Yellow Emperor's Classic of Internal Medicine*, which fully conforms to traditional Taoist thinking. According to this work, the body has some 12 vertical channels or meridians, along which are 365 points where the insertion of a needle will have a physiological effect. (Over the years, Chinese physicians have increased the number of points to 800.) Through these meridians flow the forces of *yin* and *yang*, the negative and positive aspects of our universe. These forces, which are stored in three regions of the abdomen, must be perfectly balanced to maintain one's health. Specific organs of the body correspond to each of these storage sites, and the puncture of a particular organ will remove the bodily obstruction that is disturbing the balance in one region. It does this by permitting harmful secretions to escape. Such needling also relieves tensions and can deaden from pain entire areas of the body, allowing surgery with the patient wide awake. The needles used may be either hot or cold (*yang* or *yin*), between 1 in. and 10 in. in length, and made of silver, gold, iron or steel.

Such explanations, however, could not satisfy Western physicians, who wanted to know exactly how a needle inserted in the toe could – and did – stop pain during abdominal surgery. In 1965, the Canadian Ronald McGill and Patrick Wall of England formulated the 'gate theory' of acupuncture; but like all Western attempts at describing just what occurs in the body during acupuncture treatment, the theory was not all-inclusive. Taoism is simply not amenable to the traditions of Western rational thinking.

E.W.

False teeth

The Etruscans, the best dentists of the ancient world, made partial dentures, with gold bridge work, as early as 700 BC. The teeth themselves were carved from bone or ivory, or came from other human mouths. Medieval dentistry, based on the theory that worms in the gums caused decay and toothache, rarely attempted to provide any kind of denture. Queen Elizabeth I, concerned that her face had sunk inwards from the loss of her front teeth, appeared in public with her mouth stuffed with fine cloth. But by the end of the 17th century the rich were able to obtain dentures. Impressions were impossible, so the mouth was measured with compasses. Teeth were tied to their natural neighbours with silk ligatures, and full lower sets were carved by hand (great difficulties were experienced in keeping any upper sets in place). Human teeth, pulled from the poor for a fee, were set in ivory gums, but sometimes, at court, the teeth were ornamental – in silver, mother of pearl, or, as with Lord Hervey in 1735, of Italian agate.

All these full sets of teeth came out when their owners wanted to eat. A

Parisian dentist, Fauchard, however, made some significant advances in the early 18th century. Horrified by fashionable women who would even have their gums pierced with hooks to keep dentures in place, he fastened lower and upper sets together with steel springs. At last it was possible to keep the upper teeth up, although constant pressure was needed to keep the mouth shut.

Although transplants were popular for a time in the 18th century – the teeth of a 'donor' were pulled out and rammed into a prepared socket in the recipient's jaw – false teeth steadily advanced.

One problem to be faced was that teeth made of bone or any other organic substance would be decayed by oral fluids. George Washington had terrible trouble with his teeth, as a glance at his distorted mouth (as seen on a dollar bill) will show. He was always looking for a good set of false teeth. Ivory dentures smelled and tasted unpleasant after a while, and Washington tried to make them more agreeable by soaking them overnight in port.

Just before the French Revolution, a Paris dentist introduced all-porcelain teeth, baked in one piece. From about 1845, a much improved type of single porcelain tooth was available, to be set individually in plates. This originated in the United States, as did most of the important dental innovations in the 19th century. The new tooth was the work of Claudius Ash, who disliked handling dead men's teeth.

For 'Waterloo teeth', pulled from the dead after the battle, sat in the mouths of the fashionable but ageing Regency bucks. Teeth from the dead of the American Civil War were shipped to England by the barrel. But to the Ash teeth now was added another American invention – vulcanite. This was a composition of sulphur-hardened rubber, invented by Charles Goodyear, which was cheap and easy to work. From an impression of the mouth, the teeth were built up on vulcanite facsimiles of the gums. As the fit became better, top sets stayed up by themselves, and 19th-century devices such as the suction pad and the coil spring went out. Dentures were also made of celluloid, which on one recorded occasion resulted in the teeth of a London clubland smoker catching fire.

G. N.

Spectacles

Transparent objects were used to help defective eyesight in antiquity: the best-known case is that of the Emperor Nero, who watched performances in the arena holding in front of one eye a jewel with curved facets; perhaps concave facets to correct shortsightedness. But such reports are rare and without scientific foundation.

Lenses placed close to the eyes were first certainly used towards the end of the 13th century. Nothing is known of the inventor, or of the date or place of the discovery, as it was the work of an illiterate artisan, but Edward Rosen fixes the period of the discovery to the five years after 1280. The inventor was probably a glazier, who made ornaments and glass discs for windows.

Lenses were given their name because of their resemblance in shape to lentils (the Italian for lens is *lente*, for lentils *lenticchie*) and for more than three centuries they were called glass lentils.

All lenses at this time were of the converging type. Despite the scientists' condemnation, artisans continued to make them as a remedy for long sight. They found that the older people became, the more accentuated the curve of their lenses needed to be.

While there are no exact records of the change brought about by the invention of spectacles, it must have played its part in revolutionizing man's attitude to the world. In its simplest aspect, it extended the working life of scholars and copyists; and in a more complex way it helped to demonstrate that man could be partially independent both of his physical limitations and of natural periods such as the length of daylight.

V. R.

False limbs

'One can imagine the patient more highly instrumented . . . than a Cape Kennedy rocket on the launching pad – but such engineering feats would be a clinical failure.'

This recent comment from Alex Godden, working on artificial limbs in the Engineering Department at Oxford University, epitomizes a key problem faced by doctors and engineers who design mechanical prostheses to replace missing limbs. How can one manufacture a limb that is both versatile and sensitive, *and* relatively uncumbersome and simple to use?

The first person to tackle this task in a concerted way, to establish the feasibility of artificial limbs and make them respectable in medical circles, was Ambroise Paré. He was born at Bourg Hersent, Mayenne, France, in 1509, and died in Paris in 1590. Though he came from a humble family (his father was a valet and baker), he became one of the principal founders of modern surgery.

Paré began his medical career as apprentice to a barber-surgeon, but in 1536 he started work as a military surgeon and it was in this field that he made many of his most important contributions. He established his reputation in 1545, when he reported his discovery that gunshot wounds contained no mysterious 'poison', as had

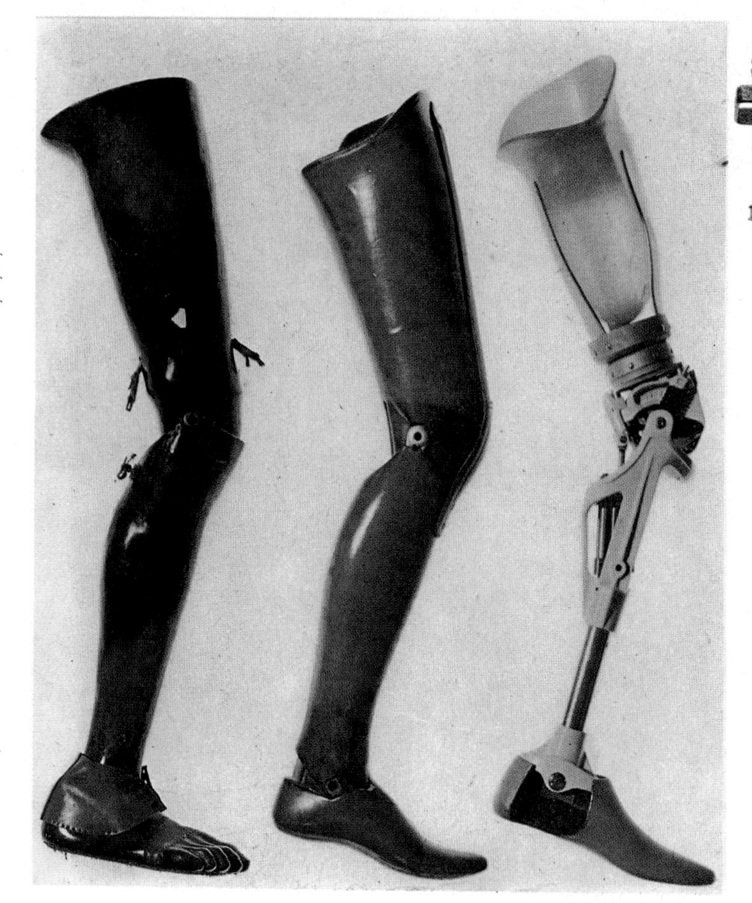

Above, left to right, *Belgian limb, First World War; wooden leg for above-the-knee amputees,* c. *1930; internal mechanisms of the latest design.*

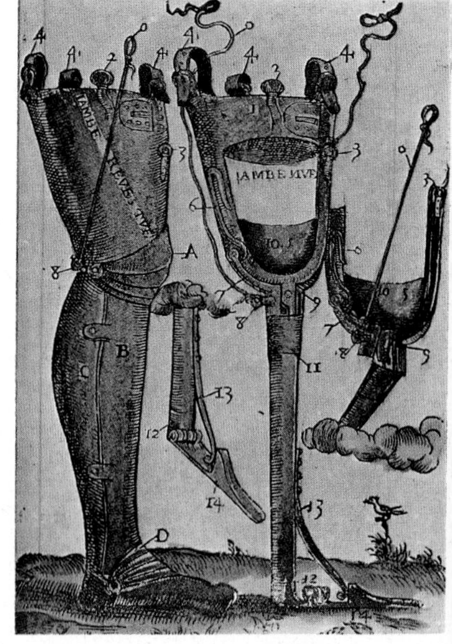

Right, *artificial limbs designed by Paré in the 16th century.*

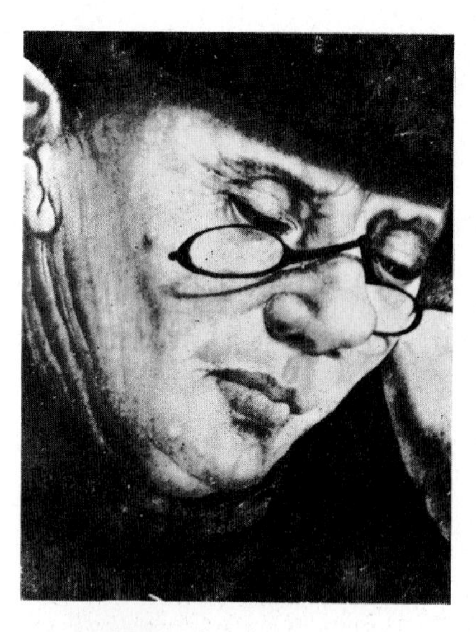

Earliest spectacles were held in the hand (left, below, *1493*)*; later, they rested on the nose* (left, above, *detail from* The Moneylenders *by Massys*).

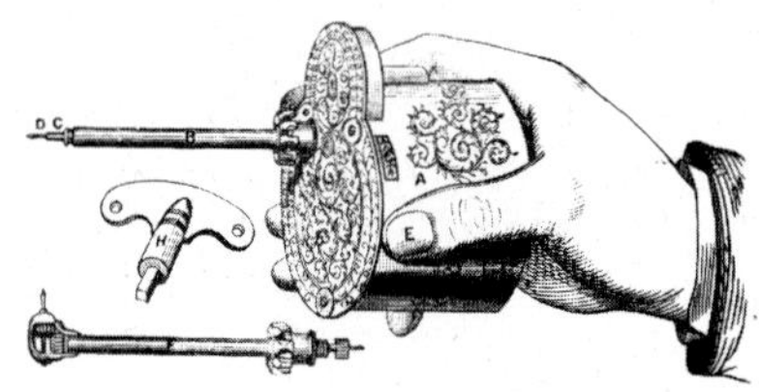

Above, *Harrington's clockwork dental drill.*

Right, *dental surgery of* c. *1890, with foot-operated drill.*

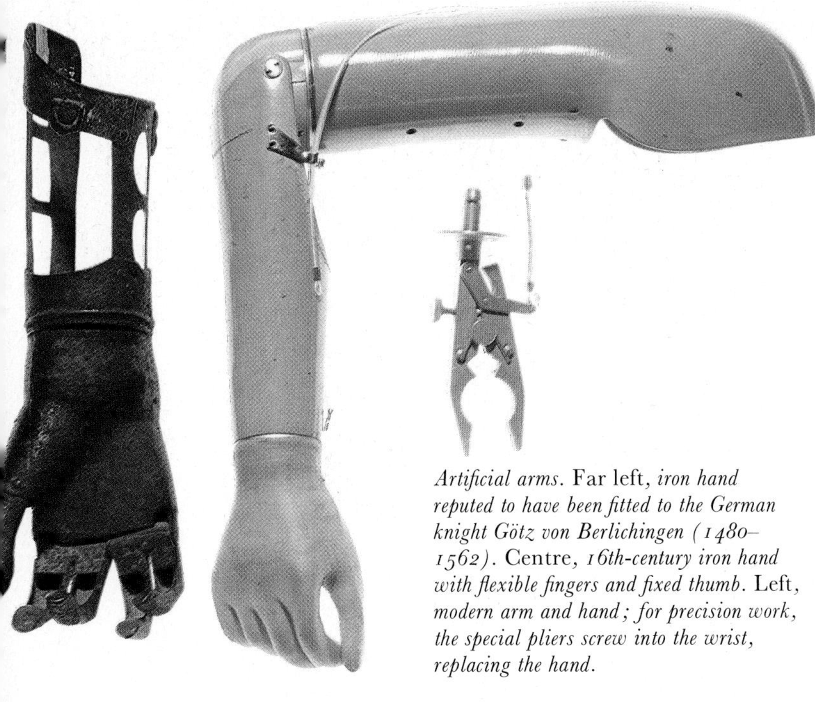

Artificial arms. Far left, *iron hand reputed to have been fitted to the German knight Götz von Berlichingen (1480–1562).* Centre, *16th-century iron hand with flexible fingers and fixed thumb.* Left, *modern arm and hand; for precision work, the special pliers screw into the wrist, replacing the hand.*

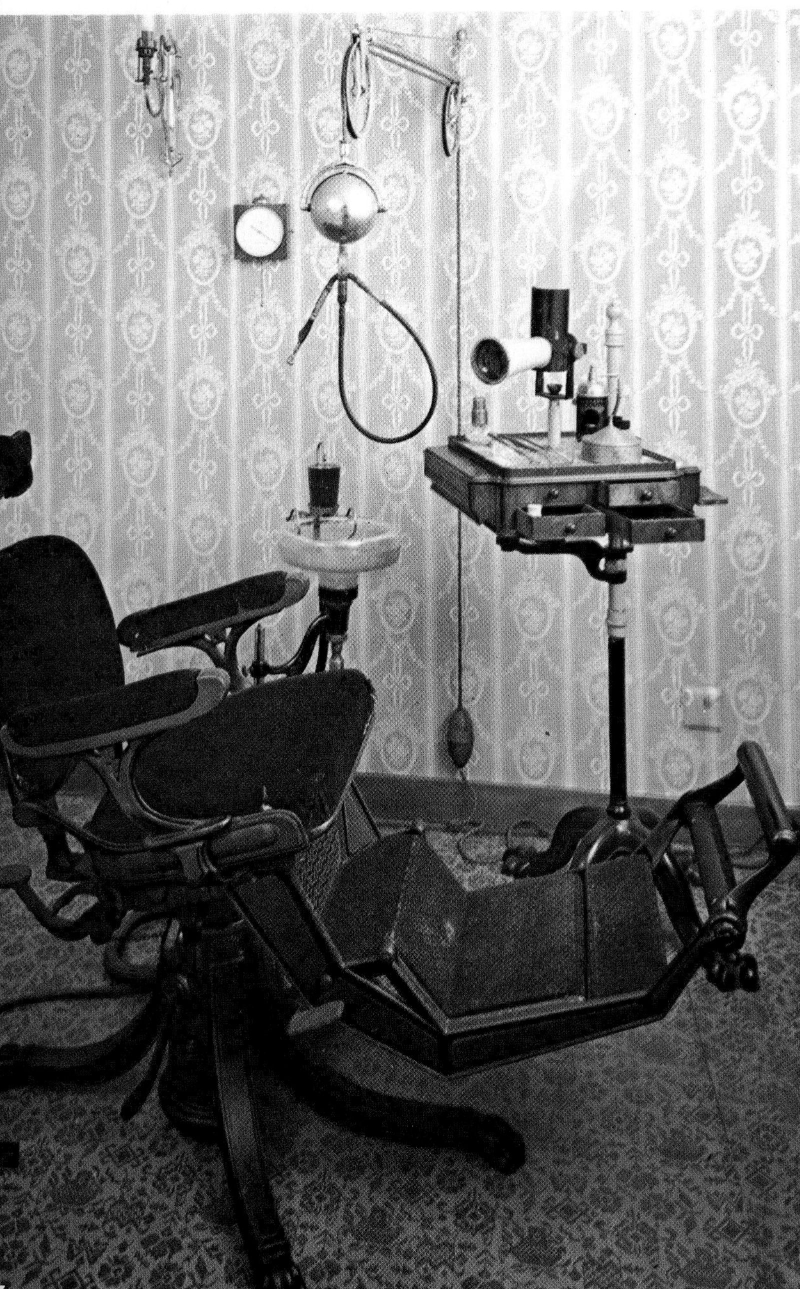

been thought previously. He urged doctors to stop treating them with boiling oil and to use a soothing dressing instead. Likewise, he discovered that ligature was a more effective – and more humane – method of stopping bleeding after amputation than a red-hot cautery.

But it was in replacing amputated limbs with man-made substitutes that Paré made one of his most decisive contributions to human welfare. Various earlier practitioners had sought to patch up the body in this way, but with only occasional success. Paré was unique for the ingenuity he used in trying to simulate natural functions with mechanical gadgetry. Much of this work was done for the benefit of wounded soldiers, and Paré devised several artificial arms and hands which he fitted successfully to his military patients. One of the simplest was a reproduction of a human hand, complete with a holder for a quill pen. One of the most complex was a sophisticated hand with fingers moved individually by tiny cog wheels and levers.

As we know only too well today, however, it is not necessarily the first person with a revolutionary idea who makes the greatest impact in his field. Enthusiastic drive and the ability to convince other people of the value of one's discovery are just as important. It seems that Paré's eminence as a surgeon – he was appointed chief surgeon to Charles IX in 1562 – was a crucial factor in getting his ideas on artificial limbs widely accepted.

Another important force in shaping the course of invention is social need. In the case of Paré, this came from the large number of soldiers injured in war. Since his day, of course, artificial limbs have become much more sophisticated. The latest of them ('myoelectric devices') operate by detecting the tiny electric currents produced by the remaining nerves and converting these into the appropriate mechanical movement. Yet social need still plays an important role – the need to help the hundreds of deformed children who were born as a result of the thalidomide tragedy in 1960–62 continues to stimulate further research in this area.

A major dilemma today is whether to produce increasingly versatile prostheses, in an effort to mimic exactly a natural limb, or to manufacture a series of separate devices. For example, a person lacking one arm could use separate attachments for eating or dressing just as we use a pen or spoon for different tasks.

B. D.

Dental drill

Torquemada was *not* the inventor of the dentist's drill. The first 'dental foot engine' was made by George Washington's dentist, John Greenwood (1760–1819), a Boston-born ex-privateersman who drifted into dentistry in New York and fashioned terrifying teeth of hippo and elephant tusk for the Father of his Country. In 1790 Greenwood adapted his mother's spinning-wheel to rotate a drill. Earlier dentists had operated drills with bowstrings, like jewellers – their use called for a high degree of dexterity in the operator and great fortitude in the patient. Greenwood's invention was ignored by all but his dentist son.

At that time most dentists used joiner-type drills operated by rotating a handle at the side, sometimes with adjustable drilling heads, and the great leap forward came only in 1829 when James Nasmyth, Scots inventor of the steam hammer and much else, devised a way of making rotary power turn corners, using a coiled steel spiral in a sleeve.

Charles Merry in America patented a hand-operated drill with flexible cable in 1858; and an Englishman, George Fellows Harrington, invented in 1864 the first motor-driven drill, which was a hand-held clockwork device with a drill attached. The next few years saw many types of motor, from pneumatic to electric; but foot-operated drills continued to be used in the 20th century.

Efficient drilling, obviating the loss of teeth, is said to have helped bring in the golden age of American dentistry, Americans being famous for their fiery golden smiles. In recent times drills have been speeded up; more speed initially meant more heat and thus more pain, but now high-speed drills with spray jets have rendered drilling virtually painless.

E. S. T.

Vaccination

'I shall never have smallpox for I have had cowpox. I shall never have an ugly, pock-marked face.'

This old wives' tale, repeated by a Gloucestershire milkmaid, was the starting point of the scientific development of vaccination and its use to eradicate one of man's most deadly diseases. Dr Edward Jenner, a country medical practitioner, subjected this piece of folklore to rigorous scrutiny, proved its validity, and thereby developed a safe and effective method of vaccination.

Jenner was the son of the vicar of Berkeley, Gloucestershire, and was born there in 1749. He began studying medicine at 13, and went to London at 21 to work for three years under the great surgeon John Hunter, from whom he learned the value of experiment. Later, choosing to practise in his native village rather than pursue a promising career in London, he decided to test the belief that people who had contracted cowpox – a disease of the udders of cows – were thereafter protected from smallpox. On 14 May 1796, after studying, hesitating and reflecting upon it for 20 years, he inoculated the arm of an eight-year-old boy, James Phipps, with pus from a cowpox sore on a milk-maid's arm. The boy developed a similar sore, but was not ill, and when he was inoculated with pus from a small-pox sore six weeks later, and again several months after that, there was no reaction. The mild cowpox infection had protected him from smallpox.

Several further experiments confirmed this first simple and dramatic proof, including tests on people who had suffered cowpox. Jenner then began a series of human arm-to-arm transfers of cowpox pus, and showed that this gave continued protection against small-pox. In 1798 he published his results (at his own expense) and, often in the face of hostility, began a campaign to persuade other doctors to use his technique. Within two years, vaccination was being widely used and in 1802 Jenner was voted a parliamentary grant of £10,000, as a token of his country's gratitude. Before his death in 1823, vaccination was being practised in civilized communities throughout the world.

Vaccination replaced the earlier and far more risky technique of *variolation* (introduced from Turkey earlier in the century by Lady Mary Wortley Montagu), in which pus from a vesicle on the skin of someone with mild smallpox was introduced into a healthy person's skin, in the hope that the mild form would follow and afford protection against a more severe attack. Though in practice this frequently caused the full-blown disease, the fear of smallpox was such that people were prepared to take such a risk. We now know that vaccination works by stimulating the production of antibodies which are effective not only against cowpox but also against the closely related smallpox virus.

B. D.

Anaesthetics

Surgery was transformed during the 19th century by two major discoveries – general anaesthesia and antisepsis. Though alcohol and other means had long been used in an attempt to combat pain during surgery, the era of real anaesthesia began in 1799 when Sir Humphry Davy suggested using nitrous oxide (laughing gas). The exhilarating properties of the gas quickly became well known, and led to a craze for parties at which it (and later ether) were used to promote conviviality. Then, in 1842, William Clark, a chemistry student in Rochester, USA, gave ether to a woman who had a tooth extracted painlessly under its influence; in the same year Crawford Long in Jefferson, Georgia, used ether for a surgical operation. However, the first conclusive demonstration that ether could be used to create general anaesthesia for a major operation came on 16 October 1846, when William Morton, at Massachusetts General Hospital, removed a tumour from the neck of a patient under ether anaesthesia. After this date, the practice quickly spread in both the USA and Britain. Sir James Simpson in Edinburgh introduced chloroform as a more pleasant replacement for ether, and used it in obstetrics and general surgery. John Snow, a London practitioner, became the first professional anaesthetist, and he and his successors devised special inhalers.

The surgeon now had time to do his work carefully; and more drastic operations also became possible, as anaesthesia greatly reduced the degree of shock suffered by the patient.

B. D.

Ultra-violet lamp

It was only in 1672 that Isaac Newton's famous experiment with a glass prism demonstrated the spectrum of sunlight, with its overlapping sections of red, orange, yellow, green, blue, indigo and violet – in that order. Sir Frederick William Herschel discovered in 1800 that there are invisible rays beyond the red sector in the spectrum: he called them infra-red rays. The German physicist J. W. Ritter thereupon scrutinized the other end of the spectrum and found that here, too, invisible rays existed beyond the violet. He called them 'ultra-violet rays'. Neither infra-red nor ultra-violet 'light' produces any sensation of sight in the human eye, though infra-red rays are heat rays and can be sensed by the skin. The presence of ultra-violet light can be shown only in roundabout ways, such as by the effects it has on certain chemicals.

It was only some decades after Ritter's discovery that the biological and biochemical effects of these rays began to be assessed; overlong exposure to the sun, which emits them along with the visible light, is injurious to health largely because of them – but if they are shut out from the spectrum for any length of time, the human body seems to miss them, and children growing up in sunless surroundings are particularly prone to rickets.

The explanation came with the discovery of the vitamins. The sun's ultra-violet rays react with the skin, making it produce vitamin D, which the body needs. But over-exposure to ultra-violet rays causes sunburn and inflammation of the skin – first, however, tanning it a beautiful, healthy-looking brown. The eyes may also suffer damage, but they can be protected by ultra-violet-absorbing dark glasses.

In climates with only a modest amount of sunshine, strictly controlled administration of ultra-violet radiation from artificial sources, from 'lamps',

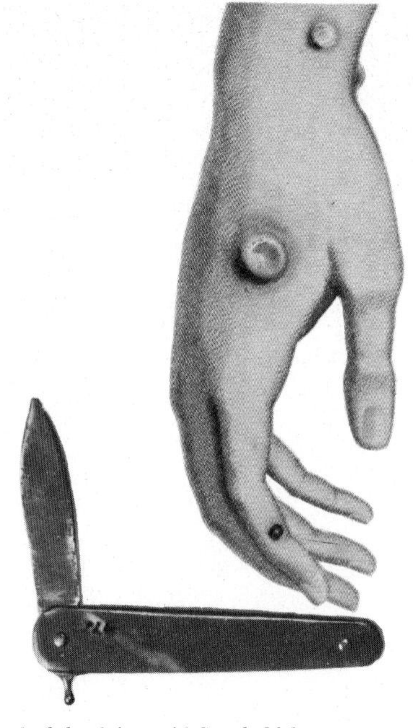

Hand of the dairymaid Sarah Nelmes, showing cowpox pustules from which Jenner made the first smallpox vaccine in 1796. Below it, a lancet of the type used for vaccination, c. *1800.*

Above, *William Morton – front left, looking down – demonstrating surgical anaesthesia in Boston in 1846.* Left, *Snow's inhaler, facepieces and drop bott*

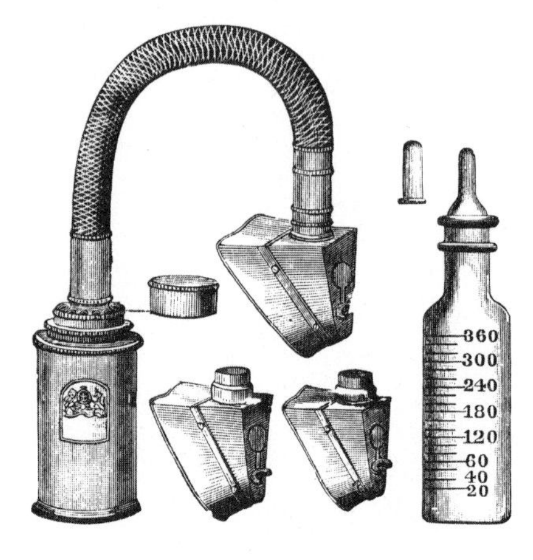

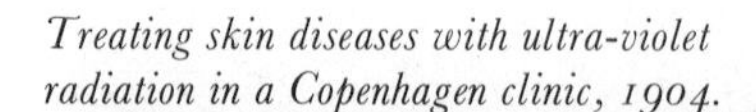

Treating skin diseases with ultra-violet radiation in a Copenhagen clinic, 1904.

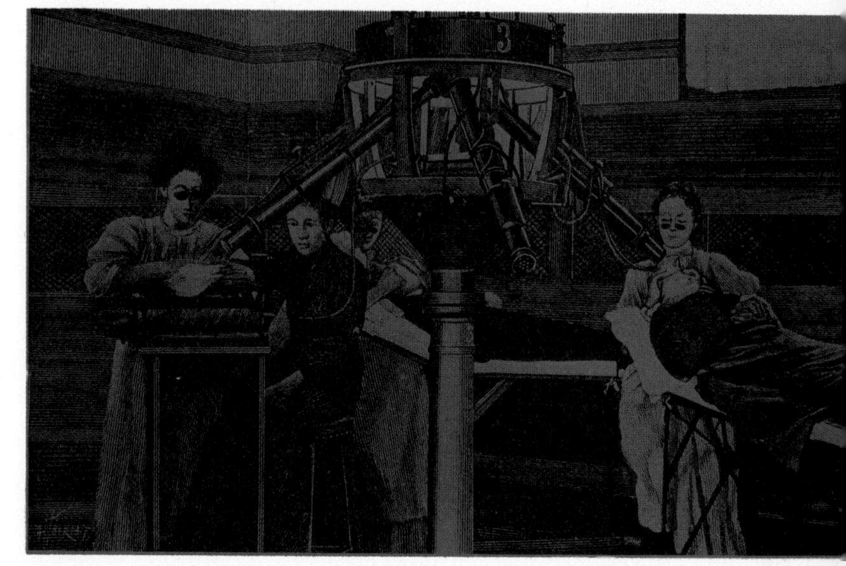

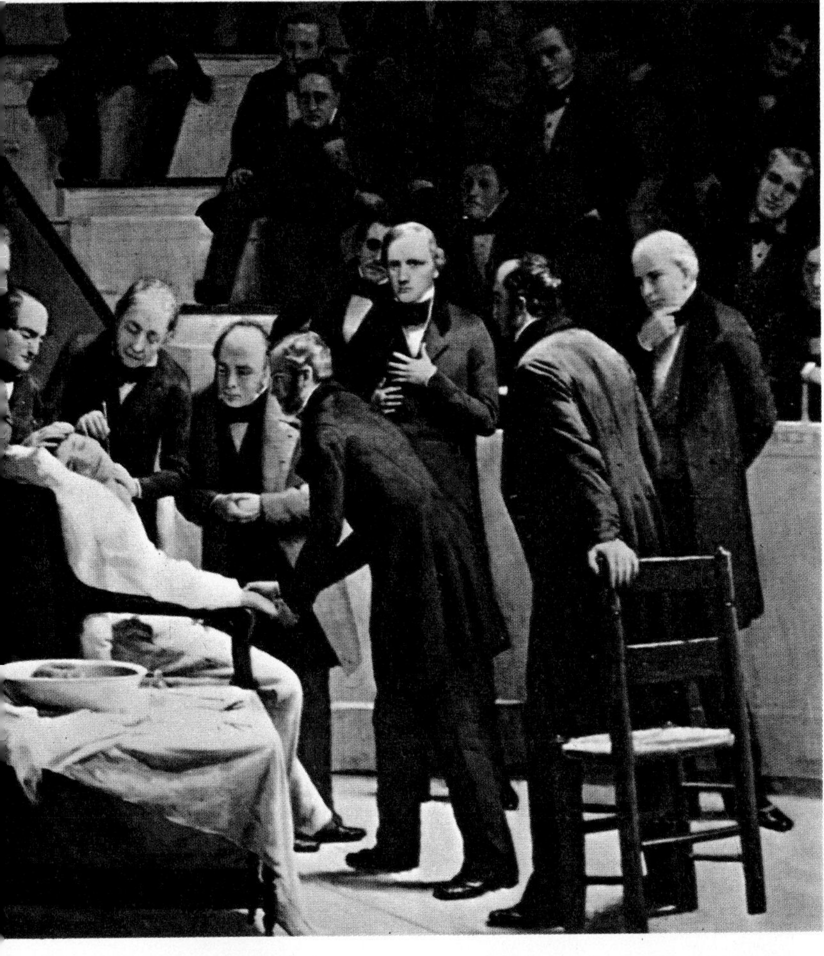

has been found beneficial to combat vitamin-D deficiency and in treating certain diseases. The best-known source is the arc-lamp, first designed around 1810 by Humphry Davy, in which an electric arc is maintained between carbons; this is still the system used in the cheaper types of ultra-violet 'health lamps'. More efficient types are those with iron-cored carbons. In medical practice, mercury-vapour tubes have been in use since the 1930s; when the current is switched on, the mercury begins to vaporize and an arc emitting ultra-violet radiation is maintained between the electrodes.

There is an extensive range of ultra-violet waves of various lengths between the visible violet and the X-ray spectrum; most of them are unable to penetrate ordinary glass, but quartz lets them through, and mercury-vapour tubes are therefore made of quartz.

Ultra-violet radiation is also widely used in the production and treatment of food, drugs and other goods. Two Englishmen, A. Downes and T. P. Blunt, found in 1877 that ultra-violet rays kill micro-organisms, a discovery which has led to the development of an entirely new technique of sterilization. Fruit and vegetables, milk and ice-cream, cheese and fish, tobacco and wine are now frequently treated by ultra-violet radiation; air and water, too, can be made germ-free by it.

E. L.

Stethoscope

This brilliantly simple device was the invention of René Théophile Hyacinthe Laënnec (1781–1826), the son of a lawyer in Quimper, Brittany. A slight, timid figure, Laënnec won a considerable reputation as a teacher and became an expert pathologist. In Paris in 1816 he stumbled on the idea of the stethoscope (which he subsequently named from the Greek *stethos*, 'chest') by chance. One day, walking in the court of the Louvre, he saw some children with their ears glued to the two ends of some long pieces of wood which transmitted the sound of the little blows of a pin, stuck at the opposite end. He immediately thought of applying this device to the study of diseases of the heart.

On the following day, at his clinic at the Necker Hospital, he took a sheet of paper, rolled it up and tied it with a string, making a central canal which he then placed on a diseased heart. This was the first stethoscope. After numerous experiments, for he was an expert wood-turner, he produced a cylinder of wood, either cedar or ebony, 30 cm. long and 3 cm. in diameter with a central canal 5 mm. in diameter. The cylinder was divided in the middle, so the two parts could be separated and more easily carried.

The stethoscope transformed everyday medicine by bringing a simple and reliable means of auscultation within the reach of every doctor. Ironically, Laënnec developed pulmonary disease himself shortly after its invention, dying of it in 1826. Laënnec's work tidied up the whole subject of pulmonary disease, until then a heterogeneous assortment of fact and fable; he wrote one of the great medical books, and with it clinical medicine entered a new era. Despite this, however, Laënnec was bitterly opposed by another Breton doctor, François Broussais, who was responsible for many deaths through his popularization of bleeding by leeches and who called Laënnec's book 'a collection of undisputed facts or useless discoveries'.

The modern binaural stethoscope, in which a tube runs to each ear from above the bell, and which has the added advantage of flexibility, was a later development.

G. R. T.

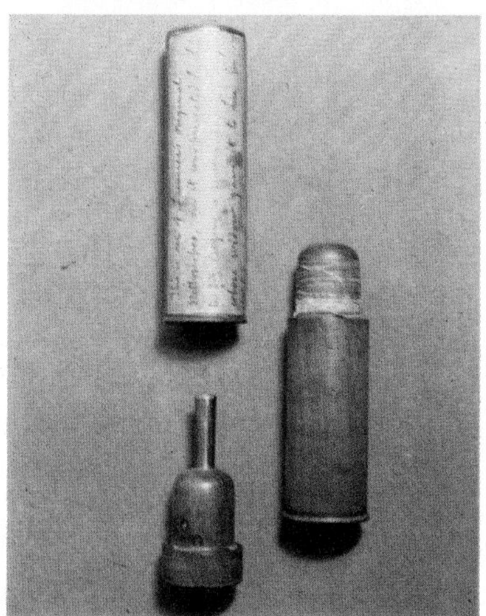

Above, left, *René Théophile Laënnec (1781–1826)*. Above, *one of Laënnec's early stethoscopes, which he presented to a French army surgeon.*

Left, *experimental blood transfusion, from a lamb to a man, in the 17th century – 250 years before the concept of compatibility was understood.*

Blood transfusion

Experiments were first carried out by Jean-Baptiste Denys (1625–1704), although it is said that the Australian Aborigines have practised blood transfusion for thousands of years. Denys first made experiments with animals, and on 15 June 1667 he transfused a 15-year-old boy with blood from a lamb. In 1668 one of his patients died after a transfusion, and although it was later proved that the patient had been poisoned by his wife, there was a great scandal about it and transfusion was not tried again until the early 19th century.

The first successful transfusion was made by James Blundel at Guy's Hospital, London, in 1818, using a syringe. At first, he worked only with hopeless cases, but in 1829 a patient with post-partum haemorrhage was saved as a result of a blood transfusion. Blundel invented two special pieces of apparatus to enable blood to be transferred from donor to patient with the minimum of physical interference. Blood transfusion was much used in the Franco-Prussian War (1870–71) to help wounded soldiers, and now various other difficulties were encountered, chiefly coagulation of the blood. Defibrination was tried, but as this also removed most of the valuable constituents of the blood it was not satisfactory.

In 1909 Karl Landsteiner, an Austrian pathologist, investigating why donors' blood sometimes – but not always – caused the recipient's blood to clot, established the existence of different categories of blood, which are now known to be four in number. The concept of 'compatibility' of blood groups was what eventually made transfusion the safe procedure it now is.

Most early transfusions were carried out directly from donor to patient, but nowadays almost all are done with stored blood. Unless this is used within three to four weeks the stored blood has its red cells removed by centrifugation; the remaining liquid is reduced to powder and kept for emergency use, the dried plasma being reconstituted with sterile water. During the Second World War, a blood plasma substitute was developed at Birmingham University by large-scale fermentation of sugar.

T. C.

Ophthalmoscope

The ophthalmoscope is a device for shining a narrow beam of light through the pupil of the eye in order to illuminate the retina. It incorporates a selection of tiny lenses, which bring the retina of the subject into focus for the eye of the observer. The choice of lens is made according to the 'strength' or curvature of the natural lens system in the eye, and by noting the lens needed the observer can, for example, work out the degree of short-sightedness or long-sightedness present. But the retina is also the one situation in the body where small blood vessels can be examined easily and in detail, so that many diseases which affect blood vessels (such as diabetes and high blood pressure) can be detected. And since the retina is almost touching the brain, it also allows a view of the nervous system.

An ophthalmoscope was invented in 1847 by Charles Babbage, the Englishman who built a mechanical 'calculating engine'; he handed his invention over to a surgeon friend, asking him to try it out. However, the dull fellow put it aside, so an additional claim to fame for Babbage was lost.

Shortly afterwards, in 1851, another was independently invented by Hermann Helmholtz. This time the instrument was properly exploited.

DD. G.

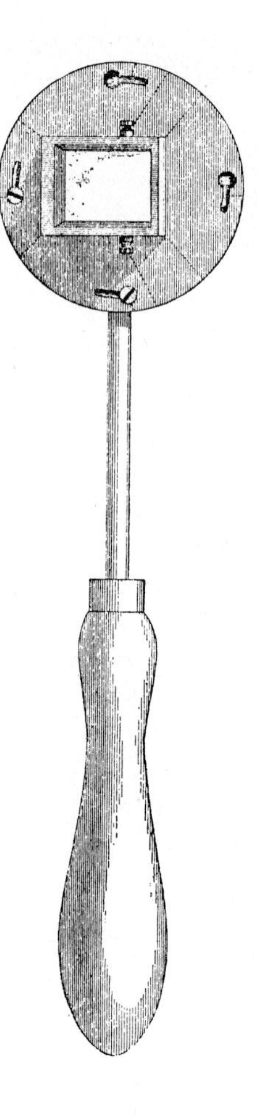

Diagram of Helmholtz's ophthalmoscope, and a picture of how the ophthalmoscope sees the retina.

Lister's spray

Until just over a century ago, surgery was a desperate remedy which killed at least as often as it cured. Lister's spray was part of an effort to give patients a much better chance of recovery.

In the 1860s the germ theory of disease was beginning to be accepted by some scientists (including Louis Pasteur). Joseph Lister, then Professor of Surgery in the University of Glasgow, wondered whether contamination by invisible organisms might not be the cause of the suppuration which then almost always developed in operation wounds, and so often led to the patient's death. He set out to fight those organisms.

In August 1865 a boy called James Greenlees was brought to Glasgow Royal Infirmary with a broken shin beneath an open wound. Lister set the bone and applied a dressing soaked in carbolic acid to the damaged flesh. The whole thing healed rapidly and without festering, and the boy walked out of the hospital six weeks later alive and well.

From then on Lister used carbolic dressings on all surgical wounds with the result that in his operations suppuration became the exception instead of the rule.

He developed his famous spray some five years later, on the assumption that some of the germs which invaded operation sites floated in from the surrounding air. Early models were cumbersome affairs on tripods, driven by a handle, and looked like the village pump. They were nicknamed 'donkey engines'. Lister's definitive model, brought out in 1875, was a small, readily portable instrument powered by a steam kettle. Placed in the corner of the operating theatre, it filled the place with an irritating mist of disinfectant which made the surgeons and nurses weep and splutter.

The spray was not popular; nor, in fact, was it needed, for while dust specks in the air do carry a rich variety of germs we now know that most of these are harmless. The dangerous kind enter wounds from the skin and clothes of the patients and their attendants, and these were being very adequately dealt with by Lister's original carbolic swabs.

The general lack of enthusiasm for the idea of airborne infections was expressed in a famous professional joke common among doctors at the time: 'Shut the door quickly, or Professor Lister's germs will get in.'

More seriously, a German surgeon, Victor von Bruns, published a paper in 1880 entitled *Fort mit dem Spray* (Away with the Spray). Finally Lister himself admitted in 1890 that he was ashamed of ever having proposed the need for destroying microbes in the air.

He need not have felt ashamed, however, for with the knowledge available at the time the spray was a iogical product of his inspired (and correct) belief that germs are everywhere, and must be kept away from damaged flesh – an insight which has undoubtedly earned Joseph Lister the right to be considered the father of modern surgery.

DD. G.

Right, *the carbolic ointment which helped to reduce the death-rate from post-operative infection.* Below, *a Lister spray – a good idea, but ultimately a dead end.*

Inoculation

That brilliant, impetuous, sarcastic and egotistical old French chemist, Louis Pasteur, when he was already 58, stumbled accidentally on the principle of warding off infection by inoculation.

He had been working on a disease of hens called chicken cholera, and had

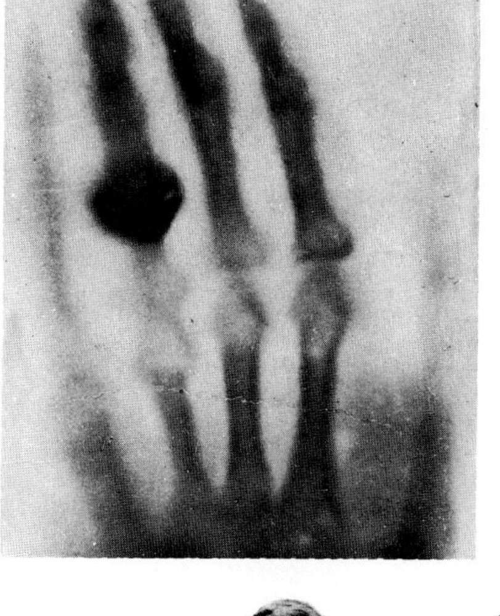

Left, *Röntgen's first X-ray photograph – his wife's hand, with wedding ring.*

Below, *the earliest type of X-ray apparatus, which dates from 1896.*

been cultivating the germs in bottles. One day he fed some of the germs to a group of birds, expecting them to fall sick and die. But they simply went off colour for a while and then recovered.

The germs had been taken from a culture which had been growing for about six weeks; Pasteur decided that they must be stale, and fed a fresh crop to the same birds. Nothing happened. But when the fresh germs were fed to a new set of hens, all the birds sickened and died in the expected way.

Pasteur realized that the six-week-old germs had somehow changed their nature, so that they were no longer capable of causing serious disease. In the jargon now used by bacteriologists, they had become 'attenuated'. Secondly, he realized that although the stale germs could no longer make his chickens sick, they did make them resistant to further infection.

At first he believed that he had discovered a panacea for all manner of infectious diseases, and that his attenuated chicken cholera germs would protect the birds, and perhaps men, against the attacks of other organisms. But he soon found that the attenuated germs will only immunize against infections by their own kind.

DD. G.

X-ray

'Probably the greatest landmark in the whole history of diagnosis.' This was the accolade given by Thomas Hunt, a distinguished British physician, to a scientific paper delivered on 28 December 1895 to the Würzburg Physical-Medical Society. In that paper Wilhelm Konrad Röntgen described his discovery of a new type of ray, which he named the X-ray. It could not be detected by the human eye, and could pass easily through paper and wood, though not through metals and other dense materials. This startling technique soon revolutionized diagnosis.

Röntgen, born in the Rhineland in 1845, became Professor of Physics and Director of the new Physical Institute at Würzburg in 1888. His great discovery was based on a fortuitous happening which he followed up with expertise and determination. On 8 November 1895 he was studying the electrical discharge produced in the vacuum tubes recently devised by the English physicist Sir William Crookes. Röntgen noticed that, when a current was passing through the tube, a nearby piece of paper which had been painted with barium platinocyanide fluoresced brightly. Moreover, this occurred even when he covered the tube with black cardboard. Within a few weeks, Röntgen had proved that the effect was caused by an invisible ray.

At first, doctors used X-rays only to reveal foreign bodies in the tissues, and to diagnose bone fractures. But a spate of subsidiary discoveries soon helped the new science of 'radiography' to probe into the deeper, soft tissues of the body. The now routine barium meal was introduced. The barium compound is opaque to X-rays, and as it passes through the alimentary tract its progress can be followed with X-ray pictures. Similar techniques are now available for studying the heart and brain. All have contributed enormously to the art of diagnosis, and several have also added to the success of pre-symptomatic screening, in which apparently healthy people are scrutinized for signs of incipient ill-health. This is particularly important in relation to cancer, where early diagnosis greatly increases the chances of successful treatment. But the role of radiography in the virtual defeat of tuberculosis in civilized communities, a battle in which mass chest X-ray units played a decisive part, would alone serve to place Röntgen's discovery among the most important in modern medicine.

B. D.

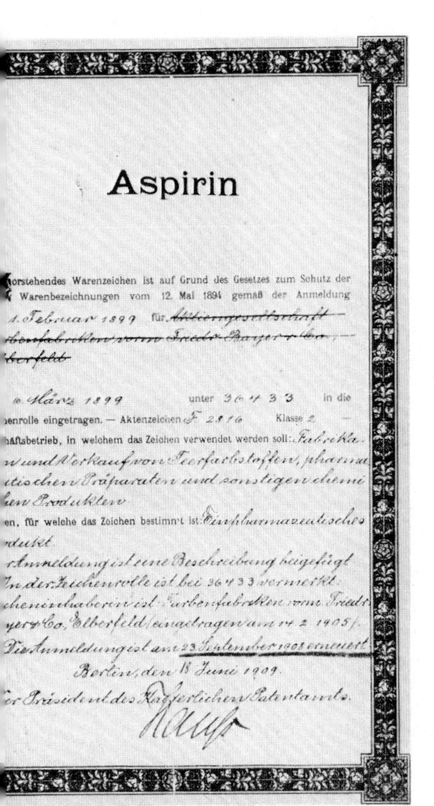

Aspirin

orstehendes Warenzeichen ist auf Grund des Gesetzes zum Schutz der
Warenbezeichnungen vom 12. Mai 1894 gemäß der Anmeldung
1. Februar 1899 für ~~Aktiengesellschaft~~
~~Farbenfabriken vorm. Friedr. Bayer & Co.,~~
~~Elberfeld~~
März 1899 unter 36433 in die
enrolle eingetragen. — Aktenzeichen F 2816 Klasse 2 —
häftsbetrieb, in welchem das Zeichen verwendet werden soll: Fabrika-
n und Verkauf von Teerfarbstoffen, pharma-
tischen Präparaten und sonstigen chemi-
hen Produkten
en, für welche das Zeichen bestimmt ist: Ein pharmazeutisches
dukt
r Anmeldung ist eine Beschreibung beigefügt
In der Zeichenrolle ist bei 36433 vermerkt:
cheninhaberin ist Farbenfabriken vorm. Friedr.
yer & Co. Elberfeld (eingetragen am 14. 2. 1905).
Die Anmeldung ist am 23 September 1908 erneuert
Berlin, den 18 Juni 1909.
er Präsident des Kaiserlichen Patentamts.

Patent granted to Friedrich Bayer & Co., Elberfeld, in March 1899 for the first synthetic acetylsalicylic acid, aspirin.

Aspirin

Aspirin has become possibly the best-known and most widely used of all drugs because of its value in relieving pain and fever, particularly rheumatic fever. Properly known as acetylsalicylic acid, it is found naturally in certain plants and tree barks, and has thus been known for centuries for its therapeutic properties. In 1899 Hermann Dreser introduced synthetic aspirin into medicine; owing to its acidity it is nowadays stabilized by the addition of a neutral salt, calcium glutamate.

Aspirin works in alleviating the symptoms of fever by lowering the body temperature, in particular by acting on the body's 'thermostat', the hypothalamus in the central nervous system of the brain. During fever this organ increases the amount of heat produced by the body; when aspirin is taken the effect is to maintain this heat production, but the loss of body heat is dramatically increased by sweating and alteration of the blood flow. Aspirin also acts on the central system in the relief of pain such as headache. The mechanism is not fully understood but it is known that aspirin has a selective depressant effect on this part of the brain. Rheumatic disease, in which the connective tissues are damaged and inflammation occurs, can be treated with aspirin, which is thought to work by suppressing the molecules that are released in the connective tissue as a result of inflammation.

In the last few years aspirin has been shown to have further potential uses. It interacts with the naturally occurring compounds called prostaglandins, which cause strong muscle contractions and are released during labour. It may perhaps be used to prevent unwanted abortion or miscarriage. Recent clinical trials on patients suffering from heart disease indicate that their condition improves when given a controlled dose of aspirin.

Although aspirin is a relatively weak pain-killer in comparison to the narcotics, such as morphine, it offers enormous advantages over other similar drugs, being cheap, easy to administer and producing minimal side-effects when sensibly prescribed. As a result, however, it is grossly abused. Over 30 million lb. of aspirin are consumed annually in the USA, indicating that the drug is indiscriminately employed for every conceivable ailment. An overdose of aspirin can cause ulcers, asthma and haemorrhage. Several derivatives of aspirin have similar therapeutic properties, but it seems unlikely that aspirin will be replaced as the basic household remedy for pains.

R.J.

Surgical transplants

Although surgical transplants did not really capture public attention until quite recently, man has been pondering the feasibility of such medical treatment for thousands of years. It is even possible that these operations were performed long before the development of modern surgery. In Egyptian manuscripts dating as far back as 2000 BC, there are references to 'a caste of tile makers', men whose task it apparently was to perform skin grafts. However, nothing approaching a clinical record of these surgeries survives. There are also numerous legends and folk-tales, some of early Christian origin, which speak of successful transplants involving noses and even entire limbs from one person to another. As might be expected, these stories are disbelieved by the modern medical community in the West.

The first clearly recorded attempts at transplantation are in the field which has since become known as plastic surgery, and they appear to have been singular failures. In 1597, the Italian surgeon Gaspare Tagliacozzi wrote of trying to 'restore the appearances of patients who had lost their noses' by using the flesh of other persons. He blamed his lack of success on the force and power of human individuality. Then, in 1823, the German surgeon G. Bunger described how he successfully reconstructed part of a woman's nose by grafting skin from her own thigh, a technique known as 'autograft'. Forty years later, Paul Bert, a French physiologist, showed how in allografts, which utilize skin or tissue from a donor, the transplanted material was regularly rejected by the recipient. To this day, the rejection of alien tissue and organs represents the major difficulty in transplant surgery.

At the turn of this century, there were several attempts made at whole-organ transplants. The American Claude Beck experimented with the kidneys, while in Germany, E. Ullmann tried and failed to treat a woman with uraemia by grafting the kidney of a pig to her arm. During the period 1902 to 1912, the surgical team of Alexis Carrel and C. C. Guthrie, working in America, effectively laid the technical basis for present-day transplants through their studies of vascular surgery. They also performed successful transplant operations on animals, using limbs as well as kidneys. Similar animal experimentation was later favoured by Soviet research doctors, and during the 1960s Moscow displayed a two-headed dog for the world to gawp at.

In 1954, a team headed by Joseph Murray of the Harvard University Medical School performed the first successful kidney transplant for the treatment of patients in the terminal stages of kidney disease. Since then, more than 5,000 such renal allografts have been carried out, mostly in the United States. Heart transplants, which are perhaps the most widely publicized allografts, did not come on the scene until late 1967, when Christiaan Barnard placed another man's heart into the chest of South African dentist Philip Blaiberg. Although many more cardiac allografts have been performed since, the survival rate is still faily low. All in all, modern medicine reckons there are 21 different transplantable human organs and tissues, including the liver, lungs, pancreas and the cornea of the eye (which last has a higher success rate because it contains no blood vessels).

E.W.

Electrocardiograph

'Those who, like myself, experimented with this instrument were thought to be rather dangerous backroom boys, unfit to be trusted with the welfare of patients.' When Sir Ian Hill, one of Britain's distinguished heart specialists, began his career in medicine in the late 1920s, the electrocardiograph was still something of a new-fangled toy. But it was soon to transform our understanding of how the heart works, in both health and disease. Today it is a routine instrument for examining people with heart disease, assessing the severity of their illness, and monitoring recovery after a heart attack.

Willem Einthoven, who invented the electrocardiograph in 1903, was born in 1860 in the Dutch West Indies and qualified in medicine in 1885. His first invention was the string galvanometer, in which a slender silver-coated quartz wire is suspended between the poles of a magnet. When an electric current passes along it, the wire (or 'string') swings

Late 16th-century cosmetic surgery: Gasparo Tagliacozzi probably had better success with homografts like this, grafting from the patient's own arm to replace lost tissue on the nose, than from the donor grafts which he also tried.

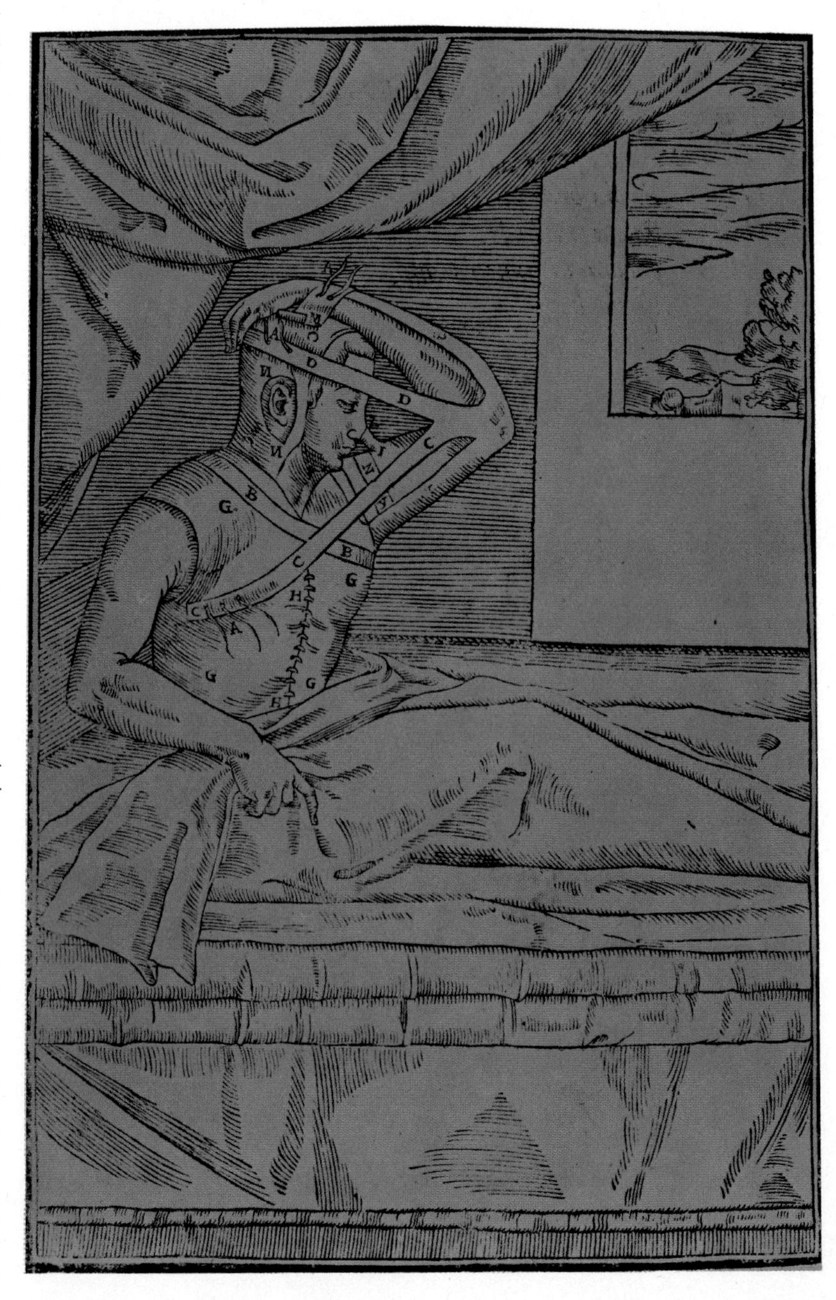

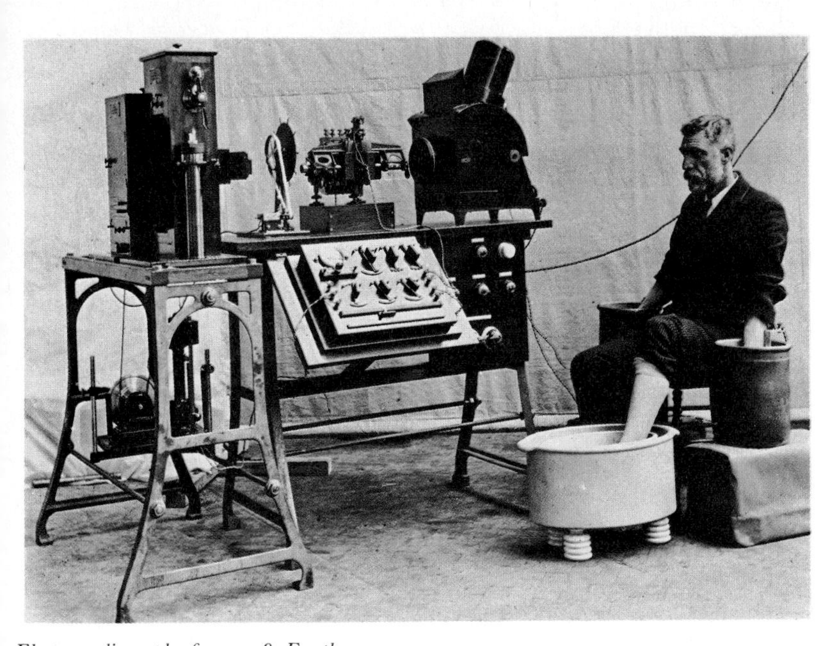

Electrocardiograph of c. *1908. For the invention of this incredibly sensitive instrument, Willem Einthoven won a Nobel prize.*

towards a position at right angles to the magnetic lines of force. The delicate apparatus is particularly suited to measuring extremely small currents – those produced when a muscle contracts, for example. About 50 years earlier, two German scientists had discovered that the frog's heart creates an electric current, and Einthoven decided to study the electrical activity of the human heart.

By placing electrodes on a patient's arm and leg, he discovered that he could, indeed, detect the pulse of electricity passing through the heart muscle as it pumps blood round the body. Moreover, he devised an ingenious way of making permanent records of the current. He arranged for the galvanometer string, as it deflected, to interrupt a beam of light, producing a shadow on paper. By using a long strip of light-sensitive paper, and moving it continuously, he was able to produce an electrocardiogram (ECG) – a continuous record of the electrical activity accompanying the heart's muscular activity. In 1924, three years before his death, Einthoven was awarded a Nobel prize.

B. D.

Primitive respirator of 1876. With a manual pump, and used solely for resuscitation, it was nevertheless the direct ancestor of today's iron lung.

Therapeutic drugs

The practice of medicine is, by its nature, a hit-and-miss affair, and for virtually all its history the hits have been a matter of coincidence. Some folk medicines – quinine, salicyclic acid, for example – were effective, but the prospect of a rational approach to curative medicine was very limited until there was some notion of what caused disease. Pasteur's discovery that some diseases were produced by micro-organisms gave the clue, but it was not easy to follow. The micro-organisms could be killed in test-tubes with heat or carbolic acid, for example, but these agents could not be used on patients, except externally.

Paul Ehrlich solved the problem. His first attack was based on the knowledge that micro-organisms could be dyed: if dyes were absorbed by the organisms, with luck some dye might be poisonous enough to be mortal. Unfortunately he did not find one, or at least not one that could cure infected rats. So after five years he tried another approach.

Medicine at the time used poisons in some of its more heroic medicaments. Ehrlich set out to find a way of producing a relatively non-poisonous compound of arsenic that might be used to attack the spirochetes that caused syphilis, then a widespread disease. His compound no. 592 was effective on syphilitic mice, and Ehrlich's colleagues decided to make a purer and more soluble variant. This, which he numbered 606, was found in 1910 to be successful in humans. 'Salvarsan' cut the incidence of syphilis in England and France by 50 per cent in five years. It was the first rationally produced medicine.

Chemotherapy, the technique that Ehrlich had invented, moved only slowly from this auspicious start, and there was no major leap forward until 1932. When it came it resulted from a return to Ehrlich's idea that dyes could be the basis of treatment. Two Germans, Mietch and Klarer, resolved to find a drug effective against streptococci and staphylococci. They tried making chemical derivatives of azo dyes, and finally sent one that they numbered 4145 to Gerhardt Domagk, who was conducting the animal experiments. 'Prontosil', the first of the sulphonamides, was eventually shown to be effective in humans.

The accident that was responsible for the next major advance has often been described, although the inventor may not have remembered exactly how it happened. In 1928 Alexander Fleming, a bacteriologist at St Mary's Hospital in London, found that a mould that had probably been introduced by another research worker had started growing on the same jelly as some bacteria he was cultivating, and the bacteria had disappeared from where the mould was growing. Fleming published an account of the work in May 1929 but lost interest in his astonishing and promising discovery. It was ignored for 10 years, until Howard Florey and Ernst Chain took it up and found a way of producing a pure penicillin, as the mould's product was called. Since then millions of lives have been saved by drugs descended from Salvarsan, Prontosil and penicillin.

T. O.

Iron lung

The iron lung was the first of the machines able to undertake one or other of the vital functions of the body so that people who would otherwise have died could now be kept alive. The iron lung takes over the job of the muscles responsible for respiration.

It was invented in 1929 by an American called Philip Drinker. It was not a particularly clever invention. The regular entry of air into the lungs depends upon the regular and sufficient contraction of the muscles involved. Sometimes these muscles fail, commonly (in the past at least) when certain cells in the brain or spinal cord are attacked by the virus of poliomyelitis.

Drinker saw that the necessary regular expansion of the chest wall could be achieved by putting the patient into an airtight box (all save the head which emerged from the box through a soft, airtight collar), and by connecting the box to a pump. The pump was arranged to produce a rhythmical drop of air pressure in the box. This meant that the outside of the patient's chest was exposed to this pump-induced low pressure, while the inside of his chest was still connected, through his air tubes, throat, nose and mouth, to the pressure of the atmosphere. Under these circumstances the weight of the atmosphere forced air into the chest each time the pump made a negative pressure in the box of the iron lung.

DD. G.

Kidney machine

When Willem Kolff produced the first kidney machine in 1944, he probably had little idea of the impact his inventiveness was to have on the world of medicine.

Kolff, a Dutchman, developed his device during the Second World War while the Germans were still in Holland. From the very first, the machine was involved in dramatic situations, for Kolff used it secretly to save the lives of partisans.

Kidneys filter several unwanted substances from the blood and push them out as urine. When the kidneys stop working, waste materials accumulate. To deal with this, kidney machines can be used to wash the blood. Blood from the patient moves through a tube or is spread across a membrane, the opposite surface of which is bathed by a fluid. The tube or membrane is made of a material which small molecules (like those of salt or water) can penetrate. Such substances cross the membrane (or tube wall) when their concentration is higher on one side than the other.

For example, blood contains a chemical called urea, which is derived from proteins. There is no urea in the fluid used to charge the artificial kidney. So urea passes from the blood into the bathing fluid as blood flows through the machine. By altering the composition of the bathing fluid and making various other adjustments, any desired quantity of materials handled by a normal kidney can be induced to move into or out of a patient's body during the course of a treatment.

Kolff's original machine functioned extremely well, and later refinements were simply to make it easier to operate. However, using a kidney machine involves inserting quite large tubes into an artery and a vein in order to draw blood from the body, pass it through the apparatus, and finally restore it to the patient's bloodstream. At first, this involved a considerable surgical operation every time a machine was connected, and the machine was obviously most useful for helping acutely ill patients through a crisis. A lot of people whose kidneys stop working die quickly from a build-up of waste products in their blood, and if this poisoning can be held at bay for a few days the faulty kidneys – and the patient – often recover. The real impact of Kolff's brainchild was felt only after 1960, when Dr Belding Scribner of Seattle worked out how to put tubes into a large artery and a fat vein in such a way that they could be left in place for months or even years. It then became possible to use artificial kidneys to treat people whose own organs had been permanently damaged.

DD. G.

Contraceptive pill

'God's will be done and if He decrees that we are to have a great number of children, why, we must try to bring them up as useful and exemplary members of society', said Queen Victoria. The Pill has already done more than any other contraceptive technique to free women from the burden of unwelcome childbirth and banish the type of fatalism illustrated by this remark.

There is no single inventor of the contraceptive as such. All societies at all times in history seem to have had their own birth-control methods and equipment, often of dubious efficacy. Yet until very recently there has also been a widespread reluctance to discuss the matter. John Marten, writing in 1702 in England, referred to an impregnated linen sheath, used to prevent venereal infection, but he refused to disclose the composition of the wash, 'lest it give encouragement to the lewd'. In 1869 a speaker at the British Medical Association's annual meeting condemned 'beastly contrivances for limiting the numbers of offspring', and even in the 1920s the level of debate in the medical Press was appallingly low.

The American counterpart of Britain's Marie Stopes as a crusading enthusiast for family planning was Mrs Margaret Sanger. It was she who established the world's first birth-control advice centre – in 1916 in Brooklyn. Though the New York Police soon closed it down as a 'public nuisance' and Mrs Sanger went to prison, she persisted with her mission. In 1927 she organized the first World Population Conference; in 1948 she launched the International Planned Parenthood Federation; and in 1951 she visited Dr Gregory Pincus, a doctor at the Worcester Foundation at Shrewsbury, Massachusetts, and encouraged him to try to develop an oral contraceptive, following his experiments on fertility in animals. Dr Pincus and his colleagues decided to investigate the possible contraceptive action of certain hormones, and, after many dead ends and disappointments, they reported their first successful results in 1955. They had discovered that a hormone called 'norethisterone' was an effective con-

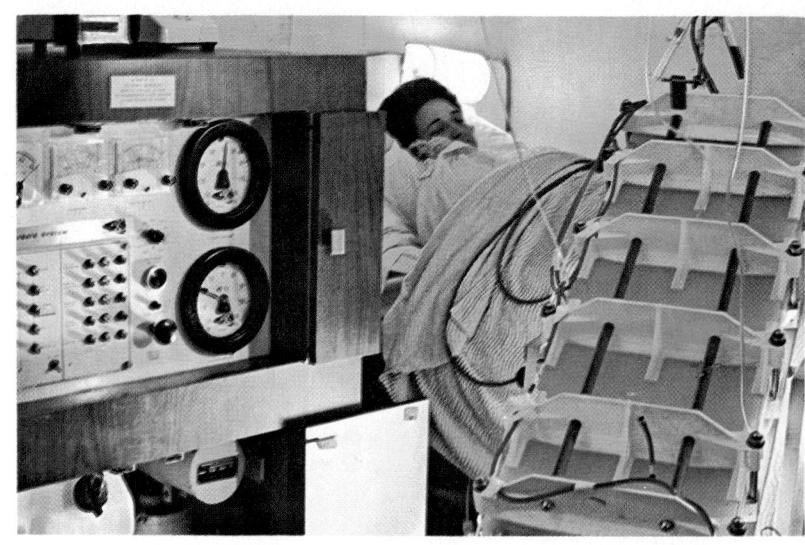

A growing number of patients who would once have died from kidney failure are now kept alive and active by regularly attaching themselves to a kidney machine, in which the cleansing of waste products from the bloodstream is taken over from the failing kidneys and carried out by dialysis through a membrane.

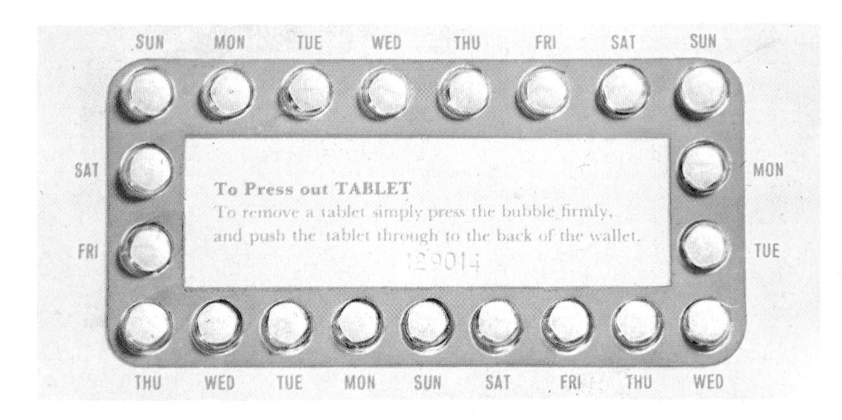

A 21-day course of Ovulen contraceptive pills.

traceptive in women. Because one side-effect was that the endometrium – the inner layer of the uterus – was shed at irregular times during the menstrual cycle, oestrogen was added to control this 'breakthrough bleeding'.

Since then, several variants of the contraceptive pill have appeared on the market, and have been shown to be effective in wide-scale trials. They all appear to work by inhibiting the release of hormones necessary for ovulation in the same way that hormones produced naturally during pregnancy prevent further conception. They may also thicken the mucus in the woman's cervix, making it impenetrable to sperms. Surveys have shown a failure rate of only one pregnancy per thousand women per year.

B.D.

A modern cardiac pacemaker. First seriously suggested in 1862, the first working model, made 70 years later, weighed 7·2 kg.

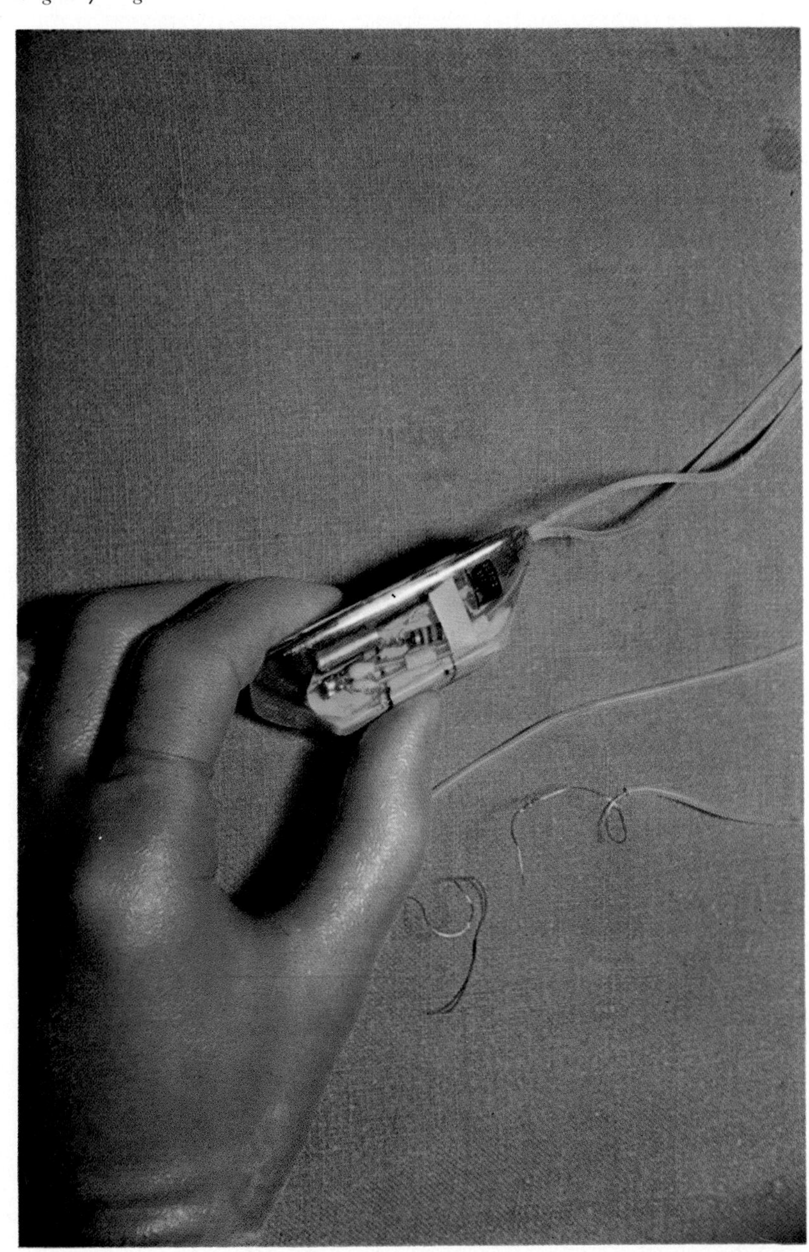

Cardiac pacemaker

Normally, the heart carries out its job of pumping the blood by means of a built-in system of rhythmic electrical impulses which are carried through the heart by nerves that lead to its muscle fibres, making them contract. Two main nerves lead to the ventricles which are responsible for the blood-pumping; if one of them fails to function properly the heart beats irregularly, and if both nerves fail for a few seconds, the brain gets insufficient blood and the patient faints. As a rule, the system soon begins to work again, but there is always the danger that the brain may be left without its blood supply for a few minutes and suffer permanent damage, or even that death may occur. The heart has a secondary, stand-by impulse system which may take over in an emergency, but it produces only half the necessary heart-beats per minute, not enough to keep the whole organism in action.

The first serious suggestion of using 'faradaic' (inductive) electric stimulation in cases of cardiac arrest was made by an English surgeon, W. H. Walshe, in a treatise which he published in 1862. Ten years later his French colleague Duchenne de Boulogne described, in his paper *Électrisation localisée*, some successful experiments with his apparatus which he called *le main électrique*, the 'electric hand': one electrode was placed on the skin of the patient suffering from cardiac arrest; the other was held by the doctor in his right hand while he placed his left hand rhythmically on the patient's thorax. This produced contractions of the heart muscle.

It was, however, an American heart specialist working for the US Navy, A.S. Hyman, who developed the first effective cardiac stimulator for clinical use in 1932. He called the apparatus, which weighed 7·2 kg., an 'artificial cardiac pacemaker', thereby introducing the term into medical language. Technical development during and after the Second World War made it possible to reduce the size of the pacemaker so much that the patient could carry it permanently in his body, and after 1950 pacemakers of almost a dozen different types were rapidly developed.

The pacemaker is not an artificial heart, nor does it take over the job of pumping the blood. It merely generates electric impulses. Some types do this all the time, others only when the natural system fails. The pacemaker is a miniature electronic unit implanted, as a rule, immediately under the skin of the chest so that it can easily be taken out and replaced. It has a battery and one or more transistors which amplify the weak current from the battery. As with Dr Hyman's original large-scale model, a wire runs from the pacemaker to the surface of the heart, or through a vein leading to its interior, ending in the right ventricle.

Thanks to the amplifying action of the transistors, pacemaker batteries have to supply very little current and therefore last for years before they have to be renewed; the latest development is nuclear batteries in which the heat from pellets of the radioisotope plutonium 238 generates an electric current. These batteries are expected to have a life-span of 10 years.

E.L.

Spear

Wooden spears have survived from the Lower Palaeolithic period (before 100,000 BC – one found at Clacton, Essex, for example, is of yew) and flint spearheads for attachment to wooden hafts date from *c.* 30,000 BC (Upper Palaeolithic). Whereas, before this, man would have hunted animals by throwing rocks at them, he was now able to arrange hunting bands, probably with beaters – and this implies an important step in his development.

The spear-thrower – a stick with a crook at the end in which the end of the spear-pole fits – gave added impetus to the weapon; it appears in the Dordogne *c.* 10,000 BC. The harpoon, a multi-barbed spear, appears at about the same date.

As a weapon of war rather than a hunting tool, the spear was favoured much earlier than the sword (which cannot, of course, be made from stone): Egyptian foot soldiers carried them, made of copper, then bronze, during both the Old and New Kingdoms (2700 BC onwards), and the Sumerians before them (3000–2350 BC) gave them to infantry and to charioteers. For the Persians also the spear was the main weapon, as shown by the procession of 'Immortals' – Darius' personal body-guard – on a stone relief in the Royal Palace at Persepolis.

It continued in service among the Greeks: javelin-throwing was one of the events of the Olympic pentathlon, and this introduces the question of how spears were used. The 'hoplite' infantryman was equipped with a Macedonian pike, with a tip at either end in case of breakages, and he clearly thrust with it. And in Homer, Patroclus cannot use Achilles' spear because it is too heavy for him – so this too would be for thrusting; but Homeric heroes often set off into battle with two spears and arrive with one, the implication being that one was thrown, and the heavier one kept for close work.

The lance – the cavalry version of the thrusting spear – was perhaps first used by the heavy cavalry of Alexander the Great; his cataphracts wore armour, and carried sword, shield and lance. From this time on, lancers are common, in Rome as in Islam and Christendom. In medieval Europe jousting, intended to perfect knights in the arts of war, degenerated into a deadly sport – Henry II of France, for example, died in a tournament in 1559. The lance, however, remained a favoured weapon until surprisingly recent times.

C. M. B. G.

Shield

Essential equipment in hand-to-hand combat, the shield was probably the first piece of armour to be invented. The earliest ones would have consisted of a wickerwork frame covered with skin, for metal shields would have been too expensive to issue to the rank and file of any organized army – leaving aside the question of weight. This early type was well established in the 3rd millennium BC: a Mesopotamian relief of *c.* 2500 BC shows soldiers in leather caps, their bodies protected by full-length square shields, carrying spears and advancing in a phalanx. The Egyptian armies were equipped in the same manner, as tomb-models show us.

Such shields could not be expected to survive in the ground. What has survived, however, is a quantity of expensive shields buried as prestigious grave-goods, such as that from Sutton Hoo in England (7th century AD), which were surely no more than ceremonial parade armour – although there are signs that the Sutton shield was in its day damaged and repaired. But to take such items into battle would have been akin to stock-car racing with a Rolls-Royce – a rough financial equivalent to the cost of a full suit of armour in the Middle Ages. The inside of the Sutton shield shows how it was carried: the forearm was passed through a leather strap, while the hand grasped a bar at the shield's centre.

The decoration of shields with precious or semi-precious stones and metals is probably as old as the shield itself. In the *Iliad* the shield which Hephaestus made for Achilles, 'large and powerful, adorned all over, finished with a bright triple rim of gleaming metal . . . fitted with a silver baldric', was decorated with scenes encompassing the whole of Heaven and Earth. The Vikings, whose shields were round and of wood, also decorated them, hanging them in rows on the sides of their long ships; Bragi the Old in the 9th century AD describes the sight as resembling 'leaves on the trees of the sea-king's forest', and the same poet composed the first of a long line of shield-

Above, *spear and shield are among the oldest weapons of war. This model of a body of Egyptian infantry comes from a Twelfth Dynasty tomb at Asjut.*

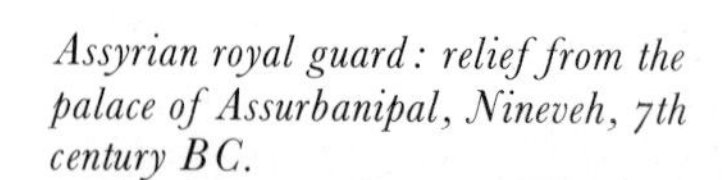

Assyrian royal guard: relief from the palace of Assurbanipal, Nineveh, 7th century BC.

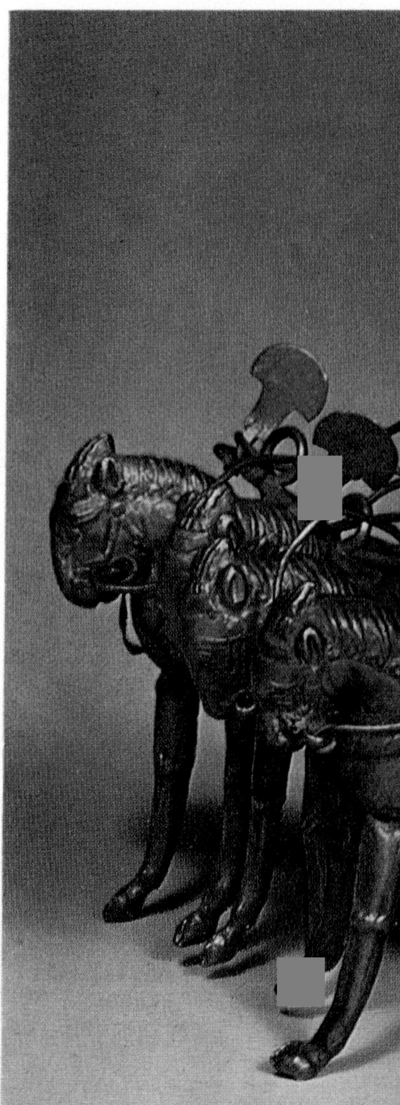

Menelaus, in his fight with Ajax depicted here on a Greek red-figure cup, uses a short sword equally useful for cutting and thrusting.

Below, *gold model of a Persian war chariot of the 6th century BC.*

songs, in which the singer gives thanks for receiving a finely decorated shield.

From such decoration to a fully developed system of heraldic devices – originally intended to announce one's identity in battle – was a short step; and the shield, the largest object presented to the enemy's view, was the main medium for the display of such emblems.

C. M. B. G.

Chariot

The first wheeled vehicles probably appeared in Mesopotamia *c.* 3500 BC, but these were clumsy contraptions for freight carriage, with four solid plank wheels; they were pulled by oxen or, for the lighter sort, onagers. The war chariot seen on the Standard of Ur (between 3000 and 2500 BC) is of this basic type. The Amorites, who followed the Sumerians, appear to have known the horse, but only used it for carriages, not for military vehicles.

It is only when a two-wheeled chariot appears, light and manoeuvrable, with spoked wheels and a horse or a pair of horses placed between the shafts, that we can begin to talk of the war chariot as a devastating force in Middle Eastern warfare. This happened *c.* 1800 BC, and it was the Hittites who were the first people to owe their power to chariotry. The archives of Boghazköy have yielded notes on the training and acclimatization of horses, and we can see them and their chariots in action against the Egyptians at the battle of Kadesh, in 1300 BC, on reliefs at Abydos (reign of Rameses II, 1290–24) and at Karnak, Abu Simbel, Thebes and Luxor. The hunting chariot is, of course, of the same nature, also the Egyptian state chariot (examples of which were found in the tomb of Tutankhamun).

Chariots made battles faster and general communications much easier, and therefore enabled the great empires of Egypt, the Hittites and Assyria to maintain their sway. These early examples would have been made from bent wood, the body perhaps covered in wicker or oxhide, with a basket attached to the 'dashboard' containing arrows or short spears – for the main weapons used from the chariot were bow and spear: one man would drive, and one or two others shoot.

Chariot-racing was probably a military sport from early times; the names of winners at the Olympic Games are recorded from 776 BC, but it was the Romans who first featured it in their Circus entertainments (the gladiators were a later introduction, and only marginally more bloody). But chariots for war were already obsolete in Homer's day. Though Alexander the Great was the first to show what could be done with cunning cavalry tactics, the introduction of mounted troops had much earlier served to make the chariot redundant.

Contrary to legend, there is little evidence that the ancient English Queen Boudicca used knives attached to her chariot wheels; in any case, she would have been far from the first to mount 'mowers'. Curtus Rufius describes their use by Darius III of Persia against Alexander the Great.

C. M. B. G.

Sword

No other artefact occupies such a firm place in literature as the sword; in medieval epics, the hero often feels an emotional attachment to his weapon. Swords commonly received names as if they were persons: Roland's sword (in *The Song of Roland*) was Durendal, Charlemagne's was called Joyeuse, and King Arthur's Excalibur. They received such veneration because making such long objects sufficiently durable required great skill.

True swords – as opposed to daggers which might only be a quarter of the length – appear only in the Bronze Age, and we find some early examples in the shaft-graves at Mycenae (*c.* 1650 BC), in one of which Schliemann found over 90 swords, many so sumptuously decorated and inlaid that they must have been for ceremonial purposes only. The majority were large thin rapiers, with blades often 3 ft. long – used of course for thrusting, not cutting. A few examples of a much more serviceable shape and size, however, were found there as well: shorter, flanged blades with enlarged tang, the flatter blade suggesting that they were used for cutting as well as for thrusting. By the Palace Period of Crete (*c.* 1350 BC), structural strength had been added by increased ribbing, and by making 'horns' at the hilt, which also acted as a hand-guard; this type was imitated in the Near East and in the Danube region. The Greeks used several types of sword, and while there are some single-bladed examples for slashing with, most swords were probably used for jabbing – by analogy with the spear, the Greeks' preferred weapon; Homer, though he speaks of silver-studded swords – no doubt a reference to Mycenean types – does not elaborate on how ordinary swords were used in battle. Much later, Tacitus tells of the advantage the Romans had over the Britons because they possessed swords with points and knew how to use them.

With the development of plate armour in the later Middle Ages, heavy weapons nearly 5 ft. long, and often requiring both hands, were developed for thrusting as much as for cutting; most fearsome was the German type seen wielded in the 1512 *Triumph of Maximilian* by Dürer. There was also the *cinquedea*, an Italian sword – or dagger, for it is found in all lengths – five fingers broad at the hilt. A lighter,

Man Living

basket-hilted broadsword appeared in the 17th century, together with a single-bladed variety known as a backsword; later military swords are descended from this latter, and also from the Eastern scimitar.

Japanese long swords are often masterpieces of the swordsmith's craft; the long process of manufacture involves encasing a soft iron core in a cutting edge of high-carbon steel; often tested for efficiency on prisoners, it is said that a good Japanese sword, placed in a stream down which a lily is gently floating, will cleave it cleanly in two.

C. M. B. G.

Siege engine

Siege engines were, in a sense, the ancient equivalent of the modern tank: some engines threw projectiles, while others were moved close to heavily defended positions – usually city walls – to destroy them. The type of warfare involved in attacking heavily defended and static positions was in existence by the time of the Hittites (from *c.* 1380 BC) and lasted until after the time of Sebastien de Vauban, Louis XIV's chief military engineer, several of whose defences (e.g. at Briançon and Neuf-Brisach) remain in good condition.

Even with great engines, sieges could last a long time: Alexander the Great had great trouble at the seven-month siege of Tyre, in 332 BC, but his eventual success, as at Metone in 354 BC, was due to his use of the catapult supposedly invented by a Sicilian general called Dionysius of the early 4th century BC.

But it was the Romans who made the most deadly use of siege engines; as well as towers and rams, they had throwing machines. Their catapults would fire javelins of up to 200 lb. in weight to a distance of nearly 500 yds., and the *ballista*, worked by torsion of ropes, could hurl a 500 lb. boulder slightly further. Josephus gives a graphic account of Roman tactics at the siege of Jotapata in the Jewish War of AD 68: surrounded on three sides by precipices, the Romans were only able to reach the fourth side after five days of battle. Hurdles were built to protect soldiers building an earth bank to the top of the walls, and 160 catapults kept up a covering fire of boulders, arrows and fire-darts – but the Jews only heightened their defences. A battering-ram, its bearers protected by horizontal hurdles, was brought forward, but the Jews attacked it with boulders and pitch, and it broke. Scaling-ladders eventually won the day, although the men underneath the 'tortoises' – an armour plating made by interlocking their shields over their heads – were badly burned by the boiling fat and pitch poured upon them.

The techniques for raising a siege were unchanged until well after the introduction of gunpowder – for gunpowder made possible even more deadly siege engines, if we count the cannon and mortar as such. When Leonardo da Vinci wrote to Ludovico Sforza il Moro at Milan (*c.* 1483) asking for employment, he stressed his attainments and projects in military engineering; painting and sculpture come low down on his own list of his abilities.

C. M. B. G.

Umbrella

The umbrella has been in use for at least 3000 years as a shade from both sun and rain. It was known in China in the 11th century BC and subsequently in Assyria, India and Persia, as archaeological finds have shown. From there it travelled to Egypt, where it was – as everywhere in the East – an emblem of rank and distinction; it still is in tribal societies, particularly in Africa, where the umbrella-bearer walks behind the great man. In ancient Greece and Rome, however, the umbrella was regarded as effeminate and rarely used by a man.

Little is known of the umbrella's use in the Middle Ages until the 12th century, when a doge of Venice had a ceremonial one made for himself: later the umbrella found its way into the regalia of the Catholic Church. The Spanish invaders of Mexico found the Aztec kings using umbrellas, and the English saw Indian princes walking under them – and, if some reports are correct, even going into battle with them.

England was probably the first European country in which, for climatic reasons, the rain umbrella became a utility for everyday use. Literature gives us the references: Drayton mentions the umbrella as early as 1620, Swift in 1710 (in *City Showers*) and John Gay in 1716 (in *Trivia*). Still, the general acceptance of the umbrella did not begin until the second half of the 18th century; the pioneer was Jonas Hanway, who first appeared with one in Pall Mall around 1750. He was a businessman who had made a fortune from trading with Russia and the Far East, where he had seen a good many umbrellas; at the early age of 38 he had amassed so much money that he could retire and devote himself to charity, founding hospitals and orphanages. He has a monument in Westminster Abbey. One of his ambitions was to introduce the umbrella for the common man and woman.

For the last 30 years of his life he invariably took his daily walks with his umbrella, undeterred by the street urchins' scornful whistling and the hackney coachmen's protests – they splashed him with mud and water by deliberately driving through the nearest puddles when he was about, for they

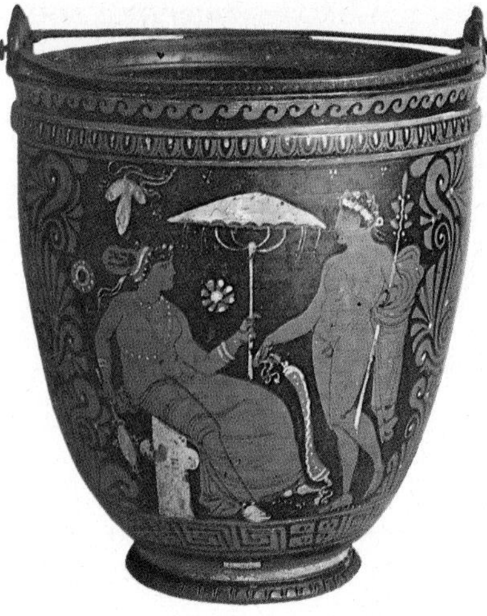

The umbrella as a ceremonial object, on a Greek urn of c. *340 BC, from Apulia.*

Right, *15th-century picture of a catapult with a lever arm, at the siege of Jerusalem.*

Below, *the warship* Mary Rose. *Her gunports were too near the waterline for seaworthiness.*

feared for their livelihood if ladies and gentlemen took to using umbrellas instead of hiring coaches in the rain.

Hanway's umbrella, like most others made at the time, had cane ribs, which made the opening and shutting rather cumbersome: steel ribs were invented by Henry Holland and first used by British umbrella-makers in 1840. Mass manufacture of umbrellas began in Manchester, Glasgow and London, offering modestly priced versions to the citizenry – who adopted them gladly, as Hanway had hoped they would, not only for protection but as a new emblem of respectability in an age where gentlemen no longer carried their swords.

E. L.

Armour

Armour made to protect all or some parts of the head and body has been used since early classical times. Mycenean troops wore leather corselets and leggings, but the 'hoplite' infantrymen who had revolutionized warfare in Greece by the 7th century BC used metal armour. The Greeks and Romans used plate armour for the torso, back and front, often together with thigh pieces and helmets. Bronze and iron were in common use, and the metal was worked as sheet or plate. The main body protection, the cuirass, consisted of breast and back pieces fastened with straps or buckles.

An important development in Roman times was the introduction of two kinds of mail armour – chain and plate mail – consisting of interlinked metal rings, sometimes sewn to strong linen or leather. The hauberk, for example, was a complete suit of mail, made sometimes in the form of a coat, fastening at the front, and sometimes as a shirt. This sort of armour was in widespread use throughout Europe at the time of the Norman Conquest and for a couple of centuries after. The disadvantage of mail was that it lacked a rigid surface capable of deflecting blows. Gradually, therefore, in late medieval times, plate armour of iron or steel came back into fashion as it became technically possible to produce fine-quality metal to fit fairly close to the body. By the time of the Hundred Years War, mounted knights normally wore plate armour. But armour of this kind was expensive, and the common foot-soldier wore much lighter protection.

Armour remained in use long after the introduction of gunpowder, although it became increasingly decorative and less utilitarian. It persists, however, in the steel helmet and in the very light body armour for infantry used, for example, by American troops during the Korean War and by British troops in Northern Ireland.

N.G.

Warship

The Greeks were probably the first people to build ships specifically for war. By the time of the Peloponnesian War (431–404 BC) they had evolved the trireme war galley, distinguished from the slow and heavy merchant ship by its slender shape and lighter draught, with considerations of seaworthiness and comfort coming well after those of efficiency and manoeuvrability. Triremes were simply constructed and it happened several times that a state which had previously lacked a fleet altogether was able to build numbers in a short time and challenge a hitherto well-established naval power.

Other early maritime peoples made less of the distinction between ships of war and those for travel, exploration or trade, and it was not until the invention of the cannon, which in fact wrought its most radical changes at sea, that the concept of the warship as such became universal. It was impossible to make wooden ships invulnerable to gunfire, and sailors could not escape by digging into safe earth. Henry VIII understood the importance of both cannon and sailing-ships and in the early 16th century he financed a new navy with money taken from the Church. Until then naval battles were fought by modified merchant ships and had mostly been a matter of grappling alongside and fighting hand to hand. Improved cannon enabled ships to stand off from the enemy and sink him by gunfire.

Speed and seamanship became the deciding factors in naval war. The old notion of a floating fort was abandoned. No longer did ships need the absurd tall castles from which soldiers could shoot their arrows, although the word 'forecastle' still remains. Indeed, soldiers were no longer carried.

At first naval architects tried to fit the heavy cannon into the castles, but these were flimsy structures and the weight of the guns there made the ships unstable. It was probably a shipwright named Baker who thought of placing the guns as near the waterline as possible. Such daring design sometimes proved foolhardy and in 1545 the *Mary Rose* – a fine example of the new kind of warship – sank in a light squall when her gun-ports took water. Low gun-ports needed carvel construction for strength, so clinker-built ships began to go out of favour.

The crude 'one-sail, one-mast' design had all but disappeared half a century before the *Mary Rose* was built, but there were still new developments in sail-making. New fabrics, smaller sails and additional masts made sailing a more subtle craft and men discovered how to sail close to the wind. Sailors learned to cope with bad weather and great oceans, and used charts and instruments to navigate through them. It was the warship that was most suitable for the voyages of discovery that marked this age, for the object of exploration was conquest.

From this time onwards the broadside was to dominate naval tactics. Admirals led their fleets into battle sailing in line instead of in disorganized clusters. Such ships of the line were to undergo only slight changes right until the coming of steam, and guns changed even less. Those of the *Mary Rose* were salvaged from the sea bed and found to be almost identical to the guns used at Trafalgar.

L.D.

Bow and arrow

Bows are among the oldest known projectile weapons in man's history. The simplest form is no more than a stave, probably slightly curved by heating, fitted with a string. Sometimes bows of this sort were in one uniform curve; sometimes they consisted of two symmetrical arches connected in the middle by a straight 'grip'. Over the centuries two broad types of bow were developed in Europe – the longbow and crossbow – both of which remained in use until the 17th century, long after the introduction of firearms. The longbow was a larger, heavier version of the simple single-stave bow, requiring more strength to operate and lethal at greater distances. It was probably found originally in Wales, and was taken into use among English troops probably in the 13th century. It was about 6 ft. long and in the hands of English archers in the 14th and 15th centuries it became a battle-winning weapon, partly because it was simple and quick to shoot – a good archer could shoot five or six arrows a minute at ranges of 150–250 yds. – and partly because archery was so encouraged and protected by legislation that it became a relatively widespread and highly skilled art. An Act of the latter part of the reign of Henry VIII, for example, provided that the inhabitants of all cities and towns were to make butts and to keep them in repair, and to exercise themselves in shooting at them on holidays, under a penalty of 20*s*. per month.

The crossbow consisted of a small bow fixed to a staff or stock of wood which the bowstring crossed at right angles. In its simplest form it was shot by hand, but tension was increasingly produced by some form of winding mechanism. The heaviest crossbows were sometimes worked by a windlass attached by a metal cap to the butt of the bow-stock. Depending on the mechanism, the crossbow could fire anything from small arrows up to heavy metal bolts or quarrels. The bolts were normally shorter than arrows and had heavier points or heads; they were winged with feathers or even with metal, and sometimes the metal wings were set spirally to rotate the bolt. It was very accurate, but its arrows did not carry so far and it took much longer to load and fire.

N.G.

Gunpowder

'It is madness to commit a secret to writing, unless it be so done as to be unintelligible to the ignorant and only just intelligible to the best educated.' Thus wrote Roger Bacon (1214–92) and secreted the secret of gunpowder in the cryptogram: Sed tamen salis petre LURU VOPO VIR CAN UTRIET sulphuris. Rearranged this becomes R. VII. PART V. NOU CORUL. V. ET and thus the secret is revealed as 'but take 7 parts saltpetre, 5 parts young hazelwood [charcoal] and 5 sulphur': and he claims that such a mixture will explode and either blow up an enemy bodily or put him to flight in terror.

Bacon was born near Ilchester in Somerset and studied in Paris and Oxford. After becoming a Franciscan monk he devoted his time to experimental philosophy and science, including chemistry, optics and physics. His writings were condemned by a Franciscan council, however, and he was jailed for 10 years for dealing in magic.

Roger Bacon does not enter history unchallenged. Berthold Schwartz – the black monk – has a well publicized claim to be the inventor of gunpowder that many continental schoolbooks still endorse. A statue to him in Freiburg im Breisgau dated his invention to 1353. In view of the lateness of this invention some of his champions modified his achievement to the equally unlikely invention of the cannon. It is now widely doubted whether Schwartz ever existed.

Certainly other men recorded the formula for gunpowder. The Chinese have been credited with the discovery and in Asia they had been experimenting with mixtures that contained saltpetre long before Bacon was born, but these may well not have been explosive mixtures. They may have been for the manufacture of bigger and better fire-arrows and rockets.

Even when the formula was widely known the manufacture of gunpowder depended upon the quality of the ingredients. Even in Napoleonic times the Paris Académie offered a prize of 4,000 livres for a process that would give good-quality saltpetre.

Gunpowder ended the permanent medieval world in a way that no other invention – except perhaps printing – could do. Gunpowder killed both knight and serf and in doing so proved them equal.

L.D.

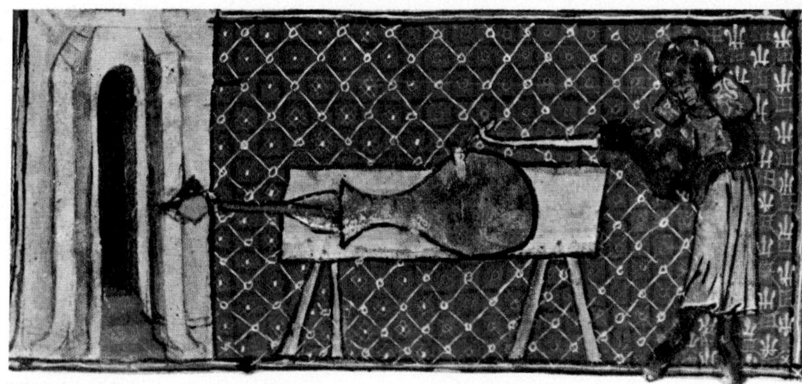

The 'dart-firing vase' – an artillery forerunner of 1324.

A 16th-century soldier taking aim with the crossbow. Detail from Martydom of St Sebastian *by G. Lopez.*

Testing a sample of gunpowder – seemingly an alarming and hazardous operation. From a 15th-century Viennese manuscript.

Grenades, emitting incendiary flames and poisonous smoke, depicted in a 17th-century Chinese book.

Artillery

It was very soon after the discovery of gunpowder that men thought of firing projectiles through a tube. However, the first practical gun and its inventor have not been recorded. Some say it was a bamboo tube in China while others insist it was more like a bucket used by the Arabs in Spain. In Europe it was the town of Amberg that in 1301 first encountered the gun fired in anger. In 1326 the first undoubted reference to guns is found in a command by the Council of Florence to prepare iron bullets and metal barrels. In that year a coloured drawing was made for Edward III by his chaplain Walter de Milemete. It provides us with our earliest picture of a gun. Its gold colour is probably because it was made of cast brass. Such a weapon was used at the siege of Metz in 1324. This 'dart-firing vase' was called a *pot-de-fer* and it is significant that its *gynour* (gunner) stands well away from it. The following year Edward III records using a 'few small crakys of war' which probably looked like this. A century later James II of Scotland was 'dung in two by a piece of mis-framed gune that brake in the shuting, by the which he . . . died hastilie.'

As well as being dangerous the new devices were expensive. Their cost was one more step along the road to national armies, for monarchs saw artillery as a way of breaking the power of the nobility by destroying their forts. Upon his cannons Louis XV had engraved 'Ultima ratio regum' – the ultimate argument of kings. However, the cost was eventually to put kings in the power of the bankers.

At first guns were only marginally more effective than the trebuchets, ballistas and mangonels that continued in use beside them. At Crécy and Agincourt the guns made no difference to the final victory. But gradually the science of gunnery improved. The military architect had to revise his craft accordingly: curved walls and lower silhouettes began a design trend that did not stop until fortifications became subterranean.

But artillery also improved. From its original place on the battlefield – ahead of the fighting men – better range enabled it to move further back. In retrospect the history of the bombing aeroplane might be regarded as just an intermediate stage in the history of artillery, and intercontinental missiles now enable the gunner to be an ocean or two away from his target.

L. D.

Shell

All early projectiles (except for burning torches and the like) relied on momentum to damage or destroy their single target. The advantages of the exploding shell over simple cannon-balls were two-fold: because its effect does not reside in the speed at which it leaves the barrel, or in its weight, the powder charge and the piece of ordnance can be fairly light; and, of course, it can hit several targets at once.

How old is the shell? It is known that shells fitted with fuses were used at the siege of St Boniface (Corsica) in 1421, and possibly the Venetians used them at Jadra in 1376. Firing such contraptions must have been an extremely risky business for the gunners: a bronze or iron canister had to be filled with powder, and some kind of fuse attached and set burning – after which the shell was placed gingerly down a cannon barrel to sit on top of the firing charge: many barrels burst.

From the earliest times until at least the late 18th century, shells were known as 'grenades', from the French for 'pomegranate' – because the grains of powder inside the shell looked like so many pomegranate seeds. What we now call grenades were certainly in use in the 16th century, and grenadiers were first formalized into units by the French in 1668–70, in the Régiment du Roi. The device was quickly adopted by the English, and John Evelyn could write in 1678 of some military manoeuvres he had watched: 'Now were brought into service a new sort of soldiers, called *Grenadiers*, who were dexterous in flinging hand grenados, every one having a pouch full. . . .'

The shrapnel shell, invented in 1784 by the English Lieutenant Henry Shrapnel, was a great advance over earlier types, because it only needed a low powder charge: earlier designs had held a lot of powder, because the idea was that the explosion should shoot shell fragments in all directions. But Shrapnel's idea was that only just enough powder to open the shell-case was needed, to let the balls inside travel forward at the same velocity as the original shell. It was first used in action against the Dutch in 1804, at Fort Amsterdam, Surinam; but difficulties were encountered with pre-ignition, due either to the bullets driving out the fuse on discharge from the gun, and the blast igniting the powder, or the friction of the balls doing the same. From 1852, Colonel Boxer produced a refinement in the shape of a sheet-iron diaphragm to separate charge and fuse from balls; his shell was approved for use in 1864, and called the 'diaphragm shell'.

Other types of shell included the incendiary, packed with molten pitch or resin (known from at least the 15th century); the molten-iron filled shell, invented by a Mr Martin in 1855, and adopted for use, but lasting less than a decade for obvious reasons. The gas-filled shell perhaps has the longest history of all: it was first envisaged by the Emperor Leo VI (AD 886–911), who suggested the use of asphyxiating fumes derived from wet quicklime which

Explosion in a mine caused by fire-damp. To prevent this, Sir Humphry Davy (left) invented the miner's safety lamp.

Opposite, *diagram of the Davy lamp: 1, bonnet; 2, outer gauze; 3, lamp standard; 4, lamp glass; 5, wick; 6, oil vessel.*

could be thrown in earthenware pots at the enemy. Gas shells were avoided by the 19th century as too horrible, but were used in the First World War.

Great improvements in the standard shell began in 1867, with the introduction of Boxer's much more accurate time fuses; the notion of regulating burning time had been current since the 17th century, and by 1819 there were 21 varieties in use for shrapnel shell in the British Army. Then in 1882, black powder was replaced first by picric acid (soon abandoned because it attacked the casings) then TNT, and then, in 1891, by cordite – which is still in use.

C. M. B. G.

Naval mine

When, at the battle of Mobile Bay, David Glasgow Farragut bellowed 'Damn the torpedoes! Full speed ahead!' he meant mines – not torpedoes. The change of name came a little after the American Civil War, but before it, because they were a 'sleeping death' (*torpid* – according to the philologists of the 19th century), they were known as 'torpedoes'.

The first recorded use of mines was at the siege of Antwerp in the 16th century, but they were then only a primitive surface weapon. David Bushnell, the submarine enthusiast, invented the first more or less effective underwater device – a keg of gunpowder supported by floats and fired by a trigger mechanism when it struck the target. Like so many ingenious American devices it was used for the first time against a British ship, the frigate *Cerberus*, on the Connecticut river. It missed the *Cerberus*, but it sank a nearby schooner with great efficiency.

The mine reached its first stage of effectiveness in the late 19th century with Heinrich Hertz's invention of the Hertz 'horn', a protuberant triggering device which made it simple and effective. By the Russo-Japanese War mines were important factors in naval strategy, and in 1914–18 more than 250,000 were laid in the shallow waters of the world by the navies of both sides.

D. D.

Lightning conductor

Benjamin Franklin is certainly the originator of the lightning conductor or lightning rod, although the Czech Academy of Sciences has attempted to establish the claims of Father Procopius Diviš.

In 1749 the Academy of Bordeaux had offered a prize for an answer to the question 'is there any analogy between electricity and thunder?' The prize was won by an essay by a Dijon doctor, M. Barbarette, who claimed that they were the same but offered no experiments.

In the meantime in America, Franklin with three friends had been conducting experiments showing that electricity was attracted and emitted by sharpened pieces of metal. Among them was the electrifying of a metal crown surmounting an engraving of the King of England. Anyone who touched it received a shock, and thus electricity showed its loyalty to the monarchy.

Franklin performed his famous experiment of flying a kite in a cloud, showing that the key attached to the thread became electrified, in 1752. To turn this into a lightning conductor it was necessary to think of joining the conductor to the earth. In 1749 Franklin had also suggested erecting a pointed iron rod on the top of a hill or a tower to see if it were electrified.

It was in Franklin's *Poor Richard's Almanack* of 1753 that he wrote a detailed description of the lightning conductor itself. The same year the Royal Society presented him with the Copley medal, and the Society for the Encouragement of Arts elected him a member.

Franklin now wrote a descriptive practical essay, 'How to Secure Houses, etc, from Lightning', and soon lightning rods were being set up in the cities of America. The clergy opposed this on the grounds that lightning expressed the wrath of God, and Franklin's collaborator, Ebenezer Kinnersley, gave lectures to prove its lawfulness. One doctor, the Reverend Theodore Prince, argued that if the lightning were conducted to the earth the earth would become increasingly electrified at that point and earthquakes would be more likely. 'Oh!', he exclaimed, 'there is no getting out of the mighty Hand of God! If we think to avoid it in the air, we cannot on the earth. . . .'

In France opposition was strong, and in Saint-Omer the citizens filed suit against a certain M. Vissery de Bois Vale, who had put up a lightning conductor, and were frightened that Heaven would punish this blasphemy. (Robespierre was the defending lawyer.)

Diviš erected his lightning rod in Moravia on 15 June 1754, clearly later than Franklin's, though he claimed to have had a plan completed by 1752 or early 1753. However, by this time Franklin's work was already known in Europe.

G. R. T.

Lock and key

Like so many things which our Western civilization takes for granted, the lock was invented by the Chinese, at least 4,000 years ago. It was basically a bolt which could be fastened and unfastened from the other side of the door or gate to which it was fixed, by means of a sickle-like hook – the key. The Egyptians improved the device by giving the key some pins corresponding to the retaining

The first bolt-action, breech-loading rifle.

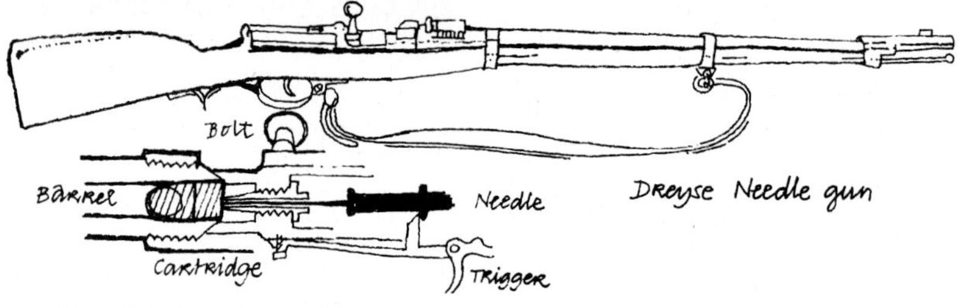

The birth of the lightning conductor: Franklin's experiment (1752).

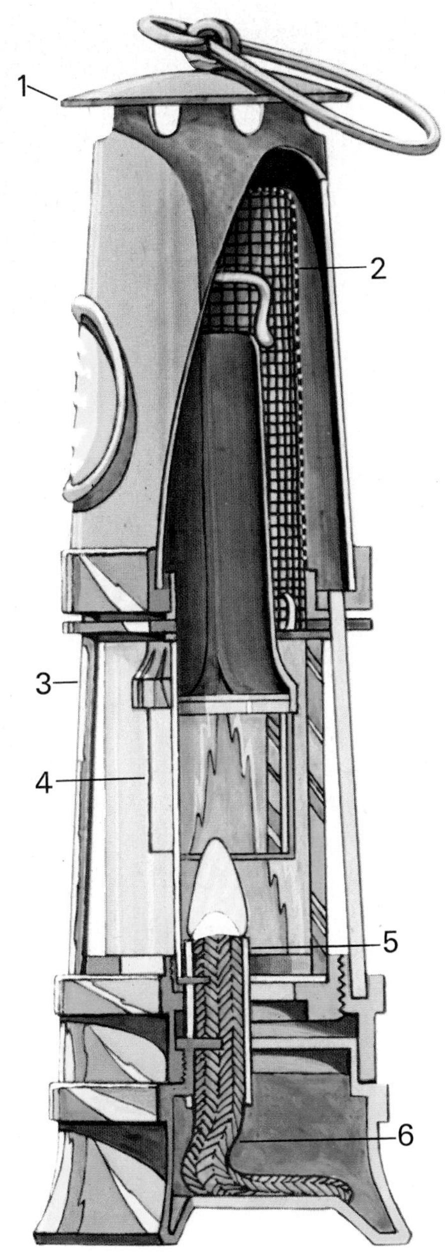

pegs, and made the part of the hole into which the latter were inserted hollow. Variations of the Egyptian locks were used by the Romans, who brought them to all parts of Europe which they conquered; however, the Roman locks had smaller bolts, and a spring was used to press the dropping pins down.

The medieval nobles, merchants and other people with money and valuables needed locks and keys not only for their doors but on their coffers, chests of drawers and wardrobes, thus stimulating the inventive minds of the craftsmen. Large, often artistically embellished padlocks were in fashion, and instead of dropping pins a pivoted tumbler came into general use in the Middle Ages. Another improvement which made the 'picking' of a lock more difficult was the interposition of a number of so-called wards, or impediments, between the key and the bolt; the key bit is designed so as to escape them.

The modern age of lock and key may be said to have begun with the Bramah lock of 1778, with half a dozen 'sliders', thin metal plates with notches, which have to be brought into certain positions before the key opens the lock. Its inventor, Joseph Bramah (1748–1814), was one of the most fertile technical minds of his time in England; among his other inventions were the hydraulic press, a water-closet, and machines for printing banknotes, making paper, and aerating water.

Then came the Chubb lock, which was invented in 1818 by a Portsmouth ironmonger, Charles Chubb (1773–1845), who later founded a factory for fireproof safes. His main improvement of the lock was the 'detector', an extra lever which fixes the bolt if an attempt at picking the lock is made.

The Yale lock, invented by an American, L. Yale (d. 1868), came into use in the 1860s; it is a tumbler lock with a small, flat key which interlocks along its length with the keyway. Its safety factor lies in the infinite number of variations of the key profile so that practically no two Yale keys are identical; but a master key will fit a series of locks, for instance in a hotel. Then there are combination locks for safes, which cannot be opened without knowledge of a code word or series of figures, and time locks for strongrooms, with one or more built-in clocks, which cannot be opened before the pre-set hour.

E. L.

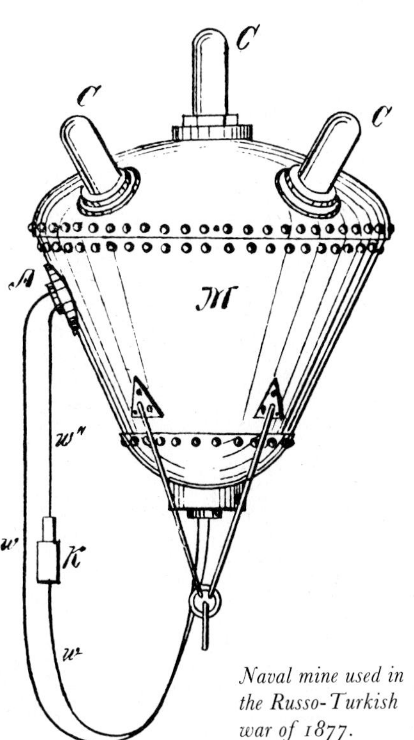

Naval mine used in the Russo-Turkish war of 1877.

Miner's safety lamp

As the industrialization of Britain escalated, so the coal mines had to go deeper. They also became bigger. In Northumberland and Durham in particular, this increased the risk of the danger from fire-damp, an inflammable gas (mainly methane) which oozed from the coal seams.

Fire-damp explosions used to kill hundreds of miners every year. Pits were transformed by a single spark into something like an enormous gun. The scorching blast seared through every gallery, and erupted out of the mine in a great inverted cone of débris.

At Felling Colliery, Durham, a terrible fire-damp explosion took place in 1812, in which 92 miners died. The local gentry decided to do something if they could, and in the following year they formed a society for preventing accidents.

The foremost chemist of the time was Sir Humphry Davy, a Cornishman from Penzance, who, though largely self-taught, displayed such inventive genius that he became professor of chemistry at the Royal Institution at the age of 24. In 1815 he was asked for his help. Davy analysed the fire-damp and found it to be a combination of hydrogen and carbon which would not explode if mixed with less than six times, or more than fourteen times, its volume of air.

He also discovered that the gas, in whatever proportion it might be mixed with air, will not explode in a small tube the diameter of which was less than $\frac{1}{8}$ in.

Bearing in mind these and other properties of the gas, he tested prototypes of four different kinds of lamp, all of which rested on different arrangements, and all of which were safe. The simplest however was the Safe Lamp. Davy's own description is best: '. . . a candle or lamp burns in a safe lantern which is air-tight in the sides, which has tubes below for admitting air, a chamber above, and a chimney for the foul air to pass through; and this is as portable as is a common lantern, and not much more expensive.'

Davy found a little afterwards that wire gauze served the same purpose as his little tubes, so long as the diameter of the holes equalled their depth. Iron wire gauze with 740 holes to the square inch was adopted as standard.

The wire-gauze lantern had an additional advantage in that the presence of fire-damp augmented the flame, both warning the miner that it was in the atmosphere and giving him light to seek safety. In the first models, the light would have dimmed and gone out. Later, Davy found that by twisting the gauze further safety factors were introduced.

Davy refused to patent his invention. That it was his was at one point challenged by supporters of a lamp made by George Stephenson the engineer. The overwhelming evidence of Davy's contemporaries supports his claim. The miners themselves called their lamps 'Davys', and testimonials, a set of plate and a baronetcy were just a small part of the proof of general public esteem.

G. N.

Fire extinguisher

'A small quantity of water, well directed and early applied, will accomplish what, probably, no quantity would effect at a later period.' Thus Captain Manby stated the purpose of his invention, the fire extinguisher.

George William Manby was born at Denver, Norfolk, in 1765 and after a military education became a captain of militia and ultimately in 1803 Barrack Master at Yarmouth. He was responsible for several inventions for saving life, including the rocket line for the rescue of sailors from shipwreck.

In 1813, watching a fire in Edinburgh where the firemen were unable to get the hose to the fifth floor in time to save the building, he was led to invent the portable extinguisher. It consisted of a strong 4 gal. copper vessel with a stopcock and jet-pipe screwed into the neck, the pipe extending nearly to the base of the vessel. Three gallons of water were put into the vessel and air was compressed into the space above by means of a pump; the stopcock was then closed and the pump removed. When the stopcock was turned back the compressed air ejected the water in a jet. As the appliance was heavy, it was supplied with a leather case and strap so that it could be slung from the shoulders, and a handcart was designed with six compartments to enable the fire watch to bring a supply of these 'engines' rapidly to a fire.

Manby sold several hundred of these extinguishers, but he failed to get them generally adopted. At the end of the century, chemical extinguishers came into vogue, in which the gas pressure was produced by carbon dioxide generated chemically. After the last war, however, it became the practice to operate water-type extinguishers by means of CO_2 under high pressure in a cartridge. The latest type of water extinguisher, the 'maintained-pressure' type, is pressurized with air by means of a pump, and is thus an unacknowledged reversion to Manby's original extinguisher.

K. G.

Rifle

The American War of Independence demonstrated what concealed riflemen could do to 'toy soldiers' marching to the formalized football matches that Europe knew as battles. A breech-loader could be reloaded more quickly than a muzzle-loader, and the reloading could be done by a man in the prone position. Already rifling, and the Minié bullet that expanded into its grooves, had made the infantry deadly; breech-loading could make them virtually invisible.

Breech-loading guns had been invented regularly ever since firearms first appeared. Some of them – complex, expensive prototypes handmade for

The first guided missile? Whitehead's torpedo of 1866.

Gas sentry ringing the alarm, June 1916.

sportsmen – had even worked. The greatest problem was to prevent the explosive gas from blowing back into the face of the user. The eventual solution used a self-contained cartridge – detonator, powder and ball sealed together.

Since the discovery of fulminate of mercury (a sensitive explosive that could be used to detonate gunpowder) in 1800, new ideas had followed rapidly. Such unlikely people as a Scottish clergyman named Forsyth and a landscape-painter named Shaw had contributed percussion ignition and copper percussion caps and a Swiss engineer, Pauly, invented a self-contained cartridge. It remained only for J.N. von Dreyse, a Prussian gunsmith who had once worked under Pauly, to add his ideas of bolt action and the long striking pin that was cocked by it. He called his gun a *Zündnadelgewehr* or 'needle gun'. It used a combustible paper cartridge with a pellet of fulminate between charge and bullet. When the trigger was pulled, a long needle went right through the base of the cartridge and into the powder to pierce the detonator. The name of his gun shows how much importance was given to the firing-pin, but the bolt action was the more important part of the design.

In fact the Dreyse gun, because of its large calibre (and the resulting low muzzle velocity) and because of gas leaks through the bolt action, was far outraged by the Austrian muzzle-loading Lorenz rifle. When war came, however, rapid loading in the prone position made the Dreyse gun the superior weapon and gave the Prussian Army their devastating victories in the 1860s.

Prussia's success with the Dreyse gun made others anxious that their own infantry should have its equal or something better. The French got more than its equal in the Chassepot design. This was a smaller-calibre gun with a rubber ring to reduce the escape of gas. The British got something very much better in the thin brass cartridge that expanded to make a gas-tight seal and thus solved the problem of blow-back.

The bolt-action rifle – direct development of Dreyse's gun – became of standard design. It had a high muzzle velocity and a range of between 2,000 and 3,000 yds. with a rate of fire of about 10 rounds a minute. It meant the end of cavalry or infantry manoeuvring in daylight on a battlefield. In 1870 at Sedan the last great cavalry charge of history suffered terrible carnage from a single volley. Such rifles are still the most widely used infantry weapons.

L. D.

Camouflage

It seems reasonable to assume that man first learned to camouflage himself by looking at the animal world, and seeing how certain insects, animals and birds blended in with their surroundings, thus securing protection from attack. Its value to man was for attack rather than defence, and no one knows whether prehistoric man used such methods. The painting of the so-called 'sorcerer' from the cave of Les Trois Frères, Ariège, France (15,000 BC or later) shows him wearing a complete stag-suit with antlers – and antlers pierced at the base for some kind of fastening had been found in excavations at Star Carr in Yorkshire, indicating that the practice was widespread. But whether such suits were for some magic purpose, or for use as hunting decoys – like rubber pigeons today – is not known.

But we today associate camouflage with military activity, and its history here is short. Until the late 19th century, war was conducted according to rules: there were deeds a well-born man would not stoop to, and weapons he would not carry. Fighting was done in the summer months, in daylight, and either by siege or by pitched battle – with banners waving and drums beating.

But as methods of war changed, so did uniforms; before the 17th century, armies were usually composed of private retainers dressed in livery, and of hired mercenaries. In that century, the regiments in the standing armies were usually dressed by their colonel, who vied with his fellows in expenditure, and often in colourfulness.

Khaki (from the Urdu *khak* = dust) was first used by the Corps of Guides in India in 1848, under the command of Sir Harry Lumsden. In 1858, a captain in the Bengal engineers commented that 'European regiments here wear white clothing in the hot weather; but white is not well suited to campaigning and most of them have dyed their coatees the well known khakee or dust colour.'

During the Boer War (1899–1902) all troops sent from England and the colonies were equipped with sand-coloured clothing, supplemented by a heavy serge for the mountain regions. Adoption of such 'service dress' (the finery being retained in 'full dress') had much to do with the increasing accuracy of the army rifle. Austria-Hungary abandoned white in 1866, and went over to dark blue, while the French clung on to their red trousers until the First World War.

C. M. B. G.

Torpedo

Unlike that of the mine, the genesis of the torpedo is clear, if scarcely simple. In 1866 Robert Whitehead, British manager of an engineering firm at Fiume on the Adriatic, developed the idea of an explosive-stuffed small boat, steered on the surface by wires, into the first guided missile – the underwater torpedo, self-contained, depth-controlled by a hydrostatic device, and eventually gyroscopically steered.

Exactly 100 years later, American submarines went to sea armed with Subroc. Fired as a torpedo deep below the surface, Subroc emerges, becomes a guided missile, moves on an interception course towards the target, submerges again, becomes a computerized homing torpedo, and finally explodes as a nuclear depth bomb. It has the added advantage of being impossible to track back to its source.

So much for progress.

At the half-way mark of its development in 1914–18, the traditional torpedo sank almost 12 million tons of Allied merchant shipping and narrowly failed to win the war for Germany. In the Second World War the torpedo sank more than 20 million tons – including more than 200 warships. For the second time the Allied powers came to within a wave-splash of disaster.

D. D.

Barbed wire

Anybody could invent barbed wire and many people did, among them Lucien Smith, who lodged an American patent in 1867; but it took a smart man to invent a reliable machine to manufacture it by the mile. The need grew urgent on the Great Plains of North America, where for every dollar spent on cattle another was spent on fencing.

In 1873, 60-year-old rancher Joseph Farwell Glidden, born in Charlestown, New Hampshire, went with his friend Jacob Haish to the county fair at De Kalb, Illinois, where both were impressed by a type of barbed wire exhibited by Henry Rose. Glidden, saying nothing, worked out a machine for threading in the spurs, and took out a patent that year. Haish, saying nothing, also devised a manufacturing machine, challenged Glidden's priority but lost (in 1875 he marketed wire with 'S' barbs and later founded Barb City Bank).

Glidden grew rich as mills in eastern states turned out vast tonnages of 'his' barbed wire. The cattle barons, used to driving great herds over everybody's lands, resented the new enclosures and tried to uproot the wire and even hired thugs to shoot settlers; but in the end barbed wire helped to open up the West by giving the settlers security.

Lawyers welcomed barbed wire, since it ushered in a quarter of a century of patent legislation. A patent taken out in Britain (by an American, William Hunt) in 1876 envisaged wire with rotary spurs and 'wheels'. In 1888 British military manuals advocated use of barbed wire (old-time defences: palisades, planted spears, *chevaux-de-frise*); Roosevelt's Rough Riders in the

Dazzle-painted merchantman in the First World War.

Below, *types of barbed wire from the First World War.*

Bottom, *the future King Edward VII firing a trial burst with Maxim's repeating gun, in 1888.*

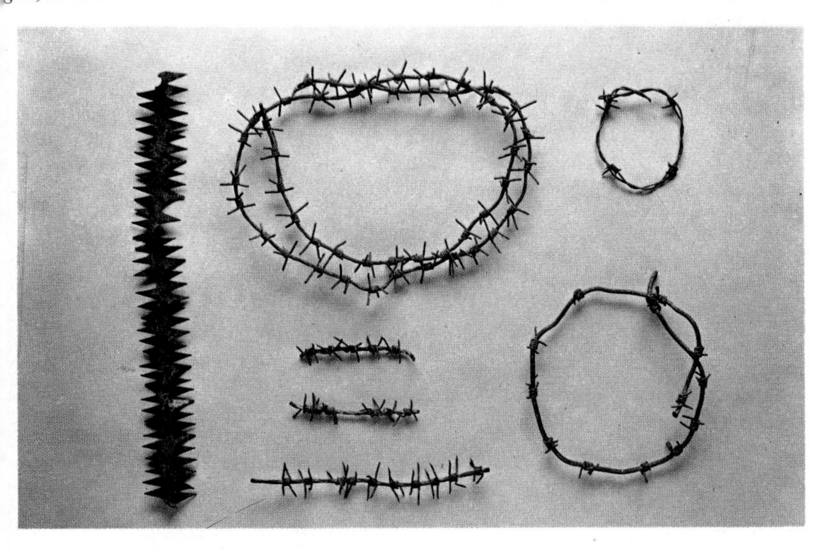

Spanish-American War used barbed wire as camp defences; on the South African veld Kitchener linked block-houses with five-strand fences to curtail the movements of Boer guerrillas; in 1914–18 barbed wire stretched from Switzerland to the English Channel. A corpse on wire became the symbol of war.

In 1930 an artist in barbed wire, Marshal Graziani of Italy, built a curtain 30 ft. wide, 200 miles long, between Egypt and Cyrenaica, to contain the Senussi. It played a crucial part in the desert campaigns of Hitler's war. Coiled wire was subsequently devised in the hope of fouling tank tracks. The symbol of the Second World War became concentration camp inmates clutching wire.

E. S. T.

Machine-gun

Mentions of repeating action guns are almost as old as the history of guns themselves. The ribauld, first mentioned in 1339, had many barrels mounted together. By waving a light across the touch-holes it could be fired like a machine-gun. By 1387 a more ambitious version had 144 barrels. The trouble with these guns was the time it took to reload them. The first attempt to solve the problem of reloading used funnels to supply measured amounts of powder and one missile into the breech. Considering the danger of the explosive supply exploding, it was just as well that most of them got no further than the drawing-board stage.

One repeater that did get built is today in the Tower of London. Patented by James Puckle in 1718, it had interchanging cylinders each with half a dozen chambers. One remarkable aspect of this design was the way in which it fired round bullets at Christians and square-sectioned ones at infidel Turks. As a gun it was a failure.

The first successful repeater was the Gatling, beloved of Hollywood movie villains. The bullets were gravity-fed from a drum. Ten barrels were cranked around, firing one at a time to make 350 rounds per minute. Its designer happily demonstrated the gun in person at battles during the American Civil War. It was also used by the British Army in South Africa. It jammed rather too easily and, being hand-cranked, was not a truly automatic gun.

In 1881 Hiram S. Maxim was in Paris when a fellow American advised him: 'If you want to make a pile of money, invent something that will enable these Europeans to cut each other's throats with greater facility.' Within two years Maxim had patented his gun.

Hurt when a child by the kick of a rifle, he remembered this energy and used each bullet's recoil force to eject the spent cartridge, insert the next and fire it. Thus the gun would fire until the entire belt of bullets was used up. It was Maxim's good fortune that while he was working on the engineering, a new explosive was invented that gave chamber and bore pressures that would enable the breech action to work.

Maxim's water-cooled gun was used in the Russo-Japanese War. It made his fortune. In the First World War it killed countless Europeans in all the armies, and not a few Americans, too.

However, Gatling was to have the last laugh, albeit from the grave. When recently the ultimate machine-gun – G.E. Vulcan 30 mm. linkless feed – achieved a rate of fire of 6,000 rounds per minute, it used six barrels and a Gatling action.

L. D.

Gas-mask

When the Hague International Peace Conference was convened in 1899 it was still 16 years before gas would be used as a weapon in armed conflict. But though it was proposed that gas be outlawed as a military weapon, the majority of the delegates, and most notably the Americans, were not impressed and the recommendation was rejected.

In a salient near Ypres, Belgium, in 1915 the German Army knocked out British and French troops with a massive gas attack. Within a few days Allied soldiers were back in the field with chemically treated cotton pads tied over their mouths and noses. The gas-mask had been born.

As the First World War progressed, so did the use of chemical gas increase, necessitating the development of something a little more durable and protective than cotton pads. The first 'respirators', as these devices were called by the British, were cumbersome affairs that severely impeded a soldier's efficiency. They consisted of a mask proper, which was tightly fitted round the face, connected by a large tube to a canister that hung in front of the wearer's body. The canister contained charcoal, through which poison gases were filtered. Unfortunately, the soldier could breathe only through the mouth, since a nose clip prevented inhalation via the nostrils. It was also difficult for him to hug the ground while advancing against machine-gun fire.

Second World War gas-masks were lighter than their predecessors and better fitting, and permitted a clearer and wider range of vision. They also eliminated the nose clamp and mouth inhaler, and the canister strap was slung over the shoulder instead of round the neck so that the filtering contraption could be carried at the side.

In the 1960s, the United States military achieved a major breakthrough with the adoption by the Army, Navy and Marine Corps of the M-17. This prototype of the modern gas-mask has neither a hose nor an external canister and yet affords protection against gases and aerosols of chemical, bacteriological and radiological agents. Air is filtered through pads of pliable material that are enclosed in cavities moulded into the facepiece of the mask.

Today, gas-masks are commonly used in industry, especially in mines and chemical plants, by firemen, rescue squads, and of course riot police. In 10 years' time, say many environmental scientists, they will be standard street wear for the citizens of Tokyo and other large industrial cities.

E. W.

Tank

In the second half of the 19th century the farm tractor and barbed wire transformed America's Wild West. Even the War Office had seen the value of the tractor as a 'mechanical horse', and had ordered them as early as 1899 for the South African War. In 1900 a motor-car engineer had built and demonstrated an armoured war car. In 1912 an Australian named de la Mole submitted to the War Office his design for a machine that was remarkably like the one that went into battle agonizing years later. His design was filed and forgotten. Lt-Col. Swinton – a historian who had studied the Russo-Japanese War – investigated the possibilities of making an American Holt tractor into an armoured weapon. Kitchener was against the idea on the grounds that it would be vulnerable to artillery fire.

It was a design for an artillery tractor that prompted Winston Churchill to ask if such a machine could be designed to cross trenches. He refused to take no for an answer and eventually a committee was formed. On 17 February 1915 they watched a Holt tractor working. Its performance was impaired by foul weather, and the committee pronounced the machine worthless. But in spite of this Churchill ordered some machines to be manufactured, thus taking personal responsibility for an expenditure of about £70,000.

It would be hard to imagine any service less likely to contribute to the birth of the tank than the Royal Naval Air Service. Yet the armoured cars they used in Flanders in 1914 had already influenced Churchill and it was an RNAS lieutenant, W.G. Wilson, who designed the tracks for the new machine. He worked with Mr W. Tritton – of Fosters of Lincoln – and, using tracks from the USA and boiler plate as armour, they put together a practical vehicle. Their second design was named 'Little Willie' and had longer tracks. It used the gears and differential from a pre-war Foster-Daimler tractor.

These same two men now returned to the drawing-board and produced a radically different design, the lozenge shape that was to become the symbol of the tank for all time, and the tank men's badge. Rejecting the name 'Big Willie' they called it 'Mother' instead. The army experts watched it crawl across Lord Salisbury's golf course at Hatfield and even 9 ft. wide trenches did not halt it. The army ordered 100 of them. For security purposes the cumbersome metal machines needed a code name: 'water carriers' was rejected in favour of 'tanks'.

The men who had supported the tank through its precarious birth-pangs asked that their surprise effect be used to the full by a large concentration of them deployed on a relatively undamaged part of the front. They asked that there be no artillery preparation, for this would make the ground more difficult for them. Haig ignored their pleas. In September 1916, although there were only 60 tanks available, he put them into the attack after a three-day bombardment by 1,200 guns.

Of 60 tanks ordered to the front, only 49 arrived and all but 18 of these bogged down in craters. Those 18, however, achieved remarkable results; one captured a village and another a trench, complete with 300 prisoners. In spite of this remarkable result, the value of this new weapon was not realized by the high command.

Eventually at the battle of the Somme in September 1916 the tank specialists were allowed to stage a tank attack in the way they had always advocated. Nearly 400 tanks attacked on a six-mile front. They were in groups of three tanks leading two platoons of infantry. The attack captured 7,500 prisoners and 120 guns, and penetrated more deeply into the enemy lines than any previous attack.

The tank had arrived.

L. D.

Napalm

An acronym for naphthenic acid and palmetate, this inexpensive and readily available derivative of petroleum and extracted palm oils was developed at Harvard University in 1942, specifically in response to a military call for a thickener that would slow down the rate of burning and increase the effective range of existing flame-throwers. Napalm not only met all of the Army's requirements, it also gave them some added benefits, the most notable being an ingredient which greatly raised the temperature at which flame-thrower fuel would burn. So adaptable was this new incendiary that it was soon being used as a successful canister-filler for air-to-ground fire-bombs.

Simple flame-throwers have been standard military hardware since the days of ancient Greece and were especially useful during the long years of the Peloponnesian Wars. During the First World War, both Germany and the Allies made use of flame-throwers fuelled by raw gasoline, although in most cases these primitive inventions burned themselves out before causing much damage. The armies of the world clearly needed a more devastating chemical killer.

The efficiency of napalm was first significantly demonstrated by the US Marines during the Second World War when it was employed against Japanese bunkers on Guadalcanal in the Pacific. Later, the chemical was used in the incendiary bombs rained by Allied air forces on Japanese industrial cities; by the war's end these bombs had burned out 40 per cent of the area of all target metropolises.

Napalm, which has a jelly-like liquid composition, is considered especially suited for attacking 'soft' targets, such as human beings and most kinds of primitive dwellings. It is equally useful against armoured vehicles, since the armour acts as a heat conductor allowing the napalm to broil the tank crew alive. Unlike explosive bombs, which must usually land directly on target, napalm devices can score a kill even with a near miss. The chemical is virtually useless, however, against so-called 'hard' targets, such as bridges, and its transport often proves problematical for aircraft since its low density creates a bulk which is greatly disproportionate to its weight. The Korean War and the American conflict in Indo-China brought an increased popularization of napalm, and in both theatres it was most commonly used against snipers, large troop concentrations and anti-aircraft batteries. Nor is there any reason to suppose that its use will be discontinued in any future hostilities.

E. W.

Ballistic missile

The missile is one of man's oldest inventions, conceivably even pre-dating the discovery of fire. The latest nuclear ICBM (Intercontinental Ballistic Missile) with a range of over 7,000 miles and a speed capability of some 15,000 m.p.h. has come a long way from the caveman's hurled rock, but the principles of motion involved, as well as the basic human urges, are very much the same.

Missiles were first put on a scientific footing with the development of the ballista, an ancient military engine which most often came in the form of a crossbow, from which the modern science of projectiles derives its name. In 1232, when the Mongols laid siege to the city of K'ai-feng, its Honanese defenders fought back with weapons described as 'arrows of flying fire'. Since no account of this battle makes

Fuelling a V2 – first of the ballistic missiles.

First conception of the tank – the 'battle engine' conceived and drawn by Leonardo da Vinci.

A 1915 prototype tank, complete with camouflage paint. Lord Kitchener, British War Minister, called it 'a pretty mechanical toy' and prophesied that 'the war will never be won with such machines'.

The 'Greek fire' used by the armies of Islam at the siege of Constantinople – a 7th-century precursor of napalm.

The first hydrogen bomb, October 1952.

mention of any kind of bow, and since the Chinese were also dropping bombs perceived as 'heaven-shaking thunder', most students conclude that gunpowder had already been discovered and was being used to make both explosives and propulsive charges.

From then until the present day, the science of rocketry has witnessed a continuous and almost steady development, contributed to by countless scientists and military experts from many nations of the world. Some names stand out, of course, from the 13th-century German scholar Albertus Magnus, who wrote us a series of powder-charge formulae for rockets in his *De mirabilibus mundi*, to the modern 'whiz kid' of missiles, Wernher von Braun.

Perhaps the most monumental contribution to missile technology was Isaac Newton's Third Law that 'every action is accompanied by an equal and opposite reaction'. Overthrowing 20 centuries of faulty Aristotelian theories of motion, this simple postulate gave man his first real understanding of just how and why a rocket operated. That was in the late 17th century, more than 250 years before the German armaments industry produced the first truly operational ballistic missile of the modern genre, the now legendary V-2. This devastating projectile, which became a familiar sight to Londoners during the closing months of the Second World War, could reach speeds of up to 3,500 m.p.h. and hit targets 200 miles from the launch site. The V-2 was propelled by a liquid-fuel rocket motor and guided by gyroscopic stabilizers. By contrast, up-to-date ICBMs such as the United States Minuteman and Polaris missiles, use pure inertial-guidance systems, which are immune to electronic countermeasures.

Today so many nations antagonistic to each other possess sophisticated nuclear-equipped ballistic missiles that the main research emphasis is on the anti-ballistic missile, with its obvious progeny, the anti-anti-ballistic missile.

E.W.

Nuclear bomb

The atomic bomb did not take long to come to fruition once the nuclear reaction had been controlled. The nuclear explosive had either to be made in a nuclear reactor or extracted from natural uranium by a new process of unprecedented expense and complexity, but by July 1945 a bomb team working at Los Alamos under Robert Oppenheimer were ready for their first test. As the fireball blazed over the New Mexico desert, a line from the *Bhagavad Gita* of the Hindus flashed through Oppenheimer's mind: 'If the radiance of a thousand suns were to burst into the sky, that would be like the splendour of the mighty one.' And: 'I am become death, the shatterer of worlds.'

The atomic bombs dropped on Hiroshima and Nagasaki brought the war with Japan to an end. They also led to a crisis of conscience on the part of the scientists involved. When the Soviet Union exploded its atomic bomb in 1949 and a powerful lobby in America began to press for the development of a weapon a thousand times more powerful still – the hydrogen bomb – many of the scientists, including Oppenheimer, opposed it.

But the pro-hydrogen bomb forces proved irresistible, and the first device was exploded at Eniwetok atoll in the Pacific in October 1952, showering a small Japanese fishing vessel several hundred miles away with radioactive dust. The Hungarian-American physicist Edward Teller was one of the most active propagandists for the H-bomb, and also made a major contribution to its development. He earned the title 'Father of the Hydrogen Bomb', as Oppenheimer before him had earned the title 'Father of the Atomic Bomb'. But where Oppenheimer was a prey to agonizing doubts, Teller had nothing but pride for his offspring. His testimony was probably the deciding factor in the withdrawal of Oppenheimer's security clearance during the McCarthy era.

For a time it looked as though the combination of hydrogen bomb and intercontinental ballistic missile had produced a nightmarish kind of stability in the armaments race, but now that stability is beginning to break down.

On the one hand there is the anti-ballistic missile (ABM), with its own nuclear warhead, capable of destroying an ICBM in flight. On the other there is the Multiple Independently-targeted Re-entry Vehicle (MIRV), a cluster of warheads which can be carried by an ICBM and then steered to separate targets. Strategic thinkers are becoming frightened that the other side might be considering a first strike, with MIRVs to knock out their missiles before they can be launched and an ABM screen to deal with those that escape.

B.S.

Coins used in the Roman Empire for the payment of taxes: Gallo-Roman relief of the early 3rd century AD, found at Neumagen, Germany.

Coinage

In many societies small quantities of precious or base metals arose naturally as a medium of exchange, since they were durable, portable and could be seen to have value according to their size and purity. Doubtless the beginnings of coinage can be traced back to a simple desire to save time, by which a man would make some mark on a piece of metal he found acceptable in weight and quality so that he would know it again. The importance of weight in early currency is, of course, preserved linguistically in such words as 'pound', 'livre', 'lira' and 'rouble' – all units of weight.

There is conflicting evidence as to where the first coins were issued. Certainly, neither the Old Testament nor Homer mentions coins and we are probably right to follow Herodotus in ascribing its invention to the kings of Lydia, a kingdom in Asia Minor, some time in the 8th century BC (though it is possible that the Lydians borrowed the idea from the Hittites). The Lydians minted ingots and coins of electrum (a natural alloy of silver and gold) and their use quickly caught on among the mercantile Ionian Greeks along the Asia Minor coast, who in turn brought it to mainland Greece (where the island state of Aegina was believed to have been the first city to mint coins – her famous 'turtles'). It is thought that the Chinese, as seems so often to have happened, invented coinage quite independently a little later than the Lydians.

The use of coinage spread round the Mediterranean and into Europe and Asia. In Britain at one time there were seventy mints (more than now exist in the entire world) and in medieval Europe enormous numbers of coins were issued, with not only nations distributing their own currency, but even cities and individual families.

Standards of coinage, however, waxed and waned; copper was added to gold under the Roman Empire and in the last two centuries before its fall very base alloys were used, sometimes with only 2 per cent of gold or even less. Alloys were discarded by the Middle Ages and the small amounts of zinc and tin now added to most 'copper' coins are said to date from the French Revolution, when church bells containing these metals were melted down to eke out the diminishing copper supply. Cupronickel, now widely used for 'silver' coins, was introduced in Belgium in 1861 and several other alloys have since been developed in various parts of the world. The first coining machine was set up in Paris in 1553; previously coins had usually been struck between an engraved plate recessed in an anvil and a punch on which the design for the other side had been engraved, though large coins were sometimes cast.

Paper money, too, represents an important step, in which real assets began to be supplanted by 'debt' money, the 'promise to pay', symbolized by words like 'bill' and 'note'. The bank note probably evolved in Europe from the receipts goldsmiths would give for coin left in their charge. During the 18th century many small private bankers issued notes which would be met so long as the bank remained solvent. It was not until 1833, however, that the Bank of England's notes were made legal tender, and 11 years later an Act gave the Bank the monopoly of banknote issue and banned the issue of any notes not backed 100 per cent by gold. Apart from the odd crisis 'suspension' the Bank of England continued to honour its 'promise to pay' in gold sovereigns until the beginning of the First World War.

J. G.

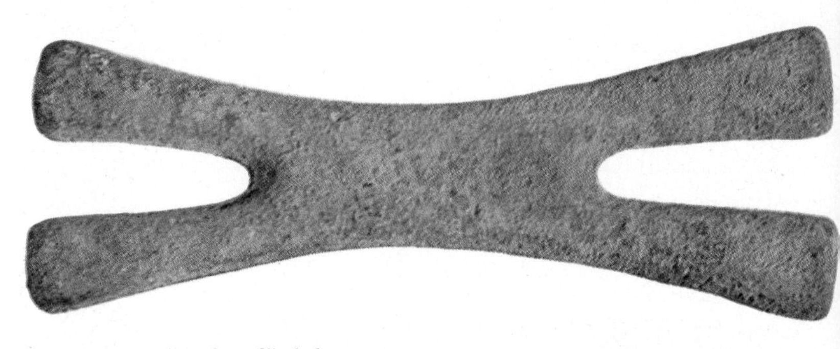

Copper currency bow from Zimbabwe, Rhodesia. Early coins were not necessarily in what we now recognize as coin shape.

Chess

The origin of chess has been much disputed, but nowadays the general opinion seems to be that its inventor was a Hindu living in north-west India in or soon after the late 5th century AD. Later Moslem tradition supplied a name for the indisputably brilliant inventor – or 'philosopher' as they called him – responsible for it: Sissa son of Dagir, who invented the game for an Indian king who had grown tired of the then most popular game, a variety of backgammon.

In fact the game 'Sissa' invented differed considerably from the one we know today. Called *chaturanga* ('four parts'), it was a sort of war game which set out to represent the four units – elephants, horses, chariots and infantry – that comprised a contemporary Indian army. What distinguished the game

Gold stater from Carthage.

*Woodcut from Caxton's edition (*c. *1482) of Jacopus de Cessoli's* The Game and Playe of Chesse.

Credit cards in a London shop window – the credit system made accessible to the man in the street.

from others of the type, however, was the inventor's ingenuity in devising moves for each piece that roughly symbolized those of the fighting unit it represented: thus our castle (rook) was originally a chariot, capable of 'rolling' in a straight line in any direction, while knights could jump over obstacles but could not arrest themselves in mid-bound.

From India the game spread in two directions. It had reached China, via Kashmir, by the end of the 8th century and thence invaded Korea and Japan. Westwards it quickly spread to Persia, where it was called *shatranj*. The Arabs encountered it there when they overran the country in the 7th century and soon it was known all over the Arab sphere of influence. Presumably the game reached Europe through Spain, perhaps in the 11th century. Certainly it was known to the cultured classes throughout Europe by the Crusades.

The astonishing universal popularity of chess has never really failed in the centuries since then. The development of the game continued with a variety of modifications and changes until the 16th century, when the last important innovation, castling, was added. In 1475 Caxton issued *The Game and Playe of the Chesse* – an indication of its popularity in Britain – and in the 16th century there began a series of great treatises on chess, starting with the Portuguese player Damiano and continuing strongly with Ruy Lopez in Spain. Thereafter the centre of the game switched to Italy, especially Venice, but it has frequently moved since then. England, with the legendary Howard Staunton, was supreme for part of the 19th century, but since then champions have been found in many other countries, including Russia, Cuba, Poland, Scandinavia, Eastern Europe and the United States.

J.G.

Credit system

The use of credit, in one form or another, is almost as old as the use of money, and historical records indicate the existence among various ancient civilizations of arrangements for performing the credit function. In ancient Athens, a financing set-up not very dissimilar to the modern mortgage was used, although it is in the field of trade that modern credit forms find their true origin. The Greeks especially had developed a specialized system of private manufacture and also had a flourishing overseas trade, thereby necessitating the growth of counterpart financial institutions. Even then, credit was needed to pay producers in advance of the ultimate sale of their product abroad.

In time, Greek financial practices spread throughout the Hellenistic world and were eventually adopted by the Romans. The major credit deals made in Rome were designed to promote domestic agriculture and industry, develop foreign trade and finance military operations. The might of the Army was considered the best possible collateral, guaranteed to bring back sufficient spoils of war to make the investment safe and profitable.

After the fall of Rome, with Europe in the hands of less domesticated barbarians than the Romans, financial practices favouring the use of credit disappeared for a time. Then, in the Middle Ages, despite the general disdain for commercial pursuits and certain religious prohibitions against charging interest, the urgency and profitability of trade forced the revival of a credit system. Most contracts, however, were disguised as partnerships or foreign exchange transactions to avoid the stigma of usury. The most prominent users of credit during this period were kings and nobles who needed more money than they themselves had in order to finance their courts and military campaigns. Not that the nobility always represented the best risk for a moneylender, since in most cases the debts were simply refinanced as they fell due (as with modern governments); but admittedly it was difficult to say 'No' to a king. Occasionally, however, a creditor could take action against the delinquent noble and there are cases on record where a prince could not leave his castle until he paid up.

The practice of mortgaging one's person as collateral for loans, and then being condemned to peonage or jail upon default, was common in medieval times, and even continued in just that way until late in the 19th century. In less blatant forms it still exists today. But the real expansion of all forms of credit systems came with the Industrial Revolution, which saw not only the building of new factories and numerous other business concerns, but also the growth of the large international cartels that fostered the development of modern international trade.

Today, credit is no longer confined to the businessman but is virtually a household word, especially since the advent of the credit card in 1938. Originally popularized by the oil companies, this scheme was really given free play in the mid-1950s when electronic computers allowed for fast and accurate billing and accounting. And now, in the United States alone, there are some 140 million credit-card owners, buying anything from a washing machine to an aeroplane ride on the never-never.

E.W.

Card games

No one knows for certain either when or where playing cards originated. One account has it that they were invented by a Chinese emperor to amuse his favourite concubine, a story which seems as likely as any other thus far advanced. In any event, cards were most probably used in parts of the Orient prior to AD 1000 and introduced to Europe around 1300. If so, it was another 100 years or more before they worked their way to England. The first playing cards actually made in the West were tarots, collectively referred to as 'the Tarot'. These were picture cards, 22 to the set, as opposed to the suited and numbered cards of Eastern origin. Tarots were, and still are, used for divination, but in the late 14th century they were combined with the 56-card Oriental set. And it was with this new deck of 78 cards that the earliest recorded form of card game was played, an interesting little affair known in Italian as *tarocchi*, in German as *tarok*, and in French as *taran*. The game was apparently never played in England.

It was the French who first adopted the modern 52-card deck (along with a 32-card set still in use on the Continent) and who also gave the four suits their present-day symbols. The English, however, changed the names, so that the symbol referred to in France as a 'square' is known to them as a 'diamond'. The first card game with any lineage to be developed in Britain was whist, which appears to have been an outgrowth of various other games such as 'triumph' that were played as early as the 16th century. In 1742, Edmond Hoyle codified the rules of whist and later turned his regulatory instincts to other card games.

Bridge was developed from whist some time in the 19th century, although contract bridge, which is now the most popular form of this game, was only introduced in 1925 by Harold S. Vanderbilt. Fifteen years later, Charles H. Goren introduced the point count system to the game and this subsequently replaced the honour count. Poker, which was first played widely in the United States more than 100 years ago, seems to have evolved in Asia and is a good example of a traditional cut-throat gambling game which is now socially acceptable. Twenty-one, which is variously known as blackjack, pontoon and *vingt-et-un*, is one of the oldest gambling games and the most popular card game in casinos the world over.

E.W.

Postage stamp

The first postage stamps, as every schoolboy knows, were the 'penny black' and 'twopence blue' issued by Great Britain in 1840. But the invention was not the result of a sudden stroke of inspiration, rather a rational attempt to introduce order and simplicity into a system which had already been developing over at least 2,000 years (if we count the state courier services of the Persian and Roman empires).

The first major nation to set up a state postal service in post-classical Europe seems to have been Louis XI's France with the Edict of Luxies of 1464. In England Charles I consolidated the efforts of his predecessors by establishing the post as a royal monopoly in 1635. Under Thomas Witherings, its first director, the English postal service maintained a frequent day-and-night service which took six days to get a letter to Edinburgh and the reply back to London. Witherings also established a regular tariff based on a single sheet of paper, which cost 2*d.* per 80 miles.

Governments everywhere were delighted by the opportunities offered by the postal service as a source of revenue. In England, improved roads and the introduction of the first mail coaches under Pitt's administration in the late 18th century enhanced the speed and efficiency of the service, but tariffs increased constantly, particularly with the onset of the Napoleonic Wars.

Thus it was due as much to discontent at high postage rates as to anything else that in 1836 Rowland Hill urged that the system of rates based on distance and number of sheets should be abolished in favour of a uniform rate based on weight, payment for which would be made in advance by means of 'labels' attached to each letter with 'cement'. These proposals were put into practice on 6 May 1840, when the first two 'labels' were issued by the British government. The system was an immediate success. In 1840 twice as many letters were posted as in the previous year, and the rest of the world, led by Brazil, Mauritius and the United States, hastened to adopt Hill's system. Nevertheless the uniform rate meant that government revenue from postage plummeted. In Britain it took 35 years for the 1839 level to be regained.

Many changes were made to the prototype stamps in the early years. Henry Archer's invention of a perforating machine in 1854 removed the need for the stamp-user to reach for a pair of scissors every time he wished to post a letter, and very soon a number of methods of defrauding the government had been devised by ingenious citizens. The 'penny black' did not last long – it was too easy to hide the signs that it had already been postmarked once – and 'penny reds' were introduced which were often printed from a number of separate plates to prevent forgery, and marked with check letters in the corners to discourage another favourite trick, that of cutting two used stamps in half and joining the unpostmarked half of each together to create a 'new' stamp.

The formation of the Universal Postal Union in 1874 demonstrated the rapidity with which Hill's system had been adopted all over the world.

J.G.

The 'Penny Black', the world's first adhesive postage stamp, issued on 6 May 1840.

Left, *a 'modern pork packing and canning establishment' in the US, which claimed to exploit 'all the pig except the squeal', was an early user of the conveyor belt. (Courtesy, Chicago Historical Society).*

An early one-armed bandit, technically simpler than those in use today, but just as elaborate in décor.

Left, *whether cheating is as old as card games is an interesting speculation. Detail from* The Sharper *by Georges de la Tour, c. 1625.*

Slot machine

The weight of coins has been used to operate simple mechanisms, ranging from toys to holy-water dispensers, since ancient times. But turning this elementary idea into the modern slot machine has proved surprisingly difficult. A Mr Denham patented a 'sale apparatus for postage and receipt stamps' in London in 1857, where a penny rolling down a chute triggered a spring which ejected a stamp. His successors soon found they needed more elaborate devices for checking whether the machine was being given a genuine coin or a washer. Little progress was made until the 1880s, when the need for metering supplies of gas and electricity, and the newfangled telephone, set off a burst of activity. The most successful inventor was R.W. Brownhill, who in 1887 patented the first useful coin-in-the-slot gas meter, allowing customers to buy a penny- or a shillingsworth of gas at a time. Automatic vending machines of all kinds became widely used over the next few years, but some of the tougher problems were not solved until the 1920s.

T.J.

One-armed bandit

One version of the invention of the fruit machine concerns a certain H.S. Mills, who began life as a Chicago newsboy. In the course of time he built up a chain of lemonade stands, and to increase profits he set up in each one an invention of his which he called the 'Kalamazoo'. This contraption consisted of a slot and three tubes. One put a penny into the slot and it fell into one of the three tubes. From two of them it emerged accompanied by two or three more pennies. From the third it did not emerge at all.

Another version is that it was invented by the young Charles Fey, of San Francisco, in 1895: he called his machine the 'Liberty Bell', and rented it to a local saloon on a 50–50 basis. However, it was Mills who was to make the most machines: he is reputed to have set up his factory in 1889, and by 1932 the Mills Novelty Co. were making 70,000 machines a year. What might seem a harmless attraction is in fact big business: according to the *Encyclopaedia Britannica* (1967) the town of Las Vegas (population 70,000) has 10,000 licensed slot-machines, on each of which the owner pays a tax to the Internal Revenue Service of $250 p.a., plus an income-based tax to the state.

In 1932, in an article entitled 'Plums, Cherries and Murder', *Fortune* exposed the profits made by racketeers from respectable businesses which only sold the machines: it instanced the turnover from slot-machines for 1931 as $20 million for greater New York alone, and $150 million for the whole of the USA. On the average machine, it was calculated that 1,000 games at a total cost of $250 would return $61·75, leaving a profit of $188·25.

Early models were technically simpler than those in use today, though Charles Fey's 'Liberty Bell' was swathed in *art nouveau* decoration, and stood on lion's-paw feet. The mechanism consisted of three reels, each with a timing device actuated by the insertion of a coin; the handle of the machine was pulled, the reels span, and the timing devices interposed rods at certain intervals against teeth on the reels to stop them. Anyone who could work out the timed intervals got rich, but a simpler idea was to block up the pay-off slot.

Modern machines are more sophisticated, for they incorporate the coin-tester, which perhaps first appeared around 1931; a newspaper report of that year writes of one firm marketing 'the Electrojector, which is a slot-machine rejector of slugs and bad coins. The makers . . . are now working on the most important: a nickel rejector . . . [which] is said to reject any bogus coin but a *marka*, a Russian imperial coin which will pass for a quarter, but is much more valuable anyway.' Modern coin-testers check diameter, weight and metal content, and reject unsatisfactory specimens.

C.M.B.G.

Conveyor belt

Modern mass production in industry largely depends on the conveyor belt. Henry Ford designed the first large-scale 'assembly line', as he called it, for the mass production of his famous 'Model T', nicknamed 'Tin Lizzie', which began in 1908. The conveyor belt was the heart of his new factory, built especially for that purpose at a time when most European cars were still being handmade by craftsmen; it was one-fifth of a mile long. The concept underlying it, however, was that of interchangeable parts, which goes back to the early days of the Industrial Revolution.

In 1798 the American teacher Eli Whitney, inventor of the cotton gin, accepted a government contract for the delivery of 10,000 muskets within 15 months. They could never have been made by the traditional methods, with one gunsmith producing the complete musket from start to finish. So Whitney had the idea of dividing the work into a great number of small jobs: each workman had to make just one component part in a continuous operation, and all the parts had to be made so accurate as to be interchangeable.

It was, at the time, a completely new system, requiring new tools. Whitney designed them: the 'filing jigs' which guided the workman's file; the stencils with up to a dozen holes, to be bored in exactly the right places; the mechanical stops fixed to the lathe, which prevented the piece of work from being made too long or too short; the dies and moulds for various component parts. Whitney demonstrated the system most effectively by taking a number of separate parcels, each containing 10 samples of the same part, to the Treasury Office in Philadelphia, where he asked officials to pick one part from each parcel at random; from these he assembled a complete musket before their eyes. Even more astonishing than the 'perfect fit' he demonstrated in this way was the fact that the parts had been made not by trained gunsmiths, but by semi-skilled mechanics, and in record time.

Another important factor which helped Henry Ford was the development of the band conveyor, which had been going on throughout the 19th century, particularly for the transporting of bulk materials in the ports. The first big band conveyor for grain was installed at Liverpool in 1868. Ford, however, used his assembly line for transporting the automobile-to-be from one worker to the next in a continuous movement. Each worker, supplied with tools and components, had to do just one job in a certain time while the work was rolling by.

E. L.

Supermarket

'I would be the "miracle man" of the grocery business.' That was the dream of the American Michael Cullen, who managed a store for the giant Kroger Grocery & Baking Co. grocery chain in the late 1920s. He was setting out his plan for 'monstrous stores . . . away from the high-rent districts'. He wanted aggressive price-cutting to attract customers in the depression years: 'I could afford to sell a can of milk at cost if I could sell a can of peas and make two cents.'

Cullen wanted the president of Kroger, William H. Albers, to finance the new idea, but his letter was intercepted by a Kroger executive who turned it down. Cullen resigned immediately and in August 1930 'King Kullen, the world's greatest price wrecker – how does he do it?' opened his first self-service store in Jamaica, New York.

It was an instant success and by 1932 Cullen had eight stores with a turnover of $6 million. Cullen died – or worked himself to death – in 1936 at the age of 52 but before that an even more successful imitator had opened up.

Robert M. Otis and Ray O. Dawson set up the first 'Big Bear' shop in an empty car factory at Elizabeth, New Jersey. 'Big Bear, the Price Crusher' offered shoppers 50,000 sq. ft. of groceries, meat, fruit and vegetables, radios, car accessories and paints laid out on cheap pine tables. Shoppers put their purchases in baskets and paid at a cashier's booth. The cash outlay by the promoters was $1,000. After one year their net profit was $166,000.

Big Bear provoked war with the traditional retailers, who persuaded newspapers to refuse Big Bear's advertisements and got a law passed in New Jersey against selling at or below cost. But Big Bear prospered and was followed by Giant Tiger, Bull Market, Great Leopard – and by the man who was not allowed to read Michael Cullen's proposal. Albers resigned from Kroger to set up Albers Super Markets Inc. in 1933, utilizing a word that had been coined in California, the home of the 'super colossal'.

By the mid-1930s supermarkets had become respectable. The philosophy was still 'pile it high and sell it cheap', but they were now purpose-built shops rather than converted factories. In 1936 shoppers were enabled to carry away even more when S.N. Goldman, president of Standard Humpty Dumpty Food Markets, invented the basket on wheels.

The final stage of supermarket development came shortly after the Second World War when the markets began pre-packaging dairy produce, meat and vegetables.

D. F.

Customer at the cash desk of a New York supermarket in November 1939.

Automation

Automation is not a single invention, but nevertheless it can be said to have come into being in 1946. In that year, Delmar S. Harder coined the name for a system he had devised for the Ford Motor Company to manufacture motor-car engines. The process was completely automatic – the first complete system applied to manufacturing – and could produce one engine every 14 minutes. Previously it had taken 21 hours. The system, as is the case with all automation systems, involved an element of 'thinking' – the machines regulated themselves, without human supervision, to produce the desired results.

The major characteristic of automation as a system is this built-in capability of self-adjustment, designed to produce the best performance in constantly changing circumstances. This concept grew out of advances in the last 15 years in instrumentation, electronics and, particularly, electronic computing. It is when they are combined into a single system that these techniques produce capabilities so startling as to be largely responsible for much of the publicly debated disquiet, perhaps even fear, of our rapidly changing world, and the calls now so frequently heard to 'control science'.

L. B.

Automation in a steelworks: six-stand hot finishing mill at British Steel's Llanwen works. The process is computer-controlled and entirely automatic.

MAN WORKING

Tools

Instruments

Mental aids

MAN WORKING

Tools

It is a far cry from a hand-held piece of stone used as an axe by early man about one and three-quarter million years ago to a cyclotron three miles in diameter and producing 300,000 million electron volts to smash atomic nuclei. Yet the basic process is the same.

Whenever my two-year-old son wants something done and cannot do it himself he takes my hand, puts it on the object (for instance, a locked box which has to be opened) and says, 'Do it'. For him I am a composite tool that can be employed to do a variety of things. There are already robots which can amplify man's muscle power to a huge extent and others which can allow him to deal with matters too fine and delicate for his fingers. These are really duplicate men, but apart from robots there are ordinary tools used by ordinary men.

The hand-held piece of stone was probably the first tool. It could be used for smashing something (like a shell), for pulping something (like vegetable fibres), for banging a stake into the ground, and for a variety of other uses. The purpose of an axe is to develop energy which can be concentrated at a chosen point. Because the edge of the axe is small, the whole energy generated by its weight and speed is concentrated in this small area. Axes are, basically, used for smashing things. And that is exactly the function of the cyclotron, which uses magnetic fields to accelerate nuclear particles so that they can be used to smash atomic nuclei.

The first axe was probably a piece of stone of a size that would fit comfortably in the palm of the hand. Later, a piece of stone with a narrow edge (for instance, a piece of flint) might have been used. The next stage, presumably, was to actually sharpen a piece of stone to give it an edge, by chipping it with another stone. Then the stone would be tied by thongs to a stick; this would allow more acceleration of the axe-head and so more energy would be built up in the head. Later axe-heads would be cast from metal and later still they would have sockets so that the haft could be inserted directly into the head instead of just being lashed to it.

Once the axe had an edge it could now make a cut. Some Stone Age axes found at a burial site were actually used in a demonstration to cut down trees, and they proved quite effective. The cutting and hammering functions would now start diverging. Different and more controlled methods of cutting would develop (knife, saw, oxy-acetylene torch). Hammering would be used for forging and shaping, for pulverizing, and also for knocking in nails, stakes or piles. We can follow the hammering path of development.

One line of development would be the mangle-type rolling mills which used a pair of rollers to squeeze seeds or even metal. Rolling mills for metal were developed in the 16th century and were much used for making coins. For softer metals they could replace extensive hammering and give a smoother effect.

But hammering by repeated blows was still necessary in many cases. Trip hammers activated by a rotating wheel and powered by water were widely used. The rotating wheel would raise the hammer, which would then be released and fall under the effect of gravity. The next stage was to replace water by steam power. The hammer was still roughly in the shape of a hand-held hammer and was swung through an arc. This is a good example of how a line c development may continue apparently progressively but in realit still stick in the same channel. It was not until 1839 that this ancien channel was escaped from, by James Nasmyth. As with so man inventions there was an actual problem to be overcome: forging th paddle shaft for the *Great Britain*, a large paddle vessel about to b built by Brunel.

The idea is obvious, in hindsight. Forget about raising the hamme on a handle: simply raise the hammer head itself and then let it fall In a way this is going right back to the first primitive axe, when th head was raised directly. Nasmyth designed a steam hammer i which a heavy head was raised by steam up a sort of tower and the allowed to fall on the forging placed below, the fall guided by th tower. The machine was not made, but a Frenchman glanced a his designs and actually made a machine, which he then showed t Nasmyth. (This illustrates how an invention may actually be a excellent idea but may not get off the ground unless the circum-stances are quite right.) Nasmyth then started to make his ow hammers, which were a great success. For instance, the sam principle applied to pile driving did in four and a half minutes wha it took a team of horses twelve hours to do.

For some purposes the hydraulic press eventually took over fron the steam hammer. This press was first developed by the ubiquitou Joseph Bramah, the inventor of the water closet, and it is possibl that fiddling around with plumbing gave him the idea.

Another example of how a development may take one line and then have to be shifted from that line is the drill. In a bow-drill a string attached to a bow-like frame is looped around the drill stick As the bow is moved back and forth the drill rotates first one way and then the other. This is a moderately efficient device which wa in use for many centuries around the Mediterranean (and by the Romans) and is still in use in many parts of the world today. The great advantage is that for each arm movement there are several revolutions of the drill and the arm movement is an easy one which can be performed by a person sitting down. In contrast, the brace type of drill is slow and requires an awkward position. The crucial difference, however, is that the brace rotates in one direction whereas the bow-drill rotates first one way and then the other. Thus the bow-drill does not really drill but erodes its way through the material. Inconvenient as the brace was, it nevertheless allowed the development of true drill bits with a spiral cutting edge.

The fascination of tools is that they are concerned not with what man has achieved but with how he has achieved it. Tools are the extension of man's arms; instruments are the extension of his senses. It is man as a tool-using animal who has achieved so much.

Instruments

Tools are active: man does something with them. Instruments are passive: they provide information and then man does something ith that information. Just about the only instrument that has an tive function is the thermocouple, which is used to measure mperature but can also be used to generate electricity and is in ct so used in satellites.

Instruments fall into two broad groups. The first is concerned ith measurement. We know that time passes but we need some strument to measure it. We know that something is hot but we ed some instrument to measure it. The second group deals with nplification. There are many things we cannot see with our ordiry senses: we need instruments to convert them into a form which e can see. It may be a matter of amplification as in a microscope or transformation as in a galvanometer. Amplification instruments n often be measuring instruments as well, but the amplification nction is the more important one.

Clocks would not be necessary if people did not have to do things fixed times. Often this necessity is self-imposed. It seems probable at ancient priests and more recent monks were the original timeongers. The priests used time as a basis for their astronomical edictions and possibly as a base for their power – for they could edict things other people could not. Monks liked an ordered life ith prayers at stated intervals, and so clocks were necessary. There probably only one situation in which a clock is absolutely necessary: hen it is used to measure not time but position, as in the use of a arine chronometer to determine longitude. Sundials, hour-glasses, ater clocks, candle clocks gave way to the pendulum clock as soon the time-keeping effect of the pendulum was discovered. There as one situation, however, in which the pendulum would not work – sea, where the rolling of a ship made the pendulum unreliable. nfortunately this was an area where clocks were really vital. It is traordinary to think that the British Navy was using hourasses up until 1839, though in fact there was an accurate marine ronometer available by that time. In 1714 the Admiralty offered prize of no less than £20,000 (a huge sum in those days) for the velopment of a marine chronometer, but they were rather reluctnt to pay out the money when a superb chronometer was actually ade for them. It was the work of John Harrison, a self-educated orkshire carpenter, and it was so good that after its first voyage of days it was only 5 seconds slow.

There are a variety of things we want to measure. The simple ot-rule for measuring distances does not seem to have been in use fore 1683. In addition to distances we need to measure angles, and rious early methods of doing this culminated in the theodolite. or temperature there was the thermometer and for atmospheric essure the barometer developed by Torricelli, a pupil of Galileo. or measuring electricity there was the galvanometer.

Instruments for amplification have rather more intrinsic interest an those which measure. In the case of the galvanometer the two nctions go together. It is easy to forget that without a detection strument discovery becomes impossible. For instance, electricity ould have been difficult to develop, without a galvanometer. milarly, nuclear physics would not have got very far without cloud ambers, bubble chambers and photographic detection layers. nce you have some measurement device, you can experiment and e what affects what and what happens when you do something ecific. Using the recently developed spectroscope as an analytical vice, Kirchhoff in 1861 discovered two new metals: caesium and bidium.

The story of the development of the cloud chamber is fascinating cause the inventor, C.T.R. Wilson, started off in meteorology, d it was his determined pursuit of cloud phenomena that led him raight into nuclear physics. The interesting thing is that he himlf followed the idea right along – it was not a matter of someone else taking it over. With a cloud chamber or bubble chamber we can actually see tracks made by the most minute particles possible. A Geiger counter, which is a very simple device indeed, allows us to detect and measure radioactivity.

Lenses had been in use for three hundred years before a Dutchman, Hans Lippershey, thought of putting one in front of the other and discovered the telescope effect, which was taken up and developed by Galileo. The microscope is attributed to an individual (Leeuwenhoek) who was not only superb at grinding the lenses and making the instrument but also at using it. Where instruments differ from most other inventions is that one single device is enough. You can invent a motor-car, sewing-machine or threshing-machine but if no one else wants to use it the invention is lost because it depends for its effectiveness on use by others – on having a market. An instrument, on the other hand, does not necessarily require market value: it is enough that one person uses it to discover things which no one else can discover.

As usual, war technology has had its influence on the development of instruments. Sonar was developed to detect enemy submarines but is now used extensively for locating fish (much as the dolphin uses it). Radar was developed to locate enemy aircraft. Although the possibility of radar had been suggested by Nikola Tesla in 1900 the actual development seems to have been triggered off when someone sent a suggestion for a 'death ray' to the British Air Ministry. Outrageous though this idea was, it served as a stepping stone to the practical development of radar.

The chromatography technique for the detection and isolation of organic chemicals is an incredibly simple technique which has completely changed the nature of biochemistry. Simplicity is a feature of many instruments which have changed the course of science and technology, as witness the telescope, galvanometer, Geiger counter, clinical thermometer, and stethoscope.

Mental aids

Man uses his senses and his instruments to gain knowledge about the world around him. But what does he do with this knowledge? How does he sort it out and organize it in order to produce more knowledge and to solve his problems? To do this, man has invented and developed a number of devices. Although they are not mechanical, like spinning-wheels or jet engines, they are just as real.

The basic way man organizes his knowledge is to 'externalize it'. This means putting it out again into the external world. Language is a communicating device: though talking to other people may be the primary purpose, it is also an important thinking device. In order to use language a person has to identify things in the external world. Sometimes he has to chop up a complex piece of the world into separate concepts. At other times he has to group different things together to give a single concept. This method of procedure creates what might be called 'units' of thought. So although language is a communication system it is also a display system: having put out his thoughts as language, man can then react to them as if they were part of the external world.

Most of the mental aids developed by man are ways of translating thought into an external form which can then be reacted to and manipulated in order to produce more thought. Written language is an obvious example. The different lines of development of language (ideographic, phonetic, alphabetic) each have their advantages and disadvantages for thinking and for communicating. For instance, a Chinese typewriter is vastly more cumbersome than an English one – on the other hand, the use of ideograms may provide a greater conceptual flexibility.

A calendar is a way of externalizing time so that it can be looked at. It is a sort of map of time. The history of the development of calendars shows how different cultures adopted different approaches, and also how much political influence was involved. A map is a display device for distance and position. A person can build up a map and then use it. The map typifies all the mental aids, in so far as it is something constructed by man and then used by man to give him an advantage. For instance, the Mercator projection has the advantage for navigators that there is a constant compass bearing between any two points joined by a straight line on the map.

Musical notation is a display device and a communicating device. Its communicating function is more important than the display function, but notation itself is not merely a communicating device, for it can exert a huge influence on the development of other ideas. For example, mathematics is almost impossible with Greek numerals and not much better with Roman numerals, which are suitable only for tallying. But the Arabic numerals (originally devised in India) open up whole new areas. The invention of the zero was also a great step forward in mathematics; so, from a point of view of convenience, was the decimal notation device.

In addition to display devices, man has also developed concept systems for helping him to organize his knowledge. Binary arithmetic is both a concept system and a display device. Algebra, geometry, trigonometry, calculus and topology are pre-work systems with their own rules. They are packages of knowledge a procedure which can be used as tools. The brilliant invention of t logarithm served to pre-cook numbers so that they were easier deal with – multiplication could be treated as addition, and divisi as subtraction.

The development of the slide-rule, a remarkably effecti calculating device, followed directly from the invention of logarithn The abacus was a way of displaying numbers so that they could used more easily in calculations. With the invention of the calcul ing machine man could actually delegate some of the calculatio themselves. The first such machine was devised in 1642 by Blai Pascal at the age of 19. As so often happens, there was a particul reason for his invention: his father was a clerk who had to car out a large amount of calculating work. The next stage was t computer, which can carry out any conceivable mathematic function (and some inconceivable ones). The basic principles of t computer were laid out in 1823 by Charles Babbage, but for the implementation they had to await electronic technology. The fi operating computer was built in 1946 and contained over 18,0 triode valves. Within two decades the power of the computer a the speed of its calculations had increased colossally. The inventi of the transistor made the machines very much smaller and mo reliable. The development of micro-circuits made them ev smaller.

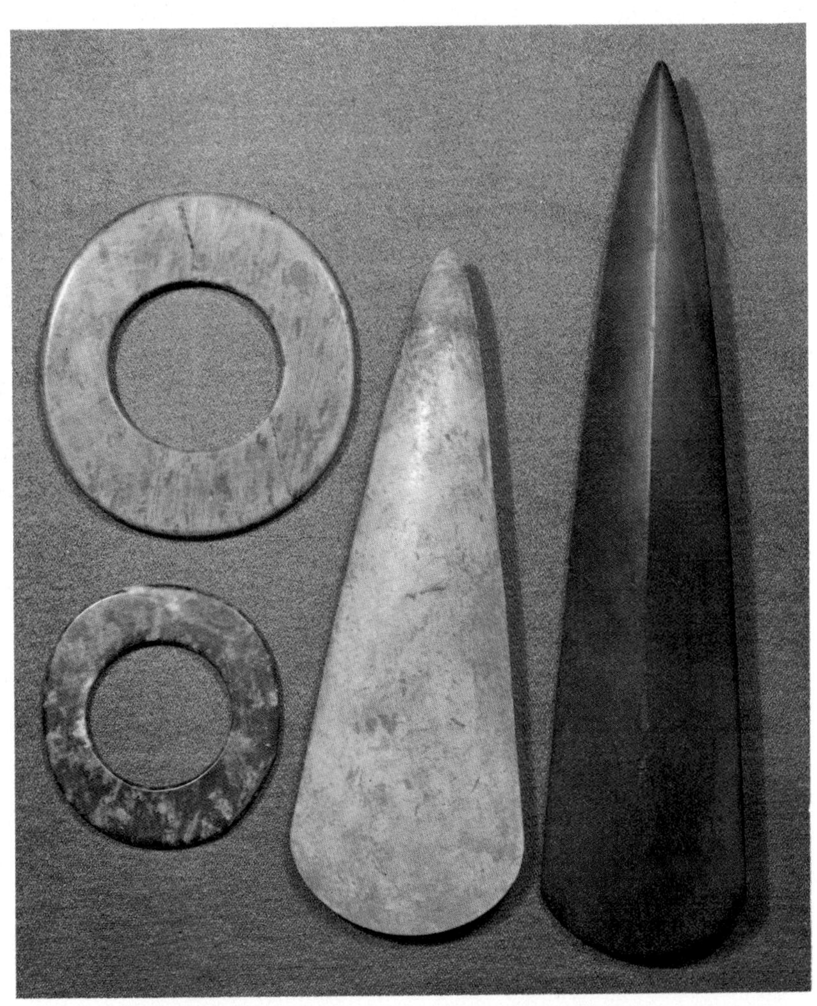

These jadeite axes and rings from Brittany date from the dawn of the Bronze Age.

A winch, turned by projecting bars. From the 13th-century 'Book of St Alban'.

Axe

The first axes were hand-axes: first of all, pebbles would have been used, then crudely chipped stones (*c.* 500,000 BC), and next more finely formed axes of flint (which can be made by striking flint with a stone, whereupon sharp-edged flakes become detached) of the Acheulian period, *c.* 250,000 BC. The progress of man in this respect can be followed by the Leakeys' finds in the Olduvai Gorge, in Tanzania, where stone tools dating back about 1¾ million years have been found.

By 3000 BC, in the Near East, we find axes of cast copper, subsequently hammered to make them harder; a sophistication in the casting – the introduction of an additional core – made it possible to produce socketed heads into which a handle would fit, instead of, as previously, being bound to the head with thongs.

Although axes have been found all over the Near East, they do not seem to have been favoured as weapons: the Sumerians, the Persians and the Assyrians all preferred spear and bow, the Greeks and Romans sword and spear. But luxury weapons, obviously ceremonial in purpose, are frequent: the double-axe is a Minoan religious symbol (the 'Labyrinth' is supposedly derived from the Lydian *labrys*, double-axe), and examples have been found made of thin sheet-metal. Socketed axes of gold have been found at Tufalau, in eastern Transylvania (*c.* 1800–1450 BC). Not only the metal, but also the curlicued shape of two of these makes them impractical as weapons, though the Mycenaeans certainly carried axes into battle and the Minoans did likewise: the Palace of Mallia, Crete, has revealed a mace-head in the form of a leopard, and a battle-axe in grey schist.

Axes are not used by Homeric heroes – except in Odysseus' sharp-shooting demonstration as described in the *Odyssey*: he performed the feat of shooting an arrow through a line of socket-holes in the hafts of double-headed axes set up in line almost at ground level. True, a minor Trojan called Peisander wields 'a fine bronze axe with a long smooth haft of olive-wood', but it did not take Menelaus long to dispose of him. Could it be that the axe was not quite gentlemanly? Certainly it was the northern tribes who favoured the weapon – the Cimmerians, Scythians, Bactrians, Franks, Vikings and English (for the English two-headed axe, see the Bayeux Tapestry) – and a recent author has suggested that 'only Saxons and Scandinavians considered it fit for any save churls' in the 11th century, although it became respectable later.

C. M. B. G.

Winch

Without the winch the architectural achievements of the Greeks, who did not have the great armies of cheap labour that their Egyptian and Babylonian predecessors had enjoyed, could never have been built. But the moment when the roller and lever, which had been used with heavy loads probably since Neolithic times, came to be combined in the 'axle in a surrounding wheel' to obtain continuous leverage is still obscure. A winch has been found at a Bronze Age copper mine in Austria, said to have been abandoned by 600 BC, but there are no illustrations before the 1st century AD.

By Hellenistic times, the winch is frequently mentioned; Pseudo-Aristotle found its operation a problem in the 3rd century, and Heron treated it as the first of his five simple machines some 300 years later. By this time the winch was furnished with a ratchet and pawl to prevent it slipping back, and had been turned on its side to form a capstan or windlass, so the workers could lean on the handspikes all the time as they walked around it. Presumably it was used at first just for traction along the ground; with the aid of a pulley as fulcrum, however, the winch would become a builder's hoist. Transferred to shipboard, the capstan could be used to haul in the anchor, while small winches controlled the rigging. Thus a simple device made possible the great ships of the late Middle Ages, with their massive anchors, whose cables would have been too much for naked hands to master.

All these winches were worked by projecting bars or handspikes, as is made clear by both the descriptions and the illustrations in ancient sources. Not until the 14th century do we find winches turned by cranks; often depicted in building operations, this type is represented in our museums in the form of the windlass crossbow.

Man Working

About the same time horses and oxen began to be used with the vertical capstan. If the job involved hoisting rather than traction, as it usually did, the effort would have to be transferred through a right angle, either through a secondary winch – as in the horse whim so much used in mine haulage – or else through a face wheel and pinion.

There was the additional difficulty of reversing, in which the horse would have to be released and then reharnessed. To solve these problems Brunelleschi invented a famous hoist while working on the cathedral at Florence; it also had three drums of varying diameters to enable him to have a light winch for raising lighter loads rapidly coaxial with the thicker ones for the heavy stones. Although much admired, and sketched even by Leonardo, his engine did not pass into general use; but right-angle gearing was sometimes mounted on winches to enable the prime mover, man or beast, to walk round a capstan, instead of men turning a winch – the same method is recorded by Agricola in the mines of 16th-century Germany.

Parallel gears to obtain mechanical advantage featured in Heron's 1st-century 'weight-lifter', the *baroulkos*, but it was not until Simon Stevin's 'Almighty', which he designed for hauling canal boats over dykes or slipways in 1586, that they entered normal practice. By the Golden Age of the canals they seem well-nigh universal, even for relatively easy jobs like opening lock gates and sluices.

Before Stevin's day, winches and capstans were always used with cables; only in the last years of the 16th century were chains introduced for mine haulage, and they are shown opening lock gates in Zonca's woodcuts of the canalized Brenta (1607). With mines penetrating ever deeper underground, the 19th century found its chain adding quite considerably to the load the winch had to lift; the answer was iron wire rope, invented in the 1830s.

A. K.

Carpenters' tools

In the Middle Ages, since nearly everything was made of wood, the basic hand tools were those of the carpenter. These had changed little since Roman times; indeed, it is possible to regard the Middle Ages in northern Europe as the gradual recovery to a point where the Romans had left off about 1,000 years before. This does not necessarily imply any special skill on their part; there seems to be a sort of Parkinson's Law about skill: that it will always manifest itself when and where there is a demand for it.

The two curious objects on the workbench and the window-sill in St Joseph's workshop (from the 15th-century *Merode Altarpiece* by Robert Campin) could be small carpenter's planes, but it is now fairly certain that they are mousetraps. It has always been usual for carpenters to amuse themselves in the winter by making these things. The one on the bench looks like the common type, where the animal nibbles at the bait and trips a spring which snaps the U-shaped arrangement down, either trapping it or breaking its back.

The brace St Joseph is using is one of the few tools the Romans did not know, and it is doubtful whether they had the other boring tool on the bench, the auger. Nothing resembling an auger occurs on any known Roman monument, but there are any amount of bow-drills, the standard boring tool in the Mediterranean area since Egyptian times. Pliny speaks of the *gallica terebra*, which the Romans apparently borrowed from those notable wheelwrights the Celts, and it thus seems likely that the auger was a northern European invention. Its main disadvantage is its 'stop-go' action; but although the bow-drill makes several complete turns first one way, then the other, the bit merely abrades its way through the wood.

The wimble or brace was the first boring tool with a continuous rotation in one direction. Like the auger it first appears in northern Europe, some time about 1425.

There are some more ancient tools on the bench, such as the pincers, which use the pivoted action of scissors to pull instead of to cut.

The sword-like handsaw for cross-cutting is practically the same tool as the Romans used, but narrow-bladed framed saws, also developed by the Romans, were used for ripping down and for bench work. (The stamp in the form of a cross matching the three stamps on the axe blade was an early form of toolmaker's guarantee.)

The basic carpenter's tool was the axe, with special types from very early times for felling the timber and dressing it to size. The adze was, of course, indispensable to cooper and shipwright, but rarely used by anybody else. The tool here is a small version of the side axe, used for final trimming.

W. L. G.

Screw-jack

'These be very much used in Germany and in Dutchland, to lift up the side of a great Dutch wagon, when that it is laden', explains an Elizabethan technologist. Screw-jacks as we know them were invented toward the end of the 15th century: Leonardo da Vinci got a German workman to make him one. From the 13th century there are also allusions to the use of a rising screw as a hoist to haul up a load with a rope and hook. All these, if they really existed, were very heavy and cumbersome

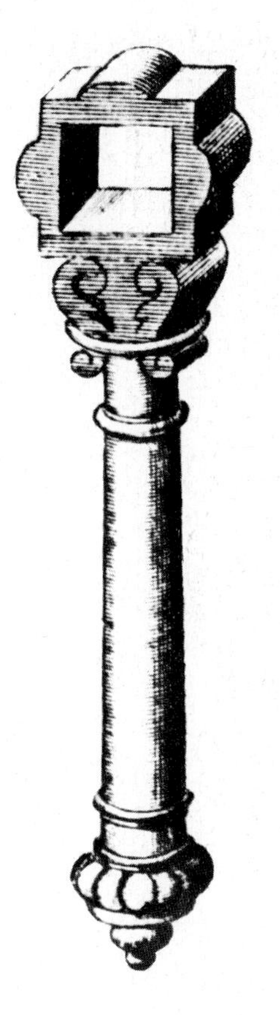

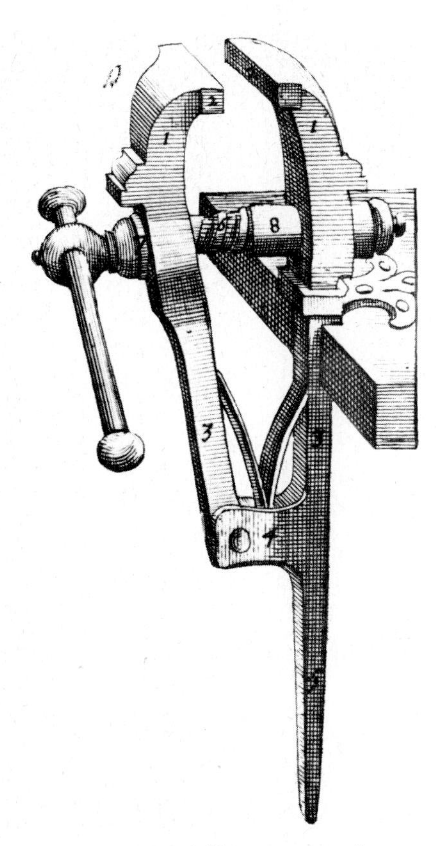

A decorative form of box spanner, and a metalworker's vice, both illustrated by André Félibien in his Principes de l'Architecture, *1676.*

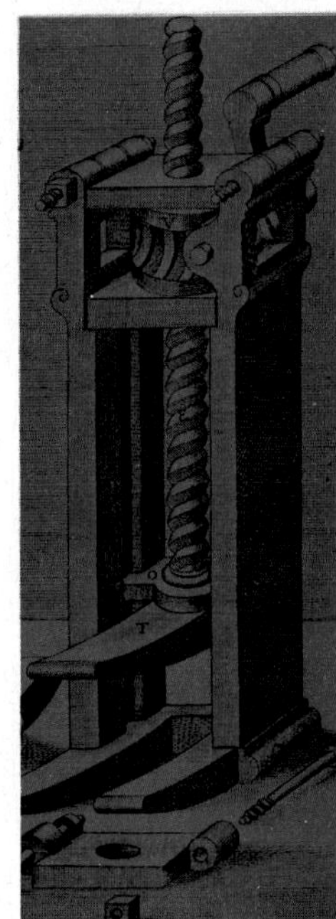

The screw-jack (left) *came into its own in Renaissance warfare, for training cannon, but is essentially the same tool as is used today for changing a car wheel.*

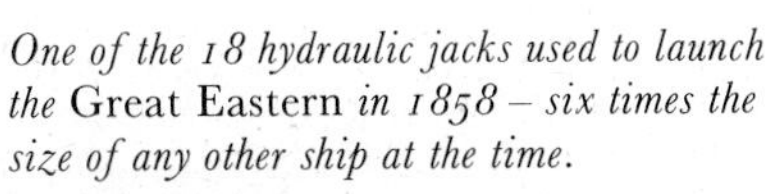

One of the 18 hydraulic jacks used to launch the Great Eastern *in 1858 – six times the size of any other ship at the time.*

Opposite, above, *the* Merode Altarpiece *(by Robert Campin, c. 1430) shows St Joseph at his workbench. Among the tools visible are a brace, pincers, auger, hammer and handsaw.*

Opposite, below, *a 16th-century rolling mill in Lorraine, the earliest picture of one in action.*

machines. But the screw-jack, which some unknown artisans devised in the Low Countries in about 1480, was the same simple handy device we know today.

It was used to lift a vehicle clear of the ground if it had fallen on to one side, as often happened on unmade roads, or to hold up the axle while a wheel was changed. For both purposes, it was steadier and easier to work than a lever, and would not slip. Unlike earlier screw-lift devices, it pushed from below but was operated from above, lifting by means of the projecting shoe, and so was useful in a great many more situations. Elizabethan English visitors to the Low Countries were very struck with this 'Engine that goeth with a screwe'.

Although these jacks were normally used for wagons, they could also help train cannon by raising the tail, and so were included in the regular issue of artillery supplies. Besides, jacks could be used to lift doors off hinges, which was very useful in street fighting or for breaking into the back doors of castles. That explains why they were so often illustrated in Renaissance books on fortification, and why the only surviving jacks of the period are so ornate; as with the guns of the day, craftsmanship was lavished on anything that appealed to princes as a military tool.

A. K.

Rolling mill

'At the Mint of Segovia in Spain there is an engine that moves by water, so artificially made that one part of it distendeth an Ingot of gold into that breadth and thickness as is requisite to make coyn of: it delivereth the plate that it hath wrought unto another that printeth the figure of the coyn upon it, and from thence it is turned over to another that cutteth it in due shape and weight.'

The earliest machines to use a pair of rollers were simple squeezing machines, like old-fashioned mangles, which were used to crush organic materials, possibly cotton seeds. They were employed in medieval Egypt to press the juice from sugar-cane. In metallurgy, however, rolling mills were not to be found before the 16th century. At first they could not be used for iron, but only for softer metals with lower melting points which were easier to work. The original version rolled copper strips and plates for prints and instruments; it produced a more even surface than could be achieved by hammering. But the version that aroused the most interest was the coining mill, described above by an English visitor to Spain.

This machine was invented somewhere in Austria or southern Germany. The King of France was negotiating for one in 1550, the Spanish coining engine was installed in 1583 and they soon spread across Europe, powered by water or horses or, in smaller models, turned by human hands. The roller-press for printing copperplate engravings worked in the same way. Another type of rolling and slitting mill was the lead-drawer, which was used to roll, slit and extrude grooved lead strips for window-panes.

These developments represented a great advance over hammering and shearing by hand. The first application to iron came a little later. Early versions were restricted to making nails and thin strips; it was not until the steam engine was applied to Cort's grooved rolling mill of 1783 and a way was found to work iron at much higher temperatures that it became possible to roll and shape the massive, elongated bars on which modern construction methods depend.

A. K.

Vice (Vise) and spanner

The vice is said to have been invented by Archytas of Tarentum, a friend of Plato, but no details are known, and although the Romans used wooden screws for wine- and clothes-presses, there is no evidence that they ever got round to applying this idea to the vice. The earliest vices were of metal and used by smiths and other metalworkers from the 16th century onwards. No woodworkers' benches are shown with vices before the end of the 17th century and they only came into common use in the 18th. Carpenters and joiners could always sit on their work to keep it down, but this was not possible for blacksmiths, even if only one end of the rod was red-hot.

The early history of the spanner is obscure, but various types were developed to deal with nuts and bolts from about 1550 onwards. The 17th-century French architect and author André Félibien illustrated a form of box spanner in 1676. Spanners with movable jaws date from about 1700 and are sometimes called 'French' spanners, which presumably indicates their country of origin.

W. L. G.

Hydraulic jack

The theoretical ideas necessary to use fluids to lift weights were established by Simon Stevin in the 16th century and Blaise Pascal in the 17th.

The invention remained latent for another 150 years, until it was developed by an English cabinet-maker, Joseph Bramah. The installation of water-closets in new houses led him to study hydraulics; the theory of the subject inspired thoughts of using it in machines.

In 1795 he patented the first hydraulic press and, shortly after, a hydraulic balance. The press was used for compressing cotton bales, tearing up trees and similar jobs.

It was Bramah himself who first suggested using this system in a telescoping hydraulic jack, in a patent of 1812. But it seems that hydraulic jacks were at first limited in application to slow lifts of heavy loads. Stephenson and Brunel both made use of them in the 1840s and 1850s to test the chains intended for their suspension bridges. A hydraulic crane was devised by Armstrong for Newcastle docks in 1847; it proved a great success, and soon spread to other ports. But the first public triumph of the hydraulic jack was in launching the great weight of Brunel's *Great Eastern* in 1858. When every other technique had been defeated, he brought up 18 large jacks which did the trick.

A.K.

Steam hammer

On 24 November 1839, James Nasmyth, machine-tool and locomotive manufacturer at Manchester, received a letter from his friend and customer Francis Humphries, the Great Western Railway's engineer. Emboldened by the success of their steamship the *Great Western*, the company had decided to build another vessel, very much bigger, the *Great Britain*. But, said Humphries, 'there is not a forge hammer in England or Scotland powerful enough to forge the intermediate paddle-shaft of the engines for her. What am I to do? Do you think I might dare to use cast iron?' Nasmyth, well trained in the theory and practice of machine design, sat down to work out an answer on the spot. He decided that the root of the trouble was simply that tilt-hammers had retained the geometry of the old hand hammer of the smiths. Steam had replaced water to drive the shaft, and the hammer head was certainly much larger. But it was still raised like a lever, and then fell through an arc of quite acute angle, so it 'suffered from want of compass, of range and fall, as well as want of power of blow', as he wrote later. If the object to be forged was of any great thickness in itself, it took up much of the available space between anvil and hammer and so 'when the forging required the most powerful blow, it received next to no blow at all'. The old design also limited the size of hammer, for if it were too heavy it would snap its helve. So he must find a way to get the hammer a sufficient height above the work, and then let it fall *vertically*.

That same day he sketched a machine in which the hammer block was joined to the rod of a piston raised by steam admitted beneath it in a cylinder, and then let fall under gravity once the steam was allowed to escape. The cylinder was to be held above the anvil between the massive shoulders of iron supports, which also provided the guides to direct the block in its fall. Apart from the advantages of longer fall in a straight line, and a potentially heavier block, the control of admission of steam through a valve would let the operator 'think in blows', regulating the speed and frequency of strokes to the requirements of the job. One of his favourite tricks when showing off the hammer was to crack an eggshell placed in a wineglass without damage to the glass.

In the end a decision was taken to have the *Great Britain* propelled by a screw instead of making her a paddle-boat; then there was a slump in heavy engineering, and the steam hammer was not built. A full 30 months after his original conception Nasmyth was in France, and visiting the Le Creusot ironworks, when Bourdon, the manager, invited him to 'come and see your own child', and showed him a steam hammer the works had built, which he had based on the quick glance he had been allowed at Nasmyth's Scheme Book the year before. The inventor could hardly complain, for he had not patented the design or kept it secret, and it was much to Bourdon's credit that he had been able to build his prototype from such a crude sketch without making any attempt to gloss over his intellectual debt. But no sooner was Nasmyth back in Britain than he hastened to patent his idea, and start manufacturing steam hammers in earnest. They were an immediate success.

His earliest version had a 30 cwt. block, but he was soon producing hammers of 4 or 5 tons. In 1843, Nasmyth's assistant Robert Wilson improved them by a self-acting drive, in place of hand gears, and the hammer was altered to admit steam above the piston, to give push down as well as up.

A.K.

James Nasmyth's steam hammer at work: painting by the inventor.

Metal plane

To every woodworker, the name Stanley suggests the adjustable metal plane. Yet it was not invented by anyone called Stanley: it was the work of Leonard Bailey, who manufactured planes in Boston, Massachusetts, in the late middle years of the 19th century. Bailey's plane was easier to adjust and quicker to sharpen than earlier, wooden planes, and was cheap enough for the average craftsman to buy. The wood-lined metal planes made by Spiers of Ayr and Norris of London cost three times as much.

Bailey's first plane was wooden, with metal fittings, but it incorporated many of the features which are still used. There was a lever cap with cam action, instead of a wooden wedge, to hold the

Below, *early Stanley-Bailey plane, with the cutter seating screwed to a wooden base – halfway to the all-metal plane.*

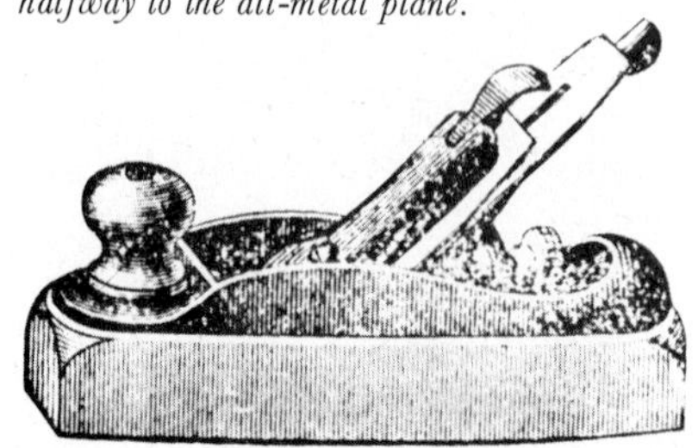

Right, *early advertisement for the blowlamp patented in 1887 – if not actually invented – by the Swede Max Sievert.*

Medieval pole lathe, worked with the foot. By its aid, such things as table legs were considerably easier to make.

cutter in position, and the cutter itself was of a uniform thickness instead of wedge-shaped. The cutter could be adjusted by turning a nut instead of having to use a hammer. The next development was to incorporate the cutter seating in a metal casting and screw it to a wood base, and then came the all-metal plane that we know today. Leonard Bailey sold his business and patent rights to the Stanley Rule & Level Company of New Britain in 1869, but all Stanley metal planes still bear his name and are of the same basic design.

C. H.

Blowlamp

The blowlamp's deceptively simple construction includes only a few basics – a strong fuel reservoir, connected by a metal tube to a small nozzle which is situated in a perforated combustion chamber. In some lamps, petrol or benzene is used as fuel, but paraffin is more commonly employed and requires initially to be forced from the reservoir by a small pump. When operating a blowlamp, one must 'prime' or preheat the fuel in the metal tube, so that it issues from the nozzle in a gaseous jet. This evaporated fuel mixes with air drawn in through the holes in the combustion chamber and, when ignited, forms a clean, hot flame of considerable force.

Often, the supply tube is designed to pass through or near the flame, whose heat provides constant self-priming, the result of which is that the lamp continues to burn until turned off.

As early as 1862, James Rhodes, of Cincinnati, Ohio, had developed and patented the *idea* of a mineral-oil vapour lamp, but he apparently never actually made one. It was not until 1880 that someone finally got a recognizable blowlamp to work, and to complicate matters, there were three claimants to the discovery. It has sometimes been said that Max Sievert, the Swedish manufacturer and exporter, was the first inventor, but the credit for the original petrol blowlamp should actually go to his engineer, C.R. Nyberg. Sievert patented the design in 1887, although sales of the new instrument had begun earlier. While Nyberg was working on his petrol lamp, however, two other Swedes, the Lindqvist brothers, produced a paraffin model. The final candidate for the invention of the original apparatus was another Swede, Ludwig Holm, and it has even been hinted that both Nyberg and the Lindqvists got their ideas from him.

Whatever the truth may be, it is clear that the invention took place in 1880, and that two of the claimants to priority became rivals in trade. Since the paraffin blowlamp is safer and more popular than the petrol variety, one should perhaps give real credit to the Lindqvists.

J. G.

Machine tools

The earliest ancestor of the lathe can probably be dated to the 4th millennium BC, in ancient Syria; by 'lathe' we understand a device for rotating any material to be worked upon by a tool for the purpose of cutting, shaping, polishing and so on. The inventor took his cue from the bow, his principal weapon; he made the string, instead of speeding the arrow, rotate the work-piece he wanted to shape. The bow-lathe and its near relative, the pole-lathe, did not produce a continuous rotating movement, but the revolutions were alternately forward and backward.

However, the wheel-driven lathe providing continuous rotation – hand- or foot-operated – was not used until the 14th century, and then only in very few trades. The first lathe for cutting screw-threads appeared in France in 1568; though the work-pieces and cutting tools were made of metal, the lathe itself was built from wooden parts. Most early lathes were of the 'dead centre' type, that is, the work-piece was inserted in a wooden block; the 'two-point centre' type, with the work-piece rotating between the 'chuck' in the headstock of the lathe and the pin-like centre in the tailstock, was an 18th-century innovation.

So was the 'slide-rest', an essential part of the modern lathe: a saddle which moves the cutting tool, fixed in its tail-post, horizontally along the rotating work-piece. Its inventor was either the prolific Joseph Bramah (1748–1814), who also invented a patent lock, the hydraulic press, and the water-closet, or his employee Henry Maudslay (1771–1831), or both of them together. At any rate, 1794 was the year of birth of the slide-rest. Other improvements of the lathe followed as England's Industrial Revolution got into its stride, with increasingly pressing demands for machinery of greater efficiency and accuracy; one of the most important machines was Maudslay's screw-cutting lathe, invented in 1800, a vast improvement on the primitive French device of 1568.

But now much of the industrial technological scene shifted to America where, early in the 19th century, Eli Whitney pioneered the system of interchangeable parts as the basis of mass production. In the 1840s, an engineering firm which still bears his name, Pratt and Whitney, developed the capstan lathe in which the tools required for various kinds of operations are held in a tool-post resembling the conventional ship's capstan; the tools take the place of the capstan handles, and by revolving the tool-holding turret they can be

Henry Maudslay's original screw-cutting lathe, invented in 1800.

brought into the required positions. By that time, of course, steam power had mechanized a wide sector of industrial production, and most of the machinery developed throughout the 19th century was designed to be mechanically operated as a matter of course.

Today, the lathe is the most widely used machine tool, particularly suitable for automatic operation and mass production of identical parts. Its main types of operation are turning, facing, screw-cutting, and boring. The work-piece is rotated by an electric motor, and various tools held by the sliding toolpost are brought up to it in proper sequence; the speed of rotation can be varied according to the requirements of the operation.

Another important machine tool is the milling machine, which has a rotating tool for cutting the work-piece while it moves past it: in contrast to the lathe, therefore, it is the tool which rotates and the work-piece which is 'fed'. The original idea seems to have occurred to an English scientist, Dr Hooke, in the 1660s, but there was no call for it in those days. In 1839 a Swiss mechanic, Georg Bodmer, invented a milling machine which, however, turned out to be much too expensive, and again the Americans came to the fore. As early as 1818, Eli Whitney had been experimenting with the rotating-tool principle, and F.A. Pratt designed what he called the Lincoln milling machine in the 1840s; but what the factories needed was something more adaptable for a wide variety of operations, and at last Joseph R. Brown, an American engineer, designed the 'universal' milling machine which he exhibited with great success at the Paris Exhibition of 1867.

This was the direct ancestor of the sophisticated modern 'miller', of either the horizontal or the vertical type. It uses any kind of cutters, some almost like circular saws, others capable of cutting square or round grooves in the work-piece; or they 'mill away' flat surfaces. The plain helical cutter is the most widely used type.

The history of the boring machine dates back to 1765 when one of Britain's first great civil engineers, John Smeaton (1724–1792), designed a cylinder-boring machine, which was in operation for ten years or so. It was powered by a water-wheel; however, as the end support of the boring tool travelled on small wheels within the cylinder, it was not very accurate and the bore was uneven. In 1775, another English engineer, John Wilkinson, improved the machine with an efficient support for the tool – a boring-bar passing through the cylinder, firmly mounted on two bearings. Still, boring machines for general use – apart from boring by lathe – were not developed until the middle of the 19th century, when Georg Bodmer designed the prototype of the modern machine, capable of boring at various speeds; but he was not a good businessman and failed to persuade the manufacturers to buy his machines. Again it was left to the Americans to market drilling and boring machines on a commercial scale, though this was not before the 1890s, when electric motors for powering them became available.

The industrial expansion in Britain, the European Continent, and the USA during the 19th century changed the way in which inventions and innovations were achieved. The 'loner', the technical genius with a new idea, became a rare figure – Edison was the last of the great all-round inventors – and his place was gradually being taken by the team, usually working for, and within, some big company which could afford to maintain a research laboratory. Thus it is impossible to say exactly when or where or by whom any of the important machine tools and techniques were created – for planing, shaping, casting, cutting, grinding, sheet-rolling, pressing, extruding, arc-welding and other metal-working processes. Some of them were offshoots of other technical developments, such as welding by the laser beam, an achievement of the 1960s. There are, however, some exceptions, cases in which the origin and development of industrial processes can be traced.

The 'universal' milling machine, invented by Joseph R. Brown and shown at the Paris Exhibition of 1867.

The continuous casting of steel, for instance, was developed first by the American engineer B. Atha between 1890 and 1910; he produced file steel in a continuous casting machine in which a crucible of the molten metal was poured through a vertical mould. But Atha was unable to devise an efficient cooling arrangement, and the whole idea remained dormant for two decades. The continuous casting of non-ferrous metals became commercially successful in the late 1920s as the result of the work of Dr Siegfried Junghans in Germany, a member of the family who owned a large watch- and clock-making company of that name. Another German firm engaged Dr Junghans for the development of his process for the casting of all metals, with which he succeeded in 1935 by introducing the reciprocating mould, designed to reduce the risk of the molten metal clinging to the mould as it solidifies – this had been a major problem. But it was an American, Irving Rossi of New York, who marketed the process all over the industrialized world.

Another most important innovation of modern times in the steel industry is the continuous hot-strip rolling process. In this case the first impetus came from Teplice, Bohemia, then part of the Habsburg empire; here, sheets of up to 50 in. width and 60 ft length were successfully rolled for the first time in 1892. The steel mill at Teplice abandoned the process after a few years because of technical difficulties, but

The shock of a chemical explosion, transmitted through water, can be used for the shaping of metal parts – a recent invention but not, so far, widely adopted.

High-speed cutting of sheet tantalum with a laser. Advantages of this technique are the absence of any cutting tool and the elimination of cutting pressures.

Charles W. Bray designed a continuous-sheet mill for the American Sheet and Tin Plate Company in 1902. Again, technical snags forced the work to be abandoned, and then a paper manufacturer of Middletown, Ohio, John B. Tytus, assembled a team which succeeded in developing the first viable hot wide-strip continuous-rolling process for Armco, the American Rolling Mill Company, which built a special continuous-rolling mill in 1923, patented the process, and licensed other steel-makers to use it. The Tytus process was based on the discovery that the hot steel must have a slightly convex cross-section; this is gradually 'ironed out' by successive rolling operations until the sheet leaves the mill flat at the end of the process.

Shell-moulding was the first major innovation in foundry technology for a long time. The shell mould consists of a thin sand mould held together by phenolic resin. This ensures that the castings have a smoother surface and more accurate dimensions than could be achieved by conventional methods; the advantages are that machining is reduced to a minimum and the moulds can be made on automatic machines by unskilled labour. The process was invented in 1941 by Johannes Croning, a Hamburg foundry owner, after years of experimenting. At that time the second World War prevented countries outside Germany from knowing about Croning's invention; it was discovered by the Allied technical teams searching in occupied Germany for new developments after the war. Croning succeeded in having the process patented in the USA, and the Ford Company was the first automobile firm to introduce shell-moulding in 1947; a year later, Ford already had a pilot moulding-shop which produced 1,000 exhaust valves a day.

A particularly interesting story is that of the use of explosives for forming metals, although it is a story of team work and not of individual inventors. Probably the first patent in this field was a British one of 1898 for the explosive expansion of metal tubes in making bicycle frames. In 1909 we find the first US patent for the explosive forming of sheet metal, and between the two World Wars gun emplacement shields were explosively formed in the French armaments industry. In 1950, an American company in Missouri used the process for making fan-hubs, but only on a small scale.

Then, a few years later, the US space programme got into its stride, new shapes and materials were required – and new techniques of forming unusual rocket components had to be developed. A major problem was that of producing large segments for the 'domes' of the Saturn rocket booster; the domes had a 33-ft diameter and were made of a very strong aluminium alloy. The North American Aviation Company undertook to make the segments by explosive forming while the Aerojet Corporation used the same technique for producing 54-in. diameter rocket cones.

There were failures, but they added to a store of experiences with the new process. By the early 1960s, the NASA technologists began to recognize that the scientific basis of explosive forming was not yet solid enough, and a small army of scientists was asked to work out the necessary theoretical and practical guidelines. Explosive metal-working uses the short, sharp release of energy caused by the detonation of a chemical explosive instead of shaping the metal by machine tools or casting. The high-energy stresses are, of course, not let loose directly on the metal, but transmitted as impulsive waves by some medium such as water or air under strict control so that the metal is shaped and not shattered. The process is cheap, versatile, and can be applied at any energy level. However, ordinary factory facilities are not safe enough for high-explosives; new buildings have to be put up, new apparatus devised. Also, the process does not lend itself to automation – each explosion has to be done, as it were, by hand. An attempt at using ice dies for the explosive forming of 10-ft-diameter high-strength steel domes did not come off as well as expected as the process had not been fully developed by trial and error within the limitations of time and the NASA budget.

As the NASA programme ran out, so did the allocations for further work with the explosive-forming technique; it was left to individual American firms to continue with research and development, but the initial enthusiasm seemed somewhat on the wane in the early 1970s. In Britain, leading firms like Rolls-Royce, ICI, de Havilland, Lucas, Napier and the Production Engineering Research Association carried out some preliminary work, though their approach was rather cautious as there appeared to be no immediate financial rewards. The College of Aeronautics carried out a series of experiments with light alloy, stainless steel and titanium. However, the general feeling among British engineers seemed to be that an explosive is an explosive and not a machine tool, and who wants a potential bomb in his factory?

So it may well be that explosive forming will, at least for the time being, rank as one of those 'also ran' ideas which add so much drama and flavour to technical history, and without which eventual progress cannot be achieved.

E. L.

Ultrasonics

Until the last century, zoologists had always been puzzled by the way bats can 'navigate' in the dark, flying around without ever bumping into obstacles. Now we know that they use a kind of radar, sending out high-pitched squeaks which are reflected from objects in their flight paths, giving warning echoes which guide the bats in complete darkness. Human ears cannot hear those squeaks because of their very high frequency. Some other animals, too, can hear sounds of higher frequencies: there are, for instance, dog-whistles to which they react while we cannot hear them. These high-frequency sounds are in the so-called ultrasonic range.

The way to produce ultrasonic waves was shown for the first time in the 1880s by the research work of the French physicist Pierre Curie, husband of Marie Curie, with whom he worked on the discovery of radium. Pierre Curie discovered the 'piezo-electric effect', a property of certain asymmetric crystals such as quartz. When such crystals are subjected to pressure, positive and negative electric charges are produced on the opposing faces of the crystals; conversely, if these crystals are subjected to an electric potential, they alter their size, generating air vibrations. Thus, if a suitably cut plate of quartz is subjected to an electric current of a frequency equal to that of the natural vibration of the quartz, it produces very high-frequency waves – ultrasonic waves.

Ultrasonic waves travel like a beam through many kinds of matter, without spreading very much, producing echoes which can be registered and recorded by a receiving set. This has led to the development of a whole range of practical applications. Ultrasonic waves are used today in industry to discover faults in opaque objects, for instance flaws and cracks in steel, which show characteristic echoes in the receiver. But ultrasonic waves can also generate high-frequency vibrations in softer materials, and can produce emulsions by breaking up particles of one liquid suspended in another, such as oil in water; they can also 'liberate' gas bubbles trapped in liquids, sterilize milk and other liquids by killing bacteria, or clarify the molten material from which artificial sapphires or rubies are made. Laundry research technicians have also been experimenting for some time with 'ultrasonic washing' as these waves can shake out dirt and grease particles from fabrics. High-frequency sound waves are also extensively used at sea for echo-sounding.

E. L.

Oxy-acetylene welding

Welding with flame, or gas welding, was introduced following the development of the Linde method of producing oxygen from liquid air in 1893, which made oxygen generally available; the oxy-acetylene torch was devised by Fouch and Picard in 1903.

Acetylene gas gives the hottest flame (3,250 °C.) and has the highest combustion intensity and thus is favoured above such other gases as propane and hydrogen since it permits higher welding speeds. It is mixed with oxygen because this gives a hotter flame than burning in air. The oxy-acetylene torch is the medium for mixing the combustible and combustion-supporting gases, and provides a means of applying the flame at the desired location. The gases are mixed in nearly equal volumes and the chemical characteristics of a flame can be altered to suit the requirements of the welding process by changing the ratio of acetylene to oxygen. For most applications the so-called neutral flame is used, with roughly equal proportions of acetylene and oxygen. An oxy-acetylene flame has a blue cone where the gas comes out of the torch, at the tip of which is the maximum temperature.

Oxy-acetylene welding is used for steel in sheet thicknesses or where good control is required. The equipment is cheap and the process can be applied to a wide variety of metals. The considerable heat spread with gas welding results in less severe heating and cooling cycles than with arc welding, because flames are less efficient than arcs, and not such compact sources of heat. Gas welding has been widely used for welding hardenable materials such as alloy steels. On sheet metal the process can often match or even exceed the speed obtained with arc processes. It is less tiring because lighter viewing glasses can be used, and control of the torch position is less critical.

T.C.

An experimental ultrasonic aid for the blind. Sound waves reflected back from an obstacle are transformed into an audible 'beep' in a small receiver or into a buzz that can be felt in the hand.

Welding in a French workshop in 1904, soon after the oxy-acetylene torch had been invented by Fouch and Picard.

Antique sundial in quadrant form, a development of the hemispherical quadrant invented in Babylon, c. 300 BC.

Hispano-Moorish astrolabe, dating from c. 1026.

Sundial

Even as the earliest of our forebears observed the most obvious visual manifestation of the earth's diurnal rhythm, the rising and setting of the sun, he was already falling into a routine that would eventually result in modern man's obsession with time. The coming and the passing of the dark of night clearly signified a division, a period of change in the course of an unavoidable cycle. But while ancient man, and especially the early farmers, did need to develop some understanding of the longer stretches of what we call time, an understanding of the seasons, a precise breakdown of the day into seconds, minutes and hours was hardly called for. Most people simply rose with the sun, did their work by its light, and left off work around dusk. People did not ask each other what time it was; they looked up at the sun, they felt it in their bones, or they simply guessed. Then, sometime around 2000 BC in places like Egypt and Mesopotamia, there evolved a class of men who wondered about the sun and its movements. These were the astronomers and they invented the sundial.

The first device used for 'telling time' was a simple gnomon, usually a vertical stick or pillar that indicated the sun's position by the length of its shadow. Obelisks and pyramids were also used in this way. By the 8th century BC, more precise sundials had been developed and the oldest of these still extant is an Egyptian shadow clock with a straight base and a crosspiece set at a 90° angle. Such instruments are still in use in primitive parts of modern Egypt.

In about 300 BC, the Babylonian astronomer Berossus invented a hemispherical sundial known as a hemicycle and, according to a 10th-century Arab star-gazer, this was still a popular tool in most Moslem countries 1,200 years later. The Greeks, meanwhile, and especially such noted geometrists as Ptolemy and Apollonius of Perga, were building truly complex sundials; and the Tower of the Winds at Athens, an octagonal structure erected about 100 BC, contained eight dials facing in different directions. All these early dials, however, including those built by the Romans, employed the use of 'temporary hours', whereby the day was divided into 6 or 12 segments that would vary in duration according to the time of year. Later, the advent of trigonometry permitted greater accuracy in marking dials and stimulated the growth of gnomonics, or dial-marking. And the invention of mechanical clocks in the 14th century prompted the eventual adoption of an equal-hour marking system.

During the Renaissance sundial-making flourished, and instruments of diverse form and great artistic merit were produced. The pursuit was highly profitable as well, and for many years methods of design and construction remained closely guarded secrets. Then, in 1531, the German cosmographer Sebastian Muenster published his *Compositio Horologiorum*, which even contained the details of a moondial, and the secrets became more common knowledge.

E. W.

Astrolabe

Called a 'mathematical jewel' in the Middle Ages, and a prized possession of Chaucer's Clerke of Oxenforde, the astrolabe has become a symbol of early astronomy. Usually, 'astrolabe' means a planispheric astrolabe, a computing and observational instrument used from Hellenistic times until the 17th century in Europe and later in Islamic countries; but there are also spherical astrolabes, sea or mariner's astrolabes and prismatic astrolabes.

The planispheric astrolabe is based on an ingenious method of drawing a circular map of the heavens so that although distances between the stars and various reference points are distorted the angles between them are not. This way of making a star map may have been known to the Greek astronomer Hipparchus (2nd century BC); it was certainly known to Ptolemy of Alexandria in the 2nd century AD.

In constructing an astrolabe, a star map of this type is usually cut out of a thin sheet of metal, and the brightest stars are indicated by pointers which spring from a tracery incorporating a circle marked with the positions of the sun throughout the year. The whole is surrounded by another circle representing the Tropic of Capricorn, and is pivoted about its centre to imitate the apparent rotation of the stars about the Pole. Below the star map (called *rete* from the Latin for 'net') is a circular metal plate of the same size, engraved with lines representing the horizon, the

Right, *early 16th-century clock mechanism. In the earliest clocks a shaft from the weighted cross-bar is fitted with two teeth that alternately engage the teeth on the bevel. This system, the first escapement, regulates the fall of the driving weight. Later escapements worked with a pendulum.*

zenith and circles of altitude between them, all drawn for a particular latitude. By correctly adjusting the *rete*, it is possible to show the positions of the stars and the sun at any time. Most astrolabes have a series of plates for use in different latitudes, which are stored below the plate in use. All the plates and the *rete* are contained in a recess (the *mater*) in a thick metal plate equipped with a suspension ring at the top and, on the back, a sighting-rule (alidade) moveable over a scale of degrees for measuring altitude. *Rete*, plates, *mater* and alidade are held together by a pin and wedge so that the *rete* and the alidade are free to turn. The back of an astrolabe and parts of the plates were often engraved with scales for use as a surveying instrument, and as a means of finding the time by various systems of measuring the hours, and with astrological tables. Primarily the astrolabe was a demonstrational instrument used in the Middle Ages for teaching astronomy and for simple computations in astrology.

The spherical astrolabe is similar in principle to the planispheric, but it involves no projective geometry since it is in the form of a globe with a superimposed spherical star map. The mariner's astrolabe is a practical altitude-measuring instrument, similar in overall shape to the planispheric, but without the star map, plates and scales. The prismatic astrolabe is a comparatively recent instrument which is employed in precision surveying.

F. M.

Hour-glass

If you can get an accurate enough one made, an hour-glass is probably a better instrument for timing boiling eggs than the average wristwatch – providing, of course, you don't want your eggs three minutes one morning and two and a half minutes the next. In any event, the hour-glass is certainly the more restful of the two timepieces to concentrate on, as its fine grains of sand gently pour through a narrow glass neck from the top bulb to the lower. But aside from egg-boilers, practically no one uses an hour-glass these days.

The earliest type of hour- or sand-glass, which seems to have been used in ancient times in various parts of the world, consisted of two ceramic or, later, glass bulbs pressed against either side of a small metal disc with a hole in its centre. A quantity of sand was included in one bulb and the whole affair was bound together with a thread. During the 14th century hour-glasses were in common use, especially at sea, where mariners referred to them as 'log-glasses'. They were used as an aid in dead reckoning and also, in conjunction with a specially knotted log-line, for determining a ship's speed. Hour-glasses were also employed for timing ships' watches, or duty shifts, and until late in the 16th century were considered the only remotely accurate means of telling time away from shore. Whenever the sand, or occasionally mercury, completed its passage from one bulb to the next, a time notation would be made in the log and the glass turned over. The British Navy made regular use of the hour-glass until 1839.

On land, hour-glasses were most often used in churches for timing sermons and fine examples of these instruments with their elegant workmanship can still be seen in some English churches. The amount of time described by the most common hour-glass varied from 15 seconds to 4 hours and it may be assumed that not a few preachers had several such devices to suit their lecturing moods.

By the end of the 18th century, glass-blowing techniques permitted the development of single-piece hour-glasses but such an achievement by the glass-blowers came a bit late in the day – except for egg-boilers.

E. W.

Clock

The origins of the invention of the mechanical clock are shrouded in mystery. The first mechanical clocks of which we have reliable information were public clocks, the earliest being set up in Milan in 1335. Recent researches have shown that a possible precursor of the mechanical escapement of these medieval European clocks may have been a device described in a Chinese text written by Su Sung about the year 1090. This was an elaborate water-clock powered by a waterwheel which advanced in a step-by-step motion. Water was poured into a series of cups which emptied (or escaped) every quarter of an hour when the weight of water was sufficient to tilt a steelyard. The mechanism was then unlocked until the arrival of the next cup below the water stream, when it was locked again. An astronomical check on timekeeping was made by a sighting tube pointed to a selected star. The timekeeping was governed mainly by the flow of water rather than by the escapement action itself, and so it may perhaps be regarded as a rather remote first link between the timekeeping properties of a steady flow of liquid and those of mechanically produced oscillations.

Indeed the fundamental difference between a water-clock and a mechanical clock is that, in the latter, timekeeping is entirely governed by a periodic mechanical motion. The type of motion employed in the earliest mechanical clocks, known as the 'verge' escapement – probably from the Latin *virga* (a rod, or twig) – was one in which a heavy bar or 'foliot', pivoted near its centre, was pushed first one way and then the other by a toothed wheel which advanced through the space of one tooth for each double oscillation of the bar. The wheel was driven by a weight suspended from a drum. Since the bar had no natural period of its own, the rate of the clock was mainly controlled by the driving weight and was greatly affected by variations in friction of the driving mechanism. The accuracy was very low and clocks could not be relied on to keep time more closely than to about a quarter of an hour a day at best, and an error of an hour was not unusual. Until the middle of the 17th century mechanical clocks usually had but one hand and the dial was divided only into hours and quarters. The earliest clocks had no dials, and time was recorded purely by the striking mechanism. This is reflected in our word 'clock' which is derived from the French word *cloche*, meaning a bell.

The development of modern science based on the idea of mathematically measurable sequences was for a long time hampered by the lack of any accurate mechanism for measuring small intervals of time. Thus, in his famous experiments on the rate of fall of bodies rolling down an inclined plane, Galileo measured time by weighing the quantity of water which emerged as a thin jet from a vessel with a small hole in it. For this and other reasons, it is not surprising that he obtained a value for the constant of acceleration that was less than half the correct amount. Nevertheless, a new era in the history of time measurement was due to Galileo's discovery that the time of swing of a simple pendulum depends only on its length, provided the arc of swing is small. To obtain a satisfactory timekeeper the pendulum had to be combined with clockwork to keep it swinging and to count the number of swings. In his old age, Galileo (who died in 1642) worked on this problem with his son Vicenzio and left drawings of a pendulum clock with an original form of escapement.

The first to construct a successful pendulum clock was the Dutch scientist Christiaan Huygens, whose original model was made in 1656. The next step in time measurement was the invention of a new type of escapement. Huygens's clock incorporated the verge type, but about 1670 the anchor type was invented, probably by the London clock-maker William Clement. The great advantage of this escapement is that only a small angle of swing is required.

G. J. W.

Theodolite

During the 16th century surveying enjoyed an unprecedented boom. Estate owners wanted to know the true dimensions of their properties, cities had to lay out new fortifications as defences against

An 18th-century sand-glass.

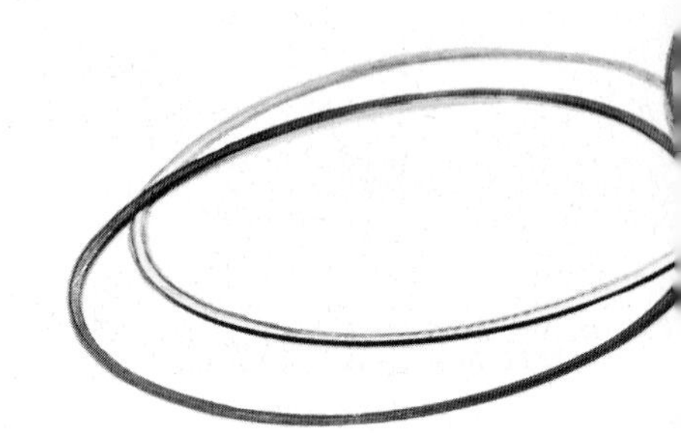

artillery, there were the new lands to be mapped overseas. The new science of trigonometry suggested methods whereby the measurer of land could calculate distances and areas without needing to tread out his boundaries. But the only instruments available were those that had been evolved by the astronomers of former times, usually one instrument for each specific task, and often demanding more advanced mathematical knowledge than the surveyors in the field could muster.

As a result, a welter of new instruments made their appearance, some of them simplified versions of those of astronomy, others combining several established instruments in one so in the end they were frequently as complex as any of their predecessors. All these multi-purpose instruments were quickly weeded out; the theodolite is almost the sole survivor.

As an instrument for measuring horizontal and vertical angles subtended by distant objects, it combines the medieval star-gazer's geometrical quadrant with the horizontal graduated circle that was occasionally used by early Renaissance architects for cartographic purposes. Such a circle, with a built-in compass, was used to map Rome about 1510: a horizontal plate that lacks a compass but does have a central pillar carrying a vertical graduated semicircle containing two inset shadow squares – in effect a double geometrical quadrant – is illustrated in a popular work of 1512, where it is called a *polimetrum*. Horizontal and vertical limbs alike are graduated in 90° quadrants, which suggests how the concept was developed. Although the woodcut is crude, this is a clearly recognizable theodolite.

Some other German handbooks of applied mathematics portray similar 'altazimuth' instruments, but the first detailed description is in Leonard Digges' *Pantometria*, which his son Thomas published in 1571; it was Digges, too, who coined the word 'theodolite'. But he attached the name to the horizontal circle only, which otherwise differs from the *polimetrum* in having four shadow squares rather than two and in being graduated as a full circle. A stand carries the vertical semicircle; this is itself graduated and is used to mark off degrees on the vertical limb.

Two incomplete Elizabethan theodolites with compasses exist, both engraved by Humphrey Cole (who had some connection with Digges), within a few years of the publication of *Pantometria*, for the earlier is dated 1574. Some theodolites were made by members of the famous German family of instrument-makers, the Habermels, but the oldest is from 1576. Although a few may have been constructed, and the device was certainly recommended by the best textbooks of the time, most surveyors continued through the 18th century to rely on separate apparatus for taking horizontal and vertical angles, presumably because a theodolite is harder to carry under the arm.

The first theodolites to have telescope sights as part of the fittings were those of Jonathan Sisson, about 1730. He also replaced the old plumb-line with a bubble-level, and constructed his vertical limb with a toothed edge so that it could be turned through a pinion, whereas all former theodolites had had to be adjusted by touch and held in position by hand. The invention of achromatic lenses meant that telescopes could be much shorter, so as to be mounted conveniently on a portable instrument. The great theodolite built by Ramsden in 1787 for what were to be the first stages of the British Ordnance Survey had microscopes as a further refinement, with micrometers for readings down to 5 sec. of arc, apart from which they could be regarded as much larger and stabler versions of Sisson's. It was Ramsden's success, however, which finally made the theodolite one of the essential tools of the surveyor.

A.K.

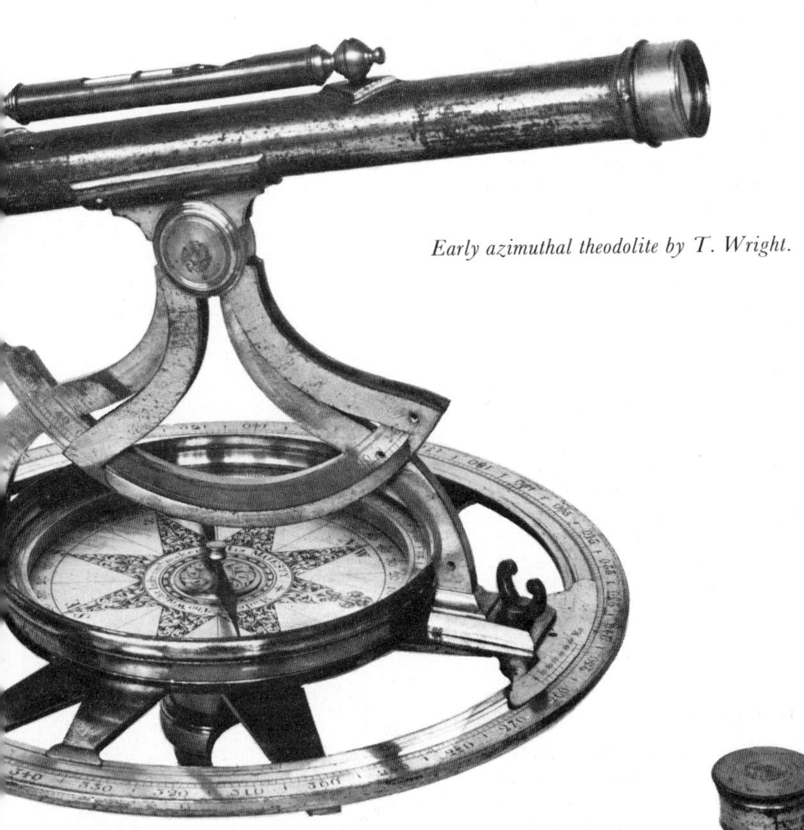

Early azimuthal theodolite by T. Wright.

Microscope

The microscope was invented, apparently by a coincidence much like that involved in the telescope, in about 1590, by Zacharias Jansen, a spectacle-maker. It was a relatively crude device, using, as did the first telescope, a concave and a convex lens, and there is no evidence that Jansen made any significant observations with it. Galileo improved the device and used it as a scientific instrument – he described the complex eye of an insect. However, Anthony van Leeuwenhoek is the outstanding figure in the history of microscopy because of his remarkable practical success. He succeeded because of his meticulous skill in grinding high-powered lenses that produced a clear image (his most powerful instruments magnified some 300 diameters) together with his uncanny persistence in using them.

That is why it was the Dutchman, and not Galileo, who first observed bacteria. The importance of this discovery cannot be overrated. By revealing the hitherto unknown and unseen world of micro-organisms, Leeuwenhoek ended centuries of superstitious speculation and set the stage for man's conquest of infectious diseases. True, it was another 200 years before the rest of science fully caught up with him, by which time compound microscopes were being much improved. But Pasteur and the other pioneer microbiologists could not have done their work without his lead. The microscope was necessary to repudiate the idea of the spontaneous generation of life, and to show that specific infections are caused by specific microbes. This led to the success of public health measures, vaccination and drugs to combat disease – the outstanding success story in the history of medicine.

Leeuwenhoek himself had an insatiable, driving curiosity. A linen merchant by trade, he was born in Delft, Holland, in 1632 and died there in 1723. He had no formal scientific training, but amused himself in his spare time with glass-blowing and fine metal work. In the course of this pastime he devised ways of grinding magnifying lenses and mounting them to form simple microscopes. Leeuwenhoek spent much of his long and industrious life in refining and improving these microscopes, which he used to study a great variety of materials – saliva, plant leaves, seminal fluid, urine, cow dung, circulating blood in the tail of a salamander, scrapings from his teeth and so on. But Leeuwenhoek was also a secretive man, jealous of his hard-won knowledge, and to this day microscopists are uncertain of how he actually used his primitive instruments to such good effect.

B.D.

Thermometer

The Greeks of Alexandria knew that air expanded as it was heated, and Philo of Byzantium, who was alive at the time of the birth of Christ, made a 'thermoscope' that must have been very like Galileo's air thermometer. But it was Galileo who, in 1592, was the first to put a scale beside the tube. This at once converted the device into a scientific instrument and made it possible to distinguish between temperature and heat. The study of heat as a form of energy depended on this distinction. Santorio Santorio, a friend of Galileo's, first adapted the air thermometer, in which a coloured liquid was driven down by the expansion of air, to measuring the body's change of temperature during illness and recovery.

T.O.

Foot-rule

The foot-rule, a wooden rod graduated in inches, and often fractions of inches, seems such an obvious device that one would have supposed that carpenters used it from very early times, but the earliest-known British instance is that described in Joseph Moxon's *Mechanic Exercises, or the Doctrine of Handiworks*, published in London in 1683. Moxon was a printer and map-maker, and Hydrographer to Charles II. He seems to have lifted most of his drawings from a French work of 1676, Félibien's *Principes de l'Architecture*. Moxon's rule shows the inches numbered up to 12, and the explanation adds that each inch was divided into eight parts, and that such rules were commonly 2 ft. in length.

There are occasional references to 'rewles' used by masons, but these were simple straight-edges. The next step was to graduate the straight-edge, often in a quite arbitrary way and without numbers. But in a picture by Mainardi, painted before 1613 and now in the Museo Civile, Cremona, part of a rule is seen: though the divisions are not numbered, the 6th and 12th are marked by a cross-line.

By 1769, graduated folding rules of a very modern type were in use, the finest in ivory and copper-gilt. The modern mass-produced brass and boxwood rule came in about 1840.

G.R.T.

Telescope

'As I stinted neither pains nor pence I was so successful that I obtained an excellent instrument which enabled me to see objects a thousand times as large and only one thirtieth of the distance in comparison with their appearance to the naked eye' (Galileo Galilei).

There is no need to explain how the telescope came to be invented – it consists only of two lenses held in line. What does need an explanation is why it took so long. Spectacles were in use in the 14th century; yet it was nearly 300 years before Hans Lippershey, a Dutch spectacle-maker, looked at the weather-vane on a distant steeple through a pair of lenses and found that it was magnified. The date was 1608.

Galileo heard of the new invention, understood how it must work, and used the new device in his battle with the accepted views on astronomy. The reception of his observations suggests why the world had to wait so long for the device. Natural philosophy before Galileo was largely treated as a branch of academic learning: the answers to problems were to be found by consulting the appropriate authorities. The Aristotelian view of the movement of the planets, as expanded by Ptolemy, was that the sun and the planets were perfect bodies and as such moved in the perfect shape, a circle, around the earth. As observations became better, these epicycles, as such patterns are called, had to become increasingly numerous so that the facts could be fitted to the theory.

Galileo, with his telescope, found that the sun was not perfect – it was marred by spots – and that at least one planet had moons moving round it, as he said our moon moved round us. But when he asked the scholars to look through his telescope to see the classical theory refuted, they refused.

The scholars were not bigoted oafs. They were simply following the doctrine put forward by Plato in particular, and classical philosophy in general, that

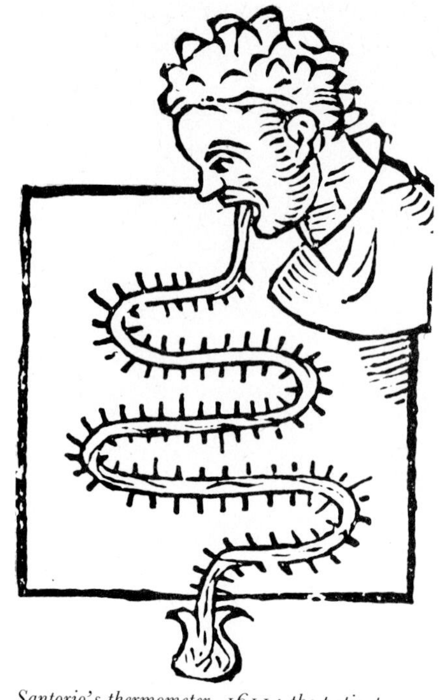

Santorio's thermometer, 1611; the patient held a glass bulb in his mouth. The higher his temperature, the further the water level dropped.

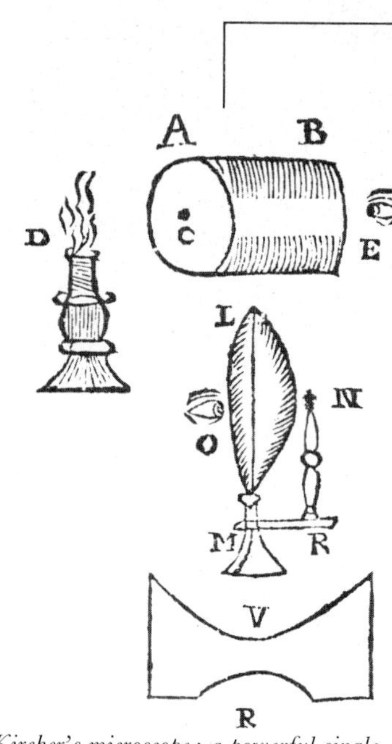

Kircher's microscope: a powerful single convex lens, mounted at the observer's end of a metal tube.

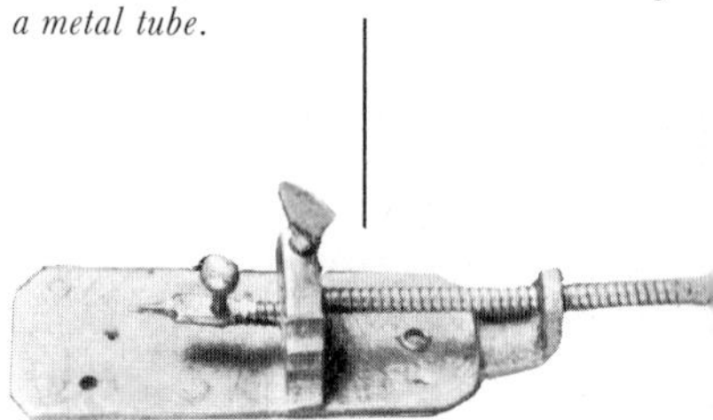

Leeuwenhoek microscope: the ultimate development of the simple lens as a magnifying device, c. 1673; the lens was mounted, and a concave mirror focused light on the object to be examined.

Chinese astronomer of the 18th century.

British Imperial standard of length placed in Trafalgar Square by the Board of Trade.

Convex lens used to correct long sight, 1280–5.

Concave lens developed to correct near sight; first mentioned 1568.

Compound microscope: in c. 1590, Jansen, a Dutch spectacle maker, aligned a concave lens and a convex one, with the concave lens as eyepiece, to magnify a small object.

Galilean telescope: 'Dutch telescope', enormously improved by Galileo.

Astronomical telescope: two convex lenses, and a wider field of view than Galileo's.

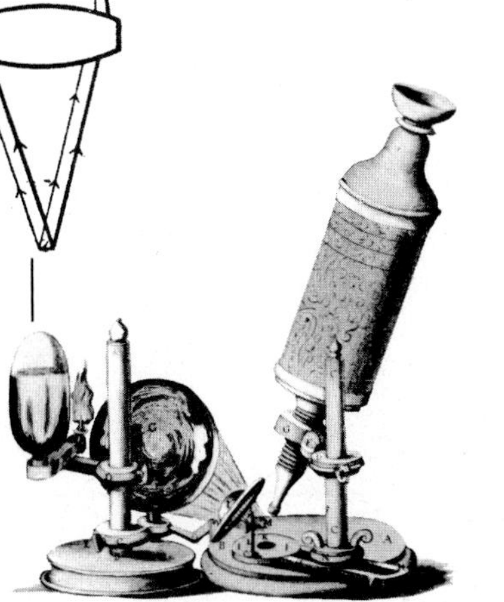

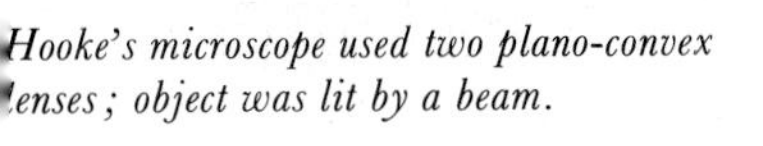

Hooke's microscope used two plano-convex lenses; object was lit by a beam.

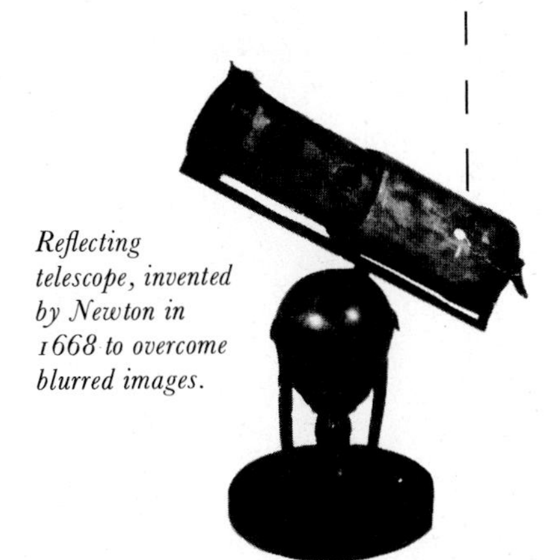

Reflecting telescope, invented by Newton in 1668 to overcome blurred images.

the senses were misleading, and that the mind was mightier. Because Galileo considered that observations were valid, and could be confirmed by others – he wrote his books in Italian instead of Latin so that anyone could read his observations and check them for himself – he has gained a slightly undeserved reputation as the founder of experimental science. And it is presumably because science was in the hands of academics that the world had to wait for the telescope until the Dutch spectacle-maker, an artisan, brought two lenses together.

Once invented, the telescope was continually found to be in need of improvement, and it has continually improved. The Dutch or Galilean telescope gave a rather narrow field of view, and because of the way it worked could not be used with an internal scale. Both of these problems were avoided in the astronomical telescope with two convex lenses that Kepler suggested and Scheiner actually built.

Scheiner used this telescope in 1611 to look at sunspots, and although Galileo had been first in the field, Scheiner was more fortunate, at least temporarily. He invented a system of using dark glass to protect his eyes from the sun; Galileo eventually lost his sight. And it was Scheiner, a Jesuit, who persuaded Pope Urban VIII to ban Galileo's *Dialogue* concerning two world systems and have the author brought before the Inquisition, condemned and punished.

Early telescopes gave coloured, blurred images. This was due to the separation of colours by glass. Newton's experiments with a prism had shown him that glass separates the colours that make up white light, and he realized that this meant that the lenses of the time focused light of different colours at different points, giving a blurred and coloured image. He thought that this problem was insoluble and invented, and personally made, a reflecting telescope, in which a mirror replaced the objective lens. As it happens he was wrong, but the greatest telescopes in the world today still use his construction.

T.O.

Micrometer

'I should not have believed the world could have afforded such exquisite rarities,' exclaimed William Crabtree after his friend William Gascoigne had showed him his micrometer and other newly invented astronomical instruments in the autumn of 1640. 'Urania's structure will grow to perfection by your assistance.' His brief account of them to Jeremiah Horrox 'hath ravished his mind quite from itself and left him in an extasie between admiration and amazement.' These flowery compliments in the grand old style of Caroline England camouflage an innovation in the spirit of a very different age, one of mechanical precision in which the most valuable scholar would be he who could carry accuracy of measurement to one more order of magnitude.

In the 30 years since the invention of the telescope it had been used primarily for qualitative discoveries. When new stars and satellites swam regularly into the observer's ken, he did not bother with subdivisions of the apparent angles of the objects he viewed, until Gascoigne 'either found out or stumbled on a most certain and easy way whereby the distance between any the least stars visible only by a perspective glass [i.e. a telescope] may be readily given, affording the diminutions and augmentations of the planets strangely precise.'

The son of a landed gentleman of Middleton, close by Leeds, Gascoigne had been dissatisfied by his Oxford education, and bored with the hunting and riding life of a Yorkshire squire, when he discovered mathematics. His enthusiasm soon took him to astronomy, the most exciting application of that science in the age of Galileo. Evidently, he was interested in the instruments rather than the observation, and with the aid of a workman on his father's estate he began to create new apparatus for himself. Possibly he began by the use of two wires held at the end of a rod: the observer looked at the sun along the rod, and by measuring the distance from his eye to the wires when they framed the sun, he could estimate its apparent diameter. This idea goes back to Hipparchus and Archimedes: the novel thing was to devise some means of doing the same with the telescope, so that those measurements which had before only been attempted on sun and moon, could be made on the planets or on the angular distance between two stars sufficiently close together to appear simultaneously in the telescope's field of view.

Gascoigne saw this could be done by changing the distance between the pointers, while that between them and the eye remained the same. Two pointers were mounted in a small case, one of which could be advanced by a screw toward the other, with a scale to register each revolution and an index hand at the end so that subdivisions down to one-hundredth of a revolution could be read off on a dial. The two pointers were placed at the focus of the object lens of the telescope and the distance between them adjusted until they just encompassed the elongation to be measured. Two hairs stretched between hooks could be used instead of pointers.

He gave an account to his old mathematics teacher and to a few others. But the times were not propitious for the calm perfecting of Urania's structure. The Civil War broke out; Gascoigne, who was a Catholic, joined the king's army and was killed in action at the battle of Marston Moor. His work was unknown to all but a very few when, in January 1667, the owner of a Gascoigne micrometer was shaken to read in the *Philosophical Transactions* that two French mathematicians, Adrien Auzout and Jean Picard, had announced a similar discovery. In their version two frames each bearing a row of slender metal rods, or hairs, slid past each other along grooves in an outer frame until the object appeared squeezed between them. One of Gascoigne's micrometers was sent to the editor, and at last, after a quarter of a century, a detailed illustration and description were published in November 1667. In fact Auzout and Picard did not print a full account of their instrument until later. But although Gascoigne's type has the priority, that of Auzout and Picard, further improved in the next decade by Olaus Roemer, was the model on which micrometers were subsequently based.

A.K.

Barometer

Science often uses human terms to explain occurrences, and the fact that pumps could lift water was explained by saying that nature abhorred a vacuum. Therefore it followed that if the air was withdrawn from the top of a tube, nature avoided the disliked vacuum by filling the space with water. Engineers had known more or less since pumps were invented that they would not lift water more than about 30 ft. – there are marvellous 16th-century sketches of a chain of pumps for emptying deep mines – but engineers and scientists did not communicate much to one another. It was Galileo who was the first to point out that the engineers' observations must mean that nature's dislike of a vacuum exerted an influence only for 30 ft.

His pupil, Torricelli, decided to see how much dislike nature would show if the vacuum was formed above mercury, a liquid more than 13 times as dense as water. In 1643 he filled a sealed tube about a yard long with mercury, put his thumb over the end, and inverted the tube so that the open end was immersed in a dish of mercury. When he took his thumb away, the mercury fell, leaving about 6 in. of what is now called a Torricellian vacuum at the top. Torricelli suspected that the column of mercury in this, the first barometer, was supported by the pressure of the atmosphere on the dish of mercury; this belief was confirmed by Blaise Pascal, who organized an experiment in which a Torricellian barometer was carried to the top of a mountain. The mercury fell as the mountain was climbed – that is, as the air pressure became lower because there was less air above the barometer.

The use of the barometer for weather forecasting started with Otto von Guericke, who built a water barometer in about 1672. Von Guericke was above all a showman, as his demonstration of the barometer well illustrates. He made a brass tube 34 ft. high with a closed glass section at the top, filled it with water and mounted it on the side of his house so that passers-by could see the little figure of a man floating high on the water in good weather, or sinking when the weather was stormy.

This helped to gain him a place in the history of science – and a reputation among his contemporaries as a dabbler in witchcraft.

T.O.

Balance spring

A pendulum is, of course, unsuited to a small portable watch that has to be carried about on the wrist or, in the 17th century, in the pocket. Besides, it soon gets into trouble at sea, where an accurate chronometer is particularly needed to help in finding a ship's position. So in the 1670s Huygens devised a clock mechanism regulated by a balance-spring, whose elasticity would cause the balance-wheel to oscillate in equal times just as well as a pendulum. But when he brought his watch out in 1675, he was taken aback to find that he had been anticipated.

He had rather bad luck with his horological inventions. This time it was Robert Hooke who claimed he had got there first, protesting that he had designed a clock with a balance-spring regulator back in 1658. Hooke explained that he had tried to form a small company to manufacture his new watch, but the project fell through because he suspected his partners, who wanted any improvements to belong to them, and not to Hooke. A few watches were apparently made, but it was not until Huygens announced his model that Hooke jumped in again with angry complaints that Huygens's spring was 'not worth a farthing', and anyway he had only developed the idea because, in the meantime, Hooke's treacherous partners had given away the secret of his original version. It is not quite clear whether Hooke's watch originally had a spiral balance-spring, like later spring-watches; Huygens's certainly did.

A.K.

Marine chronometer

Ever since oceanic navigation began, the difficulty of ascertaining a ship's position east or west had proved insuperable. It was certainly realized that what was needed was a clock that would keep such good time that a comparison between local time and the time at the Prime Meridian (Greenwich was not internationally accepted as such until 1884)

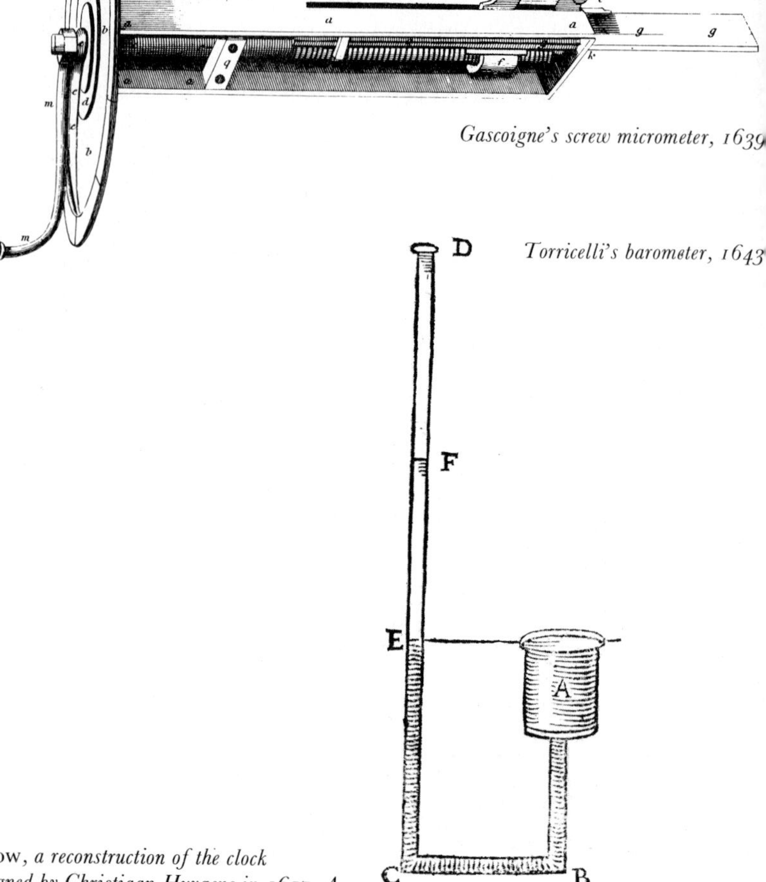

Gascoigne's screw micrometer, 1639

Torricelli's barometer, 1643

Below, *a reconstruction of the clock designed by Christiaan Huygens in 1657. A pendulum was no use for pocket watches or for clocks carried on shipboard.*

Harrison's first marine chronometer, made in 1735, with smaller versions, including (foreground) Kendall's duplicate taken by Cook on his second voyage.

Observing the solar spectrum, c. *1870*.

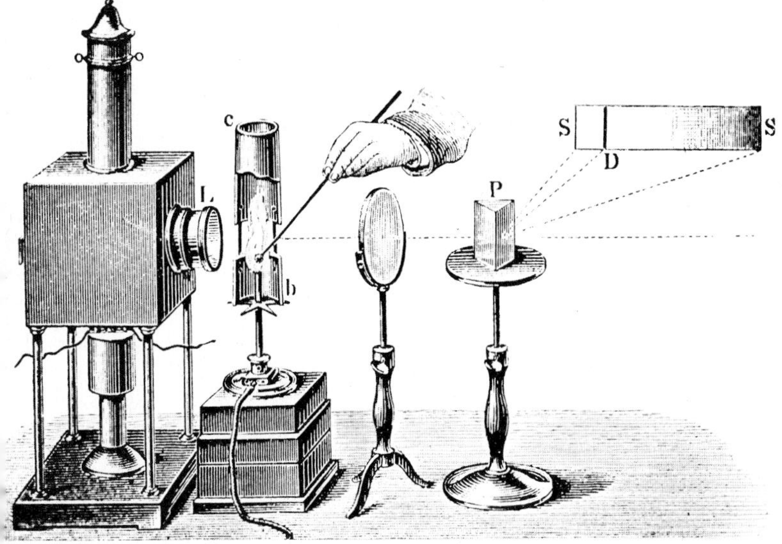

Demonstration of spectroscopy, c. *1870*, *using an incandescent light source.*

Cutaway drawing of Snap 1a, prototype miniature ($5\frac{1}{2} \times 4\frac{1}{2}$ in.) generator, producing 5 watts of power from its 20 pairs of thermocouples arranged around the surface.

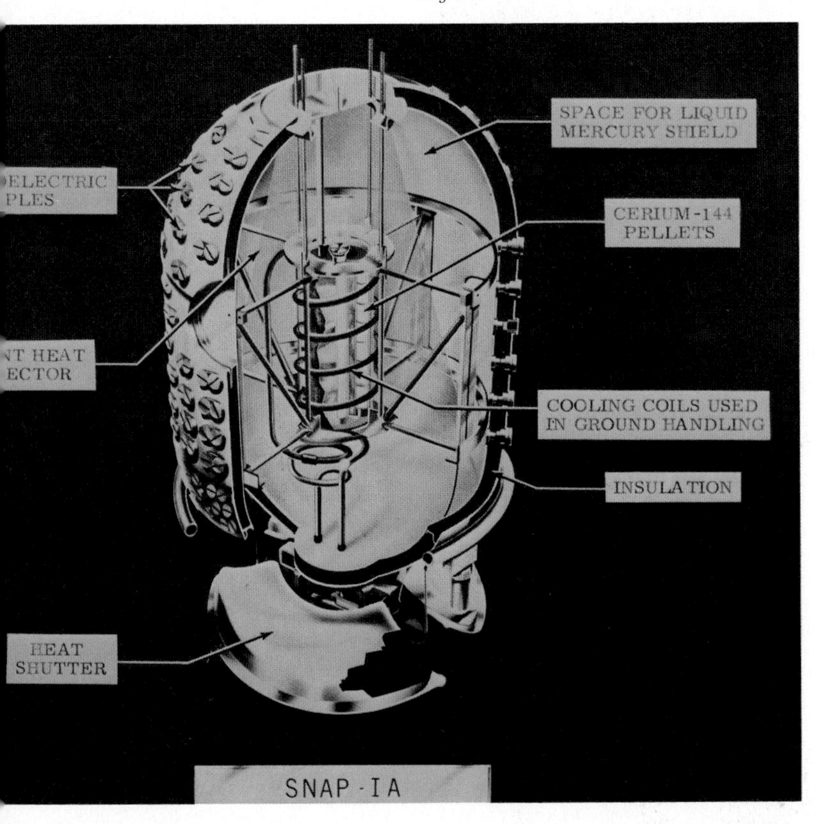

would give the correct position east or west. But no one could make a timepiece able to withstand the motions of a ship or the changes of temperature on a long voyage. Even an error of two minutes would put you 30 miles off course.

The system of using a ship's log to measure distance travelled did not allow for drift or current, and enormous errors resulted. A complicated alternative to the log, called the 'lunar method', depended on elaborate calculations. It was used by Captain Cook on his first voyage round the world.

Every government offered a prize for a solution to the problem. The French Academy was founded for the purpose, and the Royal Observatory was established at Greenwich in 1675 'to find out the so-much desired longitude of places'. In 1714, after the flagship of Sir Cloudesley Shovell had been wrecked on the Scillies (when he imagined that he was sailing up the Channel), the British Admiralty set up a Board of Longitude to offer a prize of £20,000 for a clock with an error of less than two minutes on a voyage to the West Indies and back.

John Harrison (1693–1776), a self-educated Yorkshire carpenter, entered his first clock or chronometer for the prize in 1735. He spent the next 30 years improving it and reducing its size from the 72 lb. of his first model to the handsome watch which was his fourth model in 1761. His son took this model on a test voyage under the supervision of an astronomer. When the ship was approaching Madeira, the longitude shown by the chronometer and that estimated by the pilot differed by more than 2°. The pilot begged the captain to alter course to reach the island, but Harrison asked him to wait for a few hours. The captain did so and the island came in sight as Harrison predicted. After a voyage lasting 81 days, the ship reached Jamaica, where the chronometer was found to be only five seconds slow. It was a duplicate of this model, made by Larcum Kendall, which Cook took on his second voyage in 1772–75, and which was only 7 minutes 45 seconds slow after three years at sea in temperatures ranging from tropical to polar.

The movement in Harrison's chronometer was controlled by two balances connected by springs, which were adapted to changes of temperature by a 'compensation curb' of brass and steel rods. The mainspring was partially rewound every seven seconds, with a 'maintaining power' to prevent the clock going slow during the process.

This was the greatest advance in the art of navigation ever made, but the Admiralty refused to pay the full prize money. 'By God, Harrison, I'll see you righted,' exclaimed George III when the 80-year-old inventor appealed to him. Even then, £1,000 was deducted for the cost of constructing a watch Their Lordships had ordered.

C. L.

Spectroscope

Everyone has seen nature's spectroscope, the rainbow. Sunlight is separated by raindrops into its components of varying wavelengths, and each wavelength appears as a different colour. The spectroscope splits light in the same way, allowing the different colours to be viewed separately. This dispersed array of colours is called a spectrum and the pattern of colours makes possible a chemical analysis of the light source.

Experimental science usually develops in advance of the theoretical understanding of the phenomena involved, and this was certainly the case with spectroscopy. Until Isaac Newton, colour was thought to be a mixture of light and dark, and it was believed that white light was somehow changed into other colours within the glass.

In 1666, Newton bought a glass prism 'to try therewith the phenomena of colours'. He was the first man to realize that white light is composed of all the known colours. But it was not until 1814 that a spectroscope was made and used as an experimental tool. The inventor was a German, Joseph von Fraunhofer, who built a spectroscope with lens, prism and observation telescope to measure the dispersive powers of different glasses. In doing this he observed and mapped about 700 dark lines in sunlight, now named after him. He also observed the spectra of stars, and laid the foundations of astronomical spectroscopy – the first method of analysis that could be used over distances of millions of miles.

Other important contributions to spectroscopy were made by the younger Herschel, Wheatstone, Angström and Foucault between 1820 and 1855. In 1859, Kirchhoff deduced the general law connecting absorption and emission of light, and recognized that each element has a characteristic spectrum. The use of the spectroscope as an analytical instrument dates from this discovery. In 1861, working with Bunsen, Kirchhoff was able to discover two new metals, caesium and rubidium, from investigating spectra. The work of Bunsen and Kirchhoff finally established spectroscopy as a vital scientific technique, and – extended into the ultra-violet, infra-red and microwave regions – it is now one of the scientist's most important analytical and experimental tools, widely used in investigating new chemicals.

P.J.

Thermocouple

As a rule, instruments for measuring energy cannot be turned into devices for generating or transforming energy. One exception is the thermocouple.

A 19th-century discovery was the 'Seebeck effect', so called after Thomas

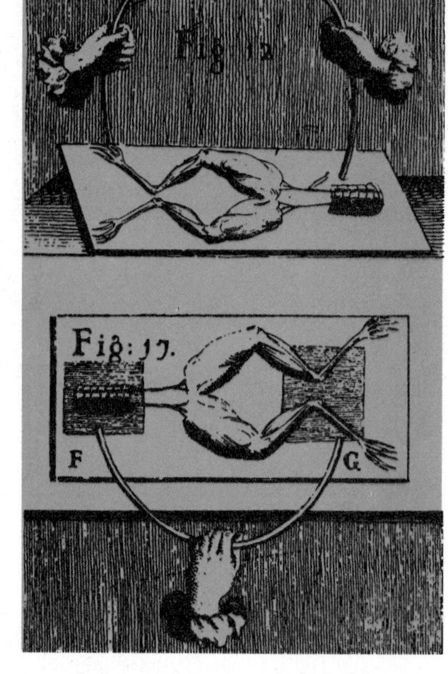

Illustration of Luigi Galvani's experiment with frogs' legs, from his De Viribus Electricitatis, *1792.*

Johann Seebeck of Berlin who found in 1821 that when two wires made of different metals are joined at their ends to form a circuit, and the two junctions are kept at different temperatures – one warm, one cold – a current flows through the circuit. Early in our century, this effect was used in reverse for measuring temperatures most accurately. Called the 'thermocouple', it is an electric thermometer consisting of two wires of different metals: one junction of the wires is at the point where the temperature is to be measured, while the other junction is maintained at a fixed low temperature. The resulting current flowing through the circuit can be measured by a galvanometer or a potentiometer, which indicates the electromagnetic force generated; in either case, the temperature can be read on a scale.

When radioactive isotopes became available after the Second World War and, at the same time, space research got into its stride, scientists began to experiment with ideas for new independent sources of energy. Isotopes seemed most suitable for small, mobile electricity-generating units, the kind of energy sources which a man-made earth satellite would need for operating its electrical equipment, such as radio transmitters. In 1959, the research workers of the US Atomic Energy Commission demonstrated a small device called 'Snap III', no bigger than $5\frac{1}{2}$ by $4\frac{1}{2}$ inches and weighing only 5 lb. It contained a tiny pellet of the radioactive isotope polonium 210, surrounded by twenty pairs of thermocouples arranged like the spokes of a wheel. The radiation from the isotope generated heat, which made a current flow through the thermocouples – about 5 watts. The descendants of Snap III, producing much more electricity, are now orbiting the earth and travelling to other planets in unmanned spacecraft. The thermocouple atomic battery, although still too expensive (and too dangerous because of its radiation) for everyday use on earth, has great advantages which are of vital importance in space: it has a very long life, it weighs little, and it needs no maintenance.

E. L.

Galvanometer

Luigi Galvani, a professor of medicine at Bologna, was lecturing to his students at his home one day in 1780 while his wife was skinning frogs with his scalpel in the adjoining kitchen. As she listened to him, the scalpel fell from her hand on the frog's thigh, which was lying on a zinc plate. The frog jerked violently. Signora Galvani screamed; the professor investigated the incident, and at once declared to his overawed students, 'I have made a great discovery – animal electricity, the primary source of life!'

He had, of course, made nothing of the sort, but he stuck throughout his life to the fallacy, making innumerable experiments with dead frogs which failed to prove anything. Nevertheless they established his name all over Europe, connecting it for ever with such words as 'galvanizing', 'galvanoscope' and 'galvanometer'. A greater scientist than Galvani, Alessandro Volta of Pavia, explained the mystery correctly: the jerking frogs, far from demonstrating the 'primary source of life', played only the modest part of electric conductors – it was the steel of the scalpel and the zinc of the plate which were the important factors. Volta showed that an electric current begins to flow when two different metals are separated by a moist conductor. The frog's thighs had merely indicated the presence of a current by contracting. From this discovery, Volta went on to invent the first electric battery, the 'voltaic pile' – but what it produced was absurdly termed 'galvanic electricity'.

That was in 1800, and from then on scientists had an efficient source of electric current for their experiments. For a long time they had suspected that there was some connection between electricity and magnetism, but it was only in 1819 that Hans Christian Oersted, a Danish physicist, found by chance that when a wire through which a current was flowing was brought near a mariner's compass, the needle swung violently. This magnetic effect produced by an electric current became the basic principle of an instrument for detecting and measuring low-voltage currents – the galvanometer.

It was developed by various researchers and in various forms during the 1820s, and applied for the first time to a practical purpose by Karl Friedrich Gauss, director of the observatory at Göttingen, Germany. In 1832, he invented one of the earliest types of electric telegraph (though the idea had previously been suggested by the French scientist, André Marie Ampère). Gauss used the deflection of a magnetic needle by the electric current to transmit signals by wire from his home to the observatory. This form of galvanometer is called the moving-magnet type; today, however, the most widely used is the moving-coil or mirror galvanometer. It consists of a coil closely wound with very thin wire, suspended in the field of a magnet; when the current is turned on, the 'torque' on the coil is balanced by the 'torsion' in the suspension wire, and the deflection angle is measured by an arrangement of a mirror, which is attached to the coil, a small lamp, and a scale for reading the units of current (amperes).

The mirror galvanometer played a decisive part in the dramatic laying of the first permanent telegraph cable between Europe and America in 1866 (a previous one, laid in 1858, had failed). The eminent British scientist Professor William Thomson, later Lord Kelvin, was on board the cable ship, the famous *Great Eastern*; it had been his idea to use the galvanometer to keep a continuous check on how much, if any, current was coming through the cable. As the current arriving from the other side of the Atlantic was of course very weak when the 2,500-mile-long cable was finally laid and telegraphic signals were exchanged, Thomson designed the most sensitive galvanometer possible. Called the 'siphon recorder', it contained an extremely thin glass siphon tube through which ink would flow by capillary action, recording the incoming signals on paper tape; the galvanometer itself, which picked up the signals, consisted of a coil of fine wire suspended between the poles of a horseshoe magnet. For many decades, Thomson's siphon recorder made it possible to maintain a reliable telegraphic service between the two continents.

E. L.

Geiger counter

The Geiger counter, the commonest and most versatile instrument used to detect radiation, is now employed routinely to measure the level of radiation from whatever source, whether nuclear power stations and research laboratories or the atmosphere. Because of the great dangers to life arising from uncontrolled nuclear radiation, reliable and quick methods of detection are necessarily of enormous importance. The Geiger counter fulfils these conditions, and like many great inventions, has, in principle, remained unaltered since it was first developed.

The Geiger counter was invented in 1910 by Hans Geiger, a brilliant young German physicist who worked with Rutherford at Manchester University. At that time Rutherford and his collaborators were working on one of the three types of rays given out by radioactive substances – alpha particles, which are helium nuclei. They were interested in counting individual alpha particles. The original apparatus consisted of a fine wire charged with a high voltage running down the centre of a cylinder in a near vacuum. When alpha particles were passed through the gas in the cylinder the particles caused the gas to form charged particles, a process called ionization. An electrical breakdown followed, and a pulse of current for each alpha particle was observed on a dial. Nowadays the particles are either heard as a 'click' on a loudspeaker or measured on a counting device.

The Geiger counter can distinguish between alpha particles and the other kinds of radiation – beta- and gamma-rays – by reducing the voltage along the fine wire in the cylinder. It is now possible to find out quite simply what sort of radiation is being emitted from a

Hans Geiger (left) with Rutherford at Manchester University, c. *1910.*

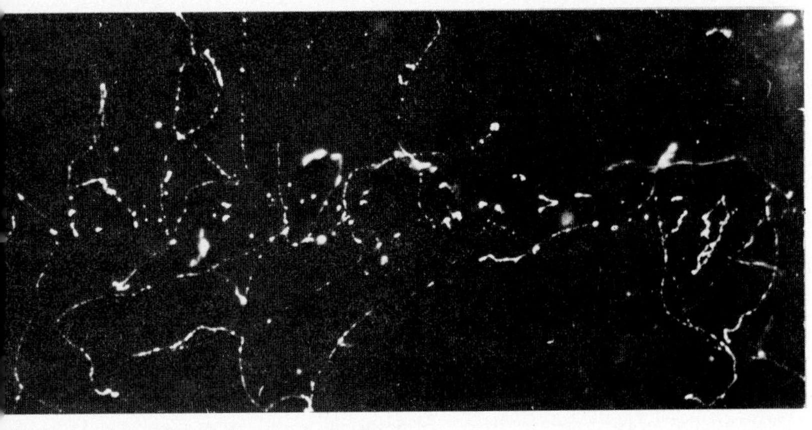

Cloud-chamber tracks of electrons produced by a beam of 'hard' X-rays.

Above, *an example of natural sonar. A blindfold porpoise finds its way between two bars by echolocation.*

radioactive source and what its energy is. A very simple variation of the Geiger counter is the dosimeter, which dispenses with nearly all the electrical circuitry of the Geiger counter. It is the size of a pencil, and is worn by people working near radioactive sources so as to give a rough measure of the radiation around them.

The increasing use of radioactive substances in the world today for many purposes, but particularly as a fuel source, is a tribute to the enormous progress that has been made this century by physicists towards understanding the atom and its particles. As a result, radiation detection methods are going to become more important as we reach the stage when the full-scale harnessing of the energy of the atom becomes increasingly widespread and safe.

R.J.

Cloud chamber

C.T.R. Wilson, a young Scottish research student then at Cambridge, decided to spend the summer vacation of 1894 helping out at the meteorological observatory which then stood on Ben Nevis. 'The wonderful optical phenomena shown when the sun shone on the clouds surrounding the hilltop greatly excited my interest', he recalled, 'and made me wish to imitate them in the laboratory.' So in 1895 he launched on a course of experiments to try to reproduce artificially the production of fog and cloud and rain.

It had long been realized that sudden expansion of air will cool it, and if that air is saturated with water vapour, the vapour will condense. So he began by devising an instrument in which his air was contained in a cylindrical glass chamber, whose floor was formed by a piston which could be suddenly withdrawn: as the volume was enlarged, the air cooled, and the water would condense. But in order to do so, it needed some material body about which the droplets could form. Normally, this is provided by particles of dust. Until Wilson took the subject up, experiment stopped there: no dust, no condensation except on the walls. But he now showed that even when the dust was removed by repeated condensations, yet if expansion passed a certain measurable limit, a 'cloud' still appeared. Presumably there were some other 'nuclei' in the air which could act as centres for this condensation. Unlike the condensation round dust particles, this kind did not diminish as the nuclei settled on the bottom. It was, he wrote, as if, although 'only present in small numbers at any given time, as fast as they are removed they are replaced by others of the same kind.' But what were these nuclei?

If the air was exposed to the newly discovered X-rays, there seemed to be many more nuclei, and he soon saw that they must be ions – atoms which have acquired positive or negative electric charge by losing one of their electrons or gaining an extra one. Over the next few years he was chiefly concerned with the investigation of thunderstorms, but in 1910 he returned to his artificial weather, influenced by the work of Rutherford and other colleagues on the alpha- and beta-rays given off in radioactivity. Would they produce similar effects to those already shown for X-rays? After prolonged work he produced apparatus based on his old principle. The new cloud chamber showed most beautifully the effects of the alpha-rays and beta-rays in creating ions as they passed through. And as the vapour condensed round each ion, the track of these particles was clearly visible and could be photographed as a trail of vapour.

Before, the existence of these alphas and betas had to be deduced from their behaviour in quantities. Now individuals could be followed as plainly as a horse can by a trail of hoof-prints. Better still, as Wilson pointed out in 1912, the pictures seemed to confirm Rutherford's explanation of the true nature of these particles, and the theory of the atom which he had proposed in consequence. In fact the Wilson cloud chamber proved the best tool in the hands of physicists exploring the world of subatomic particles in the years after the First World War. Pictures were taken that revealed the results of nuclear transformation and disintegration; cosmic rays and positrons (positive electrons) were first observed with it. So one man, aided by a couple of technicians, provided the means for a revolution in fundamental science.

Sonar

Today few commercial fishing boats are without some form of sonar device capable of searching the waters beneath or ahead for fish. Yet in spite of the acceptance of sonar for a wide range of applications the original impetus for its development came not from the needs of ocean research and exploitation but from those of the war at sea.

On 22 September 1914 the German submarine U-9 sank three British armoured cruisers in little over half an hour. A submarine of 450 tons had knocked out some 36,000 tons of the British war machine, and more than 1,200 men had been lost. The submarine menace was revealed. The need to be able to detect these underwater wolves became vital.

As early as 1912, L.F. Richardson, prompted by the sinking of the *Titanic*, had suggested using supersonic (now ultrasonic) waves as a means of detecting submerged objects by the echo. However, the idea was still little more than a

pipe-dream, even by the start of the war. It was revived by Paul Langevin and M. Chilowski in March 1915, when they began experiments in Paris.

In 1916 Langevin succeeded in producing ultrasonic waves in water that could be detected at a range of 3 km.; soon his equipment could detect an echo from a large iron plate at 100 m.

The problem was to find an effective way of producing ultrasonic pulses and receiving echoes from submerged objects in a way that enabled their position to be calculated.

Langevin decided to explore the piezo-electric effect discovered many years earlier by Pierre and Jacques Curie. They had found that when quartz is compressed it produces a very small electric current and conversely, if a current is applied to it, the quartz expands very slightly. Obviously quartz and other piezo-electric materials could provide both emitters and receivers of sound.

Followed closely by Professor R.W. Boyle, who was then working with the Royal Navy at Parkeston Quay, Langevin was the first investigator to try a plate of quartz as a receiver. Coupled to a valve amplifier, at that time a recent innovation, the receiver gave very promising results. The way was open to active sonar systems. By the end of 1918 (and the war), the Royal Navy team under Boyle was ready to try out its equipment on ships of the fleet.

Under the codename 'asdic' (for Allied Submarine Detection Investigating Committee), the British research had been one of the best-kept secrets of the war. No mention was even allowed of quartz, which they called instead 'asdivite'. In fact, submarine detection gear was known as 'asdic' in Britain until the late 1950s, when the American term 'sonar', for Sound Navigation And Ranging, was finally adopted.

T. L.

Lie detector

Early tribal cultures used primitive but remarkably effective 'tools' for reading the truth. Elaborate trials, with a great deal of ceremony, were conducted by shamans, witch-doctors or tribal chieftains, all calculated to heighten a guilty person's fear. The Bedouin of Arabia required conflicting witnesses to lick a hot iron (the liar emerged with a burnt tongue) while the ancient Chinese forced the suspect to chew and spit out a handful of rice-powder (if the powder was still dry the suspect was found guilty) and in ancient Britain if the person being questioned could not swallow the 'trial slice' (of bread and cheese) he was guilty. All three examples point up a characteristic physical reaction of fear – constricted throat muscles with difficulty in swallowing and extreme dryness of the mouth and tongue with salivation temporarily arrested. These and other physiological changes take place under the stress of telling a lie, and in theory, if these changes are accurately measured, it should be possible to tell not only if a person is innocent or guilty but at what point exactly the truth or a lie is being told.

The first scientific, instrumental method was invented by an Italian criminologist, Cesare Lombroso, in 1895; this relied on the increase in the pulse rate and blood pressure. Further research was done by Vittorio Benussi in 1914, this time using the suspect's breathing rate as another indication of innocence or guilt, and three years later W. M. Marston did some research using systolic blood pressure. In 1921 an American, John A. Larson, combined the recording of several different bodily reactions in the polygraph, which is the forerunner of the present instrument: it used blood pressure, pulse rate and breathing rate. In 1935 Leonard Keeler, of the Scientific Crime Detection Laboratory in Chicago, used a virtually similar machine (often referred to as the Keeler polygraph) to conduct the first tests in a court case. The two defendants were subsequently found guilty. Tests of a similar nature had been tried in 1924 using blood-pressure readings only, made with a Tychos sphygmometer.

Like the witch-doctor, a lie detector is really trying to plot emotional reactions through the physiological responses that accompany guilt and anxiety, such as changes in salivation, heart rate, breathing and skin temperature, which are all under the control of the autonomic nervous system. The subject is connected to a recording galvanometer by cuffs and flexible tubes called pneumographs, and other sensitive apparatus, by which fluctuations are recorded on a moving graph paper driven by a small synchronous motor. Evaluations of the electrodermal response (galvanic skin resistance), as given by the Fordham pathometer, are often included. Most importantly, the subject is seated in a relaxed position facing *away* from the machine, to eliminate bio-feedback – in other words the possible control of the results of his observed reactions.

S. M.

Cyclotron

Atomic research during the first 40 years after the discovery of radium was very much like detective work, on the tracks of a criminal no one had ever seen, for there is no microscope powerful enough to make an atom visible, let alone its infinitely small nucleus. The scientists, therefore, had to work like sleuths: the 'wanted man' had done this and that, left his marks and clues, giving the experts an idea of what he might be like and what he would do in certain circumstances. Only experi-

Top, *inside the giant cylotron at Serpukhov, USSR. The first such particle accelerator* (above), *invented in 1930 by Ernest O. Lawrence, had a chamber only $4\frac{1}{2}$ in. across.*

210-ft. radio telescope at Parkes, New South Wales.

Detail from The Justice of Otto III *by Dirk Bouts (*c. *1470). To test the truth of her evidence, the lady holds a bar of red-hot iron – an early forerunner of the lie detector.*

mental work – the setting of traps for the criminal – could prove whether the scientists' theories were right or wrong.

Rutherford had set such a trap as early as 1919 by making helium nuclei (2 protons, 2 neutrons) smash into nitrogen atoms (7 of each), transmuting a few into atoms of oxygen. After the discovery in 1932 by James Chadwick of the neutron, which proved to be an excellent 'bullet', a great amount of work was done in this new and exciting field. In 1931 the University of California had built a large machine, the 'cyclotron' designed by Professor E. O. Lawrence as a device for accelerating nuclear particles in a circular track. This was based on much the same principle as a child's swing, the movements of which are increased by a succession of pushes; in the particle accelerator, as these machines are properly called, the particles are made to pass between the poles of strong electromagnets which impart 'pushes' to them until they reach the highest possible speed – which means energy, measured in electron-volts. Finally, when the desired energy is reached, the beam of particles is deflected by a magnet and smashes into the target.

From the collision with the atomic nuclei in the target material, much can be learnt about the basic 'building-blocks' of matter. A spray of subnuclear particles, knocked out of the target's atoms by the impinging 'bullets' (usually protons) can be directed into a bubble chamber, where their tracks, lines of tiny bubbles in liquid hydrogen, are photographed and analysed. Speed, mass, charge and other characteristics can be read off from the tracks in the bubble-chamber photographs, enabling the sleuths to build up a precise picture of the 'wanted man'.

Lawrence's first cyclotron, in 1930, was an elegant little device 4½ in. in diameter accelerating particles to a final energy of 1¼ million eV. This may sound a lot, but bigger and bigger machines have been built in the intervening years, culminating in the 3-mile-diameter giant to be built across the Franco-Swiss frontier near Geneva, with which it is hoped to produce pulses of the order of 300,000 million eV. Thus, to investigate the smallest entities, only the largest 'microscopes' will do.

Nuclear research is still proceeding, in the attempt to answer some of the still unsolved questions about the structure of the atom. Strange particles have been discovered (whose 'strangeness' can be precisely determined), the behaviour and functions of which have not yet been fully explained. But none of them can be regarded as the ultimate, indivisible unit from which the nucleus, the atom, the universe itself is made. In this quest for the secret of matter the cyclotron and its descendants continue to play an important part.

E. L.

Radio telescope

The radio telescope was first made possible by the work of Karl Jansky, an American physicist who in 1931 was working as an engineer for the Bell Telephone Company in America. He was investigating radio-frequency disturbances which hindered transoceanic telephone services, and noticed that background 'noise' interfering with short-wave reception varied in intensity with an almost daily rise and fall. The period proved to be 23 hours 56 minutes, which happens to be the period of the earth's rotation relative to the stars. This was the clue from which he concluded that radio waves from outer space were bombarding the earth in a steady stream. Hearing about Jansky's discoveries, Grote Reber, an American amateur astronomer, built a 'dish' antenna with a 31 ft. diameter and demonstrated that the origin of cosmic radio-frequency radiation is not confined, as Jansky had thought, to the region of our galactic centre but is also spread along the galactic plane. A new branch of science, radio astronomy, had been born.

We now know that the earth receives radio waves emitted by the sun and some planets, by other stars, galaxies, and even by gases in outer space. Some waves are thought to be generated by the collision of matter with charged particles, others by point sources of intense radio energy, called 'radio stars'. Since the lengths of radio waves are much greater than those of visible light, the equipment that receives them must be correspondingly larger than optical telescopes. Although there are many different designs, the radio telescope is basically a huge concave reflector that receives weak radio signals from space, focuses them on a special antenna, amplifies them and records them.

In the early 1940s, astronomers made the first serious use of radio telescopes for studying the sun and its transient phenomena, and in 1942 Reber compiled the first radio maps of our own Milky Way galaxy. After 1945 radio astronomy made its most significant advances. Today, the best-known instrument is the 250 ft. steerable dish at Jodrell Bank, England, although the largest is the 1,000 ft. receiver built in a natural earth depression at Arecibo, Puerto Rico, by the US Department of Defense. This telescope, which is primarily designed for ionosphere research, is non-steerable and therefore depends on the daily rotation of the earth for making its observations.

E. W.

Electron microscope

By 1878 it had been realized that there was a theoretical limit to the resolution of the optical microscope. In other words, no matter how lenses were improved there would always be some details too small to see. Ernst Abbé calculated the smallest detail that could be resolved optically was about one thirty-millionth of an inch.

From the 1850s scientists, and especially those in Germany, had been investigating the so-called 'cathode rays'. The exact nature of these rays was unknown, but scientists were soon making use of their properties, and around the turn of the century magnetic or electrostatic fields were being used to concentrate cathode ray beams. Because cathode rays – streams of electrons – have a much shorter wavelength than light, they can resolve finer detail. In 1926–27 Hans Busch discovered the analogy between the effect of a magnetic coil on an electron beam and that of a convex lens on a beam of light. However, although Busch is nowadays generally considered to have laid the formal foundations of electron optics, the significance of his findings was not then recognized, and initially no attempts were made to develop an electron optical instrument.

In 1928, Max Knoll and Ernst Ruska set out to investigate Busch's theory by experimenting with electron beams and focusing coils. Knoll and Ruska used a coil to form an image of a small aperture at a magnification of just over 1. They then added a second stage of magnification to give a final image with a magnification of 17. This instrument is usually regarded as the first electron microscope, although it was built primarily for accurate measurement of magnification rather than for examining specimens. Knoll and Ruska's instrument was first demonstrated in 1931. By the end of 1933 Ruska had built a 'supermicroscope' with which he achieved magnifications of up to 12,000 and surpassed the resolution of the optical microscope. In law, however, the inventor of the electron microscope is Günther Reinhold Rüdenberg who filed patent applications covering the combination electron lens for microscopy shortly before Knoll and Ruska's work was published in 1931.

The electron microscope has made an enormous impact on biological research, contributing especially to the investigation of cell structures and viruses.

P.J.

Radar

Radar was made to happen by Robert Watson-Watt; others had made it possible. In the 1880s Heinrich Hertz, experimenting with Clerk Maxwell's theories of electromagnetic waves, found that the 'sparks' which he generated were reflected from the pillars of his laboratory. In 1900, Nikola Tesla suggested that moving ships could be located by radio reflections. In 1924, Edward V. Appleton and his colleague, Miles A.F. Barnett, employed signals transmitted from the ground and reflected back from an ionized layer to establish the existence and height of the 'mirror in the sky', the Appleton layer of short-wave telecommunications.

In 1925, Gregory Breit and Merle A. Tuve contributed the pulse-signal. In 1934, the British Air Ministry committee, under Sir Henry Tizard, referred to the Government Radio Research Station, of which Watson-Watt was superintendent, the perennial 'death ray' which was supposed to paralyse an aircraft in flight. Watson-Watt and his colleague A.F. Wilkins did the calculations of energy requirements which brusquely disposed of the death ray, but Watson-Watt on half a sheet of paper spelled out 'radio-location', showing that detection of moving aircraft was a practical proposition. A month later, he gave an actual, and successful, demonstration. The result was a chain of radio-location stations which protected the eastern approaches of Britain and proved their effectiveness in the first German raid of the Second World War, when an aircraft was detected, tracked and destroyed in the Firth of Forth.

Airborne systems became possible with the development by J.T. Randall, J. Sayers and H.A.H. Boot of the cavity magnetron, a device so compact that Tizard could take it to the United States in his pocket, yet capable of generating pulses of energy equivalent to a large broadcasting transmitter. With such refinements, pilots were able to detect their adversaries in the air and bombers could turn a radio-searchlight of invisible rays on to a ground target and see the details on a cathode ray screen. Bombs and shells, fitted with radar proximity fuses, could respond to the echoes of their own signals and near misses became direct hits.

R.C.

Chromatograph

Chromatography is a method for coping with one of the biochemist's most frustrating problems: separating the components of complex mixtures.

Sometimes Nobel prizes are awarded not for great advances in understanding, but for the invention of techniques which make these advances possible. This was the case with chromatography, for the development of which the British chemists A.J.P. Martin and R.L.M. Synge received the Nobel prize for chemistry in 1952.

In its most widely used form, chromatography is almost unbelievably simple. A spot of the mixture to be separated is dropped on to a piece of absorbent paper, and one edge of the paper is dipped into a solvent. As the

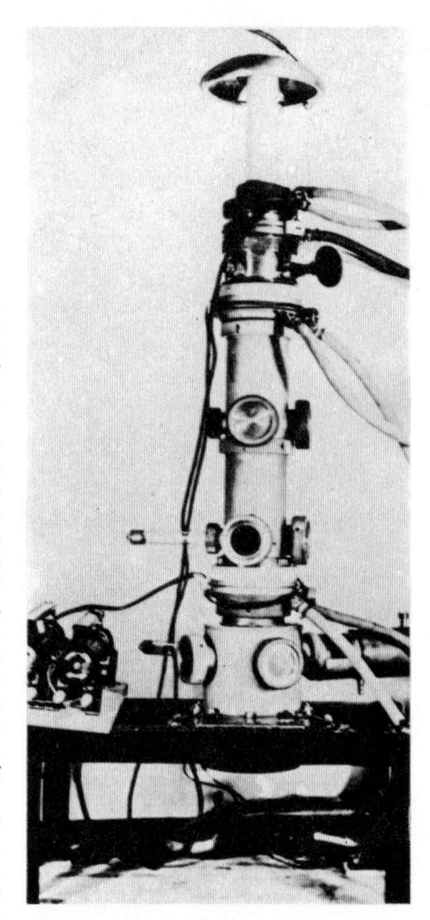

Ruska's 'super-microscope' of 1933, and, below it, 1, the point of a hypodermic needle; 2, a hair on a woman's leg; 3, a fly's eye at low-power magnifications. Modern instruments can magnify up to 200,000 times.

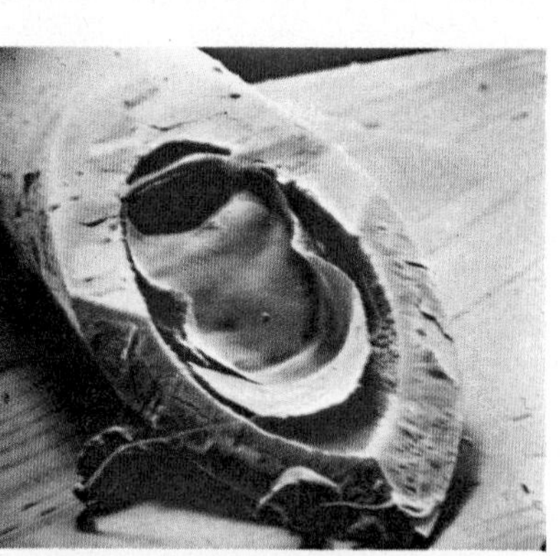

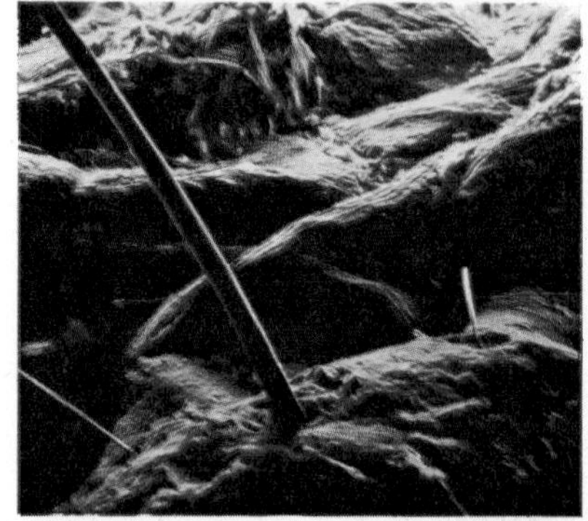

Below, *the Wash, on the east coast of England, as seen by airborne radar in September 1944.*

Below, *medical fibrescope. The viewing probe is 0·025 in. in diameter, small enough to pass through a hypodermic syringe.*

solvent spreads across the paper (just as ink spreads on blotting-paper) it carries the components of the mixture with it. But they move at different speeds. The components of the mixture end up as physically separated spots which can be cut out with a pair of scissors. The speed at which a substance moves also helps to identify it.

A primitive form of chromatography was worked out in the 1900s by Tswett in Russia, but it had very little practical use. Paper chromatography (as it is called) was invented in 1944 and is now as much a part of the biochemical laboratory as the typewriter is of the office. Two of the most important developments in biochemistry of the last 20 years would probably have been impossible without it: the first complete chemical analysis of a protein – insulin – by Frederick Sanger at Cambridge; and the elucidation of the detailed mechanism of photosynthesis in plants by Melvin Calvin in California.

B.S.

Chromatographic paper showing how the components are separated by the solvent.

Bubble chamber

The bubble chamber is one of the devices used to detect the particles within the atom, and to investigate the reactions of these particles with nuclei. It was invented by Donald Glaser, an American, in 1952 at a time when nuclear physicists were building machines such as linear accelerators and cyclotrons to produce atomic particles of very high energy. The bubble chamber fulfilled many of the exacting requirements necessary for observing complex atomic and nuclear events.

The basic idea of the bubble chamber, inspired by that of the cloud chamber developed by Wilson, is that atomic particles with electrical charge produce a trail of minute bubbles as they pass through a superheated liquid. The chosen liquid is hydrogen kept at a temperature only 20° above absolute zero, and under high pressure. When the pressure is suddenly reduced the hydrogen becomes unstable, ready to evaporate; at this moment nuclear particles are shot through the liquid, bubbles are instantly formed along the paths of the particles, and the paths are then photographed. The bubble chamber is used to look in detail at what happens to the nucleus of an atom when bombarded by, for example, a proton. All the components of the reaction will be observed by their bubble trails, and the products can often be identified by placing the chamber in a powerful magnetic field.

The bubble chamber offers many advantages over other particle detection methods; it is more sensitive, the pictures of the particle tracks are clearer and it is more efficient to use. On the other hand its construction presents many engineering problems – as may be imagined with a device containing 100 gal. of liquid hydrogen. The largest such machine in the world, 80 in. long, is at the Brookhaven National Laboratory in the USA.

R.J.

Bubble-chamber tracks. The tight spirals are mostly electrons; heavier particles swerve less in the magnetic field.

Fibrescope

A microscope that can look round corners is the picturesque result of success in making glass fibres several times thinner than a human hair.

The first problem was to make a mirror shaped like a bootlace and with its own light source. By wrapping a glass with one refractive index round a glass rod of another, the law of total internal reflectivity could be satisfied – that is to say that the light passes back and forth along them with no loss. This makes it possible to look at 'insides' from outside, no matter how long and winding the journey.

The greatest difficulty is to make the glass fibres thin enough for a bundle of them to be inserted into the narrow tunnels of the body, into the centre of a hypodermic needle, or through a leak-proof seal to look inside working machinery such as a nuclear reactor. The glass fibres are of two sorts – one carrying 'coherent' light which produces a picture at the eyepiece, and the other scrambled or 'incoherent' light which the fibres pass into the interior to illuminate the target. By forming bundles of the two kinds an image of the lighted object is produced at the surface eyepiece or camera.

The Atomic Energy Authority and, more recently, the Rank Organization have pioneered developments in Britain. By 1965 a 25-micron fibre (a micron is 0·001 mm.) was being produced. American companies, notably Bausch & Lomb, had concentrated on making the fibres smaller, and achieved a size of 15 microns. The Bausch & Lomb 'Flexiscope' was at first used for industrial inspection – as it gave off a 'cold' light it was useful for checking inside fuel tanks – but the American Cystoscope Company succeeded in sterilizing glass fibres, which opened up endless possibilities in the medical field. Bronchoscopes, which are swallowed, and introscopes, for probing the stomach, are normally bundles of fibres of 7 microns' thickness. But perhaps the millions spent on research and development will not be recovered until the image can be captured by a camera.

A.C.

Language

We will probably never know when the first language developed, what it sounded like, or who were the first people to use it. Writing is believed to have existed for 6,500 years, perhaps a little longer, but the beginnings of man's first systematic communication by the use of vocal symbols lies far back in the mists of prehistory. What caused even spoken tongues to evolve is the same factor that prompted civilizations to form and grow: there was a need. In the case of language the need was to communicate an image that referred to other than the time present. When a caveman stood up, he did not have to tell his family, 'I am standing up'. They could see that much. Similarly, when he went off in the morning to hunt, there would have been no need for explanations since he was merely following a routine to which the others had grown accustomed. But if during the course of one such expedition he chanced to discover another cave inhabited by creatures much like himself, he might later feel the need to relate his experience. And for this he would need a language.

Language, unlike the intelligence which makes it possible, is strictly a learned phenomenon. It is a form of knowledge, passed on from one generation to the next and constantly changing. How complex a language becomes – and many primitive tongues are as intricate and subtle as modern ones – again depends on the needs, real or imagined, of the persons speaking it. If Sanskrit seems more highly developed than Malay, it is because the former was used by philosophers and the latter by farmers and tradesmen. One group had a greater need of expressing abstract ideas than the other, and this itself stemmed more from environment and population conditions than from innate intelligence.

Today there are very approximately 3,000 primary languages, or speech communities, with anywhere between a few dozen and many millions of speakers. In addition, most of these languages have several or more dialects, some of which cannot be understood by other persons in the same language group. Chinese, for example, includes at least eight mutually unintelligible dialects, although all of these share exactly the same system of writing. The dialect which, over a period of time, acquires the most prestige in a given linguistic area eventually becomes the model for that people's standard or dictionary language. In this manner did official English come from London English and standard Italian from the dialect of Tuscany. What is considered correct speech today, however, may well be all wrong tomorrow.

To only a very limited extent has it been possible to trace genetic relationships between different languages, and most research in this field has been confined to the Indo-European and Hamito-Semitic families or maximal stocks. Yet these two groups represent but a very small part of the world's linguistic spectrum; and some languages – such as the Andamanese of the Andaman Islands, south-east of India – appear to be unrelated to any other language.

Despite the great variety of extant languages, only 12 primary tongues are spoken by 50 million people or more, and of these the North Chinese vernacular and then English are predominant. Occasionally certain auxiliary languages will present themselves, such as the original lingua franca, the language of the Franks, used in the Mediterranean after the time of the Crusades. This was a mixture of French, Italian, Greek, Arabic and other words. Both pidgin English and Swahili belong to this category. From time to time there have been attempts made to fabricate a universal language, such as Esperanto, but somehow the need for such a system has not been felt by enough people to make a go of it. Perhaps mathematics, despite its complexity and exclusiveness, has the best linguistic claim to universality.

E. W.

Abacus

Mankind's first attempt to mechanize calculation is of great antiquity. The word 'calculation' actually comes from *calculi*, the Latin word for the stones that were used in classical times, but the first abacuses consisted of a board or table covered with sand, on which counters could be moved. In this primitive form the abacus was used for addition and subtraction. The later form was the familiar rectangular frame with rods on

An early printed picture of the abacus, mankind's first attempt to mechanize calculation. Chinese, c. 1593.

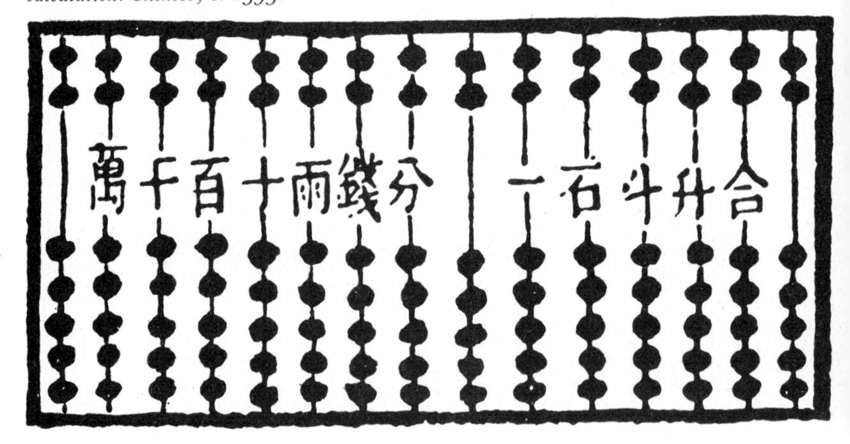

The Rosetta stone, cut in 196 BC in three scripts and deciphered in 1822, has a unique place in the history of writing.

Egyptian scribes writing – ideographs rather than an alphabetic script. Relief from the grave of Mereruka, Sakkara.

Detail from a pre-Conquest Aztec calendar. In the blue square at the top is a sign representing the year Two Reed, which came round every 52 years.

which pierced beads could slide – each rod serving as a place in decimal notation – and it could be used for multiplication, division, and the extraction of square and cube roots. It is still used extensively in the East today.

A. K.

Writing

Picture-writing dates back to representations on cave walls of the Upper Palaeolithic (*c.* 25,000 BC), in which the pictures are signs for things, and have nothing to do with speech. In this form of writing there is and can be no grammar or syntax. Ideographs are pictures which convey ideas, and it can be seen that they are an advance on picture-writing: 'legs' might mean 'walk' as well as 'legs', and the picture 'sun' might convey the idea of 'heat' or 'day'. Certain scripts, like Egyptian, Cretan and Hittite, have ideographs and also phonetic elements.

Phonetic writing is an important step because what is written represents the *sounds* of a given language – and not just generalized ideas which can be 'read' in any language. It is of two types: syllabic, in which the symbols are phonetic and represent syllables; and alphabetic, where a small number of letters represent sounds, and where each sound is represented more or less by only one letter. Alphabetic writing is, of course, easier to write and to read than, for example, Chinese, with about 80,000 possible symbols.

The oldest type of writing is perhaps cuneiform, which was in existence by *c.* 4000 BC; this is a system of 'wedges' (Lat., *cuneus*, a wedge), and was probably at first a mixture of pictures and ideographs. Its nature possibly stems from the availability of clay in the ancient Near East, and the consequent ease of impressing a wedge-shaped stylus therein. It was still in use in Mesopotamia in *c.* AD 75, but had been dwindling since the 5th century BC.

The age of alphabetic writing is hard to determine, but the prototype, called 'Proto-Semitic', has been dated to *c.* 1730–1580 BC. A few centuries later the inhabitants of Syria and Palestine were using an alphabet of 20 letters, quite similar to the later Greek alphabet, but (as in Hebrew) omitting the vowels. A family tree of the Roman alphabet would plausibly begin with Egyptian hieroglyphs, a mixture of ideograms and phonetic elements; then Old Semitic would split into two branches: Northern Semitic, the ancestor of Aramaic, Arabic, Persian and certain Indian scripts, and Southern Semitic, giving for example the scripts of Arabia and Ethiopia. From Northern Semitic derive Phoenician, Greek, Etruscan and, finally, Roman writing.

Of great importance to the historian is the decipherment of scripts once unreadable, enabling him to read laundry bills as well as monumental inscriptions. Egyptian hieroglyphs, for example, were already unreadable by the time of Christ, and *c.* AD 300 a man called Horus-Apollon affirmed that hieroglyphs were completely pictorial, each symbol representing an idea. His works were well known in the Renaissance, and spawned a whole series of pseudo-Egyptian picture-scripts. It was not until the discovery of the Rosetta Stone, with inscriptions of a priests' decree of 197–196 BC, in hieroglyphs, Demotic and Greek, that progress was made; the stone was brought to England in 1802, but it was the young Frenchman, Jean-François Champollion, working from plaster casts, who deciphered it in 1822, using comparative methods.

C. M. B. G.

Calendar

The measurement of time has been important to man ever since he began to settle and to cultivate the soil. The rise of cities and complex urban life meant that accurate measuring of time was even more necessary. The early calendars, which noted the division of time into manageable and known units, were based upon seasonal and astrological data and religious ideas. When a civilization gained a knowledge of mathematics and astronomy, accurate calendars usually resulted.

Primitive man reckoned time by the lunar calendar, by the phases of the moon and the planets. The urban Sumerians (*c.* 3000 BC onwards) divided day and night into two 12-hour periods. They seem to have been the first people to do this, but they still followed the lunar calendar. In Egypt, however, from the Third Dynasty onwards (*c.* 2780 BC), the priests produced a tropic or solar calendar based upon a year of 365 days, this being the approximate time taken for the earth to journey round the sun. Their researches, spanning a period of 50 years (modern time), formulated a calendar based also upon the frequency of the Nile flood, so vital for the Egyptian crops and for life itself. The Egyptian year was divided into 12 parts, with five days left over which acted as feast days. The writings of the Greek poet Hesiod show that by about 800 BC the early Greeks, by means of rustic botanical lore as well as astronomical data, were also using a calendar of 365 days. From 747 BC onwards, the astronomers of Babylonia reckoned time in years from a fixed point in time. Thus, for example, in the Hebrew Scriptures we read of events happening so many years after the Creation, or the Great Flood, instead of a particular year in the reign of a certain king. The Hellenistic Seleucid dynasty (*c.* 280 BC onwards) also recorded time from a fixed point.

In America the Mayas (AD 600–800) and the Aztecs (*c.* 1300–1500) linked religion closely with the calendar. The Mayas, in particular, were fine astronomers, and from Copán their priests calculated a 365-day year, consisting of 18 units of 20 days, plus five 'unlucky days'. The Mayas and the Aztecs also observed at the same time a 260-day ceremonial cycle.

As the true solar year is in fact 365 days, 5 hours, 48 minutes and 46 seconds, by Roman times the solar calendar of exactly 365 days required to be modified. In 46 BC Julius Caesar, on the advice of the Greek Sosigenes, reformed the calendar. One day was added to a year every four years, thus making the 'leap' year. The 12 months of the year were divided into units of 30 or 31 days, except that in non-leap years the month we call February had only 29 days. A little later Augustus, the first Emperor of Rome, caused the month Sextilis to be renamed after himself, and arrogantly added a day to it (making it 31 days), taking it away from February.

It was during Roman times that the first regular written almanacs, or astronomical registers of the days of the year, were produced. These were made by Ptolemy in Alexandria during the second century AD.

Even the Roman reforms did not produce a completely accurate calendar, however, and by 1582 the vernal equinox occurred on 11 March according to the current calendar, instead of on 21 March as it should have done. It being important for the Church to fix the saints' days and holy days accurately throughout the world, Pope Gregory XIII reformed the calendar again, and thus 4 October 1582 became 15 October, at least in Catholic countries. To prevent the error recurring, Gregory's reformed calendar provided for 24, not 25, leap years every century.

Many Protestant countries were slow to change to the new 'Gregorian' calendar. Britain did so only in 1752, when 11 days had to be added to bring the country into line with the new calendar: the riots this occasioned, as people demanded to be given back their 11 days, are famous. Other countries accepted the Gregorian calendar even later: Russia changed over only after the 1917 Revolution, and Thailand adopted it as recently as 1940. Traces of the old 'Julian' calendar remain in the financial field. The monetary year begins on 5 April (old Lady Day) in Britain – the start of the old-style new year.

M.H.

Arabic numerals

Man counts because he trades. An ancient farmer living alone with his family, deriving all their needs from the land he cultivates and from nature around him, may never wonder how many or how much. Living so simply, he knows what he needs to know and accepts the rest. Should he one day stumble upon another farmer, one who harvests different crops from his own, they may trade. And let several families make mutual contact, each with something new to offer the others, then sooner or later a system of barter will develop. And for this you need numbers.

Early man, like children and even many adults today, counted on his fingers and the hand is considered the simplest form of abacus. With the advent of civilization, as language in general first found itself committed to writing, some form of numeral symbology seemed called for. Various arithmetic systems quickly developed in various parts of the world, but the number symbols used by the early trading nations did not easily lend themselves to universality. We assume, though we do not know for sure, that it was in India about 2,500 years ago that what we now call Arabic numerals were first used. On the Buddhist inscriptions of Asoka, erected in the 3rd century BC, the symbols 1, 4 and 6 appear. One hundred years later, we find a 2, 4, 6, 7 and 9 engraved on the Nana Ghat monuments; and by the 2nd century of our era, the period of the Nasik caves, all of the 'Arabic' numerals – except the 8 – had been recorded. The zero, which also evolved among the Maya, began with the Hindus as a dot or a small circle and was referred to in Sanskrit by the term for 'vacant' or 'void'.

The earliest definite reference to the Hindu numerals beyond the borders of India is in a note written by a Mesopotamian bishop, Severus Sebokht, about AD 650, which speaks of 'nine signs', not mentioning the zero. By the end of the 8th century, some Indian astronomical tables had been translated at Baghdad and these signs became known to Arabian scholars of the time. In 825, the scholar al-Khwarizmi wrote a small book on numerals, and 300 years later it was translated into Latin by Adelard of Bath. Some historians believe that these number symbols came to Europe even before they arrived in Baghdad, but the oldest European manuscript containing them dates from AD 976 in Spain.

Today, whatever their true origin, Arabic numerals and the decimal system which they facilitated represent the most universal language which man has yet constructed. They are found in Chinese, Japanese and Soviet journals as freely as in the market-places of Germany, Greece or England. None the less, in the most modern of our many counting devices, the electronic computer, the decimal system itself has now been supplanted by the binary code.

E.W.

Zero

A sign for zero appeared only at a late stage in the development of symbols for numbers. It was not often necessary to write 'nothing' and so the word did not have to be abbreviated by a symbol. A zero symbol only became necessary with the invention of place-value notations and even then was not rapidly adopted.

In a place-value notation, the meaning of a number symbol depends on its position in a group of symbols: for example, in the modern decimal system, 7 means 'seven' in the group 47 but 'seventy' in the group 74. Each 'place' in the decimal system may be occupied by one of 10 symbols, including 0 which can be thought of as indicating an 'empty place'.

The earliest known place-value notation was used with the cuneiform script that was written with reeds on wet clay tablets by the Sumerians in Babylonia in the 18th century BC. The Sumerians wrote numbers in base 60 – i.e., each place could be occupied by one of 60 symbols. But only 59 symbols were used: as there was no symbol for 'empty place', a scribe would just leave a space.

This seems to have been acceptable for several centuries despite the inevitable confusion which arose from using, for example, the same symbol (equivalent to 1) for 1, 60, 3,600, 216,000 or any other power of 60.

Not until about 300 BC, when Babylonia was part of the Seleucid Empire, were the empty spaces filled with a symbol. The scribes began to use the same symbol for zero as they used to separate sentences of prose: it looked like the modern colon.

By that time, India had been much influenced by Babylonia and had adopted a modification of the Babylonian script. Indian mathematicians simplified the Babylonian number notation and changed from base 60 to base 10, thus creating the modern decimal system. Very little evidence exists of the chronology of Indian number symbols but it seems that, like the Babylonians, the Indians for a long time saw no need to write a symbol for zero. The earliest example of Indian use of the decimal system with a zero dates from AD 595.

Zero was indicated by a point; the circle notation developed some time before the 9th century when it appears in China; but before then Arabian mathematicians had adopted the decimal notation and the point symbol for zero. The circle symbol arrived in Europe about the 12th century.

The Arabians also used the Indian name 'vacant' (i.e. 'empty place') for zero. In Sanskrit this is *sunya*, in Arabic, *cifr*, which later became 'cipher' in English.

Far away from the mainstream of Western history, the Mayan culture of

Iraq c. 1000 A.D.

Arabic little changed in 1000 years

١٢٣٤٥٦٧٨٩ ·

Spain 976 A.D.

W. Europe c. 1360 A.D.

Italy c. 1400 A.D.

How our numbers have developed. Zero was originally indicated by a dot.

The 'new' mathematics, a woodcut from Margarita Philosophica, *'The Philosophic Pearl', by Gregor Reisch, 1515.*

Opposite, *the last page of Pepys's diary, in Shelton's shorthand: 'for . . . all the discomforts that will accompany my being blind, the good God prepare me. May 31, 1669'.*

Portrait (c. 1506) by Jacopo dei Barbari of the mathematician-monk Fra Luca di Pacioli, author of a 600-page work on arithmetic and geometry.

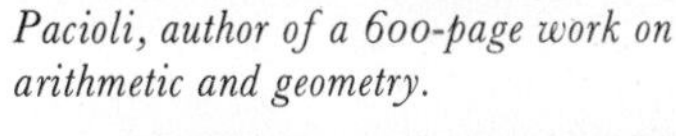

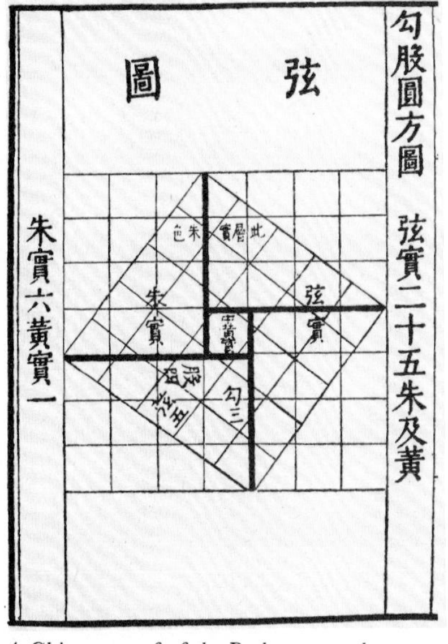

A Chinese proof of the Pythagorean theorem, on the basis that $3^2 + 4^2 = 5^2$.

A Greek papyrus fragment, bearing the statement and solution of a geometrical problem.

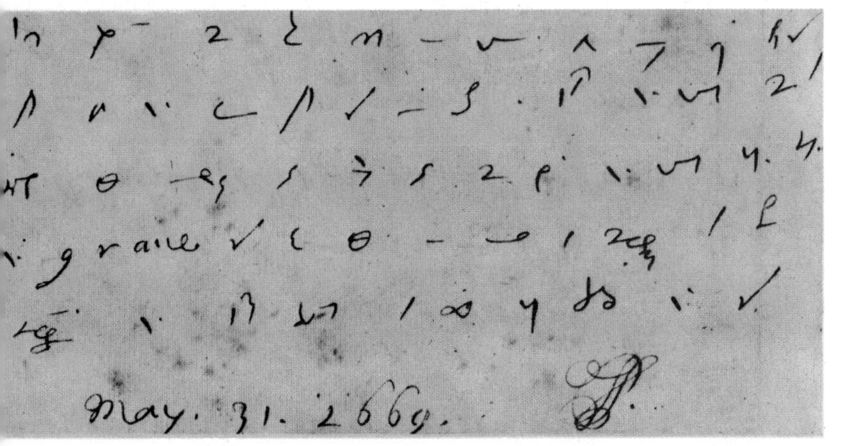

Central America, which died out at the end of the 9th century, developed a place-value system of notation with a symbol for zero. Mayan numbers were written vertically and are read from the bottom upwards. The Mayans worked in base 20 and had two sets of symbols: for everyday use there was a simple bar-and-dot notation with zero represented by a stylized shell; for formal inscriptions each number, including zero, was represented by the head of its patron god. It is conjectured that the Mayans first used their zero symbols at about the same time as the Babylonians used theirs on the other side of the earth, but the oldest Mayan numerical inscription dates from no earlier than the end of the 3rd century AD.

D.F.

Geometry

The early study of geometry is inseparably associated with the name of Euclid, yet almost nothing is known of the man except that he taught in Alexandria somewhere around 300 BC and wrote the world's most successful textbook of elementary mathematics, the *Elements*, used to introduce geometry to students for 22 centuries.

The 13 books of the *Elements* contain several hundred propositions (or theorems). These are proved by using simple rules of logical argument, starting from only five, unproved, commonly accepted statements (axioms) and some definitions. The work is a masterpiece of logical exposition which has been a model for mathematicians ever since.

Euclid appears to have been a skilful publicist for his book: there is a story that when asked by his king, Ptolemy I, whether there was a simpler way to learn geometry than by reading the *Elements*, Euclid replied, 'Sire, there is no royal road to geometry'. But it seems from another story that Euclid thought the road to geometry should be travelled for intellectual gain only: a student who had just learned his first theorem asked Euclid, 'But what shall I get by learning these things?' Euclid called a servant: 'Give him a penny, since he must make a profit from what he learns.'

The success of Euclid's *Elements*, however, was disastrous for historians, for all earlier textbooks of geometry were thrown away as useless. The best available assessment of Euclid's originality was probably made by Proclus (AD 410 or 412–85), a geometry teacher at Athens, who wrote that 'Euclid . . . put together the *Elements*, collecting many theorems of Eudoxus, perfecting many others by Theaetetus, and bringing to irrefragable demonstration the things which had only been somewhat loosely proved by his predecessors.'

Among the predecessors of Euclid were Thales of Miletus (640?–546 BC), who probably proved that the base angles of an isosceles triangle are equal; Pythagoras (died about 497 BC), who is said to have sacrificed oxen after discovering the theorem which now bears his name; and Hippocrates of Chios (*c.* 470–400 BC), the first person to write a book of 'elements of geometry'.

The Greeks were not the first to study geometry, some knowledge of which is necessary for even simple architecture. Sacred texts from the Indus Valley of about 1500 BC include geometrical rules of proportion (including Pythagoras' theorem) among their instructions for constructing altars. Pythagoras' theorem was also known in Babylonia 13 centuries before Pythagoras made his thanksgiving sacrifice. Indeed the 19th-century physicist, Lord Kelvin, remarked that 'God is a geometer.'

The Greeks made tremendous advances over the geometrical observations of earlier cultures: Pythagoras' theorem is dealt with at the end of the first of Euclid's 13 books. But the principal difference between Greek and pre-Greek geometry was that no one before the Greeks thought of proving what seemed to be facts of nature. The word 'element' in the title of Euclid's book referred originally to a link in a chain and Euclid's lasting achievement is the chain of reasoning contained in the *Elements*, even though he did not design every link himself.

D.F.

Shorthand

The high level of demand for shorthand typists in the business world of today is no more than a contemporary reflection of a discrepancy in human capacities that has vexed mankind for centuries: the inconvenient fact that writing takes longer than speaking. Ingenious men have made attempts to get round this problem ever since the 4th century BC, or so a 'shorthand' inscription on a marble slab from the Acropolis at Athens suggests. But the earliest record of an organized system dates from somewhat later, 63 BC in fact, when a Roman freedman, Marcus Tullius Tiro, invented a set of *notae* for recording the speeches of his friend Cicero and other speakers in the Roman Senate. Tiro's system, which was based on the letters of the Roman alphabet, was taught widely in schools and used by the Roman emperors. It remained in use for several centuries and was, for example, still being employed by the Merovingian King Clotaire II in AD 625, while an improved version, apparently devised by Seneca, was used by Thomas à Becket. The great advantage of Tiro's system, which many later ones have followed, was that the reader could deduce simply from the angle at which each consonant was written what

Above, *world map of Henricus Martellus Germanus, 1489: surprisingly recognizable, though the American continent is still* terra incognita.

vowel followed it, without the addition of any other sign.

All sorts of more or less standardized systems followed in the Middle Ages and later, reaching a peak in Britain: in 1588 Dr Timothy Bright published – and dedicated to Queen Elizabeth I – his *Characterie: an Arte of Shorte, Swifte, and Secrete Writing by Character*, and 14 years later John Willis brought out his *Arte of Stenographie*. Willis's was the first alphabetic system in English; it was followed rapidly by others, including Thomas Shelton's (1630) in which Pepys wrote his *Diary*, Thomas Gurney's (1750) which Dickens used, and the influential system of Samuel Taylor which was adapted for use in French, German and Spanish.

Most modern systems, however, are phonetic. The first of this kind was William Tiffin's, published in the mid-18th century, which again led to a flood of variations. But the first really scientific classification of sound was made by Isaac Pitman in his *Stenographic Sound Hand* (1837). Pitman's system uses lines, curves and hooks, with contractions and 'grammalogues' for certain whole words which occur frequently. It remains the most widely used system in Britain today. It has, however, proved less durable in the United States, where it was introduced by Stephen Pearl Andrews in Boston in 1844. There the 'Light Line Phonography' of John Robert Gregg, brought out in 1888, has become the most favoured method.

Elsewhere in the world innumerable systems, both phonetic and alphabetical, are in use, such as Fayet's French system and Gabelsberger's German one. In the late 18th century the Stolze-Schrey system, in which the length of the signs is directly related to their pronunciation, was perfected and this method has been highly influential.

Shorthand speeds of up to 280·4 words per minute have been recorded, with an accuracy rate exceeding 99 per cent; nevertheless an uneasy feeling must haunt many of the tens of thousands of people whose livelihood today depends on their skill at shorthand that future generations are going to look back with amazement on a situation where almost every business letter has to be laboriously transcribed and typed out by someone different from the person who dictated it, and that it cannot be long before some genius invents an economical and practical machine which will deprive them of their jobs while solving the problem of the gap between speech and writing speeds once and for all.

J.G.

Map

Although there is a map of northern Mesopotamia extant on a clay tablet dating from *c.* 3800 BC, we have no other maps except for town plans which antedate 2000 BC. The oldest world map is perhaps that on a clay tablet in the British Museum of *c.* 500 BC, which shows Babylon at the centre, and the universe all around it.

The Egyptians undertook journeys of exploration as early as 1493–92 BC, but there is no evidence that they produced maps before Hellenistic times. The Greeks had maps by the 6th century BC: Herodotus, for example, records that Aristagoras showed 'bronze tablets upon which was engraved the whole circuit of the sea, the earth and the rivers' to Cleomenes, King of Sparta, in 504 BC.

But it was at Alexandria, beginning in the 3rd century BC, that a more practical and scientific approach to the problems of map-making was attempted. The most influential Alexandrian scholar was Ptolemy, whose *Geography* of *c.* AD 150 is a guide to drawing a map of the world; a Byzantine manuscript of it reached Italy and was translated into Latin in 1406, and the first printed edition to contain maps as well as text appeared at Bologna in 1477. Although Ptolemy was wrong in many details, his insistence that map-making should have a scientific basis led to a renaissance in the science. His importance can be gauged by looking back to the Middle Ages, which had made little progress in cartography: their maps tended to be diagrammatic, not too different from the Babylonian clay tablet of 500 BC, and probably deriving from Roman examples. The best-known type is the T-O map, a circle with a 'T' inside it, the bar of the 'T' representing the rivers Danube and Nile, the riser the Mediterranean; Asia is at the top, Europe bottom left, Africa bottom right, and the Ocean surrounds the whole inhabited world. Another type, not too different, is the Beatus map: Beatus of Valcavado wrote his *Commentary on the Apocalypse* in AD 776, and illustrated it with a map of the world, which again divides the known world schematically with Ocean all around.

Although T-O maps survived into the 16th century, that century also saw the production of the first printed sea-charts for sailors; even earlier, with the help of Italian (mostly Venetian) cartographers, the Portuguese had begun to explore and map the coast of Africa – the earliest (*c.* 1448) of their charts stretches as far as Senegal. Mercator (1512–94) is popularly the inventor of a projection (perhaps first used by Erhard Etzlaub in 1513) which produced a map on which any drawn straight line in whatever direction is a line of constant compass-bearing. This projection was the most practical way of representing on the flat the spheroid form of the earth (although it is the more inaccurate the nearer one approaches to the poles), but it did not meet with immediate success on its publication in a world map in 1569, for Sir John Narbrough, the naval commander, could complain as late as

Below, *table from Napier's* Descriptio, *published in 1614. This was the foundation of modern logarithms.*

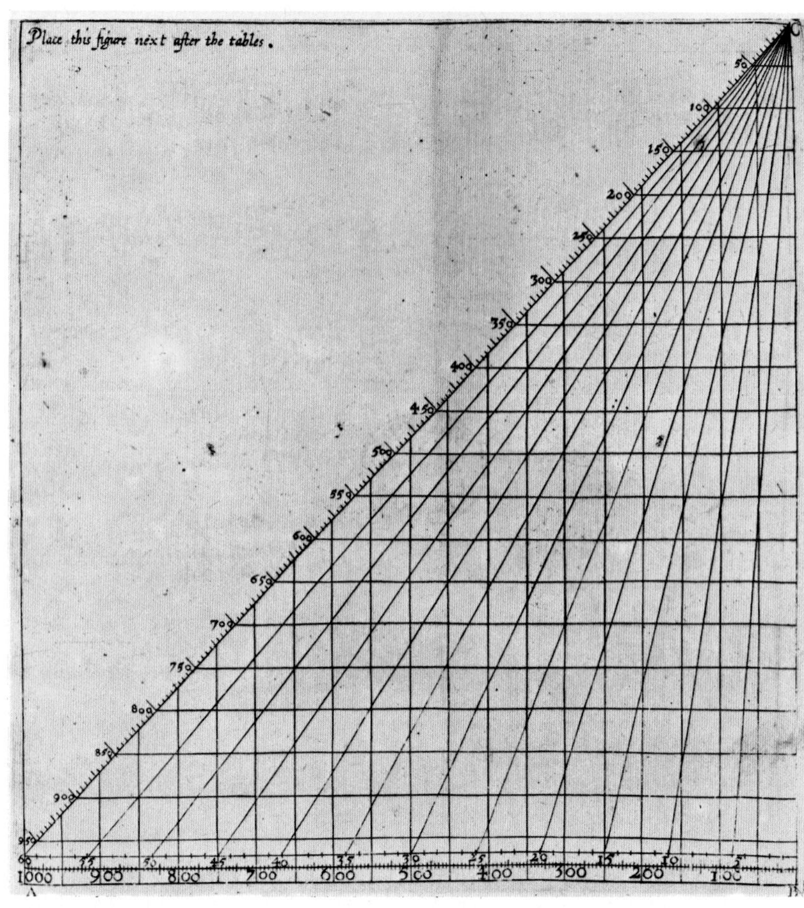

A passage from the Rhind mathematical papyrus proves that the Egyptians had an understanding of what today would be called algebra.

The Easter trope 'Quem quaeritis?' from the Sarum Processional: a stave of only four lines but recognizably the ancestor of modern musical notation.

1694: 'I could wish all Seamen would give over sailing by the false plain charts, and sail by Mercator's chart, which is according to the truth of Navigation; but it is a hard matter to convince any of the old Navigators. . . .'

The 16th and 17th centuries saw the great collections of maps published by Ortelius and the Blaeu family beginning in 1605; the latter family's *Atlas Maior* was produced by Joan Blaeu in French, Latin, Dutch, German and Spanish editions from 1663. The Latin edition is of 11 large folio volumes, containing in all 3,000 pages of text and 600 maps: it was the most expensive book available in the second half of the 17th century – a prestige gift renowned for its luxury binding and beautifully coloured maps more than for its scholarship.

But the systematic surveying of much of the world had to wait for the national surveys of the 19th century; many, like that of Britain (first in the field, in 1791) were established to help military rather than civilian communications.

C.M.B.G.

Algebra

Although its origins are not known for certain it is generally thought that algebra first developed from early arithmetic in ancient Babylonia. Clay tablets bearing mathematical tables of a kind that we should today regard as algebraic have been found dating from as early as 1800–1600 BC, although the modern term was not applied to the system until the Arabs brought algebra to the Western world and, in Baghdad in about AD 852, al-Khwarizmi wrote his *Kitab al jabr w'al-muqubala* (*The Rules of Restoration and Reduction*).

Algebra is basically no more than a system of calculation using symbols (usually letters of the alphabet) to represent numbers and signs to represent the relationships of these symbols to each other. Early mathematicians doubtless found that in a complicated calculation the use of an actual number might be inconvenient, and so introduced a symbol to represent it. From this initial step algebra gradually evolved until it laid down a set of prescribed steps by which problems or equations unmanageable by simple arithmetic could be solved. It also led to the formation of formulae in which algebraic symbols stood for an unknown quantity for which a known number could be substituted to provide an unvarying rule which, formulated as a result of mankind's long experience, required no testing. Most geometrical formulae are of this kind.

Since the 1920s an abstract algebra has been developing in which not only the numbers involved but also the calculations made are generalized.

Algebra has been widely used by scientists and engineers in many periods, most especially, of course, our own. When one considers the part it has played in working out the secrets of, for example, nuclear energy, the title 'Great Art', given to algebra during the Renaissance to distinguish it from the 'Lesser Art' of arithmetic, seems merited.

S.F.

Musical notation

Probably the earliest known musical notation was used for the Vedic hymns of southern India around 700 BC. This system contains approximately 300 symbols based on the contemporary alphabet, each representing a series of notes in a predetermined pattern. For example, the symbol known as 'cha' means that one should sing three particular descending notes in quick succession. In the 'staff' notation most common today, each symbol of the system represents one single sound, explaining its relative pitch and relative duration. Systems in which a symbol represents more than one sound are called 'polynomial' and systems in which one symbol represents one sound are called 'mononomial'. The development of the staff notation familiar today is the development from 'polynomial' to 'mononomial' systems.

Guido D'Arezzo (AD 995–1050), a Benedictine monk, is the inventor who made the most significant advances towards staff notation. Fifteen letters of the Roman alphabet had been applied to the notes used in music of Guido's time. There was also a system of 'neumes', or symbols similar to shorthand symbols, written above the words of a song to indicate the accompanying tune. A sign like a full stop, for example, meant a short note; an undulating line meant a group of two notes, and so on. The relative highness or lowness of the note depended on its respective distance from the text. One of Guido's ideas was to introduce lines, or staffs, parallel to the text, which represented certain notes, so that intervals between the notes could be more accurately determined by considering their position in relation to these lines. The first two lines drawn in this way were to represent 'F' and 'C', and these letters, in a degenerate form, remain the symbols for our 'treble' and 'bass' clefs in staff notation today.

A system of notation which Guido also developed was the 'hexachord'. This was a system whereby six consecutive notes were taken as a unit in the same way that the eight notes of the 'octave' are taken as a unit today. In Guido's time music, mathematics and magic were bound up together, and six was an appropriately significant number. Guido named his six notes after the words of a Latin hymn in which consecutively ascending notes occur coinciding with the following syllables in the lyric: '*UT* queant laxis *RE*sonare fibris *MI*ra gestorum *FA*muli tuorum *SOL*ve polluti *LA*bii reatum, sancte Iohannes.' In the 19th century, John Curwen, a Congregational minister, along with a Miss S.A. Glover, developed this principle into the 'doh re mi' notation or tonic sol-fa system familiar today as a handy aid in sight-reading music.

Perhaps because the introduction of electronically recorded music has lessened the need to rely on graphic systems of musical representation, or perhaps because of a change in the relationship between composer and performer, modern composers have been particularly adventurous about notation, and Stockhausen, Earle Brown and many others have invented their own notations for their work. Whether John Cage's notation, based on the blemishes on the paper he wrote on, was an advance in approach similar to some of Guido d'Arezzo's, time will tell. Or not tell, as Cage would probably say.

P.WH.

Logarithms

'Neper, lord of Markinston, hath set my head and hands to work with his new and admirable logarithms. I hope to see him this summer, if it please God, for I never saw book, which pleased me better, and made me more wonder.'

Henry Briggs, professor of geometry at Gresham College, London, was expressing his wonder at the book which John Napier, 8th Laird of Merchiston (1550–1617) had published in 1614. *Mirifici Logarithmorum Canonis Descriptio* was the culmination of 20 years' work by Napier. It contained a table of logarithms and a description of how to use them.

Briggs immediately realized that Napier's invention was enormously important. A significant part of scientific work at the time was astronomy and the related techniques of navigation and surveying, all of which involved a great deal of laborious calculation. The invention of logarithms had much the same liberating effect on scientists as the invention of electronic computers three centuries later.

Briggs's meeting with Napier is described by Napier's friend, the astrologer William Lilly: 'Almost one-quarter of an hour was spent, each beholding the other almost with admiration, before one word was spoke. At last Mr Briggs began, "My Lord, I have undertaken this journey purposely to see your person, and to know by what engine of wit or ingenuity you came first to think of this most excellent help unto astronomy; but, my Lord, being by you found out, I wonder nobody else found it out before, when now known it is so easy."'

And it seems that Napier and Briggs were content to leave discussion of how logarithms were invented at that. Given the state of mathematical knowledge at the time, however, Napier's inventive step was not a large one and most of his 20 years of work had been spent actually computing the logarithms.

Napier's logarithms were very different from those in use today because he had designed them specifically for trigonometrical calculation. In Napier's version the logarithm of 10,000,000 was 0 and logarithms increased as the number decreased. The logarithm of 1,000,000, for example, was 23,025,850.

Briggs stayed with Napier for a month, discussing improvements to the logarithmic method. Briggs wanted to change the base so that logarithms of multiples of 10 would be whole numbers. Napier replied that he had already thought of this and also wanted to 'reverse' the scale so that 0 would be the logarithm of 1 and logarithms would increase as the number increased – in other words, Napier had thought of the modern form of logarithms.

'I could not but acknowledge [this] to be far the most convenient,' said Briggs, who undertook the enormous task of computing a new table of logarithms to 14 places. Napier died two years later and Briggs, exhausted, gave up his calculations in 1624 when he published one-fifth of the new table in his *Arithmetica Logarithmica*. The table was completed (but to 11 places only) in 1628 by Adrian Vlacq.

D.F.

Slide rule

'I therefore devised to have another Ruler with the former: and so by setting and applying one to the other, I did not onely take away the use of Compasses, but also, make the worke much more easy and expedite.'

So in 1633 did William Oughtred, Rector of Aldbury, describe his invention of 'twelve yeares agoe', which heralded the beginning of computation using mathematical instruments. At the turn of the 17th century, the only aid to calculation was a simple abacus, which shopkeepers used to help them visualize their accounts. Only an educated few could manipulate even simple numbers.

The first step towards alleviating the drudgery of calculation came from John Napier, Baron of Merchiston, who, in 1614, published his invention of logarithms.

This prepared the way for the immediate forerunner of the slide rule. In 1620, Edmund Gunter, a professor at Gresham College, London, described his 'logarithmic line of numbers'.

One year later, Oughtred made the first rectilinear slide rule, consisting of two scales, both bearing logarithmic lines, which were held together in the hand when in use. The delay in publishing his invention caused a bitter quarrel between Oughtred and his former pupil Richard Delamain, who described a circular slide rule in 1630 and was therefore the first to publish an account of any type of slide rule. Oughtred accused him of stealing his ideas and, of course, Delamain claimed priority of invention. Due to Oughtred's high standing in society, his word is not doubted as having been the first inventor, but Delamain very probably invented his instrument independently.

Many subjects interested Oughtred. He walked in 'the pleasant and more than Elysian fields of the diverse and various parts of humane learning'. He travelled little, but delighted in filling his house with students. He enjoyed good health into his old age, which he attributed to 'Temperance and Archery'.

The earliest straight slide rule now existing was made in 1654, and resembles the modern design of a slider moving in a fixed stock. Before the end of the century, many specialized rules were designed for use by excise officers, masons, carpenters and others.

J.M.P.

Calculating machine

Arithmetic, if much of it has to be done, is tedious and dull; the results are all too often unreliable and need checking, and no doubt the 20th century will see the last of the clerks, doomed to add up accounts for the whole of their working life. The first major step in replacing the clerk with a machine was made in the 17th century.

Blaise Pascal's father was a clerk whose daily work involved enormous numbers of simple calculations – additions and subtractions – with money. To make his father's work more pleasant, Pascal invented a calculating machine. The principle used in such a machine is simple, especially when working with a decimal currency. A dial representing units is rotated; it is fitted with some device that, as it passes the number 9, engages the next dial to the left and moves it one space. This operation, which corresponds with 'carrying one' in simple addition, is quite difficult to mechanize. The problem is that, at one moment, more than one dial has to move simultaneously. When 1 is added to 9999, for example, five dials all have to move round one space in unison. Pascal's invention was a delicate and elaborate clutch device, which he called a *sautoir*, that moved the dials. He offered his machines for sale, and pointed out that no rival contained a *sautoir*, which 'made it as easy to add 1 to 9999 as to 1'. According to his sister, Pascal invented the machine in 1642 when he was 19 years old. Although the instrument-makers of the time found it extremely difficult to work to the accuracy needed for a device like this, about 70 were made, some of them as

Pascal's calculating machine. The diagram in the picture explains the operation of the sautoir.

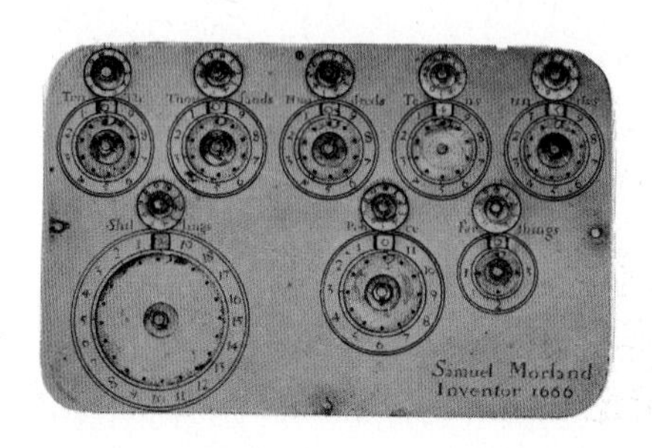

Calculating machine made in 1666 by Samuel Morland, a contemporary of Pascal.

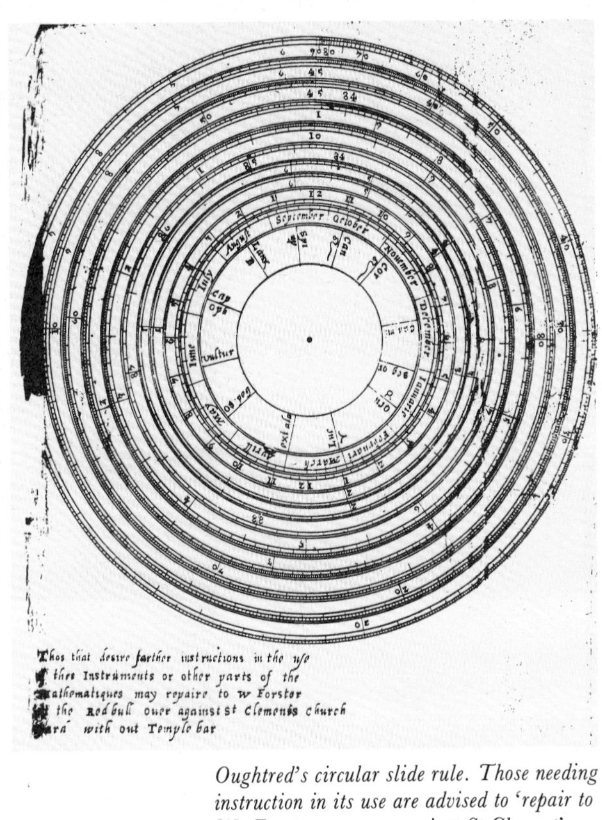

Oughtred's circular slide rule. Those needing instruction in its use are advised to 'repair to W. Forster . . . over against St Clement's Church without Temple Bar'.

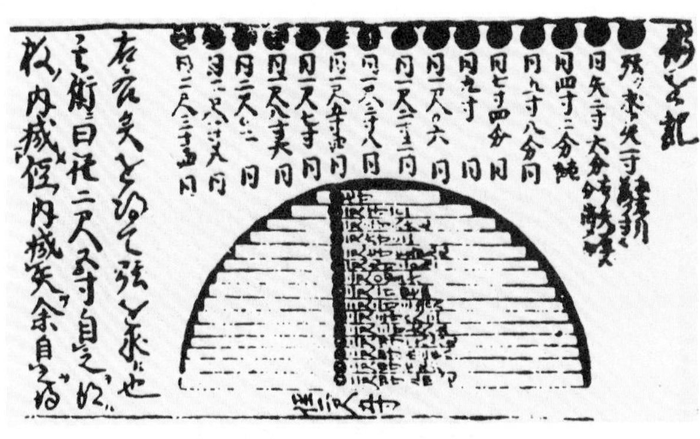

Late 17th-century Japanese diagram, illustrating area measurement by integration.

This millimetre-sized ceramic logic module (one of thousands in an IBM computer system) can distinguish only two states, 'on' and 'off'. This translates into the 1 or 0 of binary arithmetic.

presents for the King of France. It was, incidentally, not a pure decimal machine, because the French also used *deniers* and *sous*, 12ths and 20ths of a franc.

This device, like others of the time, could handle easily only addition and subtraction. Obviously, a series of additions is equivalent to multiplication, but the process is tiresome. The essential devices for speedy multiplication were embodied in the two calculating machines that Leibniz built in 1694 and 1706. The problem that has to be solved is to add the same number again and again by, for example, turning a handle. One of Leibniz's devices was the 'stepped reckoner'. This was a cylinder with nine cogs cut parallel to its axis, running successively further along it. The 'calculating cog' could be moved above this cylinder. If it was moved a short way, '2' would be added each time the cylinder was turned: if moved further, '4' and so on. This device, and the 'pin wheel', for the same purpose, is used in modern calculating machines, and was also incorporated in the first machines that had a large-scale success because they did not eventually break down owing to poor construction. These were built by Charles Xavier Thomas in the 19th century.

T.O.

Calculus

The basic problems of the calculus – finding the tangent to a curve (differentiation) and finding the area enclosed by a curve (integration) – have been studied by Western mathematicians since the 5th century BC. Solutions were found in a number of special cases by a wide variety of methods – some of them simple, some of them clumsy and laborious.

In the late 17th century two men, Newton and Leibniz, developed simple general rules – a 'calculus', or method of calculation – for dealing with these problems. As Newton put it in an essay describing his discovery: 'To resolve Problems by Motion these following Propositions are sufficient.'

Isaac Newton (1642–1727) was interested in astronomy as an undergraduate at Cambridge University. He made observations of the movements of planets and studied Galileo's work on the subject, but in order to explain planetary motion he needed to study mathematics. Higher mathematics was not then taught at Cambridge, so Newton had to proceed entirely by unguided reading. The advanced works he read contained many attempts at the tangent and area problems.

His reading began a period of intense creativity. He established general rules for differentiation, recognized that integration was just the inverse of differentiation, created a coherent basis for his methods and wrote an essay setting out his ideas. 'All this was in the plague years of 1665 & 1666. For in those days I was in the prime of my age for invention & minded Mathematicks & Philosophy more than at any time since.'

But Newton was only 24 when he devised the calculus, and he had never been introduced to the scientific Establishment which then as now controlled publication of new work. His work on calculus became public only by stages, in letters written over the next few years as his genius began to gain recognition.

Gottfried Wilhelm Leibniz (1646–1716) was, in contrast to Newton, a worldly, gregarious man, a philosopher and diplomat, whose ideas came from meeting people and from selecting what he wanted from the books they recommended. While on a diplomatic mission to Paris in 1672, Leibniz met Huygens, who fired his enthusiasm for mathematics. He visited London to give a demonstration of a calculating machine he had invented, but gave the impression of being a dilettante. No one, it seems, thought it worth while to tell him of Newton's work.

Leibniz's approach to the tangent and area problems was conditioned by the philosophical problem that had exercised him since his student days: how to systematize, and reduce to a few simple elements, all human thought. The practical effect of this grandiose scheme was Leibniz's expertise at reducing concepts to symbols. Between 1673 and 1676 he invented the $\int$ and δ symbols that are still used today and realized that the rules of differentiation and integration were simply rules for manipulating these symbols.

The independent unpublished discoveries of the two men caused bitter controversy over priority. It can, however, now be seen that each in his own way laid the foundations of the modern period of mathematics.

D.F.

Binary arithmetic

'One suffices to derive all out of nothing,' remarked a delighted Leibniz about the system of binary arithmetic that he described in 1679. For Gottfried Wilhelm Leibniz, philosopher, mathematician, logician and diplomat, the discovery that every number could be represented by using only two symbols, 0 and 1, was both a step nearer his grand design of a universal symbolism in which all arguments could be expressed, and also a mystical demonstration of the existence of God. 'All combinations arise from unity and nothing, which is like saying that God made everything from nothing, and that there were only two first principles, God and nothing.'

Leibniz was so impressed by the theological implications of binary arithmetic that he wrote about it to a Jesuit missionary in China, Father Joachim Bouvet, suggesting it could help in the conversion of the Chinese. Leibniz was amazed when Bouvet replied that he had detected a use of the binary system in the *I Ching*, then believed to be 3,000 years old. Leibniz wrote a paper crediting the Chinese with the invention of binary arithmetic and this idea has been popular ever since.

In fact, Bouvet had looked at a version of the *I Ching* compiled by Shao Yung in the 11th century AD and, although it is true that the famous 64 hexagrams (*kua*) are made up from two types of line, there is no evidence that anyone before Bouvet thought of this as binary arithmetic.

Leibniz also invented an early version of the modern desk-calculator and would undoubtedly have been pleased that his binary arithmetic was revived in the 20th century as a vital element in the design of electronic computers. Credit for thinking of this application appears to be evenly divided among a number of people.

In 1932, C.E. Wynne-Williams, working at the Cavendish Laboratory, Cambridge, designed a high-speed electronic counter for alpha particles in which the count was recorded on a 'scale of two'.

In 1938, Louis Couffignal, designer of the first French computer, suggested using binary arithmetic in calculating machines. At about the same time, John V. Atanasoff, a physicist at Iowa State College, thought of using binary arithmetic for an electronic equation-solving machine.

Atanasoff described his machine to John W. Mauchly, who later joined the team at the Moore School of Electrical Engineering, University of Pennsylvania, that produced the first designs for the modern form of general-purpose digital electronic computers. It seems likely that J. Presper Eckert, Jr, chief engineer of the Moore School team, came across Wynne-Williams's work.

John von Neumann (1903–57) wrote the report that summarized the Pennsylvania team's thinking. 'In spite of the long-standing tradition of building digital machines in the decimal system we feel strongly in favor of the binary system for our device.' Binary arithmetic has been the language of computing ever since.

D.F.

Metric system

The metric system was developed by a series of government-appointed commissions during the French Revolution. At the end of the 18th century, France had a chaotic variety of weights and measures. Each province had a different standard unit of length; each trade had different methods of measurement. The revolutionaries thought this a typical example of feudal maladministration, hindering trade and confusing consumers; scientists wanted the system reformed on an international scientific basis.

On 8 May 1790 the National Assembly, prompted by Talleyrand, decreed the reform of weights and measures. The first stage was for the Academy of Sciences to 'indicate the scale of division which it believes most convenient for all weights, measures and coins'. The Academy set up a commission to consider this question, consisting of Jean-Charles de Borda (1733–99), designer of navigational and surveying instruments; Marie-Jean-Antoine-Nicolas, Marquis de Condorcet (1743–94), secretary of the Academy, deputy for Paris in the Legislative Assembly and intellectual guide of the Revolution; Pierre-Simon Laplace (1749–1827), the 'Newton of France'; and Mathieu Tillet (d. 1791), treasurer of the Academy. The chairman was Joseph-Louis Lagrange (1736–1813), 'the lofty pyramid of the mathematical sciences', who proposed using the decimal system 'to accord with the gradation of numbers' and countered every argument against it. The commission recommended decimalization on 27 October 1790.

The next stage was to decide on a 'natural' unit of length. The same commission considered this question except that Tillet was replaced by Gaspard Monge (1746–1818), military engineer and examiner for commissions in the Navy.

The commission's second report recommended that the unit of length should be one ten-millionth of the distance from the Equator to the North Pole, measured along the meridian that passes through Paris.

This recommendation was adapted from an idea that Abbé Gabriel Mouton of Lyon had published in 1670, and was probably put forward by Borda, the surveyor; certainly it was Borda who proposed the name: mètre, from the Greek *metron*, 'measure'.

The unit of mass was to be the gramme, the mass of a cubic centimetre of water.

The Assembly accepted the proposals on 30 March 1791 but it was several years before the new system was ready for use. The problem was to measure the new units. Twelve academicians were commissioned to do this. The most important job was given to Jean-Baptiste Delambre and Pierre Méchain, who were to measure the difference in latitude between Barcelona and Dunkerque (using instruments designed by Borda) and thus determine the length of the metre.

With a revolution raging around them, Delambre and Méchain worked patiently for seven years on the task. They were arrested (surveying looked suspiciously like anti-Revolutionary activity); work was held up for a year while half the commission (including Borda) was purged by the Committee of Public Safety; Borda's instruments had been 'metricated' but the trigonometrical tables for use with them had not.

Eventually Delambre completed his report on the survey in 1810 (Méchain died in 1804). He presented it to Napoleon who remarked, 'Conquests pass, but such works remain.'

D.F.

Artificial languages

Although the Greeks had toyed with the idea of a universal language, it was not until the efforts of Francis Bacon in the early 17th century that any thorough systems were worked out: his *The Advancement of Learning* (1605) sketches a language of symbols, and we find several scholars in the following years – Douet (1627), Vossius (1635) and Lodwick (1647), to name but three – who studied several ancient and modern languages in attempts to evolve a suitable alternative.

But the main period for the development of universal systems was the late 19th century. Idealistic dreams of universal brotherhood resulting from communication between nations were often stressed, and existing 'natural' languages were put out of court because of their individual quirks and because, as the British Association Committee of the International Research Council on the subject delicately put it in 1919, 'the great international languages of the past have all borne the marks of imperial prestige which prevented them from being welcomed by alien races. The adoption of any modern language by the common consent of the chief nations is therefore unlikely.' The suggested alternatives have certain elements which are wholly invented, and others which are adaptations of existing language groups.

Volapük, invented by a German priest named J.M. Schleyer, was first published in 1880. A mixture of English and Romance languages, its structure is mainly German. It spread fast, and when the Third Volapük Congress met in Paris in 1889, there were 283 Volapük societies all over the world; but it soon wilted before Esperanto, which was first published in 1887 by the Russian, Dr L. Zamenhof, for just those idealistic motives given above. He selected the roots of his words from the European languages; some words are similar, anyway, and excluding these, his dictionary contained only 2,642 root-words. Grammar is part borrowed, part original: for example, suffix -a indicates an adjective, suffix -o a noun, and the definite article is 'la' for all genders. The beginning of the

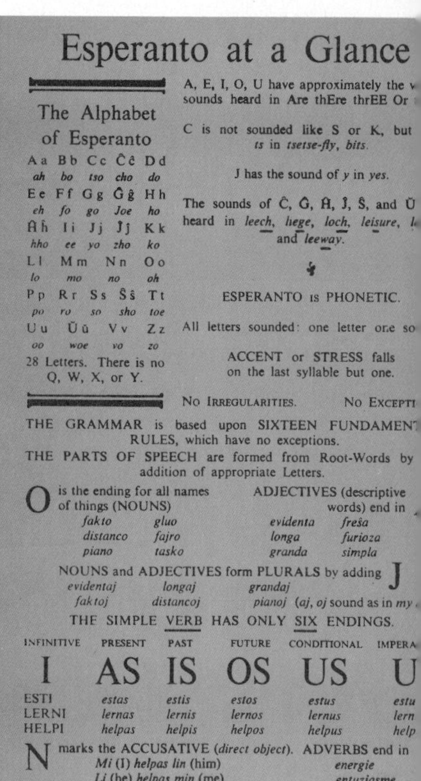

Esperanto, first published in 1887, had a simple grammar and skilful advertising.

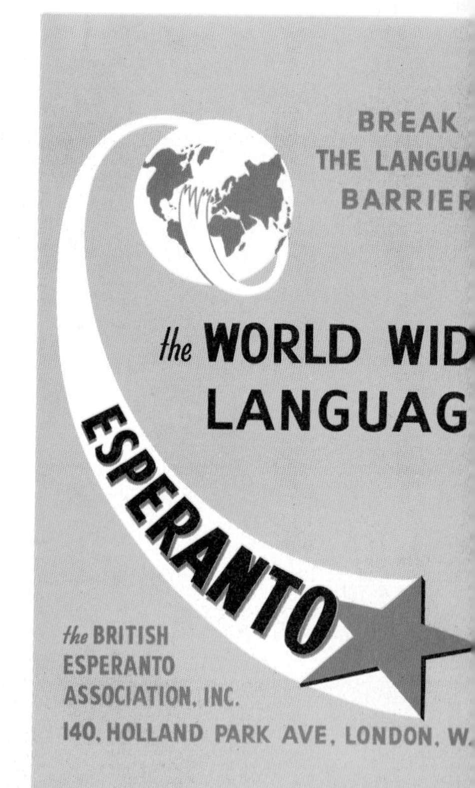

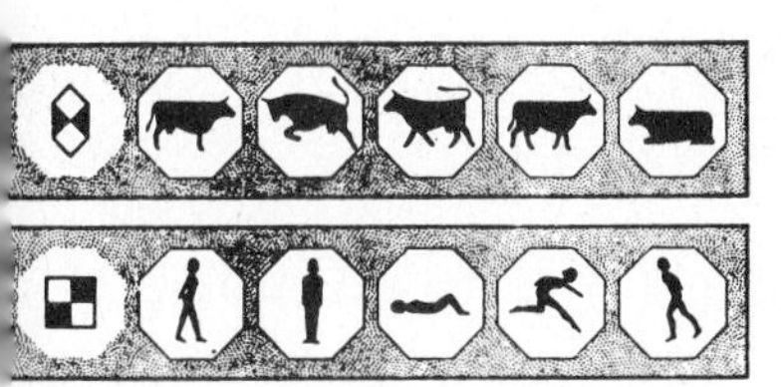
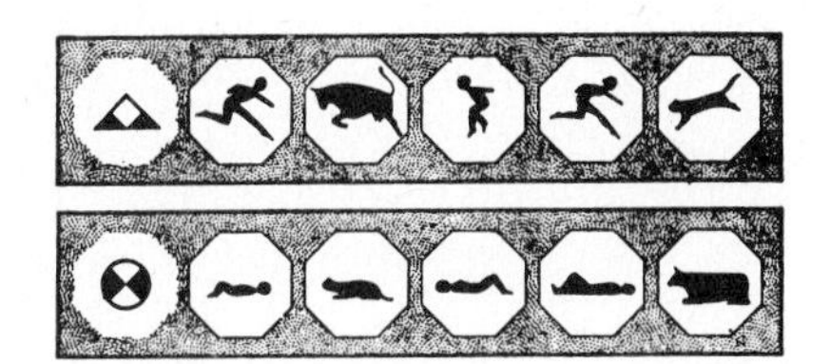

Non-verbal intelligence test, requiring recognition of shape and subject.

A frivolous French prophecy, dated c. *1900, of what teaching would be like in 100 years – in essence not far off the language laboratory of today.*

Below, *watching the screen of a teaching machine as the computer flashes the correct equation.*

Lord's Prayer will demonstrate how it works and how easily it is read: 'Patro nia, kiu estas en la ĉielo, sankta estu via nomo; venu regeco via; estu vola via, kiel en la ĉielo, tiel ankau sur la tero. Panon nian ĉiutagan donu al ni hodiau . . .' While a person familiar with some European languages will be able to read this, learning, writing and speaking it is another matter; but it survives and indeed thrives. Translations of the classics can be bought, and Esperanto has, in its early years, been used for scientific papers (the universal language for which is now English).

Ido was developed by Louis de Beaufront and Louis Couturat from 1900 on the foundations of Esperanto and its derivative, Idiom Neutral; it works in much the same way, with prefixes, suffixes and roots, and easily recognizable tenses. For example, the sentence, 'I believe that we are subject to the law of habit in consequence of the fact that we have bodies', would read: 'Me kredas, ke ni esas submisata a la lego di la kustumo per konsequo di la fakto, ke ni havas korpi.' Ido, in its turn, was revamped by the distinguished linguistic expert Otto Jesperson in 1928, the result being the New International Auxiliary Language, Novial.

C. M. B. G.

IQ tests

In 1905 Alfred Binet, a French psychologist working on a method of identifying mental retardation in schoolchildren, introduced the first scale or group of mental tests and therewith began a movement from which modern educational systems have yet to recover. Binet, along with his student and colleague Theodore Simon, was greatly influenced by the work of Sir Francis Galton, the British psychologist who since the 1890s had been devising ways for testing general mental abilities. The Binet-Simon Intelligence Scale specifically measured memory, reasoning ability and comprehension, and in 1916 was published in the United States by still another psychologist, the American Lewis M. Terman. The Binet system, which is wholly pragmatic in approach, completely dominated the field for at least 20 years. In 1939, this system was partly replaced by a series of tests designed by David Wechsler and known as the Wechsler-Bellevue Scale. Since then hundreds of psychologists and educators have collaborated to design as many different tests for nailing down the precise mental age of a schoolchild. But to this day no one has established an acceptable criterion for measuring the accuracy of the tests themselves.

The idea of an intelligence quotient, or IQ, was first suggested in 1912 by the German psychologist Wilhelm Stern, and once again it was Terman who popularized the procedure. The IQ is computed by dividing the mental age of a subject by his chronological age, then multiplying the result by 100 to eliminate the decimal. Mental ages are determined by a battery of standardized intelligence tests, whereby the average number of correct answers given by a certain age group is considered the norm for that age. If a child of 10 achieves a score that is judged normal for 12-year-olds, then his mental age is 12. The average IQ, which seems to remain relatively unchanged after the age of 14 or 16, is of course 100.

One quality of intelligence for which no intellectual testing method exists is insight, and yet it is this very quality which effectively guides the efforts of every productive genius, the Leonardos, the Darwins, the Einsteins: the men who for better or worse have created the collective mind of modern civilization. The faculty which most of us regard as intelligence may be nothing more than instinct looked at from another angle.

E. W.

Teaching machines

Programmed learning is a new technology, but it has a long history in research and a longer one in human thought going as far back as the theories of Socrates. Because machines are utilized most people flinch from what they view as dehumanized education, but in fact the difference between an electronic computer and counting pebbles placed in the sand is only one of degree – and a picture-book, a bucket of water or a barbed-wire fence are all teaching machines of a sort.

Sydney L. Pressey was responsible for the first teaching machine developed. Originally used in the U.S., where it is called programmed instruction, the machinery is known as 'hardware' and the programme, or actual instructional sequence, known as 'software'. Pressey was working at a teacher training college in the 1920s when, to lessen delays in the return of written work to the students, he had duplicated copies run off of the tests he had devised and supplied a paper answer slip placed between two pieces of punchboard. The top piece had a list of answers with corresponding holes in the punchboard, but only the correct answer had a matching hole in the hidden layer of board. The device was self-correcting in that only when the student's pencil point went right through both boards did he know that he had the correct answer.

Pressey's discoveries upheld the tenets of another American who lectured at a teachers' college of Columbia University. Edward L. Thorndike (1874–1949) stated in his 'law of effect' that learning accompanied by pleasure and satisfaction on the part of the student is more likely to be permanent than learning accompanied by dissatisfaction and frustration – and that in this case pleasure was the 'reward' which would reinforce the student's behaviour.

Having learnt that the method not only saved time but increased the rate of learning, Pressey progressed to a testing machine with correlated keys. If the wrong key was depressed nothing happened but if the right key was depressed the next question was automatically presented. A mechanical counter recorded the student's performance. Pressey was unable to market his invention because of the effects of the Depression at that time.

Dr B.F. Skinner, professor of psychology at Harvard University, is famous for his 'baby box' and his work on behavioural psychology, which involved extensive research on conditioning in animals following on from the work of the Russian psychologist, Ivan Pavlov (1849–1936).

Skinner's work in this field produced programmed learning as an accidental offshoot when, after an open-day visit to his daughter's school in the 1950s, he was distressed at the ineffectiveness of conventional teaching methods. Using his knowledge he developed a technique for presenting information to human learners. The 'linear programme' is an instructional sequence which takes each learner through the same steps gradually without deviation, and gives encouragement when the right reply is given, to reinforce the student's enthusiasm; this is known as 'operant conditioning'.

A different method of programming was devised by Norman A. Crowder to train technicians in the US Air Force to identify faults in electronic equipment, and has since been developed much further. It is a teaching sequence which allows the learner to choose an answer, even if this is a mistake, and routes him to information according to the choice he has made. Until a few years ago programmed learning was divided into two bitterly opposed camps, the 'Skinnerians' and the 'Crowderians', but now both methods are quite commonly employed together in learning programmes.

Computer

There are some mathematical calculations that actually cannot be done, but there are many that no human can do quickly enough, sometimes simply because no one lives long enough. A simple calculating machine, derived from the type that Pascal first made, speeds up individual calculations, but is extremely limited because humans have to intervene at every stage in a long calculation.

The visionary who thought of a way to solve this limitation was an Englishman, Charles Babbage, who persuaded the British government in 1823 to finance an 'analytical engine'. This would have been a machine that could undertake any kind of calculation. It would have been driven by steam, but the most important innovation was that the entire program, as it is now spelled, of operations was stored on a punched tape, a method that is still used. Babbage derived this idea from a punched-card system that Jacquard used to direct looms in making complex woven patterns. It was later taken up by Herman Hollerith, who found that he was barely able to work out one decennial US census before starting on the next. He halved the time by using punched cards to store the information.

Babbage's machine was not completed and would not have worked if it had been. The standards required were far beyond the capabilities of the engineers of the time, and in any case rods, levers and cogs move too slowly for really quick calculations. Only electrons, which travel near the speed of light, are rapid enough.

The first electronic computer was called ENIAC, and was completed by J.P. Eckert and J.W. Mauchly by 1946. It was an immense, relatively primitive device, using 18,000 radio valves. These demanded an enormous amount of electricity and in turn generated so much heat that cooling was a major problem. In addition, valves are not very reliable and the machine would not work continuously for long. The modern computer, which will characterize the new industrial age as the steam engine did the old, became possible only after the invention of the transistor.

T.O.

IBM System/360 computer. The data on which the computer, already programmed, is to work are being fed in by means of a punched card. The holes are translated into pulses of electric current.

Electric tabulator and sorter devised by Herman Hollerith to enable him to keep pace with the US decennial census.

KEY DEVICES

KEY DEVICES

Key devices are inventions or developments which embody some principle so basic that it enables man to do a number of different things. Rope is an example of such a key device. It goes back to the beginning of man's history and the first ropes were probably strands from creepers or vines. Later the idea of twisting or plaiting strands developed and to this day that remains the basic technique. With a rope one can do a great number of things. It becomes possible to lash objects together – before nails, screws or adhesive, the only method of joining. It was used for attaching an axe head to its shaft and for attaching the timbers of a boat to the frame (using strips of leather). Poles could be lashed together to build houses or traps for animals. Rope (or string) also enabled man to make fishing nets, or fish with a line and hook. It also allowed him to use captive harpoons. Rope could be used for suspending things or for holding up tent poles or ships' masts. Finally rope could be used as a way of transmitting force: in pulling rocks or trees along or in lifting. In these situations a rope provided a more convenient way to apply the force, made it possible to apply the force over a distance, and allowed several people to combine forces. Later on, with pulleys and winches, rope allowed man to change the direction of force and also to amplify it.

Adhesives were another way of putting things together. The first adhesives – some of which continue in use to this day – were derived from insects or from the skin or bones of animals. What is interesting is that some of the more modern adhesives, such as the epoxy resins, were developed when the need for adhesives would seem to have disappeared, since there were so many other methods of attachment available (screws, nails, rivets, soldering, welding etc.). War technology and, in particular, aircraft construction provided at least some of the impetus here. The more modern adhesives can be used for metals as well as wood.

The screw is another method of attachment. In itself it is an extraordinary device. Being a combination of lever, wedge and wheel, it combines three of the most basic key devices. A wedge is wrapped around an axle so that rotary (wheel) motion is used. Finally the screwdriver uses leverage to exert considerable pressure on the 'wedge' as it is driven into the wood. The fascinating thing is that the screwdriver was actually invented before the screw. It was used to force in or extract slightly twisted nails. The screw proper was developed only towards the end of the 18th century and given a point halfway through the 19th.

Solder as a method for joining metal was in use in 3000 BC. It seems likely that, as with so much of metal technology, a combination of experimentation and chance produced metal alloys with different melting temperatures which could function as solders. Riveting, soldering and, later, welding provided the means for attaching metals until the bolt was developed.

The rope, screw and solder are devices for attaching things together in a physical way, but there was another way in which man had to join things. This involved putting together a source of power and what it was to achieve. As we have seen, ropes also serve this function: they connect the man pulling with what is to be pulled. Oxen walking round a circular track have to be connected wit vertical well-wheel for raising water. This was done by gears wh translated horizontal motion into vertical motion. It is proba that from these gears developed all other gears, including th which increased speed or power instead of just changing the dir tion of motion. Archimedes himself may have invented the wo gear. The cam, which can be regarded as a gear with a single too was developed to allow a rotating shaft to activate trip hamm It acted to raise the hammer, which was then released and fell gravity. In this way the rotating motion of the shaft was transla into up-and-down motion. The crank was developed to perfo the same function, but by this means, motion could be two-w instead of relying on gravity for the return stroke. The crank probably first developed in China to use water-wheel power operate a bellows.

The universal joint was designed to allow two shafts at an an to each other to be coupled together. Its invention is a curious st of how an inventive individual faced with a minor problem can co up with a brilliant invention. The inventive individual was very fertile Robert Hooke; the problem was connected with astronomical device. He invented the universal joint and thou no more about it. Today, when every motor-car depends on universal joint, we also think no more about it but take for gran what is really a brilliant invention.

Motion can be transmitted by shafts, by gears, by belts and chains. Chains had been used for decoration for a very long ti but their use as a sort of super-strong rope had to await the devel ment of wrought iron in the 19th century. The use of a chain driving purposes had to await the invention of the 'bicycle' chain James Slater in 1864 – the invention that allowed the cumberso penny-farthing bicycle to develop into the bicycle we know toda

The next group of key devices are the ones that actually allow m to amplify effects. The importance of this group cannot be ov estimated. A very simple amplifying device is the wheel. It ampli not because it actually increases force as such but because it increa the effectiveness of a given traction force. With the wheel (and w roads) the same amount of traction could shift much larger loa Much later, the invention of ball-bearings again increased effecti ness by reducing friction.

The most obvious amplifying devices are the wedge, lever, ge and winch. In each of these a small amount of force applied ove longer distance leads to a much greater force being produced ove smaller distance. With the wedge, considerable vertical movem forces the surrounding wall apart a tiny amount but at much grea force. One advantage of the wedge is that the movement obtain is not reversible and so the effect can be cumulative. With the lev considerable movement of the longer limb gives a smaller movem but increased force of the shorter limb. With the gear a sma diameter wheel with fewer teeth driving a larger wheel with ma more teeth increases the force of rotation but diminishes the spe A winch is a sort of rotary lever which – with the rope, as we ha seen, to translate its force into useful effect – can act continua change a large, weak rotation into a smaller, more powerful one. e take for granted this exchange of distance for force and yet it is emarkably useful effect.

Storage is just another type of amplification. For instance, ccessive small bucketfuls poured into a large container can eventu-y produce a considerable weight. This could be regarded as the plification of the weight which one man could carry. The same rage effect can be seen with a spring. As in a crossbow, a spring n be wound up quite slowly but when released delivers all its power once. Springs have many other uses as well. They can operate ps (and this may have been their first use), they can absorb pact as in a carriage spring, or they can store energy and release slowly as a clock spring does. The history of the development of ings shows a variety of forms: simple elastic material obtained m animals, torsion springs as used in the Roman catapults, leaf ings, coil springs and then back to torsion bars. We take the ingginess of metal for granted, yet in 200 BC this property was noted the first time and regarded as very odd – since metals were used their hardness and rigidity. Even today we should find it difficult think of glass springs, and yet they are quite feasible.

A totally different sort of amplification was needed to make the rvels of radio communication and electronics possible. A radio ve, like light, attenuates with the square of the distance it has velled. It is easy to see that the power of the wave must become ry weak. Powerful devices are needed to amplify the received nal enough to operate a speaker. Even with telegraphy some sort amplifier is useful but not so essential: here the amplifying process a gate system. In other words, the current to be amplified simply ntrols the gate through which the energy is passing. The fluctua-ns in the tiny control current are transmitted to the energy itself d so an amplification is brought about. This process was put into ect in the triode valve, on which electronics was built. This was cumbersome device, recently replaced by the transistor, which n be made incredibly small. Smallness in itself could be regarded a negative amplification of space.

Another group of key devices includes control devices. The story Galileo and the swinging cathedral lamp in Pisa is well known. om it came the notion of a pendulum as a regulating device for clocks. Christiaan Huygens, the brilliant Dutch inventor, seems to have had a similar idea at the same time and to have put it into practical effect somewhat before Galileo. Curiously, the use of a spring as a regulating device for clocks poses exactly the same question about who thought of the idea first, and in this case the two people involved were the same Christiaan Huygens, and Robert Hooke. James Watt is well known for the invention of a governor that would control his steam engine, but it seems likely that the governor idea, in almost exactly the same form, had been in use to regulate the distance apart of millstones when a high wind made the windmill arms themselves rotate faster. Nevertheless, credit is due to Watt for applying the device.

One of the most basic control devices is the valve which controls the flow of gases or liquids in piston systems. Without the valve most of our steam and petrol engines could not exist. The valve itself started as a very simple flap that was used on bellows to allow air to enter the system and then to ensure that it did not leave by the same route but was forced through the nozzle into the fire. Valves activated by the actual flow of the liquid were later developed into valves that were activated separately by cams and other devices. Without valves only rotary engines (like turbines) and centrifugal pumps would be possible.

The gyroscope is as much a reference point as a control device. In fact its first suggested practical application was as an artificial horizon for ships; today it serves that same function in aircraft. The extraordinary thing about the gyroscope is that, alone of all devices, it steps outside man's world. A gyroscope provides a stable reference point that is outside the oscillations not only of the local environment but of the world itself. This is a property unique to the gyroscope. An interesting insight into the process of invention is given by the story of how giant gyroscopes were proposed, to stabilize ships and to prevent them rolling. This idea was superseded by the much more practical one of having quite small gyroscopes as reference systems that could activate underwater plates projecting from each side of the ship and acting to prevent roll. The concept of control as opposed to direct action is well illustrated in this example. Most of the key devices being developed today are concerned with system control rather than direct action.

Key Devices

Knife

One-edged cutting and scraping tools were among the first artefacts produced by man, and they are surely directly descended from the hand-axe. Excavations at the Olduvai Gorge in Tanzania have revealed progressively more comely hand-axes, dating from about $2\frac{3}{4}$ million years ago, in the successive layers; crude flaked stones had evolved, by the Upper Palaeolithic period (*c.* 25,000 BC) into a much more specialized range of tools – chisels, needles, knife-scrapers – made with punch and hammer to form fine blades by edge-trimming, or pressure flaking. 'Backed blades', with a blunt side to take a haft, and one cutting edge, date from this period. By the Neolithic period (*c.* 6000 BC), knives, chisels and scrapers formed an important part of everyday life. Knives and axes of flint were not only often very beautiful, but also decidedly effective: experiments in Denmark using genuine stone axes, not sharpened in modern times, showed that four men could clear 600 sq. yds. of silver birch forest in 4 hours. The backed blades mentioned above were sometimes used in sets by Neolithic Man to produce quite large tools; reaping-knives, of wood or antler-bone, have small knife-edges inserted in grooves.

When metals came to be used, the shapes cast or beaten followed those already developed in stone. Copper, bronze and stone succeed each other, with overlaps, but, more than this, we find metal weapons being recopied in stone by more primitive peoples: in Scandinavia for instance, where the Stone Age continued while the Mediterranean peoples were using bronze (*c.* 1800–1500 BC), flint knives with handles and butts have been found.

Metal blades were, however, a great improvement; they were made of copper from *c.* 6000 BC in parts of the Near East. Whereas stone blades fractured easily, copper, as long as the edge was hardened by cold-hammering and the whole blade given a hefty bulk, did not prove too soft. Hilts would be made separately, and joined to the tang with rivets.

Bronze was an even greater advance, for it was easier to cast in complicated patterns, and much harder – hence less of it was needed in each blade, which acquired a longer and more slender shape. Iron – better still – was for a long time much too expensive for anything except luxury goods, but the Hallstatt culture of the Alpine region was producing iron-bladed knives with bronze hilts from the 7th century BC, and trading with them.

Knives were used for warlike as well as for peaceful purposes; a modern development is the bayonet (included here because it is indiscriminately one- or two-edged). This weapon dates from the early 17th century, when it was realized that a dagger rammed into the muzzle of a musketeer's gun gave him a second chance to defend himself after he had fired and before he had a chance to reload: such 'plug-bayonets' had given way to ring-fixings by the end of that century. But the most fearsome war-knife ever made was surely the Bowie, first used by the American frontiersman of that name (1795–1836); this had a blade up to 2 in. wide, and anything from 9 in. to 15 in. in length.

C. M. B. G.

Lever

'Give me a fixed point outside the Earth to stand on and I will move it', said Archimedes, the greatest of Greek physicists and mathematicians. He lived in the 3rd century BC and was among the first educated Greeks to take an interest in technical problems; as a rule, such 'vulgar' occupations were left to the working class of antiquity – the slaves.

What did Archimedes mean? Like every true scientist, he was concerned to examine things which had always been taken for granted, investigating the principles and natural laws involved. Thus he studied the mechanical laws underlying the action of one of the oldest and apparently simplest devices, the lever. For many ages, men had used a long stick or the strong branch of a tree to shift heavy weights: a stone or another piece of wood was put under it at the most effective point (later called the 'fulcrum'); one 'arm', the one that was put under the weight, had to be much shorter than the other one, which the workers pulled or pressed down.

Archimedes calculated the forces involved. If, for instance, the two 'arms' have a proportion of 1:5 in length, the workers will need only one-fifth of the strength they would require for shifting the weight directly, without a lever.

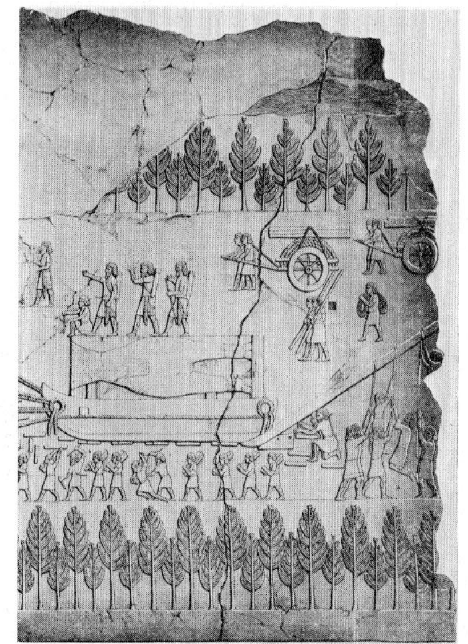

Relief representing the construction of Sennacherib's palace at Nineveh: a statue of a winged human-headed bull is being levered up a slope.

Right, *pre-dynastic flint knife from Gebel-el-Arak, Egypt, with ivory handle.*

Below, *invention of the wheelbarrow (from the Luttrell Psalter,* c. *1340) meant that one man could do the work of two with a stretcher. In China by the 3rd century AD, in Europe by the 13th.*

Below, right, *this crude representation from Uruk puts the invention of the wheel as far back as 3200–3100 BC.*

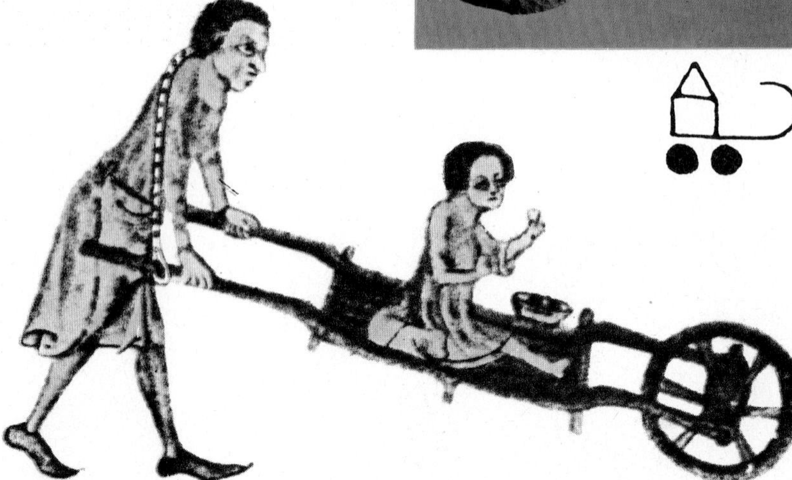

Use of the wedge as a splitting device is illustrated in an anonymous portrait of Abraham Lincoln indulging in the American sport of competitive log-splitting. This was probably painted in 1858, during the Lincoln-Douglas debates. (Courtesy, Chicago Historical Society).

Below, *spoked wheels meant lighter work for Assyrian slave gangs.*

However, what you gain in power, you lose in the time or distance over which the force has to be applied. For example, in order to lift the weight 1 ft., you have to press the long lever arm down 5 ft.; thus you exert the same amount of force which you would have to apply if you moved the weight directly, but the lever translates your action into a more practical form – it makes the job easier. Archimedes put that principle into simple terms: the longer the arm to which force is applied, the less that force need be. His often-quoted dictum about the earth and the fixed point, however, does not mention the lever itself. It would have to be so long that its working end would stick right out of our own galaxy into outer space. Nor does he take into account another force which helps the worker if he stays on earth – the force of gravity.

In modern practice, several kinds of lever have been developed. The oldest form, where the fulcrum is between the power and the weight, is still the most widely used; the crowbar belongs to this class. In another class, the weight is in the middle, as with the wheelbarrow or the nutcracker. In a third, the power is exerted in the middle as with a pair of tongs or the treadle of a lathe. All these devices, and many more, act on the lever principle, though often in a disguised form.

E. L.

Wedge

Most ancient peoples were probably expert in the use of the wedge. Since primeval times wedges have been used to move, raise or transport loads. The mechanical leverage a wedge affords offers many advantages over the lever, which requires the application of physical force, and a pair of wedges acting together can be made to exert a considerable force (as it often is on building sites today).

The caveman used the wedge because it worked. Early civilizations grasped the principle behind its function, however, and began adapting these forces. Nails, screws, axes, chisels, planes and a host of other tools are all, at bottom, adapted wedges.

Ingenious operators of the wedging principle were the American Indians, who would fill deep, wedge-shaped cracks in the rocks with water in early winter and later collect the great slabs of rock which had been forced apart by the expansion of the water when it turned into ice in the cold. The Greeks showed the same sort of imagination when they forced wedges of cork into holes drilled in marble slabs and then soaked them with water, causing them to expand and split the marble into manageable blocks.

Today, of course, this humble device is still going strong in a wide range of applications. Some of these are not at first sight obvious developments of the simple wedge, but the glass optical wedge used as a filter in photography, the cam which plays so vital a part in almost every engine, and many other key devices can arguably be said to derive from the piece of stone placed by the cavemen under an otherwise immovable boulder.

S. M.

Wheel

It seems that we should not think of the deliberate invention of 'the wheel' as an abstract concept applied to all forms of rotary motion. In antiquity much of this type of motion was utilized horizontally, as in spindles and drills (especially bow-drills) or potters' turntables and wheels, the drill being developed independently in the Old and New Worlds, while in the Old World a knowledge of wheeled vehicles does not equate with the use of the potters' wheel, or vice versa, nor do either relate to the use of spindles with horizontally rotating whorls for spinning. The vertical wheel as applied to transport is basically a device for minimizing the friction between a vehicle and the ground over which it moves, in the manner of rollers under a sledge (but not demonstrably or necessarily derived from such usage). Wheeled transport was never invented in the New World, and in the Old the idea need not have a single point of origin; archaeological evidence at the moment suggests that earlier claims for the priority of the early civilizations of Mesopotamia in the use of wheeled transport is very uncertain.

All the earliest wheeled vehicles have solid block wheels made of one or more planks split longitudinally from the tree stem: a cross-section of a tree trunk is useless, since it will split along the medullary rays which run at right angles to the concentric rings; it is also more difficult to cut from a trunk without adequate saws, unknown until later antiquity. The copper or bronze tools available to the first makers of carts and wagons were furthermore not suitable for the carpentry needed for spoked wheels.

But any wheeled transport needs first a social context in which it can be accepted as a valuable increment to existing technology and, in addition, natural circumstances of terrain where alternative means of heavy transport, such as river boats, are not feasible, and an economy in which vehicles can be used to advantage. To these basic requirements must be added adequate timber resources and efficient woodworking tools, as well as domestic animals with the necessary tractive force and staying power. These were in the first place domestic and castrated cattle used in pairs with a central draught pole and yoke. The earliest domestic horses were small but swift, and seem to have been first used in transport when the spoked-wheel war chariot was developed, probably in north Syria, early in the 2nd millennium BC. Traction between shafts is a post-classical development in Europe, apparently with Chinese origins, associated with the rigid horse-collar.

With the recent revisions in radiocarbon dating, it is difficult to prove that the earliest evidence for wheeled transport was in Mesopotamia (in the Uruk IV phase), before that from Caucasia and east and central Europe, around 3500 BC; thereafter in both areas the use of such transport continued throughout antiquity, and from 2500 BC in northern Europe. In Egypt the use of the Nile as a waterway meant that wheeled vehicles were not used until the 17th century BC, at which relatively late date the chariot was introduced from outside Egypt, together with the horse. In China, our first evidence of wheeled transport, two or three hundred years later in the Shang Dynasty, is again that of chariots and horses. In the New World wheeled vehicles were never developed (though wheeled toys were made occasionally), and here the lack of suitable domesticated mammals for traction must have been one of the reasons.

S.P.

Net

The use of the net goes back almost to the dawn of civilization. It is possible that coastal and river folk, near the present-day Riviera, used nets in Upper Palaeolithic times (*c.* 11,000 BC). Certainly later on in Mesolithic times the Maglemosian peoples of northern Europe used nets made out of 'bast' fibres from the cellulose cells of certain plants (*c.* 7500 BC). The remains of an early willow bast net, weighted by stones and supported by wooden bark floats, have been found at Antrea in Finland. Further evidence comes from cave-paintings in Spain, and from tomb-paintings of the Old Kingdom era in Egypt (*c.* 2750 BC). Almost all early fishing nets were simply knotted, though some were made which had up to three twists to each loop.

Nets were not only used for fishing purposes. Caps or hair nets, mounted with shells, have been found in Upper Palaeolithic burial sites, and woollen hair nets were in use during the Bronze Age in Norway (*c.* 2000 BC). The special net plait used, known as *sprang* in Norwegian, still survives in parts of Norway, Galicia and Croatia. Various raw materials, apart from the usual hemp and sisal fibres, have been utilized. The ancient Egyptians used flax, date, palm and papyrus fibres and even camel hair. Esparto grass was probably used in prehistoric Spain, and in Persia a fine silk netting, with decorated ends of gold and silver, was produced.

The making of nets by hand cannot have altered much through the ages. Netting consists of loops of thread ('stitches') secured, usually, by knots. These stitches are made by means of needles or gauges of bone, wood, and now steel and the ubiquitous plastic. In the early Middle Ages the Corders of the Ropery Guild were important manufacturers of ropes and netting. This has always been a handicraft industry carried out by individual outworkers or fisherfolk. Nets produced have varied from billiard-table nets to different types of fishing net. Today man-made synthetic fibres have largely taken over from the traditional bast fibres. There are, however, a few flourishing net-making centres left in the Western world, and Bridport in Dorset is the centre of a net-manufacturing industry whose products are sold throughout the world.

M.H.

Rope

Long before the advent of large factories and sophisticated machinery, what is today a mundane production operation was a highly prized skill. Such was the case in 4th-century India, where the rope-making trade was so specialized that separate classes of experts were engaged in producing rope for different purposes, one for elephants, another for horses and so on. And, unlike the modern assembly-line worker, each craftsman of that time was thoroughly familiar with all phases of the rope-making process, from cultivating, harvesting and drying the hemp plant (*Cannabis sativa*), to spinning the yarn, forming the strands and finally laying these into stout, strong rope.

Yet this slow and careful approach to rope-making would have seemed streamlined to the earliest rope-makers. They lived more than 4,000 years before that, in an age when fibres were gathered from bark and a sort of yarn was made by rolling the fibres together with the palm of the hand against the bare thigh. Archaeological studies indicate that our early ancestors first used cord and rope to attach handles to their tools and weapons. By 2800 BC at the latest, hemp was being used for fibre in China, and with the development of shipping in the eastern Mediterranean half-way through the 2nd millennium BC, heavy rope came into great demand. In both Europe and America, soft fibres such as hemp and flax were still used for most rope until about 1850, and even today the cannabis plant supplies us with the basic material for tarred hemp rope as well as certain fittings for marine purposes, although the strongest natural fibres are from the abaca plant of the Philippine Islands (misnamed Manila hemp, the abaca is actually a kind of banana plant), the agave of Yucatan and Cuba, and various

Use of a net for catching birds. From the 15th-century BC tomb of Nakht, Thebes.

Twisting rope by hand: picture from a German manuscript of 1399.

sisals from East Africa, Indonesia and Haiti.

As with many crafts, the technical aspects of rope-making never really changed until the onset of the Industrial Revolution, and less than 250 years ago yarn for rope was still being spun by hand. The next major leap in the evolution of this important trade came during and after the Second World War with the introduction of synthetic fibres such as nylon, Dacron, saran, polyethylene and polypropylene. The utilitarian advantages of these man-made materials are manifold: greater strength, increased immunity to deterioration from mildew, rot and abrasion (thereby eliminating the need for special protective treatment), and both economy of storage and ease of handling through the use of smaller and lighter ropes. The lifetime of synthetics is rated as three to four times that of natural fibres.

If, in the civilization of the Phoenicians, rope was most at home among seafarers, so it is today; and more than half of all the rope produced in the world is used by the marine and fishing industries. As for the rest, it is everywhere around us, from the common household clothes-line to the heavy-duty cable-laid ropes used in mammoth drilling operations.

E. W.

Knots

Primitive man twisted plant fibres and sinews into rope long before he learnt to spin or weave, and knots were obviously an inevitable development from this, for a single strand of rope was largely useless until it had been formed into a loop, attached to something or joined to another rope. Very early knots were used to secure the vines that held stone axe-heads to their handles, for tying housing and roofing posts and for lashing log rafts together, but the first systematically repeated knot pattern known to us was made by the Neolithic lake-dwellers of Switzerland and Denmark for the formation of their fishing-nets.

A good knot depends on friction for its success – the more a load is applied the more friction is increased and the greater the knot's holding power grows. This principle must have been discovered very early in man's history, and, so far as can be seen, knot-tying was almost a universal accomplishment in the prehistoric world. The pre-Columbian Indians in South America made crude vine-rope bridges which are the prototypes of the modern suspension bridge and which depended entirely on knots. Knotted rope was used in the construction of the Tower of Babel and the Pyramids in Egypt, where the state even employed professional rope-knotters whose job it was to tie equally spaced knots in the long ropes used for measuring. Greek sailing-ships were using knotted ropes in their rigging by the 6th century BC at the latest, and Herodotus tells how the Persian King Xerxes crossed the Hellespont in 480 BC by constructing a floating bridge of boats lashed together, over which his invading army swarmed into Greece. But knots also served as a mental aid from a very early date. Pizarro found the Incas keeping elaborate accounts by means of systematically knotted strips called *quipu* which functioned as memory aids, and the Polynesians, early sea migrants, used knots to cross-hatch an open-weave papyrus mat and form a navigational map.

The connection of the art of knots with the sea has persisted for centuries. During the great age of navigation knotting and splicing became an indispensable art and it is to this period that we owe many of the variants on the simple knot adapted for special purposes. The romance of their names rolls off the tongue: Double Blackwall and Stunner Hitch, Cow Hitch, Lark's Head and Cat's Paw, three-part Turk's Head. Other knots, too, were developed, whose use extended beyond the maritime world: the Weaver's Knot was suitable for uniting ropes of different size, such as those found on a loom, while the Thief's Knot would unravel under strain, allowing its load to go free.

The advent of steamships somewhat diminished the importance of knots, since standing rope rigging was replaced by metal cable and cable can be spliced (woven together) but not knotted. Nevertheless knots are still indispensable today in a wide variety of fields, and while on the one hand some of the earliest knots known to man, such as the Reef Knot, continue in robust health, new knots, such as the Electrician's Knot, continue to be invented.

S. M.

An engraved gem (Phoenician, from Tharros in Sardinia), depicting a carpenter shaping an anchor, also includes the anchor chain as a decorative border.

Right, *Isambard Kingdom Brunel, posed against the massive anchor chain of the* Great Eastern.

Chain

Oddly enough the manufacture of chain, a strong flexible connector of metal links, did not develop on any scale until the early 19th century, although the use of metal chain is attested in ancient times. Small chains made out of gold and silver, and used primarily for decoration, were made in classical times. The Romans used bronze chains in their galleys, and link chains were recorded as being used on water-raising machines in the Hellenistic world about 200 BC. Although the 'democratic metal', iron, was increasingly used from about 1100 BC onwards it was not at first used for making chains. Ironworking remained rather primitive. According to Agricola's *De Re Metallica* iron chains and buckets were being used in 16th-century pumping machinery instead of ropes and earthenware pots. In 1634 a blacksmith named Philip White patented an iron mooring chain, but the industry did not develop greatly until wrought iron of good quality became increasingly available in the 19th century.

The problem of iron's tendency to stretch under strain was overcome in the early 19th century by Thomas Bunton, who invented the two-piece link with a central stud, thus improving the strength of each individual chain link. Round about the same time, in 1808, Captain Sir Samuel Brown, RN, patented a design for an improved iron chain. In 1820 a village blacksmith named Noah Hingley forged the first ship's cable, using simply a forge, hammer and anvil. There eventually grew up, in England in particular, a thriving industry in chain-making. The country was rich in iron ores, and good-quality Lancashire wrought iron was always sought after, being preferred to steel for its anti-corrosive properties. When, in 1864, James Slater patented the driving chain (foreseen by Leonardo da Vinci), the idea of the modern precision machine-made chain, as used on bicycles and in industrial machines, was born.

Chains are made by forge welding, that is, by softening the metal and striking it into shape. Larger chains can be strengthened by metal studs, inserted into the links while they are still hot. Chain-making by hand has almost died out by now, although a few years ago many blacksmiths in the Midlands in England were specialist chain-makers. Exceedingly strong men were required in the chain-making shops of the Black Country, and often for manufacturing heavy chains of over 1½ in. in diameter a whole team of strikers was required to hammer the chain into shape. Workers were able only to work in shifts of four hours at a time. Nowadays the finished chain is tested for strength by a hydraulic apparatus on a chain-testing bed. Some small chains of precious and semi-precious metals are still handmade by craftsmen today, however, for example in the manufacture of jewellery.

M. H.

Adhesive

Natural adhesives have been in use since prehistoric times, when beeswax, resin, rubber and shellac (produced by parasitic insects which live on trees) were all probably known. In ancient Egypt glue for cabinet-making was made from the skin, bones and sinews of animals.

Vegetable adhesives may be made from the gum that exudes from certain trees and plants, or from rubber, or from starch obtained from potatoes, rice or wheat. But other less orthodox substances also act adhesively: Mrs Beeton suggested the use of treacle on paper to be used in the catching of flies – they simply stuck to it and could not escape, an ingenious method in the days before chemical insecticides were commonplace.

The chemical adhesives which appeared with the development of synthetic materials in many fields in the mid-1930s widened the horizon considerably. Until then adhesives had been used on a large scale mainly in the woodworking and paper industries. Now not only porous substances such as paper and wood could be bonded, but also glass, metals and plastics.

Although the natural adhesives are still used, notably in the woodworking industry, synthetic adhesives have largely taken over. Familiar types in the home include pressure-sensitive adhesives, as used on the transparent 'sticky tape' which has so many uses, and emulsion adhesives which are used for sticking rubber soles on shoes and for self-sticking envelopes.

In industry these substances are now extensively employed. They are often used in place of rivets and bolts, as their even strength over the whole bonded surface gives rise to less strain than when stress falls on one part of the agent used. In aeroplanes, where sometimes the outer covering is bonded by adhesive to the main framework, streamlining is facilitated and costs reduced. Motor bodies and boat hulls are commonly built of glass fibres bound together by a synthetic adhesive, forming a product with the important features of lightness and impermeability. From the most sophisticated of synthetic varieties down to the child's flour-and-water paste, few substances are employed as widely in man's handiwork as the adhesive.

S. F.

Solder

There are four main ways of joining together sheets of metal: by bolts or by rivets, by soldering and by welding. Bolts, depending as they do on the screw-thread, only became current in post-Renaissance times. Welding, the fusing together of metal by heat alone and without the use of solder, required higher temperatures than was the case with solder; it is really the child of the Iron Age, and an early example is the welded iron head-rest found in Tutankhamun's tomb (*c.* 1350 BC), which was probably a gift from Syria.

The two methods of joining most used in antiquity were rivets and pins, and soldering; the former is by far the older, and the rivets were not considered unsightly, but often integrated into the design. The important feature of any solder is that it will melt at a lower temperature than the pieces of metal it has to join. Probably the first solders were not consciously sought, but were

Cauldrons for boiling gelatinous glue – a primitive and malodorous trade – from an illustration of 1866.

A rag-and-chain pump, illustrated by Agricola (De Re Metallica, *1556*), *employed a primitive form of the crown wheel and pinion.*

Left, *a torc, or collar, made of electrum, a gold and silver alloy – a fine example of Iron Age craftsmanship from the 1st century BC. By this time solder would have been part of the craftsman's equipment.*

Gears would have made easier the turning of this outsize well-handle. From Agostino Ramelli's Machine, *1588.*

This design by Thomas Sheraton (1793) for a 'chamber horse' shows in detail one of the first examples of springs in upholstery.

the result of chance observation that gold from different sources (some of it naturally alloyed with silver or copper) melted at different temperatures. Experimentation would then show how the melting points of solders, as well as their hardness, could be changed by varying the proportions of metals in them (common ones in antiquity were gold/silver, gold/copper and silver/copper). Great care was necessary in the choice of solder, otherwise the work might be damaged by too much heat. Soft-soldering, at a temperature lower than that required for gold and silver work, was established in the Near East by 3000 BC: there is a copper panel of Im-Dugud, from al-'Ubaid, now in the British Museum (before 3000 BC), in which the stags' copper antlers are pieced together with soft solder. Hard soldering may be of about the same date – an authority has affirmed that by 2500 BC, at Ur, the soldering of gold and silver was well known. Brazing – the word betrays its connection with copper and brass – is much the same as soldering, but the joining of these metals with hard solder scarcely antedates 1000 BC.

Benvenuto Cellini, in his *Treatise on Goldsmithing* (first published in 1568), gives a characteristically bombastic account of making a silver statue for the French king: 'Arms, legs and body I hammered out in separate pieces, and the head in one whole piece as if it were a vase. . . . I soldered and fitted them together. The solder I used was *ottavo*, that is a solder composed of one eighth part of an ounce of copper to one of silver. To do the soldering I had fixed to the tube of my big bellows several channels of such length as I deemed necessary for the purpose of blowing from below on the beds of coal that I had placed under the back of my work. . . . I blew the bellows and gradually made the solder run, and I kept on with this, now applying it from above, now from below. . . . Then came the job of soldering the big pieces together, and that was where those great French experts failed. . . . Cellini solved it, of course – by first joining the parts with silver wire, then controlling the temperature of the solder with ash.' (Translation by C.R. Ashbee.)

C.M.B.G.

Gear

Gears first appeared as an essential component of the chain of pots, often called by its Arab name *sakieh*, used to draw water from a well; the wheel that bears them has to turn about a horizontal axle – all very well if men are to work it, but what if you want to make use of four-footed beasts? What genius saw that one wheel can turn another at right angles to it if you fix a series of pegs perpendicular to the wheel's plane round the rim?

The trail leads back to the Egypt of the Ptolemies, where it peters out somewhere in the 3rd–2nd centuries BC. This face wheel or crown wheel usually drove a 'lantern', the crudest form of pinion, just two discs joined by a cage of staves or bars, the crown wheel teeth meshing between them very roughly. The *sakieh* with its gears spread around the Old World, and in a great many communities was for centuries the only pair of gears in the village. But by the 1st century BC someone had seen how the gears could be used to combine two recent inventions, the horizontal watermill and the vertical water-lifting wheel; Vitruvius gives us the first account of gears in his description of a watermill with a vertical waterwheel, face wheel and lantern drive to the millstones. This type of mill spread more slowly than the horizontal variety but by the early Middle Ages, at least in the lowlands of Europe and some favoured parts of the Moslem world, it became the normal way to grind flour.

The worm-gear – or perpetual screw as medieval and Renaissance engineers preferred to call it – was possibly the invention of Archimedes himself and was employed in a few instruments, but only men of exceptional skill, like Heron of Alexandria, would trust themselves to make one properly. In these small scientific instruments, gears would be of metal; a unique set of brass gears has survived from Greco-Roman times, in a little computing device fished out of the sea from a wreck off Anticythera. Metal gears continued in use for some astronomical instruments, but only became common with the appearance of mechanical clocks. It was the clockmakers who first sought to make them more efficient, and created gears as opposed to wheels with attached teeth. Leonardo da Vinci spent a good deal of time working out gear ratios and ideal tooth shapes and a 16th-century successor, Turriano, is said to have made a wheel-cutting machine to form the gears for the great planetarium he built for the Emperor Charles V. But both the theory, and the machine tool for large-scale production of precisely cut teeth, had to wait for another 100 years: the wheel-cutter first came into general use in Restoration England, the mathematical analysis in the work of Roemer and La Hire in France a little later.

A.K.

Spring

Most materials are to differing degrees elastic, and if bent out of place will return to their original conformation with considerable force. It must have been noticed very early in human history that saplings and the branches of young trees are highly flexible, for many primitive cultures made traps which exploited this property by wedging a stick behind some kind of door or cage, or else by pulling down a pole attached to a noose; once the tensing element was released, the stick or pole sprang back. Indeed bows are springs which use the elasticity of young wood in the same way; the bow is pulled back and then released. In the Middle Ages, the same idea began to appear in machines, in looms, lathes, drills, powder mills, even with saws; the operator provided the downstroke by hand or by foot with a treadle, pulling down the working tool, and a pole to which it was attached by a cord then sprang up to give the return movement.

Elastic materials will resist twisting no less than bending: in place of simple springs, torsion springs, invented probably in the 4th century BC and tightened with ropes of twisted sinew or hair, added enormously to the power of catapults and mechanical crossbows under the Hellenistic empires. At this point it came to be realized that metals have even greater elastic capabilities than wood or horn or any of these organic substances. Philo (writing about 200 BC) treats this as a new discovery which he expects his readers will find hard to believe. Impressed by the resilience of Celtic and Spanish swords, some of his predecessors at Alexandria had carried out experiments to find out the reason for it. As a result his master Ctesibius had invented projectile engines that used springs composed of curved bronze plates – in effect, the first leaf springs; these were further improved by Philo himself. The ingenious Ctesibius, while he was about it, also thought up an engine that would use the elasticity of air under compression in a cylinder.

Only much later did it occur to anyone that metal springs, too, would store energy better if a twisted rod were compressed rather than a straight rod bent. According to a biography of Brunelleschi, he constructed 'a certain alarm clock, in which are various different generations of springs', and recently it has been suggested that this is depicted in a late Quattrocento notebook of machinery which has some curious sketches of clocks with helical springs. Springs of this kind also appear in contemporary mousetraps. Clocks with coil springs, compressed horizontally rather than vertically, were certainly in use by about 1460, but very much as luxury items for princes, so it was not for about another century that spring watches became the mark of the solid middle-class man.

A. K.

Valve

As a valve is a means of allowing the flow of water or some other fluid such as air to move in one direction only, it must almost certainly have occurred first as part of the one early device that required this kind of motion: the bellows. Agricola in his Renaissance survey of metallurgy explains how the bellows for the forge is to have an air-hole, with a flap, slightly longer and broader than the hole, 'a thin board covered with goat skin, the hairy part turned toward the ground' attached to the board by leather straps, and so placed that 'when the bellows are distended the flap opens, when compressed, it closes'. The flap-valve is certainly far older than Agricola, as old as the wedge-shaped board bellows itself; but its date is hard to establish, for the name was, of course, taken from the older types of leather bag bellows in which the air-hole could be closed by the operator's foot or hand. Apparently the earliest representations are in the form of bronze lamps, perhaps Hellenistic, although there is no reference to their valves earlier than the late Roman poet Ausonius, who compares the gills of a landed fish in its last gasps to 'a woollen valve that receives and halts the wind alternately through a hole as it plays to and fro in the beech-wood cavity'.

The story of the use of valves in machinery could be said to begin rather with Ctesibius' force-pump. Vitruvius and Hero give detailed accounts of it, stating that 'coins placed in the upper apertures of the pipes by a clever joint do not let that go back which is forced into the vessel'. It seems that these original flap-valves of Ctesibius' were circular (hence the name of 'coins'), and that they may have been adapted from those sometimes used to control ventilation in roofs of the time. Later, the rectangular form took their place, but the name was kept. A few Roman force-pumps have been recovered with their valves badly corroded but still recognizable. In Heron's account of a twin-cylinder force-pump to be used as a fire extinguisher, he also introduces a primitive poppet valve, little discs sliding up and down on three curved pins. Ctesibius' hydraulic organ had slide valves to control air intake to the pipes. Otherwise, until the Renaissance all pump and bellows valves were flap-valves (also called clack-valves).

A conical poppet valve invented by Leonardo da Vinci was no doubt the source of those which appear in the plates of Ramelli's book of mechanical inventions (1588). Aleotti, a contemporary of his, employed a butterfly valve to control the flow of water in a pipe in some of his automatic puppet-shows. But neither of these seems to have been widely adopted and valves altered but little between the days of Heron and the invention of the steam engine, which required a much more precise control of the sequence of inflow and outflow; this led to the development of elaborate valve gear linked to the motion of the engine itself, from the

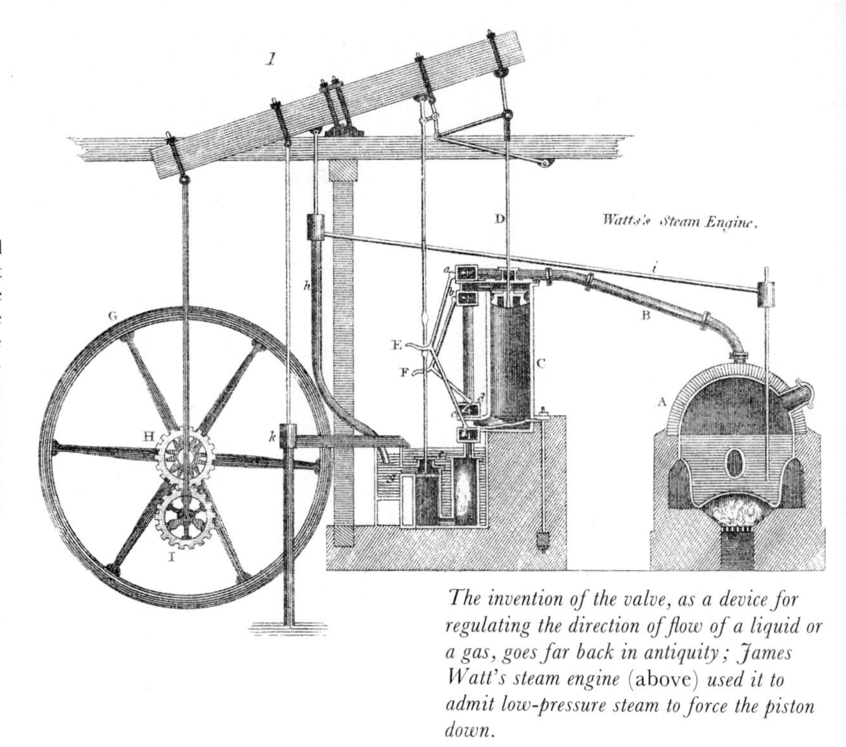

The invention of the valve, as a device for regulating the direction of flow of a liquid or a gas, goes far back in antiquity; James Watt's steam engine (above) *used it to admit low-pressure steam to force the piston down.*

Cams on a revolving shaft operating trip hammers (from Diderot's Encyclopédie, *1751–72). Later, cam and shaft were made in one piece, thus reducing friction.*

'snifting valve' devised by Newcomen to release the air that accumulated in the cylinder, to Murdoch's slide valves (1799) and the equilibrium (or 'Uncle Abraham') valve to keep the balance on each side of the piston in a double-acting engine.

A.K.

Cam

Alike in China and the West, our cam is descended from the projecting lug used to trip hammers. In the East, trip-hammers were used from an early date, certainly BC, to hull rice; the hammer was depressed by the operator's foot and fell when released. By the 1st century AD water-power had been applied to this device, first by an arrangement of spoons about an axle, which when they filled with water tipped over and spilled; the other end then dropped to depress the shaft of the hammer. Eventually the spoons became a series of buckets around a wheel, and the 'handle' a lug mounted on the same axle.

Once the handle had been divorced from the bucket, 'using one main shaft it is possible to actuate several bellows [by lugs on the shaft], on the same principle as the water-powered trip-hammers', as Wang Chen says. In one of his illustrations a vertical waterwheel works four such tilt-hammers through lugs on what is now a true camshaft.

At the other end of Eurasia, a parallel story. Little trip-hammers were part of the repertoire of Hellenistic inventors, but they were seldom used widely around the Mediterranean except in automata, until towards the end of the 1st millennium AD. This is a little curious: perhaps the lack of timber for such massive shafts as were needed may have had something to do with it, for they required only a vertical waterwheel and its prolonged axle, no gearing.

In medieval western Christendom, the single lug first appeared for the smith's tilt-hammer, about AD 1000. But a machine with several hammers, and so several cams, was first used in Europe to produce paper from rags. This product was invented in China where the rags had been beaten to pulp by hand, but in Europe the pulping was carried out from the 13th century by water-driven trip-hammers. About the same time, almost the same device was used to full cloth, and in this form it spread through Europe.

But the medieval camshaft really triumphed when it was seen that a longer drop on a confined surface could be used to great effect in any pounding or crushing operations. The hammers became a straight iron-shod wooden post (the stamp) with lugs at the side by which the cams could raise it, and then let go, so that the stamp could fall through guides. Once again, one of the first uses was for a Chinese invention – crushing charcoal for gunpowder – but a host of other jobs for the camshaft followed: pounding earths for pigment, 'falling' fulling stocks, beating rams to compress beeswax or various seeds for oil, and crushing and breaking up mined ore.

Apart from the cam used with the 'stackfeed' to regulate early spring-driven watches, it was not until the age of cast iron that cam and shaft could be made as one, and the cam correspondingly shaped to its task, with a reduction of friction loss which was as important as the gain in strength afforded by the new material.

A.K.

Detail from an etching, St Eligius in the Goldsmith's Shop, *by the Master of Bileam. Drawing the wire through the dies by hand must have needed both strength and control.*

Wire

'And they did beat the gold into thin plates and cut it into wire.' This sentence in Exodus is one of the first written references to wire. It has, however, been found in Egyptian tombs and in the ruins of Troy, in ancient China, Italy, and in some countries in the Middle East. At this early time it was made by the smiths, who hammered lengths of metal into thin strips, an exhausting and lengthy process capable of producing only very short pieces of wire. To obtain anything longer it was necessary to hammer several pieces together.

It is not known exactly when the first wire was produced by pulling a rod through a die or draw plate, but such wire is known to have existed in AD 1000 and historians have placed its probable first use some 300 years earlier. For several centuries wire made by this method was drawn through the dies by hand, but this was slow and difficult for the person doing the pulling, and produced wire whose consistency and thinness was totally dependent on his strength. The section to be drawn was hammered to a size at which it could be pushed through the reducing hole in the die, then the wire-drawer grasped it in his hands or in a pair of tongs and pulled it through; the wire then assumed the shape of the hole through which it was pulled. There were a number of modifications to this method but all were ultimately dependent on the strength of the wire-drawer.

During the Middle Ages wire was used to make the rivets for the moving parts of armour and in 1350 a man called Rudolf de Nuremberg was credited with having developed a method of wire-drawing which made use of water-power. At about this time the musical instrument known as the 'cittern' was developed, one of its major innovations being the use of wire strings. In 1854 William Humphrey established a wire mill in England using water-power.

Most wire was still of the thin single-strand sort and it was not until later that rope-wire began to be made. Although specimens of rope-wire found in the ruins of Pompeii show that methods of stranding wire were known over 1,500 years ago, knowledge of the method seems to have been lost, for there are no specimens or records extant from that period until the 19th century, when iron-wire ropes were used in the construction of a suspension bridge in Geneva. Today wire is generally made from various metals and alloys, brass, bronze, copper, aluminium, zinc, steel, gold and platinum.

T. C.

Crank

The first crank in the West, with pin and crank arm in one piece, may be seen turning a grindstone in a 9th-century psalter; thereafter cranks are illustrated only occasionally for half a millennium, and then quite suddenly, shortly after 1400, there hardly seems to be an illuminated manuscript (at least in the Low Countries) that does not have one, turning hoists, cages, spits, even instruments of execution. The same generation saw the birth of the compound crank, originally in the form of the brace, a simple hand tool. But within a matter of years of the first depictions of the brace it occurred to someone that the arm that turns the brace could be replaced by a rod; the rod would be simply an extension of the human arm, in handmills, but the linkage could also be reversed, with the rod driven by the rotating crank to work a pump, as it is portrayed in a manuscript of 1431. And so the crankshaft was born. Cranks were often used in the 15th and 16th centuries to drive bellows and sawmills, the only machines that required this two-way control. Although crankshafts were occasionally employed in pumps, two-throw and even four-throw crankshafts were envisaged, and may have been installed here and there. They did not really catch on for a long while, however, perhaps because so long as heavy-duty machinery was all of wood, crankshafts could not easily be made in one piece, and this would have put great strain on the joints.

At all events, crankshafts did not come into their own until the days of cast iron. In 1780, James Watt found himself forestalled by a patent for the use of the crank to obtain rotary motion from the reciprocating action of his steam engine – the reverse of the older chain of motion. He was furious, though he was thereby inspired to design the sun-and-planet gear for the same purpose. But as the offending patent lapsed, the crank became the normal means to do the job, and if two or more cylinders were used, or power had to be delivered on both sides – for instance to the paddle-wheels of a steamship – a crankshaft was the answer, subsequently bequeathed by the age of steam to all 20th-century piston engines, by whatever fuel they may be driven.

A. K.

Pendulum

The story goes that Galileo discovered the isochrony of the pendulum – the fact that the time of a pendulum swing depends only on its length and not on its weight at the arc through which it swings – by timing the swings of the great lamps of Pisa Cathedral with his pulse-beat, during an earthquake tremor, and so finding them all apparently taking place in equal time.

Like all favourite anecdotes, this one has had cold water thrown over it by modern historians. But however he may have come by it, there is no doubt that he was the first to discover the principle, demonstrate it (1588–91) and make use of it in his research into theoretical mechanics. He did sketch a suggestion for a clock that could employ it, but whether any such clock was ever constructed in his lifetime is another matter. The Dutch physicist Christiaan Huygens became interested in the topic for the same reason as Galileo, in order to exploit it in his inquiries into the behaviour of falling bodies. He appears to have been the first to have a clock made which applied this property of the pendulum, realizing that it must prove a more accurate regulator of the motion of the clock mechanism than any then in current use. He made the further discovery that the pendulum bob would not in fact oscillate in perfectly equal times unless it were constrained between restricting cheeks of a particular (cycloid) curve so that the bob would itself describe a cycloid arc. A clock incorporating this in its design was then made to Huygens's specifications by a local clockmaker of The Hague named Salomon Coster, and installed at Scheveningen in December 1657. The whole idea looked like being a great success, profiting Huygens's income as well as his reputation, when an indignant message arrived from Italy to say that Galileo had done it all long before.

All the same, the priority of construction and publication is Huygens's; he did eventually obtain a patent for his version, from which all later pendulum clocks are derived, and his invention gave great impetus to attempts to improve the mechanics of the clock.

A. K.

Screw and screwdriver

The wood screw, or screw nail as it is sometimes called in the USA, is comparatively modern; but in the 16th

Reconstruction of Galileo's study, with a pendulum and an experimental clock mechanism.

Below, *the screw developed, in the 16th century, from the nail – given a twist for the sake of a firmer hold. Félibien's 'tournevis' (1676), with its short, blunt blade, evolved by the late 18th century into the longer-bladed screwdriver of today.*

Left, *piston pump and crank shaft, from* De Re Metallica *(1556) by Agricola.*

century gunsmiths and armourers were already using a small tool with a blade – the original 'turnscrew' – to adjust their gun mechanisms. The mechanisms themselves were 'nailed' to the gun-stock, but it was found that the nails held better if they were given a twist. Like all other nails, they were hammered in, and were extremely difficult to get out again. The only solution was to cut slots in the heads before they were knocked in, so that they could be taken out with the 'turn-screw', which thus became the first 'screwpuller' or 'unscrewer'. Félibien's *tournevis* of 1676 is of this type.

As these twisted nails were made by hand, they were naturally expensive and were used only for special work. Towards the end of the 18th century, however, some unknown genius, possibly in Birmingham, found a way of making better, but still pointless, wood screws by machinery. This made them cheap enough to be used more frequently for fixing hinges, doors, furniture, etc. But the efficiency of their finer thread was reduced by hammering, and some longer-bladed tool was needed to screw them in. In about 1780, the joiners' tool-makers of London introduced the longer-bladed screw-driver, still known in the trade as 'London-pattern'. In about 1840, Nettlefold improved wood screws by giving them points, and the screwdriver has never looked back since.

W. L. G.

Air pump

While Mayor of Magdeburg, Germany, Otto von Guericke became interested in the disputations of scientists and philosophers about the possibility of a vacuum and, being an engineer by training, he decided to try to settle the question by experiment. In 1650 he made his first air pump – like a hand-operated water-pump but with the parts sufficiently well made to be reasonably airtight. It worked. In an evacuated vessel he showed that a bell ringing inside could not be heard, that candles would not burn, and that animals died.

His large-scale demonstrations were spectacular. One took place before the Emperor Ferdinand III and his court, in which two hemispheres 12 ft. in diameter were fitted together along a greased flange at the circumference, and the sphere was then evacuated. Two teams of eight horses pulled on ropes fastened to each hemisphere but were not able to separate them, yet when air was admitted they fell apart. In another, in 1654, a vertical open-ended cylinder was evacuated below the piston and 50 men with ropes attached to the piston were drawn up when they tried to withdraw it. In this way the piston was made to do work while a vacuum was maintained beneath it.

But could a vacuum be created without an air pump? After many years steam was found to be the answer. Thomas Savery, in 1698, was the first to pump water by steam, admitting it into a closed vessel and then pouring cold water over the vessel to condense the steam within, thus producing a vacuum. This he used to suck water up from a mine and the vessel emptied by using steam from a boiler to press it out. This cycle of events was then repeated.

Savery's device was called 'The Miner's Friend'. It had no piston or moving parts, and was not an engine – just a pump.

Before this, in 1690 in France, Denis Papin had made a model apparatus of a piston $2\frac{1}{2}$ in. in diameter in a well-fitting cylinder. With a little water inside the cylinder he was able to show by successively heating and cooling it that a vacuum was created below the piston as the cylinder was cooled. Though it was of no practical use, this was the first apparatus in which a piston was moved and work done by condensing steam.

These three achievements of von Guericke, Papin and Savery were combined to produce a practical steam engine by Thomas Newcomen of Dartmouth in 1712.

A. BU.

Universal joint

In 1676, Robert Hooke, the English Leonardo da Vinci, published a lecture on his 'helioscope', an instrument employing a system of reflecting glasses to observe the sun in safety. It was to be manipulated by means of his novel 'Universal Joynt . . . a Universal Instrument . . . for communicating a round motion through any irregular bent way, without shaking or variation'. Although Hooke describes the method of constructing his new invention in some detail and suggests vaguely that it might come in useful for various purposes, he himself meant it only for astronomical work, or in the design of clocks and sundials, and it caused little excitement at the time.

Hooke was a man of incredible fertility of mind, throwing off inventions by the score in the time he could spare from formulating revolutionary theories in physics, chemistry and geology, and from endless discussions with like-minded friends in London coffee-houses. His diaries usually hint at least as to how some new idea took shape in his fiercely active brain, and the minutes of the Royal Society report the experiments where he made known his latest discoveries.

But the diaries do not record his spending much time over the universal joint, and he did not bother to demonstrate it before the Society. That the invention was his and his alone as far

When the air is pumped out of the bell-jar the pigeon dies – to the dismay of the tender-hearted. Detail from Experiment with an Air Pump *by Joseph Wright of Derby, 1768.*

Right, *von Guericke's first air pump (1650).*

Below, *Robert Hooke's diagram (1676) of the universal joint.*

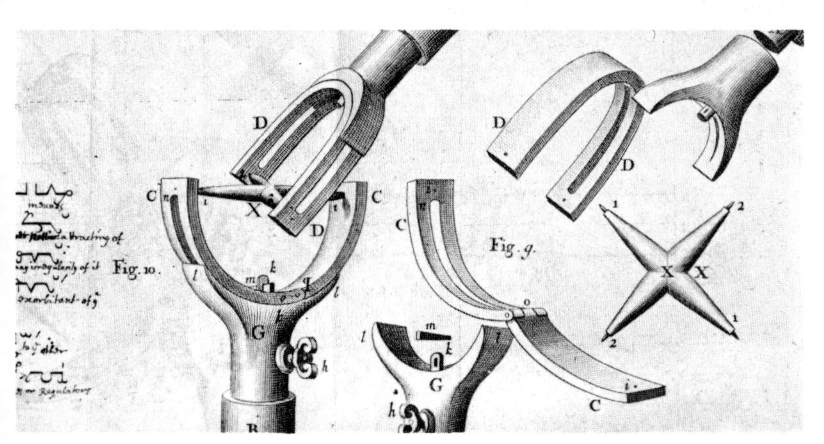

as machinery goes, seems beyond dispute. But in the transmission of power, like so many other inventions, there was little call for a joint with freedom to communicate motion in all directions before the transport revolution of the 19th century.

A. K.

Gyroscope

Any body spinning about a movable axis can claim to be a gyroscope, whether it is as small as a child's top or as large as the earth. The first practical application of the gyroscope is probably due to Serson, who in 1744 persuaded the British Admiralty to test at sea a spinning rotor which would indicate a stable horizontal reference for ships. The rotor was supported on a pivot so as to be free from disturbances caused by the ship – forerunner of the gyroscopic horizon used in modern aircraft.

Many years after Serson, a Scotsman called Sang (1836) and a Frenchman, Foucault (1852), independently seized on the idea of gyroscopic stability to demonstrate the rotation of the earth. Sang, however, had not the resources to construct a sufficiently accurate rotor. Foucault – who invented the word 'gyroscope' – succeeded in making such a rotor.

In the early part of this century a burst of activity was directed to the stabilization of ships against roll by massive gyroscopes. The primary inventor was Otto Schlick, and the first application in Britain was in 1908, but this method of stabilization was superseded by a system using underwater fins controlled by small sensing gyroscopes. At about the same time inventors like Brennan, Scherl and Schilovsky were constructing vehicles capable of running on two wheels instead of the usual four, held upright by internal gyroscopes; but the world would not accept them.

Another marine application, the gyro-compass, now an essential feature of all but the smallest ships, is generally credited to Anschutz-Kaempfe of Germany (1908), though he was closely followed by Sperry in the United States. It can indicate true north by sensing the rotation of the earth and then pointing the rotor axis in the direction of the North Pole, even when the ship is rolling and tossing at sea.

The most remarkable advances in the use of gyroscopes have been in the air, where an essential for navigation and control is an accurate vertical reference. A simple pendulum will not do, since it will oscillate wildly if its point of support is accelerated. In artificial horizons and autopilots a gyroscope with a vertical axis – like an upright top – is used. As the aircraft climbs or rolls the angle it assumes relative to the unchanging gyro-axis provides a reading which can be used either directly by the pilot or as part of an automatic control system.

During the past 20 years these airborne applications have culminated in complex 'inertial navigation' systems, at the heart of which are miniature precision gyroscopes capable of detecting turns of small fractions of a degree per hour – much smaller than the slow rotation of the earth. The systems can operate as self-contained units, recording the precise position and orientation of the aircraft independent of any external source of reference whatever, over flights of many thousands of miles.

L. M.

The gyroscope, the ultimate stable source of reference: essentially a simple device, yet it has many uses, from the stabilization of ships at sea to the control of blind flying in the air.

Governor

The centrifugal governor used in Watt's steam engine of 1789 caused only a moderate stir at the time; Watt seems to have thought of it as only a subordinate part of his engine by comparison with the motive system. Yet, as the ancestor of all feedback devices to enable a machine to regulate itself automatically, its place in the history of invention seems assured, for it was certainly the first device to control speed effectively by varying the intake of fuel. Watt's governor consisted of a pair of centrifugal pendulums, linked ultimately to the engine's revolving flywheel, and directly attached to a sleeve, which was connected in turn to the valve that admitted steam to the cylinder. When the flywheel turned more rapidly, the two balls swung outwards, which lowered the sleeve; when speed dropped they fell with it and the sleeve was forced up. As the steam valve was opened wider or narrower accordingly, so they ensured that an even speed was maintained.

The history of Watt's governor perhaps goes back to the ball-and-chain or ball-and-rod devices sometimes used instead of flywheels on medieval and Renaissance machines. However, these only performed the function of flywheels, producing a more regular movement in a drill or crank by storing energy to carry the instrument over its 'dead point'; they did not control the speed or intake of power and can at most only have suggested the shape of the governor. It was not until mechanics had grown accustomed to the behaviour of pendulums in clocks, and had learnt about centrifugal force, that anyone thought of using the ball-and-rod combination to *control*. Millwrights were always faced with the problem that they could not exploit high winds because the millstone tended to ride upwards when its shaft rotated very fast, increasing the distance between the stones so that the grain in the middle was not properly ground. Tentering the stones so that they were always kept just the right distance apart was done

In the Otto four-stroke engine the pair of centrifugal pendulums, linked to the flywheels, controlled the speed of the engine by a simple form of feed-back. As the speed of revolution increased they flew outwards, opening a valve and reducing steam pressure. As speed dropped, they fell, and steam pressure rose.

Cutaway view of the hold and upperworks of a cable-laying ship at the end of the 19th century.

by hand, until in 1787 Thomas Mead designed a method by which two pendulums, independently attached to the spur gear that drove the stones, could raise and adjust the tentering lever through chains and a universal joint. Another pair was linked to the sails in such a way as to open and close them in response to changes in speed; by altering the amount of sail exposed to the wind, the miller could similarly adjust the speed of the shaft. Two years later, a patent for a rival version with rack and sector instead of the chains was taken out by Stephen Hooper.

Meanwhile, the first steam-powered mill, the Albion Mills erected in London by John Rennie, was equipped with a governor like Mead's, for Boulton wrote to his partner Watt in May 1788 that it had a device 'for regulating the pressure or distance of the top mill stone from the Bed stone in such a manner that the faster the engine goes the lower or closer it grinds . . . and when the engine stops the top stone rises up . . . this is produced by the centrifugal force of two lead weights which rise up horizontal when in full motion, and fall down when the motion is decreased, by which means they act on a lever.' This must have been Watt's inspiration, for although this governor was obviously applied to the millstones and not to the engine, Watt adapted it to its later purpose before the end of 1788. Since he knew he could make no claims to have discovered the basic principle involved, he did not try to patent it. Still, he was sufficiently ahead of his competitors to attempt to protect his governor by secrecy and concealment.

A. K.

Ball-bearings

It seems likely that the Italian Renaissance sculptor and goldsmith Benvenuto Cellini (1500–71) first saw the potential of a circle of freely revolving balls to reduce the friction set up between two rotating bodies. In 1543, 'having finished the beautiful statue of Jupiter, I placed it upon a wooden base, and within it I fixed four little globes of wood which were more than half hidden in their sockets and so admirably contrived that a little child could, with the utmost ease, move the statue backwards and forwards and turn it round', he recorded in his autobiography.

The rotating balls in this modern ball-race allow a shaft passing through the middle to rotate freely. This method of reducing friction was first used c. *1780 in the rotating head of a windmill.*

Rolling contact ball-bearings loose in their races, however, were not used until the last quarter of the 18th century, when they were introduced into windmills, the earliest known (about 1780) being in a post mill, in which the entire mill structure revolves about the central post. Radial ball-bearings first appeared in 1794, when Philip Vaughan, an ironmaster of Carmarthen, Wales, patented them for the axle-bearing of a carriage. From then on through the 19th century, and especially in the 1850s and 1860s, a large number of patents were taken out which used ball-bearings with the axles of everything from merry-go-rounds through the propeller shafts and gun turrets of warships to armchairs and bicycles. Nevertheless, the invention was not really exploited to the full until the advent of powered vehicles, using metal parts in rapid motion with the consequent risk of great loss through friction. Ball-bearings, therefore, did not really assume the indispensable role they have today until the arrival of the motor car, and of the precise grinding machines required for the mass production of accurately spherical metal balls.

A. K.

Telegraph cable

In 1851, when England and France were to be connected across, or rather under, the Straits of Dover, the first attempts at producing a cable – strands of wire in an insulating cover – failed. Wire-drawing and insulating with a ½ in. thick envelope of gutta-percha was a haphazard affair, and when the cable was laid, the Morse messages that got through came out garbled at the receiving end. A second cable with a copper core and thicker insulation was laid a few months later and proved serviceable.

Five years later, the American businessman Cyrus W. Field formed the Atlantic Telegraph Company for the purpose of laying and operating a submarine telegraph cable between Britain and America. About 340,000 miles of fine wire were produced and intertwined; the insulating gutta-percha alone weighed 18 cwt. per cable mile. This venture, too, failed, though its scientific director was Professor William Thomson, later Lord Kelvin; after only three weeks' operation, the transatlantic cable ceased to convey messages. A new, stronger cable was laid in 1866 under the personal supervision of Professor Thomson.

These cables were, of course, meant to carry only low-tension current of a few volts, sufficient for telegraphic signals. With the advent of electric power and lighting, high-tension cables had to be designed and manufactured. The first 'concentric' cable, which made technical history, was the Ferranti power cable laid in 1890 across the Thames at Deptford. Its outer casing was a wrought-iron tube; embedded in impregnated paper were two copper tubes as conductors. This cable remained in use until 1933.

Today, impregnated paper is still the most common insulating material for underground cables, while submarine cables have layers of gutta-percha, PVC or polythene, with lead or

The Swedish chemist Jöns Jakob Berzelius, whose discovery of the element selenium in 1817 led eventually to the development of the photo-electric cell.

aluminium sheathing outside; if armouring is considered necessary to prevent mechanical damage, one or two layers of steel, or stranded steel wire, will provide good protection. Cable cores are invariably of copper because it is the best conductor; power cables can contain over 100 fine copper strands. For very high voltages, the cable may also have a filling of insulating oil.

Submarine telegraph and telephone communication cables have relays – repeaters or amplifiers – built in at regular intervals so as to transmit the electric impulses over great distances. The modern trend, however, is towards the co-axial cable; it has a core of copper tubes which can carry great numbers of telephone conversations, Telex messages and many radio and television programmes simultaneously on as many different wavelengths in the microwave band.

E. L.

Photo-electric cell

In 1817, the Swedish chemist Berzelius discovered a new element which he called selenium (from the Greek word *selene*, moon) – a non-metallic, widely distributed element. It was only much later that it was found serviceable as an electrical resistance; for instance, it was used as a rheostat at the cable station on the island of Valencia, off the coast of Ireland. Here, an engineer by the name of May noticed a strange phenomenon: the selenium bars let the electric current pass through them much more freely when the sun was shining on them than at night.

May made his discovery in 1873, but again there was no immediate practical application – though in 1883 an early pioneer of television, the German engineer Paul Nipkow, based his system of picture transmission on this property of selenium when he invented his famous 'Nipkow disc' for scanning a scene to be transmitted by electricity. A few years later, Heinrich Hertz, the first man to produce and receive electromagnetic waves, did some research on the phenomenon of 'photo-electricity', as it was eventually called. J.J. Thomson, who discovered the electron, put forward a theory that electrons were emitted from certain kinds of substances due to the incidence of light; and later still Einstein applied the quantum theory to photo-electricity to explain why the emission of electrons started as soon as light fell on sensitive surfaces, the speed of the electrons depending on the wavelength of the light.

The first technical use of photo-electricity was made in 1902 when Arthur Korn, a German physicist, developed an efficient system of photo-telegraphy; it was probably he who invented a basic type of photo-electric cell for the purpose: a small glass cell, highly evacuated, and coated with some alkali metal on the inside, except for a narrow 'window' through which light can enter. Inside the cell there is only a thin metal ring; this and the alkali coating are kept at a negative and positive potential. When light reaches the coating it gives off electrons according to the intensity of the light, and the ring attracts the electrons. This produces a weak electric current, which can easily be amplified. In Korn's photo-telegraph, the (still) picture to be transmitted is brightly lit and scanned, point by point, by a photo-electric cell; and the modulated current is then transmitted to the receiver, where the light values of the picture modulate a lamp which exposes a photo-sensitive paper or film rotating on a drum – point by point and line by line. The system is still in use for the transmission of pictures by wire or radio.

That type of cell is called the photo-emissive cell, but there are also two other types in use today: the photo-conductive cells, which still use selenium and are most useful for switching street lights on and off automatically; and the photo-voltaic cells which incorporate two kinds of substances, for instance copper and cuprous oxide, separated by a specially prepared boundary. When light falls on that boundary, a voltaic current is produced.

Photo-electric cells of these various kinds have found a wide range of applications: as photographic exposure meters; as burglar alarms which switch on bells when an intruder interrupts an invisible ultra-violet light beam; as counting units in factories, recording each piece of goods as it passes by on the conveyor belt; or as colour-sensitive cells, reacting differently to light of different wavelengths so that they 'see' whether certain kinds of fruit or vegetables, such as peas, have the right colour or should be thrown out before packing or canning.

The most important applications of photo-electric cells are, of course, those in television – the camera is just one big cell – and in sound-film production and presentation, where photo-electric cells turn speech, music, and noise from tape-recordings into optical sound-track on film, and the sound-track back into sound. Another instrument which uses photo-electricity in industry and research is the photo-multiplier or electron-multiplier for the detection and measurement of very small quantities of light; it consists of a system of electrodes acting like a chain of relays, amplifying the amounts of electrons released by the incidence of light until they are easily measurable.

E. L.

Aerosol spray

This familiar Aladdin's Lamp of the

Progress of the aerosol: the idea remains the same through all the refinements of shape and size.

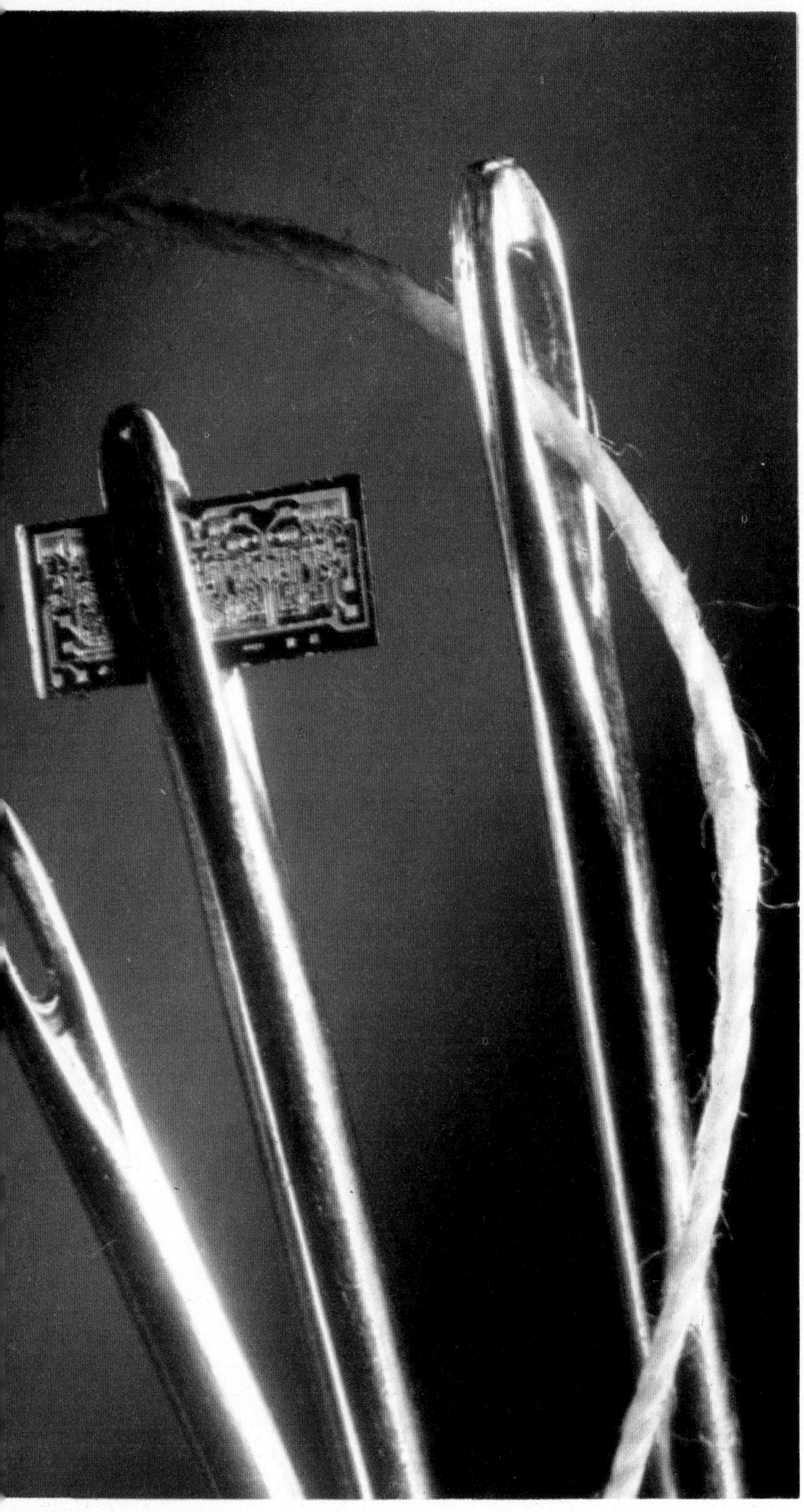

Before encapsulation in plastic, this tiny silicon chip with its complex printed circuit and over 120 components can pass through the eye of a needle. Such micro-miniaturization has only become possible with the invention of the transistor.

Opposite, *helium-neon laser apparatus, set up to test different ways of using laser beams.*

20th century was invented in 1941 by the American scientist Lyle David Goodhue. Since then it has come to be used as a dispenser of many hundreds of household commodities. It consists of a canister containing fluid packed under pressure. The pressure is maintained by gas in the upper portion of the tin and on pressing a valve button the fluid is released in the form of a very fine spray or mist. Alcohol is usually used as the dispensing medium so it is important that the fluid to be dispensed should dissolve in it.

Briefly, what happens is that the appropriate ingredient is dissolved in the alcohol, though sometimes, as in the case of aluminium chlorhydroxide, the main ingredient in most anti-perspirants, the substance remains in suspension. The concentrate is measured into the canister and the valve and cap fitted and squeezed tightly over the orifice. The propellant gas or liquid, usually a fluorinated hydrocarbon, is then pumped in under pressure through the valve in a measured quantity to give a regular pressure. Finally the spray nozzle is fitted to the valve stem. This type of aerosol is very useful in a pharmaceutical context as it gives the chemist a new dosage form and it also has the advantage of being totally hygienic.

There are three main types of aerosol, based on particle size: the first is the 'space spray' which dispenses the active ingredients as a finely divided spray of particles no larger than 50 micro-millimetres in diameter. Those with larger particles are known as 'surface-coating sprays' and as their name suggests produce a wet, rather coarse spray suitable for coating a surface with a film of material. Finally there are the 'aerated foam' systems such as are used in the aerosol shaving preparations. Perfumes, cleaning agents, insecticides, anti-perspirants and paints are among the most common commodities now dispensed in aerosols; among the more unusual must be London fog – for film effects.

T. C.

Transistor

Today it is possible to build into a single chip of silicon, no bigger than a full stop on this page, many times the number of components that can be packed into a conventional radio set. This revolution in electronics is above all the result of a single invention, the transistor, first demonstrated by William Shockley, John Bardeen and Walter Brattain at the Bell Telephone Laboratories in the US in 1948.

Transistors can do virtually all the jobs of conventional radio valves. And since transistors are more reliable, more rugged, smaller and require only a fraction as much power and no warm-up time, the valve has been on the way out since 1948.

The transistor exploits the curious electronic properties of materials like silicon and germanium, which are neither electrical conductors nor resistors and are therefore known as 'semi-conductors'.

Shockley's key discovery was that the properties of semi-conductors could be modified in important ways by 'doping' them with traces of certain impurities.

However, early transistors still could not perform one of the vital functions of the valve: amplification. But by combining semi-conductor materials doped in different ways Shockley and his collaborators succeeded in making a device that did.

B. S.

Laser

The laser, an intense, narrow beam of light with all the light waves exactly in step, was discovered by Theo Maiman at the Hughes Laboratory, Malibu, in 1960. The laser beam is so concentrated that a beam 1 cm. in diameter at its source on earth spreads to only a few kilometres when it reaches the Moon, 250,000 miles away.

The story of the invention of the laser goes back to 1917 when Einstein declared that it should be possible, by allowing light of the right frequency to fall on an excited atom, to stimulate the atom into releasing its excess of energy in the form of light. Following the successful invention in 1954 of a microwave amplifier based on this principle, Schawlow and Townes calculated in 1958 that a 'laser' (Light Amplification by Stimulated Emission of Radiation) could be built. Their attempts, using potassium vapour, failed, but Maiman, who used a dilute form of ruby, succeeded in generating a laser beam. Subsequently a vast number of laser materials, from gases to solids, have covered the spectrum from the near ultra-violet through to the far infra-red. Many of these are feeble emitters and have been rejected, leaving a mere half dozen of promise. These range from the low-powered helium-neon red laser, which costs £40 and is readily available for school demonstration, to high-powered systems emitting more than 1 kW., which are useful for metal cutting at high speeds. Special techniques enable the laser to deliver its energy in very short bursts when it can produce powers of the order of 600 million kW.

The laser's most dramatic use is as a death ray – now being explored but not yet used by the US army. Its main use at present is for delicate welding, including refixing the retina in the human eye: its most exhilarating potential use is for holography – a technique for producing pictures that contain all the visual information necessary to reconstruct a three-dimensional image.

DK. G.

CHRONOLOGY

The Pattern of Invention

Any attempt to chart the rising tide of invention is bound to be arbitrary. The wheel and the plough, pottery and the abacus, the boat and the lever cannot be attributed to any inventor or date. Some of the most important early inventions, particularly among those made before 1500, when this table starts, were made anonymously in the East and percolated to the West. The principle used in compiling the table is to list the invention that is recognizably the origin of the present-day version. Usually, this will be the first reliably working version. Sometimes, as in the case of Benjamin Franklin and the lightning conductor, we have taken the first clear description of how to make the device; turning his description to use was not difficult. But this is not the case with, say, the automobile or the aeroplane, and here we give the first successful version.

If no date is given, this is because precision is not possible: the approximate date of the invention can be deduced from its position on the page.

Chronological Table

	Contemporary Events	Man Moving	Man Taking
1500	1492 Columbus lands in America		
		Pencil Dredger	
	1534 Henry VIII founds Church of England		
	1571 Battle of Lepanto		
		1580 Newspaper	
	1588 Spanish Armada		
1600			
	1616 Pocahontas comes to London		
	1649 Beheading of Charles I		
	1660 Louis XIV assumes power		
	1693 Last witch burnt at Salem		
1700		1709 Piano	
			1712 Steam engine
	1720 South Sea Bubble	1716 Diving bell	
		1731 Sextant	
			1772 Papier mâché
	1776 Declaration of American Independence		
		1783 Balloon	
	1789 Storming of the Bastille		
		1792 Semaphore 1797 Parachute 1798 Lithography	
1800		Steamship	1800 Battery
		1812 Printing technology	
	1815 Battle of Waterloo		
		1816 Camera	
			1820 Elastic 1821 Electric motor 1827 Matches
		1829 Braille 1830 Railways	
			1831 Dynamo Transformer
		1832 Tram and trolleybus	
			1833 Water turbine
	1837 Accession of Queen Victoria	1837 Screw propeller Diving apparatus	
		1837 Electric telegraph 1839 Bicycle	1841 Vulcanized rubber

Man Living	Man Working	Key Devices	
1509 Wallpaper	Rolling mill; Theodolite		1500
Naval mine	Vice and spanner		
1560 False limbs			
1589 Knitting frame	1590 Microscope	1588 Pendulum	
	1592 Thermometer		
	1600 Foot rule		1600
	1608 Telescope		
	1614 Logarithms		
	1621 Slide rule		
	1640 Micrometer		
	1642 Calculating machine		
	1643 Barometer		
	1658 Balance spring	1659 Air pump	
Cigarette	1666 Calculus		
		1676 Universal joint	
1679 Pressure cooker	1679 Binary arithmetic		
			1700
1733 Flying shuttle; Seed drill			
	1735 Marine chronometer		
1752 Lightning conductor			
1770 Spinning jenny			
1771 Factory farming			
1775 Water closet			
1779 Crompton's mule		Screw and screwdriver	
1789 Beehive		1789 Governor	
1790 Dental drill	1791 Metric system		
1792 Cotton gin		1794 Ball-bearings	
1796 Vaccination			
1799 Bleaching powder; Anaesthetics			
1801 Gas lighting			1800
1810 Ultra-violet lamp	1812 Hydraulic jack		
1815 Miner's safety lamp	1814 Spectroscope		
1816 Fire extinguisher; Stethoscope			
1818 Blood transfusion			
	1821 Thermocouple		
1826 Corn-reaper			
1830 Food canning; Sewing machine			
1830 Lawn mower; Plywood			
1830 Threshing machine; Crop rotation			
Artesian well	1832 Galvanometer		
1834 Refrigeration	Steam hammer		
1840 Postage stamp			
1844 Paper patterns			
1847 Ophthalmoscope			

	Contemporary Events	Man Moving	Man Taking
			1850 Paraffin
			1856 Bessemer steel
			1860 Internal combustion engine
	1863 Battle of Gettysburg	1863 Underground railway	
		1864 Driving chains	1866 Dynamite
			1868 Plastics
	1871 Germany becomes a State		
		1873 Typewriter	
		1875 Submarine	
		1876 Telephone	
		1877 Gramophone	
		1880 Half-tone block	
		1881 Stereophony	
		1884 Motor-car Airship Fountain-pen	1884 Steam turbine
		1885 Motor-cycle	1886 Aluminium
		1888 Pneumatic tyre	
		Glider	
		1893 Carburettor	1893 Diesel engine
		1895 Cinematograph	
		1896 Radio	
		1897 Teleprinter	
		1899 Magnetic tape-recorder	
1900	1901 Death of Queen Victoria		
		Silk-screen printing	1902 Synthetic minerals
		1904 Caterpillar track	
		1905 Hydrofoil Aeroplane	
		1907 Helicopter	
			1912 Stainless steel
	1914 First World War	1914 Traffic lights	
		1926 Rocket	
			1930 Gas turbine
		1932 Parking meter	
		1934 Catseye Television	
		1935 Hearing aid	
		Juke box	
		1937 Xerography	
		1938 Ballpoint pen	1938 Stirling engine
	1939 Second World War	1939 Jet engine	
			1942 Nuclear reactor
		1943 Aqualung	
		1947 Polaroid camera	
		1948 Long-playing records	
		Music synthesizer	1952 Float glass
		1955 Hovercraft	
		1956 Videotape recorder	
		1957 Spacecraft	1959 Fuel cell
		1962 Communications satellite	

Man Living	Man Working	Key Devices	
1848 Rifle Camouflage			
Weedkillers and pesticides		Telegraph cable	
1853 Safety lift		1852 Gyroscope	
1856 Aniline dyes			
1860 Can-opener Linoleum			
Pasteurization			
1869 Torpedo	1869 Metal plane		
1870 Lister's spray			
1873 Barbed wire			
1875 Gas mantle			
1876 Carpet sweeper			
1879 Filament lamp Cream separator			
1880 Inoculation	1880 Artificial languages Blowlamp		
1883 Machine-gun Man-made fibres	Ultrasonics		
1887 Combine harvester			
1889 One-armed bandit			
1895 X-ray			
1897 Breakfast cereals			
1899 Aspirin			
1901 Vacuum cleaner			1900
1902 Safety razor Surgical transplants		1902 Photoelectric cell	
1903 Electrocardiograph	1903 Oxy-acetylene welding		
1904 Vacuum flask			
1908 Conveyor belt			
1910 Therapeutic drugs	1910 Geiger counter		
	1911 Cloud chamber		
1913 Zip fastener	1912 IQ tests		
1914 Brassière			
1915 Tank Gas mask	1916 Sonar		
Synthetic fertilizers	1921 Lie detector		
Washing machine			
1929 Iron lung	Teaching machines		
1930 Supermarket	1930 Cyclotron		
Cloud-seeding	1931 Radio telescope Electron microscope		
1934 Fluorescent lamp			
	1935 Radar		
1937 Electric blanket			
1939 Insecticides			
1942 Napalm		1941 Aerosol spray	
1944 Ballistic missile Kidney machine			
1945 Nuclear bomb			
1946 Automation	1946 Computer		
1948 Microwave cooking		1948 Transistor	
	1952 Bubble chamber		
1955 Non-stick pans Contraceptive pill			
	Fibrescope		
		1960 Laser	
1966 'Miracle' rice			

NAME INDEX

Index

Index

Index

Index

ENERAL INDEX

numbers in italic refer to main entries

Index

Index

Index

Index